Essential Concepts of Environmental Peacebuilding

This book provides a comprehensive introduction to the essential concepts of environmental conflict and peacebuilding by compiling and explaining key terms in an easy-to-navigate A–Z format.

Over the past 80 years, scholarly interest in war and peace has flourished, resulting in the institutionalization of disciplines such as international relations, security studies, and peace and conflict studies. Since at least the 1970s, research has also investigated the planetary effects of global environmental change. Environmental peacebuilding emerged at the intersection of environment, conflict, and peace, weaving together several related threads that address both environmental risks of conflict and environmental opportunities for peace. The language used to describe, explain, predict, teach, negotiate, and form policy in the arena of environmental peacebuilding is often nuanced, evolving, and contested. This book fills a gap in the literature by providing a consistent understanding and approach to key aspects of environmental peacebuilding. The book begins with a thematic introduction to contextualize the topic, written by a team of world-leading scholars. The book then follows an A–Z format, with over 150 entries from experts in the field describing and explaining key concepts and providing examples of how different terms are applied within this growing discipline.

This comprehensive A–Z guide will be useful for students, academics, professionals, and policymakers seeking a wider understanding of environmental peacebuilding. The book will also be useful for those studying security studies, peace and conflict studies, environmental studies, and environmental law.

Carl Bruch is Director of International Programs at the Environmental Law Institute (ELI) and Executive Director of the Environmental Peacebuilding Association (EnPAx), USA. His research focuses on environmental law, policy, and practice in conflict-affected situations.

Rachel Weaver holds a Master of Environmental Management from Duke University, USA, with concentrations in Environmental Economics and Policy, and Terrestrial and Freshwater Ecosystems. At Duke, she served as a Graduate Research Assistant and as a Member of the Conference Secretariat for the Environmental Peacebuilding Association.

Richard Matthew is Professor of Urban Planning and Public Policy at the University of California, Irvine, USA. He is Director of UCI's Climate and Urban Sustainability Program and the Alec Glasser Center for the Power of Music and Social Change. His research focuses on the connections among climate change, conflict, health, and displacement in highly vulnerable regions.

Erika Weinthal is John O. Blackburn Distinguished Professor of Environmental Policy, a Member of the Bass Society of Fellows at Duke University, and the President of the Environmental Peacebuilding Association, USA. She specializes in global environmental politics and environmental security with an emphasis on water and energy.

Larry Swatuk is Emeritus Professor in the School of Environment, Enterprise and Development (SEED) at the University of Waterloo, Canada. He is also Extraordinary Professor in the Institute for Water Studies, University of the Western Cape, South Africa; External Researcher, Bonn International Centre for Conflict Studies, Germany; and Vice-President of the Environmental Peacebuilding Association.

A–Z Guides for Environment and Sustainability

This series provides accessible, easy-to-navigate overviews of a range of different topics related to environment and sustainability. Following an A–Z format, each book contains entries which map out an important concept or term and illustrates how it connects more broadly to other ideas and disciplines. With related terms and further reading included alongside the entries, these innovative volumes will be of great interest to students and scholars learning, teaching and researching in this field.

Essential Concepts of Sustainable Finance
An A–Z Guide
Edited by Elisa Aracil and Ibrahim Sancak

Essential Concepts of Land Politics
An A–Z Guide
Saturnino M. Borras Jr. and Jennifer C. Franco

Essential Concepts for Implementing the Sustainable Development Goals
An A–Z Guide
Edited by Frank Biermann, Thomas Hickmann, Yi hyun Kang, Carole-Anne Sénit and Yixian Sun

Essential Concepts of Environmental Peacebuilding
An A–Z Guide
Edited by Carl Bruch, Rachel Weaver, Richard Matthew, Erika Weinthal, and Larry Swatuk

Essential Concepts of Science Communication
An A–Z Guide
Edited by Declan Fahy and Todd P. Newman

For more information about this series, please visit: www.routledge.com/A-Z-Guides-for-Environment-and-Sustainability/book-series/AZES

Essential Concepts of Environmental Peacebuilding

An A–Z Guide

Edited by
Carl Bruch, Rachel Weaver, Richard Matthew,
Erika Weinthal, and Larry Swatuk

Routledge
Taylor & Francis Group
LONDON AND NEW YORK

Designed cover image: Getty Images

First published 2026
by Routledge
4 Park Square, Milton Park, Abingdon, Oxon OX14 4RN

605 Third Avenue, New York, NY 10158
and by Routledge

Routledge is an imprint of the Taylor & Francis Group, an informa business

British Library Cataloguing-in-Publication Data
A catalogue record for this book is available from the British Library

ISBN: 9781032994390 (hbk)
ISBN: 9781032994314 (pbk)
ISBN: 9781003604143 (ebk)

DOI: 10.4324/9781003604143

Typeset in Sabon
by Deanta Global Publishing Services, Chennai, India

Access the Support Material: www.routledge.com/9781032994314

Contents

Contributors

Osama Abu-Libda is a Research Expert on Natural Resources and Environment at Questscope for Social Development, Amman, Jordan

Rami Abi Ammar is a Program and Research Officer at the Arab NGO Network for Development, Beirut, Lebanon

Tareq Abu Hamed is the Executive Director of The Arava Institute for Environmental Studies and Dead Sea Arava Science Center, Southern Arava Valley, Israel

Shaadee Admadnia is the Director of Jurisdictional REDD+ at Conservation International, Arlington, Virginia, USA

Billy Agwanda is a PhD Student at the Carter School for Peace and Conflict Resolution at George Mason University, Fairfax, Virginia, USA

Saleem H. Ali is a Professor at the University of Delaware, Newark, Delaware, USA

Jeremy Allouche is a Professorial Fellow at the Institute of Development Studies, Brighton, England, United Kingdom

Bashar Alshawwa is a Water Diplomacy Specialist at Swim Drink Fish, Toronto, Ontario, Canada

Gabriela Nagle Alverio is a J.D./Ph.D Candidate in Environmental Policy at Duke University Sanford School of Public Policy, Nicholas School of the Environment, and School of Law, Durham, North Carolina, USA

Randall Amster is the Co-Director of the Environment and Sustainability Program at Georgetown University, Washington, D.C., USA

Eileen M. Angelini is a Renee Sachs Holocaust Education Fellow at the Crane Center for Mass Atrocity Prevention, Syracuse, New York, USA

Fakunle John Aremu is a Policy Advisor at the International Food Policy Research Institute, Washington, D.C., USA

Beatriz Arnal-Calvo is a Doctoral Researcher at the School of Humanities and Social Sciences, University of Brighton, Brighton, England, United Kingdom

Theresa (Isa) Arriola is an Assistant Professor of Anthropology at Concordia University, Montreal, Quebec, Canada

Jeremiah Ogonda Asaka is an Assistant Professor at Sam Houston State University, Huntsville, Texas, USA

Jill Baggerman is a Transboundary Water Cooperation and Peacebuilding Expert

Mohammed W. Bah is a PhD Student in Urban and Environmental Planning and Policy at University of California, Irvine, Irvine, California, USA

Megan Baker is an Assistant Professor of Anthropology at Northwestern University, Evanston, Illinois, USA

Timothy Balag'kutu is a Lecturer and Researcher at the Centre for Peace and Security Research, University of Professional Studies Accra, Accra, Ghana

Aleksandra Balinskaia is a PhD Candidate at Kanazawa University, Kanazawa, Ishikawa, Japan

Elsa Barron is a PhD Student at the University of Notre Dame, Notre Dame, Indiana, USA

Sinéad Barry is an Analyst at adelphi, Berlin, Germany

Luis Paulo Batista da Silva is a Professor, Department of Geography, Federal University of Bahia, Salvador, Bahia, Brazil

Pauline Baudu is the Director of the Climate Security Programme at the Conference of Defence Associations Institute and a PhD Candidate at Université du Québec à Montréal, Montreal, Quebec, Canada

Sarah Bebb is a PhD Researcher at the University College London, London, England, United Kingdom

Alexander Belyakov is an Independent Consultant, Toronto, Canada

Ambreen Tour Ben-Shmuel is a Research Fellow at the Harry S. Truman Institute for the Advancement of Peace, Jerusalem, Israel

Philippe Le Billon is a Professor at the University of British Columbia, Vancouver, British Columbia, Canada

Michelle Bonatti is an Associate Professor at the Leibniz Centre for Agricultural Landscape Research, Müncheberg, Brandenburg, Germany, University of Berlin, Berlin, Germany, and University of Vila Velha, Vila Velha, Espírito Santo, Brazil

Nancy Boyer is an Adjunct Assistant Professor, Department of Geography, and Visiting Scholar, Department of Political Science and International Relations, University of Delaware, Newark, Delaware, USA

Meryam El Bouhati is a PhD Researcher and Teaching Assistant at the University of Lille, Lille, Hauts-de-France, France

Gidon Bromberg is the Israel Director of EcoPeace Middle East, Tel Aviv-Yafo, Israel

Leif Brottem is an Associate Professor of Development Studies at Grinnell College, Grinnell, Iowa, USA

Oli Brown is an Associate Fellow at Chatham House, London, England, United Kingdom, and the Director of Alp Analytica, Samoëns, Auvergne-Rhône-Alpes, France

Carl Bruch is the Director of International Programs at the Environmental Law Institute, Washington, D.C., USA, and the Executive Director of the Environmental Peacebuilding Association, Washington, D.C., USA

Róisín Burke is an Assistant Professor of Public International Law at the Netherlands Institute of Human Rights, Utrecht University, Utrecht, the Netherlands

Chloe Canavan is a PhD Student at Murdoch University, Perth, Western Australia, Australia

Ana Elisa Cascão is a Consultant and Researcher

Tom Cavanagh is a PhD Candidate at the University of New South Wales, Sydney, New South Wales, Australia

Gülsena Çevirken is an LLM Candidate in Public International Law at Utrecht University, Utrecht, the Netherlands

Elizabeth Chalecki is an Associate Professor at the University of Nebraska, Omaha, Omaha, Nebraska, USA

Aili Winstanley Channer is a Graduate Student at the Warburg Institute at London University, London, England, United Kingdom

Alan Channer is a Senior Fellow at Global EverGreening Alliance, Melbourne, Victoria, Australia, and the Co-Director of OikoDiplomatique, Nairobi, Kenya

Esther C. Chinweuba is a Programme Officer for the UN Environment Programme World Conservation Monitoring Centre, Cambridge, England, United Kingdom

Jennifer Clapp is a Professor at the School of the Environment, Resources, and Sustainability, University of Waterloo, Waterloo, Ontario, Canada

Alexandre Corriveau-Bourque is a Lead Consultant at the Tenure Beat Consultancy, Montreal, Quebec, Canada

Gabriel Gomes Couto is a Programme Coordinator of Climate and Conflict at the Peace Dividend Initiative, Geneva, Switzerland

Alec Crawford is the Director of Nature for Resilience at the International Institute for Sustainable Development, Winnipeg, Manitoba, Canada

Cormac Cullinan is the Director of the Wild Law Institute, Cape Town, South Africa, and Co-Founder of the Global Alliance for the Rights of Nature, San Francisco, California, USA

Geoffrey Dabelko is a Professor at the Voinovich School of Leadership and Public Service, Ohio University, Athens, Ohio, USA

Simon Dalby is a Professor Emeritus at the Balsillie School of International Affairs, Wilfrid Laurier University, Waterloo, Ontario, Canada

Daniëlla Dam-de Jong is a Professor of International Sustainable Development Law, Grotius Centre for International Legal Studies, Leiden University, Leiden, the Netherlands

Silvia Danielak is an Assistant Professor at the Carter School for Peace and Conflict Resolution, George Mason University, Fairfax, Virginia, USA

Lynette de Silva is a Senior Instructor and Co-Director of the Program in Water Conflict Management and Transformation, Oregon State University, Corvallis, Oregon, USA

Susannah Dibble is a Staff Attorney at the Environmental Law Institute, Washington, D.C., USA

Elise Doumergue is an Environment, Fragility and Conflict Consultant at the World Bank, Washington, D.C., USA

Holly Dunn is an Assistant Professor at the Government Department, Colby College, Waterville, Maine, USA

Kendra Dupuy is an Assistant Professor at the Department of Political Science, Western Washington University, Bellingham, Washington, USA

Janet Edmond is the Senior Director of Peace and Development Partnerships at Conservation International, Arlington, Virginia, USA

Nico Edwards is a Doctoral Researcher at the School of Global Studies, University of Sussex, Brighton, England, United Kingdom

Kelechi Eleanya is a Cameron Speth Fellow at the World Resources Institute

Tom Ellison is the Deputy Director for the Center for Climate and Security, Washington, D.C., USA

Annika Erickson-Pearson is a Facilitator and Community Organizer

Abdullah Fahimi is a Research Fellow at DGAP's Center for Climate and Foreign Policy, Berlin, Germany

Rebecca L. Farnum a Lecturer in Environment and Security at Queen Mary University of London, London, UK

Bryan Fok is a Kellogg International Scholar at the University of Notre Dame, Notre Dame, Indiana, USA

Noah Fritzhand is a Research Fellow at the Center for Climate and Security, Washington, D.C., USA

Gina Gilson is a Postdoctoral Researcher at the University of Nevada-Reno, Reno, Nevada, USA

Sarah Glaser is a Senior Director at the World Wildlife Fund, Washington, D.C., USA

Sherri Goodman is the Secretary General of the International Military Council on Climate and Security, Washington, D.C., USA

Julia Gorricho is a Programme Coordinator at the Blue Action Fund, Berlin, Germany

Jacobo Grajales is a Professor of Political Science at Université Paris 1 Panthéon-Sorbonne, Paris, Île-de-France, France

Christina Greene is the Research Scientist at the Arizona Institute for Resilience, University of Arizona, Tucson, Arizona, USA

Silja Halle is a Programme Officer at the United Nations Environment Programme, Geneva, Switzerland

John Hardman is a Senior Instructor at the Department of Educational Leadership and Research Methodology, Florida Atlantic University, Boca Raton, Florida, USA, and Founder of Regenerative Organizations, Boca Raton, Florida, USA

Tracy Hart is the Lead Environmental Specialist and Global Lead on Environmental Risk Management and Fragile and Conflict States at the World Bank, Washington, D.C., USA

Nataliia Hendel is a Leading Researcher at the State Scientific Institution, Institute of Information, Security and Law of the National Academy of Legal Sciences of Ukraine, Kyiv, Ukraine

Anne Hennings is a Leading Researcher at the Peace Academy Rhineland-Palatinate, University of Kaiserslautern-Landau, Germany

Laura Higgins is an LLM Candidate in Public International Law at Utrecht University, Utrecht, the Netherlands

Thomas Homer-Dixon is the Founder and Executive Director of the Cascade Institute at Royal Roads University, Victoria, British Columbia, Canada

Larissa Jane Houston is an Energy and Climate Law Researcher at the University of Graz, Graz, Styria, Austria

Elaine (Lan Yin) Hsiao is an Assistant Professor at Kent State University, Kent, Ohio, USA

Karen Hulme is a Professor at the University of Essex, Colchester, England, United Kingdom

Saher Husain is an LLM Candidate in Public International Law at Utrecht University, Utrecht, the Netherlands

Tobias Ide is an Associate Professor of Politics and International Relations at Murdoch University, Perth, Western Australia, Australia

Machane Daniel Isaac is a PhD Research Scholar at the Department of International Relations, South Asian University, New Delhi, Delhi, India

Maria Ivanova is a Professor of Public Policy and Director of the School of Public Policy and Urban Affairs at Northeastern University, Boston, Massachusetts, USA

Marie G. Jacobsson is a former Ambassador and Principal Legal Adviser on International Law at the Swedish Ministry for Foreign Affairs, Stockholm, Sweden, and Associate Professor at the University of Lund, Lund, Skåne, Sweden

Martina Jaskolski is a Climate, Peace, and Security Policy Scientist at CGIAR and an Adjunct Faculty of Sociology at the American University in Cairo, Cairo, Egypt

Natalia Jiménez Galindo is an Independent Consultant

David Kaelin is a Head of Expert Group WASH at the SDC/HA, Swiss Federal Department of Foreign Affairs

Ousseyni Kalilou is the Founder and CEO of the Gum Arabic Institute of Poverty Alleviation (GAIPA), Greensboro, North Carolina, USA

Utchang Kang is a Programme Management Officer for the Global Mechanism of the UNCCD, Bonn, North Rhine-Westphalia, Germany

Mawa Karambiri is a Policy and Technical Engagement Specialist for the Sahel at the Center for International Forestry Research—World Agroforestry, Nairobi, Kenya

Anna Kasperovich is an LLM Candidate in Public International Law at Utrecht University, Utrecht, the Netherlands

Elena Kavanagh is a Lecturer in Law at the School of Law, University College Cork, Cork, Munster, Ireland

Rina Kedem is the Co-Director of the Jordan-Israel Center for Community, Environment, and Research at the Arava Institute for Environmental Studies, Southern Arava Valley, Israel

Ilan Kelman is a Professor of Disasters and Health at University College London, London, England, United Kingdom, and Professor II at University of Agder, Kristiansand, Agder, Norway

Erica Key is the US Global Hub Director of Future Earth, Montreal, Quebec, Canada

Shahab Khalid is a Climate Security Consultant at the Alliance for Bioversity and CIAT, Rome, Italy

Atieh Khatibi is a Research Project Assistant in Climate Law and Human Rights for the Research Center for Climate Law (ClimLaw: Graz) at the University of Graz, Graz, Styria, Austria

Elizabeth Koch is the Process Support Team Lead for the Women in Water Diplomacy Network

Florian Krampe is the Director of Studies, Peace and Development and Acting Director of the Climate Change and Risk Programme at the Stockholm International Peace Research Institute, Stockholm, Sweden

Thomas De Lathauwer is an LLM Candidate in Public International Law at Utrecht University, Utrecht, the Netherlands

Michael Lawrence is a Research Fellow at the Cascade Institute at Royal Roads University, Victoria, British Columbia, Canada

Ann J. Le is a PhD Candidate at the University of California, Irvine, Irvine, California, USA

Marja Lehto is the Ambassador for International Legal Affairs at the Ministry for Foreign Affairs of Finland, Helsinki, Finland, and Adjunct Professor of International Law at the University of Helsinki, Helsinki, Finland

Matti Lehtonen is a Senior Advisor at the United Nations Environment Programme, New York, New York, USA

David Lehrer is a Lecturer and the Director of the Center for Applied Environmental Diplomacy at the Arava Institute for Environmental Studies, Southern Arava Valley, Israel

Laurel Levin is a Junior Research Fellow at the Climate and Community Institute, and a graduate student at the School of International Service at American University, Washington, D.C., USA

Veronica Limeberry is an NSF Post-doctoral Fellow at the American University School of International Service, Washington, D.C., USA

Katharina Löhr is a Professor for Socioecological Forest Governance, Eberswalde University of Sustainable Development, Eberswalde, Brandenburg, Germany, and Leibniz Centre for Agricultural Landscape Research, Müncheberg, Brandenburg, Germany

Madeleine Loll is the Gender Administrator for the Environmental Peacebuilding Association, Washington, D.C., USA

Stephen Lonergan is a Professor Emeritus at the Department of Geography, University of Victoria, Canada, Victoria, British Columbia, Canada

Tobias von Lossow is a Research Fellow at the Clingendael Institute, The Hague, South Holland, the Netherlands, Research Associate at the Department of Water Governance, IHE Delft, Delft, South Holland, the Netherlands, and a Lecturer at Leiden University, Leiden, South Holland, the Netherlands

Päivi Lujala is a Professor at the University of Olu

Richard Marcantonio is an Assistant Professor of Environment, Peace, and Global Affairs at the Kroc Institute for International Peace Studies at the University of Notre Dame, Notre Dame, Indiana, USA

Kimberly R. Marion Suiseeya is an Associate Professor of Environmental Policy, Duke University, Durham, North Carolina, USA

Albert Martínez is a Programme Manager at the European Institute of Peace, Brussels, Belgium

Paula Gutiérrez Martínez is a GESI and Conflict Sensitivity Consultant at Global Green Growth Institute, Cape Town, South Africa

Stephanie Martínez is a PhD Candidate at the Urban and Environmental Planning and Policy Program, University of California, Irvine, Irvine, California, USA

Richard Matthew is a Professor at the University of California, Irvine, Irvine, California, USA

Francis Massé is an Assistant Professor at Durham University, Durham, England, United Kingdom

Melissa McCracken is an Assistant Professor at The Fletcher School at Tufts University, Medford, Massachusetts, USA

Leonardo Medina is a Climate Security Research Specialist, International Center for Tropical Agriculture (CIAT), Cali, Valle del Cauca, Colombia

Adriana Melgar is a Regional Evaluation Analyst for Food and Agricultural Organization Regional Office for Asia and the Pacific, Bangkok, Thailand

Stacey Mitchell is an Associate Professor at Georgia State University's Perimeter College, Decatur, Georgia, USA

Daimon Mkwandawire is the Mission Secretary on Ecology and Economy for the Council for World Mission, Singapore, Singapore

Saeede Mokhtarzade is a student in Human Rights at the University of Erlangen, Erlangen/Nuremberg, Bavaria, Germany

David Mozersky is the President and Co-founder of Energy Peace Partners, Sacramento, California, USA

Héctor Morales Muñoz is a Senior Advisor on Climate Diplomacy and Security at adelphi, Berlin, Germany

Faith Mutegi is a Knowledge Management Officer at Amref Health Africa, Nairobi, Kenya

Valery Ndagijimana is a PhD Candidate at the Leibniz Centre for Agricultural Landscape Research, Müncheberg, Brandenburg, Germany, and Leuphana University, Lüneburg, Lower Saxony, Germany

Emnet Negash is a Postdoctoral Researcher at Oregon State University's Conflict Ecology Lab, Corvallis, Oregon, USA

Iryna Nikolaieva is an Environmental Consultant at PAX, Utrecht, the Netherlands

Sarah Njeri is a Lecturer on Humanitarianism and Development at the Department of Development Studies, SOAS University of London, London, England, United Kingdom

Cyril Obi is the Program Director of the Social Science Research Council, Brooklyn, New York, USA

Irene Ojouk is a Doctoral Researcher at the Right Livelihood College, Centre for Development Research, University of Bonn, Bonn, North Rhine-Westphalia, Germany

Shannon O'Lear is a Professor at the University of Kansas, Lawrence, Kansas, USA

Daniel E. Orenstein is an Associate Professor at the Faculty of Architecture and Town Planning, Technion, Israel Institute of Technology, Haifa, Israel

Juliet Chiamaka Osueke is a Research Assistant at the Fund for Peace, Washington, D.C., USA

Vincent Otieno is a Lecturer at the Department of Geoscience and Environment, Technical University of Kenya, Nairobi, Kenya

Sara Ouasti is an LLM Candidate in Public International Law at Utrecht University, Utrecht, the Netherlands

B. Eni Owoeye is a Master of Public Policy Candidate at Duke University Sanford School of Public Policy, Durham, North Carolina, USA

Ariel Ozdemir is a Consultant for PAX, Utrecht, the Netherlands

Gül Özerol is an Associate Professor at the University of Twente, Enschede, Overijssel, the Netherlands

Grazia Pacillo is a Senior Scientist and Co-lead of the CGIAR FOCUS Climate Security at the Alliance of Bioversity and CIAT, Rome, Italy

Stavros Evdokimos Pantazopoulos is a Teaching Fellow at the School of Law, National and Kapodistrian University of Athens, Athens, Attica, Greece

Anastasia Papadopoulos is a PhD student at the Balsillie School of International Affairs, University of Waterloo, Waterloo, Ontario, Canada

Rejoice Papka is a PhD Researcher at the University of Greenwich, London, England, United Kingdom

Christina Parandii is the Senior Project officer Humanitarian Disarmament at PAX, Utrecht, the Netherlands

Meaghan Parker is the Executive Director at the Council for the Advancement of Science Writing, Seattle, Washington, USA

Hassan Partow is a Programme Manager for the Disasters and Conflicts Branch, United Nations Environment Programme, Geneva, Switzerland

Ananya Peddibhotla is a Graduate Research Assistant at the University of California, Irvine, Irvine, California, USA

Toni-Giovanni Pegurri is a Doctoral Researcher at the University of Lille, CERAPS, Lille, Hauts-de-France, France

Caroline Pellaton is the Operations Director at the Geneva Water Hub, Geneva, Switzerland

Nancy Peluso is a Professor at the University of California, Berkeley, Berkeley, California, USA

Laura E. R. Peters is an Assistant Professor at the College of Earth, Ocean, and Atmospheric Sciences, Oregon State University, Corvallis, Oregon, USA, and an Honorary Lecturer at University College London, London, England, United Kingdom

Liliana Pimentel is an Independent Researcher and Consultant and the Co-chair of the Water Interest Group at the Environmental Peacebuilding Association, Washington, D.C., USA

Benjamin Pohl is the Head of Programme, Climate Diplomacy and Security at adelphi, Berlin, Germany

Erik Post is a PhD Candidate at the University of British Columbia, Vancouver, British Columbia, Canada, and Senior Advisor to the European Partnership for Responsible Minerals, Brussels, Belgium

Maryruth Belsey Priebe is the Co-Founder of aXXelerate

Celestine Procter is a Senior Fellow at the Alliance for Peacebuilding, Washington, D.C., USA

Maureen Purcell is a Postdoctoral Scholar at the University of California, Irvine, Irvine, California, USA

Mara Pusic is a Research Associate at the Environmental Law Institute, Washington, D.C., USA

Harifidy Rakoto Ratsimba is a Researcher at the Land, Landscape, and Development Research Lab, LLandDev, University of Antananarivo, Antananarivo, Analamanga, Madagascar

Danni Reches is an Advocacy Officer at EcoPeace Middle East, Tel Aviv, Israel

Sidra Riaz is Presidential Fellow at the Steven J. Green School of International and Public Affairs, Florida International University, Miami, Florida, USA

Lauren Herzer Risi is a Senior Fellow and Director of Environmental Security at the Stimson Center, Washington, D.C., USA

Karam F. Robeil is an Environmental Policy Researcher for the American University of Iraq - Sulaimani, Sulaymaniyah, Iraq

Zoe Rosenblum is a PhD Student at Oregon State University, Corvallis, Oregon, USA

Galeo Saintz is the Chair of the IUCN Task Force on Migration, Environmental Change, and Conflict, Gland, Switzerland

Emily Sample is the Director of Research, Learning, and Evaluation for the Alliance for Peacebuilding, Washington, D.C., USA

Asiem Sanyal is a Marine Conservation Scientist and Artist

Zerrin Savaşan is an Associate Professor of Public International Law and Department of International Relations at Selçuk University, Konya, Turkey

Adam Savelli is a Regional Lead on South and Southeast Asia for the CGIAR FOCUS Climate Security, the Alliance of Bioversity International and CIAT, Rome, Italy

Nynke Schaap is a PhD Fellow at the University of Copenhagen, Copenhagen, Denmark

Marie Schellens is a Senior Analyst Environment, Peace, and Security at PAX, Utrecht, the Netherlands

Juliane Schillinger is a Technical Adviser at the Red Cross Red Crescent Climate Centre, The Hague, South Holland, the Netherlands

E. Lisa F. Schipper is a Professor at the Department of Geography, Institute of Geography, University of Bonn, Bonn, North Rhine-Westphalia, Germany

Susanne Schmeier is an Associate Professor and Head of Water Governance Department at IHE Delft, Delft, South Holland, the Netherlands

Aliakbar Shams is an Attorney and Member of the Young Professionals in Foreign Policy

Jeannie Shroads is a Consultant

Mirali Shukla is the Deputy Chair for the IUCN CEESP Theme on Environment and Peace, Gland, Switzerland

Joseph A. Siegel is Professor at the Haub School of Law at Pace University, White Plains, New York, USA

Emmanuel Sindikubwabo is the Executive Director of We Do Green Rwanda, Kigali, Rwanda

Radhika Singh is a Climate Security Specialist, Alliance of Bioversity and CIAT, Nairobi, Kenya

Shreya Singh is an Intern on Climate, Business and Peace at the Peace Dividend Initiative, Geneva, Switzerland

Britta Sjöstedt is a Senior Lecturer at the Faculty of Law at Lund University, Lund, Skåne, Sweden

Jeannie Sowers is a Professor of Political Science and International Affairs, University of New Hampshire, Durham, New Hampshire, USA

Phoebe Girouard Spencer is an Environmental Economist at the World Bank, Washington, D.C., USA

Ella Stack is a Consultant at the Global Environment Facility Independent Evaluation Office, Washington, D.C., USA

Tim Stork is a PhD Candidate and Lecturer at Tilburg University, Tilburg, North Brabant, the Netherlands

Andres Suarez is an Assistant Professor at Universidad de la Costa, Barranquilla, Atlántico, Colombia

Larry Swatuk is an Emeritus Professor at the University of Waterloo, Waterloo, Ontario, Canada, and an Extraordinary Professor at the Institute for Water Studies, University of the Western Cape, Cape Town, Western Cape, South Africa

Alice Taberner is a PhD Candidate at the University of São Paulo and King's College London, London, England, United Kingdom

Joram Tarusarira is an Associate Professor at the University of Groningen, Groningen, the Netherlands, and a Climate Security Research Associate for CGIAR, Rome, Italy

Ty Terlaak is an undergraduate honors student at the College of Earth, Ocean, and Atmospheric Sciences, Oregon State University, Corvallis, Oregon, USA

Bejoy K. Thomas is an Associate Professor at the Indian Institute of Science Education and Research, Pune, India

Dalmas Tiampati is the Founder of the Maasai Center for Regenerative Pastoralism, Kajiado, Kenya

Mara Tignino is a Senior Lecturer for the Faculty of Law, University of Geneva, and Vice Director of the Platform for International Water Law at the Geneva Water Hub, Geneva, Switzerland

Guido Trivellini is a Freelance Expert on Ecology and Development

Jessica Troell is a Senior Attorney and Director of the International Water Program and Africa Program at the Environmental Law Institute, Washington, D.C., USA

Eugenia Tsatsos is a Senior Security Advisor and Partner at Mirades Security Inc., Montreal, Quebec, Canada

Alexandra Turgul is a Research Associate at Oregon State University, Corvallis, Oregon, USA

Juha Uitto is a Visiting Scholar at the Environmental Law Institute, Washington, D.C., USA

Bernd Unger is the Senior Project Manager at Dorsch Impact GmbH, Essen, Germany

Pauliina Upla is an Independent Researcher

Emmanuel Valax is a PhD Researcher and Teaching Assistant at the Institute for Environmental Sciences, University of Geneva, Geneva, Switzerland

Arabella Angeline Vallee is a Member of the Environmental Peacebuilding Association Secretariat

Stacy D. VanDeveer is a Professor at the University of Massachusetts, Boston, Boston, Massachusetts, USA

Alessandra Vaselli is a Climate, Peace, and Mobility Specialist at the Alliance of Bioversity International and CIAT, Rome, Italy

Emmanuelle Veuillet is a postdoctoral researcher at the School for Advanced Studies in the Social Sciences, Institute of African Worlds, Paris, Île-de-France, France

Alina Viehoff is an Advisor at adelphi, Berlin, Germany

Janani Vivekananda is the Head of Programme for Climate Diplomacy and Security at adelphi, Berlin, Germany

Natalija Vojno is the Founder of Our Future First, Ontario, Canada

Rachel Weaver is a Master of Environmental Management Candidate at Duke University, Durham, North Carolina, USA

Erika Weinthal is a John O. Blackburn Distinguished Professor of Environmental Policy at Duke University, Durham, North Carolina, USA, and the President of the Environmental Peacebuilding Association, Washington, D.C., USA

Doug Weir is the Director of The Conflict and Environment Observatory, Hebden Bridge, West Yorkshire, United Kingdom

Lars Wirkus is the Senior Management and Head of Research Infrastructure and Data at the Bonn International Centre for Conflict Studies (BICC), Bonn, North Rhine-Westphalia, Germany

Amanda Woomer is the Principal and Founder of Baobab Tree Consulting, Atlanta, Georgia, USA

Jocelyn Wright is a Professor in the Department of English Language and Literature, Mokpo National University, Mokpo, South Jeolla, South Korea

Ekrem Yılmaz is a Research Assistant for the L.L.M Public International Law and Department of International Relations, Selçuk University, Konya, Turkey

Milen Yishak is a Consultant for the Climate and Security Expert Network, Washington, D.C., USA

Aseel Zahran is a Peace and Conflict Researcher

Christianne Zakour is the Caribbean Representative for the Loss and Damage Youth Grantmaking Council

Mark Zeitoun is the Director General of the Geneva Water Hub, Geneva, Switzerland

Nick Zuroski is a Senior Manager of Policy and Advocacy for the Alliance for Peacebuilding, Washington, D.C., USA

Wim Zwijnenburg is the Humanitarian Disarmament Project Leader at PAX, Utrecht, the Netherlands

Acknowledgments

The creation of this publication represents a collaborative endeavor that was enriched by the insight, experience, and dedication of numerous individuals and institutions. This resource would not have been possible without their collective contributions.

We extend our deepest gratitude to the many contributors who devoted their time and expertise to crafting the entries in this resource. Their thoughtful scholarship and commitment to advancing the field of environmental peacebuilding have been invaluable. The diversity of perspectives and experiences they brought to this work has resulted in a resource that reflects the global, interdisciplinary nature of environmental peacebuilding.

We are grateful to Osama Abu-Libda, Angela Amaya, Fakunle John Aremu, Alexander Belyakov, Octavio Goio, Paula Gutiérrez Martínez, Suleiman Halasah, Bello Ifeoluwa Oreofe, Liliana Pimentel, Thomas Reeve, Marie Schellens, Juliane Schillinger, Alma Segoviano, Gisele Sirot, Manuel Soque, Sophia Stanger, Eugenia Tsatsos, and Wenhan Zhao, who provided the translations of terms (found in the online annex) and who made this work more accessible to a wider global audience. Their efforts ensure that this resource can serve communities and practitioners across different linguistic and cultural contexts, embodying the inclusive spirit that defines environmental peacebuilding.

Special recognition goes to the participants of the workshop, "Toward an Environmental Peacebuilding Glossary: A Consultative Workshop," held at the Third International Conference on Environmental Peacebuilding in The Hague, the Netherlands. Their engagement and constructive feedback refined the methodology, identified terms for inclusion, and advanced our collective understanding of the terminology used in this evolving field.

We also acknowledge the valuable contributions of many members from the Environmental Peacebuilding Association, who proposed terms for inclusion and provided ongoing support throughout the project's development. We are especially grateful to Jill Baggerman, Erin Czelusniak, Karishma Goswami, and Atieh Khatibi, who helped review, edit, and refine the final publication.

Finally, we extend our sincere gratitude to the Jubitz Family Foundation and the PeaceNexus Foundation for their generous financial support, which has made it possible to offer this publication as an open-access resource and support outreach and dissemination. This funding ensures that practitioners, researchers, students, and communities worldwide can benefit.

AI Acknowledgment

Some contributors to this volume used AI-based tools for proofreading, improving language clarity, and formatting citations. No AI tools were used for content generation or substantive writing.

Introduction: Fostering Dialogue of Environmental Peacebuilding

Toward a Shared Understanding

Rachel Weaver, Richard Matthew, Erika Weinthal, Larry Swatuk, and Carl Bruch

Introduction

Environmental peacebuilding is an emerging area integrating multiple disciplines and fields of practice. It brings together a variety of individuals, communities, organizations, and institutions in common cause—to explore the possibilities for environmental action to foster inclusive and durable forms of peace. Yet not everyone understands environmental peacebuilding in the same way, especially as the terminology that signifies and shapes environmental peacebuilding in theory and practice continues to evolve. Experience gives rise to new theories and practices, which in turn foster debate, discussion, revision, and adaptation. Through this interplay of research, action, and debate, a discourse of environmental peacebuilding is emerging, where discourse may be defined as "a way of constructing meanings which influences and organizes both our actions and our conception of ourselves" (Abercrombie et al. 1988, 71). By assembling a compendium of essential concepts, organizations, and related technical terms, this glossary aims to describe how different constituencies understand certain terms and to begin a conversation to improve communication and build trust. There are several potential uses of the glossary: in training, research (especially across disciplines), and in translation.

We recognize that discourse emerges from the social relations of power. As stated by Barnett and Finnemore (1999, 711):

> IOs [International Organizations] exercise power by virtue of their ability to fix meanings, which is related to classification. Naming or labeling the social context establishes the parameters, the very boundaries, of acceptable action. Because actors are oriented toward objects and objectives on the basis of the meaning that they have for them, being able to invest situations with a particular meaning constitutes an important source of power. IOs do not act alone in this regard, but their organizational resources contribute mightily to this end.

In compiling this glossary, we are conscious of the potential for this project to embody Western bias and perceptions that could support forms of neocolonialism. These concerns have long been expressed in relation to the broader domain of peacebuilding, which, especially in its original formulations, was predisposed to using Western beliefs, values, practices, and institutions as benchmarks for setting targets and measuring success. To be clear, we are not trying to force agreement on terms or fix their meaning; we recognize the contested character of concepts and terms, the complexity of peacebuilding situations, and the need for local understandings and aspirations to shape outcomes.

DOI: 10.4324/9781003604143-1

At the same time, within peacebuilding, there are important elements of continuity and opportunities to share best practices. Hence, it is worthwhile to gather the key terms and concepts that have emerged over three decades of operation to facilitate the organic development of a common understanding of mutually beneficial socio-ecological action. We thus view this glossary as a living document, and we anticipate that understanding will evolve. This glossary is an initial and incomplete entry point, not a definitive end point. We used a crowdsourcing methodology, soliciting terms from the environmental peacebuilding community, and some terms will resonate more for some than others. Several terms that were suggested were omitted for a variety of reasons, including a lack of suitable contributors, the narrowness of the term, or the more appropriate positioning of a term in the glossary of a related field of study (such as political science, environmental studies, or sociology).

We aimed for a broad, inclusive approach, and this is reflected in the list of contributors: 229 authors from 57 countries who work in diverse professions and sectors. Most entries have more than one contributor, encouraging the weaving of different perspectives and experiences. Indeed, this was the first time that some of the contributors collaborated. There is an understandable tendency for researchers to seek to clarify the range of meanings a term embodies and then use this clarity for identifying areas of agreement and disagreement, aiming to build a generalizable understanding of a concept or issue. At the same time, there is a tendency among practitioners to be highly sensitive to local "on the ground" context, attuned to terms that might be divisive, focused on co-developing meaning as part of the peace process itself. In most cases, entries reflect a combination of the generalizable and the specific. In every case, we asked contributors to reflect on how the term applies to environmental peacebuilding and to provide specific examples.

Context

The above caveats notwithstanding, the high level of interest in this volume from practitioners, researchers, and policymakers suggests broad recognition of the importance of environmental peacebuilding (EPB). It is an area of policy, action, and analysis that emerged in the 1990s as part of a transformative and interlocking set of events and initiatives including, among many others, the end of the Cold War in 1991, the launch of the first website that same year, the Rio Earth Summit in 1992, and the genocide in Rwanda in 1994 (Bruch et al. 2019; Ide et al. 2021). Influential responses to the period of intense global change that characterized the late 1980s and 1990s included widespread efforts to rethink security (Buzan 1997), broad confidence in liberal democracy as a model for political and economic success (Fukuyama 1989), and a desire to move the global economy into a new framework informed by environmental science and ethics (WCED 1987).

Within this turbulent context, UN Secretary-General Boutros Boutros-Ghali determined that the end of the Cold War created an unprecedented opportunity for the UN to address an alarming increase in armed violence, especially in Africa and the Middle East, that was reversing the global trend toward peace experienced in the years after World War II (Cravo 2018). Building on the work of Johan Galtung, who coined the term "peacebuilding" in about 1975 (Galtung 1976), Boutros-Ghali released *An Agenda for Peace* (1992), which marked a new direction for the United Nations. The key innovation in this report was the integration of Galtung's concept of peacebuilding—of moving beyond peacemaking (agreements to cease hostilities) and peacekeeping (ensuring the terms of peace agreements were implemented)—to mobilize resources with the objective of creating the conditions for enduring peace by addressing the "root causes of conflict"—its social and structural factors (UN 1992, para. 15).

In the decade following the launch of the Secretary-General's report, peacebuilding was operationalized through four categories of action: professionalizing security forces, developing an economic plan attractive to investors, strengthening governance institutions, and implementing a national reconciliation process (Cravo 2018). Critics of this approach underscored its top-down, cookie-cutter tendencies, which were implicitly guided by the idea that countries like Canada and Sweden were good models to emulate, and worried that peacebuilding was essentially recolonizing spaces coming out of war (Chandler 2017; Cravo 2018; Tanabe 2017). Their calls for greater sensitivity to local contexts, more realistic time frames for change, the broad inclusion of civil society, and reconstruction and visioning processes led by local stakeholders, together with on-the-ground experience that made it clear that peace could be supported but not imposed from the outside, have catalyzed the past two decades of reimagining and evolution. Today, according to the UN, peacebuilding:

> refers to efforts to assist countries and regions in their transitions from war to peace and to reduce a country's risk of lapsing or relapsing into conflict by strengthening national capacities for conflict management and laying the foundations for sustainable peace and development.
>
> (UN n.d.)

As the concept of peacebuilding was maturing through research and practice, a parallel multilateral effort focused on coordinating a global response to shared environmental challenges, which were not formally part of the early peacebuilding agenda. This tendency to ignore or compartmentalize environmental issues has a long history—indeed, the founding documents of the UN did not mention the natural environment at all (and made only one reference to economic development in Article 55 of the UN Charter).

Nonetheless, in 1968, alarmed by a growing body of scientific evidence, Sweden asked the UN Economic and Social Council (ECOSOC) to consider organizing a conference on the environment. In 1969, through Resolution 2398, the UN General Assembly announced that a conference on the human environment would be convened in Stockholm in 1972. The challenge that animated the Stockholm Conference, carefully articulated by Barbara Ward and René Dubos in *Only One Earth* (1972), was to protect the natural environment and facilitate development in the Global South while working in the tense security context of the Cold War. The participants' hope was that shared recognition of the somewhat objective scientific basis for environmental protection might catalyze cooperation in a deeply divided world. Since that time, many efforts have been made to link the environment to the UN's development and security priorities, including in the discussions at the 1972 UN Conference on the Human Environment and the accompanying Stockholm Declaration (Westing 2013).

These efforts were powerfully amplified when the first Earth Summit (1992) highlighted the gravity of trends in climate change, biodiversity loss, and pollution, and called for broad adoption of a paradigm of sustainable development. The release of *An Agenda for Peace* in 1992, followed two years later by the pioneering UNDP report linking its programs to the concept of human security, provided two novel pathways for mobilizing support around the importance of the environment to human and national security (Matthew et al. 2009).

Much initial focus was given to the ways in which global environmental change might cause, trigger, or amplify violent conflict (Homer-Dixon 1994). This line of analysis quickly generated widespread policy interest and support but also significant criticism and counter-arguments (Deudney and Matthew 1999; Swatuk 2000). For example, Mark Halle, who directed the International Institute for Sustainable Development's (IISD's) European office, commissioned a state-of-the-art review assessing research linking environment, security, and development (1999). When Achim Steiner became the Director General of the International Union for the Conservation of Nature (IUCN), he became interested in this work and commissioned another study with Halle, *Conserving the Peace*, which focused on exploring the potential role of the conservation community in peacebuilding activities (Matthew et al. 2002). Steiner later became the Executive Director of the United Nations Environment Programme and initiated the formal integration of environment and peacebuilding, beginning with the policy justification document, *From Conflict to Peacebuilding: The Role of Natural Resources and the Environment* (Matthew et al. 2009).

As these efforts were unfolding in the UN, Conca and Dabelko were systematically making the case for an environmentally sensitive approach to the entire spectrum of peace-related activities, presented in their 2002 edited volume on *Environmental Peacemaking* (Conca and Dabelko 2002). This influential work countered the predominant narrative of resource-driven conflicts and, combined with relevant work across the globe, generated interest in the field that this volume addresses.

Methodology

The development of this publication was highly collaborative and iterative. We began by brainstorming and crowdsourcing potential entries to draw on the expertise of scholars and practitioners in the field of environmental peacebuilding. An initial review meeting was held during the Third International Conference on Environmental Peacebuilding (June 2024), where participants provided feedback on the objectives of the glossary, format of entries, and proposed entries, suggesting additional terms.

Following this, we secured authors for individual entries, with an emphasis on fostering collaboration across regions and disciplines. Authors were identified through open announcements shared by the Environmental Peacebuilding Association as well as through nominations from colleagues across the community. Teams of one to three authors (occasionally more) were asked to draw upon the existing literature to draft a concise definition of their term and provide a contextual exploration of its significance within environmental peacebuilding. This included, where applicable, tracing the origins of the concept, highlighting varying interpretations, and examining any applications across related disciplines. For entries focused on institutions, authors were asked to provide a clear description of that institution, including its history, role, and impact within the context of environmental peacebuilding. Key literature was cited throughout each entry to support the explanations and interpretations. Where terms carried broader disciplinary meanings, contributors were encouraged to clarify their specific significance for this field and to reflect diverse interpretations of their uses. Each entry then underwent a review and revision process by the editors to ensure consistency, clarity, and rigor.

Moving the Field Forward

There has been tremendous enthusiasm for this project. The development of a common reference point has been seen as an important step in supporting communication and collaboration within a rapidly expanding field. Deciding what to include has required careful judgment, as has determining how best to link and organize related terms. Additionally, decisions about which organizations to feature reflect both the breadth of institutions working in this space and the limits of what can be captured in a single resource.

Numerous challenges remain for environmental peacebuilding. Ours is a world of turbulence, of dramatic and constant change across social, political, economic, and environmental systems. Concerns about standardizing terminology within a context of flux are especially salient. Terminology in the field has always been nuanced, contextual, and continually evolving. Even foundational concepts such as "environmental peacebuilding" have been employed with different understandings and definitions across disciplines, languages, and regions. This variety can sometimes result in confusion and fragmentation in both academic and policy settings, and yet it also reflects the diverse contexts and evolving global challenges in which environmental peacebuilding is innovated, developed, and practiced.

Accordingly, this glossary is not intended to impose a rigid set of standardized definitions. Instead, it aims to help clarify how different people think about terms, including the same terms when used with different understandings. By making these distinctions visible, the glossary fosters greater awareness of the underlying assumptions shaping how environmental peacebuilding is conceived and practiced. In this sense, it facilitates understanding across perspectives and helps strengthen linkages in an evolving and maturing field.

Open access is central to the project. This glossary is intended to be part of a broader suite of platforms, resources, and activities that make knowledge in environmental peacebuilding widely accessible. In this way, it contributes to the democratization of the field, supporting the participation of practitioners, scholars, and communities across regions and disciplines. Where possible, the publication strives toward a broader common understanding and consensus. However, the ongoing evolution of terms and institutions means that consensus must be balanced with adaptability (Weaver 2025). The case of "post-conflict environmental assessment," as well as the role of UN peacekeeping structures such as what used to be called the Department of Peacekeeping Operations and is now the Department of Peace Operations, illustrates how terminology evolves over time in response to practice and institutional engagement.

This glossary also seeks to build conceptual bridges. Environmental peacebuilding necessarily crosses sectors, disciplines, and domains of conflict and peace dynamics. It requires attention to the important interactions between practitioners and researchers, as well as recognition that the same words—such as "conflict," "environment," "sustainability," or "resilience"—can hold quite different meanings depending on a variety of factors, such as a person's lived experience, positionality, disciplinary background, type of work, and role within the work organization or institution. Some disciplines have precise definitions for terms, while some terms are used across disciplines, and there are distinct conceptual priorities for different fields. These differences, rather than being a barrier, can provide a dynamic tension that interrogates prior understandings, tests them against experiences, and generates new insights when acknowledged and brought into dialogue.

Precisely because environmental peacebuilding manifests itself differently across temporal and geographic spaces and scales, the language used can take on different meanings. We have endeavored to translate the terms into as many languages as possible. Translations were completed by native speakers of these languages who offered their time and expertise, many of whom were connected to the Environmental Peacebuilding Association. Terms were translated into Arabic, Chinese, Dutch, French, German, Portuguese, Russian, Spanish, Ukrainian, and Yoruba. The languages represented reflect the contributions that were offered, rather than a predetermined set, underscoring the community-driven and inclusive nature of this project. Generally, some terms were easier to translate across languages, while others were more difficult, as a result of conceptual, cultural, or contextual differences. Particularly where concepts were relatively new, translations were not yet agreed upon, and they often had direct/literal translations.

Conclusion

Ultimately, we hope the publication contributes to ongoing efforts to foster dialogue and deepen shared understanding of key concepts in environmental peacebuilding. Engaging with the diversity of terminology and interpretations helps strengthen the foundations of a field that is both dynamic and essential to addressing the interconnected challenges of environment, conflict, and peace. Striking the balance between conceptual clarity and adaptability is necessary for environmental peacebuilding to remain both grounded and innovative in responding to emerging challenges.

As conflict and violence continue across the world, it remains essential for scholars and practitioners to pay attention to the environmental underpinnings of such conflicts and the environmental potential for peacebuilding. Hopefully, this glossary will provide an invaluable tool for those who seek to further human security, sustainable natural resource management, and environmental cooperation.

Environmental peacebuilding is rapidly maturing as a field. Central to its development is the interchange across disciplines, geographies, languages, scales, and theory and practice. This glossary is understood as a living process. We envision an ongoing, iterative process, and we encourage you to engage. If you have comments, additions, praise, or feedback, please reach out to us at glossary@environmentalpeacebuilding.org.

References

Abercrombie, Nicholas, Stephen Hill, and Bryan S. Turner, eds. 1988. *The Penguin Dictionary of Sociology*. Penguin.

Barnett, Michael N., and Martha Finnemore. 1999. "The Politics, Power, and Pathologies of International Organizations." *International Organization* 53 (4): 699–732. https://doi.org/10.1162/002081899551048.

Boutros-Ghali, Boutros. 1992. "An Agenda for Peace: Preventive Diplomacy, Peacemaking and Peace-keeping." *International Relations* 11 (3): 201–18.

Bruch, Carl, David Jensen, Mikiyasu Nakayama, and Jon Unruh. 2019. "The Changing Nature of Conflict, Peacebuilding, and Environmental Cooperation." *Environmental Law Reporter* 49 (2): 10134–154.

Buzan, Barry. 1997. "Rethinking Security after the Cold War." *Cooperation and Conflict* 32 (1): 5–28. https://www.jstor.org/stable/45084375.

Chandler, David. 2017. *Peacebuilding: The Twenty Years' Crisis, 1997–2017*. Springer.

Conca, Ken, and Geoffrey D. Dabelko, eds. 2002. *Environmental Peacemaking*. Woodrow Wilson Center Press.

Cravo, Teresa Almeida. 2018. "Peacebuilding: Assumptions, Practices and Critiques." *ASPJ: Africa and Francophonie* 8 (1): 4–20. https://www.airuniversity.af.edu/Portals/10/ASPJ_French/journals_E/Volume-09_Issue-1/cravo_e.pdf.

Deudney, Daniel H., and Richard A. Matthew, eds. 1999. *Contested Grounds: Security and Conflict in the New Environmental Politics*. State University of New York Press.

Fukuyama, Francis. 1989. "The End of History?" *The National Interest* 16: 3–18. http://www.jstor.org/stable/24027184.

Galtung, Johan. 1976. "Three Approaches to Peace: Peacekeeping, Peacemaking, and Peacebuilding." In *Peace, War and Defense: Essays in Peace Research*, Vol. II, edited by Johan Galtung. Christian Ejlers.

Halle, Mark, Geoffrey Dabelko, Steve Lonergan, and Richard Matthew. 1999. "State-of-the-Art Review on Environment, Security and Development Co-operation." Report for the Working Party on Development and Cooperation and Environment, OECD Development Assistance Committee.

Homer-Dixon, Thomas F. 1994. "Environmental Scarcities and Violent Conflict: Evidence from Cases." *International Security* 19 (1): 5–40. https://www.jstor.org/stable/2539147.

Ide, Tobias, Carl Bruch, Alexander Carius, et al. 2021. "The Past and Future(s) of Environmental Peacebuilding." *International Affairs* 97 (1): 1–16. https://doi.org/10.1093/ia/iiaa177.

Matthew, Richard Anthony, Jon Barnett, Bryan McDonald, and Karen O'Brien, eds. 2009. *Global Environmental Change and Human Security*. MIT Press.

Matthew, Richard Anthony, Oli Brown, and David Jensen. 2009. *From Conflict to Peacebuilding: The Role of Natural Resources and the Environment*. Earthprint.

Matthew, Richard Anthony, Mark Halle, and Jason Switzer, eds. 2002. *Conserving the Peace: Resources, Livelihoods and Security*. International Institute for Sustainable Development.

Swatuk, Larry. 2000. "Power and Water: The Coming Order in Southern Africa." In *The New Regionalism and the Future of Security and Development,* edited by Björn Hettne, András Inotai, and Osvaldo Sunkel. Palgrave Macmillan.

Tanabe, Juichiro. 2017. "Beyond Liberal Peacebuilding: A Critique of Liberal Peacebuilding and Exploring a Postmodern Post-liberal Hybrid Model of Peacebuilding." *International Relations and Diplomacy* 5 (8): 447–59.

United Nations (UN). 1992. "An Agenda for Peace." Accessed January 31, 1992. http://www.un-documents.net/a47-277.htm.

United Nations (UN). n.d. "Peacebuilding." Accessed September 21, 2025. https://www.un.org/peacebuilding/.

United Nations General Assembly (UNGA). 1992. *Report of the Secretary-General Pursuant to the Statement Adopted by the Summit Meeting of the Security Council on 31 January 1992, An Agenda for Peace: Preventive Diplomacy, Peacemaking and Peace-Keeping*. A/47/277 (June 17). United Nations.

Ward, Barbara, and Rene Dubos. 1972. *Only One Earth: The Care and Maintenance of a Small Planet*. W. W. Norton.

Weaver, Rachel. 2025. "Tracing the Evolution of Terms and Concepts in Environmental Peacebuilding: A Critical Analysis." Nicholas School of the Environment, Duke University. https://hdl.handle.net/10161/32242.

Westing, Arthur H. 2013. *Arthur H. Westing: Pioneer on the Environmental Impact of War*. Springer.

World Commission on Environment and Development (WCED). 1987. *Our Common Future*. Oxford University Press.

30x30

Elaine (Lan Yin) Hsiao and Sarah Glaser

30x30 is an international agreement to:

> ensure and enable that by 2030 at least 30 per cent of terrestrial, inland water, and of coastal and marine areas, especially areas of particular importance for biodiversity and ecosystem functions and services, are effectively conserved and managed through ecologically representative, well-connected and equitably governed systems of protected areas and other effective area-based conservation measures, recognizing indigenous and traditional territories, where applicable, and integrated into wider landscapes, seascapes and the ocean, while ensuring that any sustainable use, where appropriate in such areas, is fully consistent with conservation outcomes, recognizing and respecting the rights of indigenous peoples and local communities, including over their traditional territories.
>
> (Convention on Biological Diversity [CBD] 2022)

The term "30x30" derives from the Kunming-Montreal Global Biodiversity Framework's (GBF) Target 3 quoted above, replacing its predecessor, Aichi Biodiversity Target 11, which aimed to protect 17% of land and 10% of coastal and marine areas by 2020 (CBD Secretariat 2020). Although Aichi Biodiversity Target 11 was not met in 2020, 30x30 proponents, organized as the Global Deal for Nature (GDN), wanted an ambitious goal that would continue to push State Parties, donors, and the international **conservation** community toward their end goal of 50% of the planet protected by 2050 or "Half-Earth" (Wilson 2016). The GBF's 2050 Goals are ambitious, seeking to "substantially increas[e] the area of natural ecosystems by 2050" and to significantly reduce the extinction rate of all species (CBD Secretariat 2023). In the oceans, 30x30 initiatives include the designation and expansion of marine protected areas within coastal waters. The consideration of Other Effective Conservation Measures (OECMs) in the GBF would increase the ability for protected ocean spaces to also respect and protect cultural heritage sites and other traditionally managed areas. More recently, the protection of the high seas, including important marine wildlife corridors, areas of fisheries productivity, and protection of the deep

DOI: 10.4324/9781003604143-2

sea have arisen as countries consider ratifying the United Nations High Seas Treaty (United Nations 2003).

While support for 30x30 has been widespread, critics have emphasized the importance of where and how 30x30, and even Half-Earth, is achieved. Studies examining areas of conservation interest or importance where expansion of protected areas could be established to meet GBF Target 3 indicate that 1–1.87 billion people live in these territories and would be directly affected if half the planet were set aside for area-based conservation (Schleicher et al. 2019; Worsdell et al. 2020). Notably, these people are in large part Indigenous or Afro-descendant, sustaining traditional cultures and livelihoods, and already marginalized within their countries. In a joint submission to the CBD Secretariat, 128 organizations and individuals expressed concern that even with a 30% target, they estimated "that up to 300 million people could be negatively and seriously affected" (Minority Rights Group et al. 2020). In a report by the **International Union for the Conservation of Nature (IUCN)**, experts warn that conservation goals should not further displace peoples without their consent, even in the face of **climate change** and **biodiversity** crises, and instead urged conservationists to engage in **conflict** and migration-sensitive approaches that could support peaceful transitions (Hsiao et al. 2024). Unfortunately, the violent mass evictions of Indigenous Maasai from their homelands in early 2022 because of a third expansion of the Ngorongoro Conservation Area in Tanzania set a poor example for what 30x30 would mean for **human rights** and environmental conflicts (Currier and Mittal 2021; Amnesty International 2022; Forest Peoples Programme 2022).

In places like California, however, 30x30 has been taken up and applied to return lands to Indigenous guardianship. California's Pathways to 30x30 confirmed in 2024 that the state set aside 26.1% of lands and 21.9% of coastal waters under long-term conservation, including through ancestral land returns to California Native American tribes (California Natural Resources Agency 2025). In April 2022, the Governor's Administration issued Pathways to 30x30, stating that California's initiative "aligns with broader state commitments to advance **justice**, equity, diversity, and inclusion, strengthen tribal partnerships, and sustain our economic prosperity, clean energy resources, and food supply" (California Natural Resources Agency 2022). These efforts have been facilitated by the governor's formal apology to California Native American Tribes (Executive Order N-15-19), its administration policy on Native American Ancestral Lands, and a $100 million fund for Tribal **Nature-Based Solutions** (Newsom 2020; Newsom 2019; California Natural Resources Agency 2018). Indigenous-led conservation, including through land return, remains a promoted method for achieving 30x30 in ways that align with both **environmental justice** and **environmental peacebuilding** (Hsiao 2024).

References

Amnesty International. 2022. "Tanzania: Halt Brutal Security Operation in Loliondo." June 15. https://www.amnesty.org/en/latest/news/2022/06/tanzania-halt-brutal-security-operation-in-loliondo/.

California Natural Resources Agency. 2018. "Tribal Nature-Based Solutions Program." https://resources.ca.gov/Initiatives/Tribalaffairs/TribalNatureBasedSolutionsProgram.

California Natural Resources Agency. 2022. *Pathways to 30x30: Accelerating Conservation of California's Nature.* Administration of Governor Gavin Newsom, effective April 22. https://

resources.ca.gov/-/media/CNRA-Website/Files/Initiatives/30-by-30/Final_Pathwaysto30x30_042022_508.pdf.

California Natural Resources Agency. 2025. *Pathways to 30x30: California Annual Progress Report*. Administration of Governor Gavin Newsom. https://resources.ca.gov/-/media/CNRA-Website/Files/2025-30x30-Pathways-Progress-Report.pdf.

Convention on Biological Diversity (CBD). 2022. "Kunming-Montreal Global Biodiversity Framework." Decision Adopted by the Conference of the Parties to the Convention on Biological Diversity 15/4. CBD Secretariat, December 19. https://www.cbd.int/doc/decisions/cop-15/cop-15-dec-04-en.pdf.

Convention on Biological Diversity (CBD) Secretariat. 2020. "Aichi Biodiversity Targets." In *Strategic Plan for Biodiversity 2011–2020, Including Aichi Biodiversity Targets*. https://www.cbd.int/sp/targets.

Convention on Biological Diversity Secretariat. 2023. "2050 Goals." In *Kunming-Montreal Global Biodiversity Framework, Section G.*, December 20. https://www.cbd.int/gbf/goals.

Currier, Andy, and Anuradha Mittal. 2021. *The Looming Threat of Eviction: The Continued Displacement of the Maasai Under the Guise of Conservation in Ngorongoro Conservation Area*. The Oakland Institute, June 10. https://www.oaklandinstitute.org/looming-threat-eviction.

Forest Peoples Programme. 2022. "'We Will Not Leave. Not Now. Not Ever': Tanzanian State Violently Seizes Maasai Land in Loliondo." https://www.forestpeoples.org/pt/publications-resources/press-releases/article/press-release-we-will-not-leave-not-now-not-ever-tanzanian-state-violently-seizes-maasai-land-in-loliondo/.

Hsiao, Elaine Lan Yin. 2024. "Conversations on Indigenous and Decolonial Approaches to Environmental Peacebuilding." Interview by Kelsey Coolidge. *Peace Science Digest*, Special Issue on Decolonial and Indigenous Approaches to Environmental Peacebuilding. https://warpreventioninitiative.org/peace-science-digest/conversations-on-indigenous-and-decolonial-approaches-to-environmental-peacebuilding/.

Hsiao, Elaine Lan Yin, Richard A. Matthew, Philippe Le Billon, and Galeo Saintz. 2024. *Planet on the Move: Reimagining Conservation at the Intersection of Migration, Environmental Change and Conflict*. IUCN CEESP.

Minority Rights Group, Rainforest Foundation UK, and Survival International. 2020. "NGO Concerns over the Proposed 30% Target for Protected Areas and Absence of Safeguards for Indigenous Peoples and Local Communities." 1 September. https://assets.survivalinternational.org/documents/1959/final-en-ngo-concerns-over-the-proposed-30-target-for-protected-areas-and-absence-of-safeguards-for-indigenous-people-and-local-communities-200901.pdf.

Newsom, Gavin. 2019. *Executive Order N-15-19*. Office of the Governor. https://www.gov.ca.gov/wp-content/uploads/2019/06/6.18.19-Executive Order.pdf?emrc=b13680.

Newsom, Gavin. 2020. *Statement of Administration Policy: Native American Ancestral Lands*. Office of the Governor, September 25. https://www.gov.ca.gov/wp-content/uploads/2020/09/9.25.20-Native-Ancestral-Lands-Policy.pdf.

Schleicher, Judith, Julie G. Zaehringer, Constance Fastré, Bhaskar Vira, Piero Visconti, and Chris Sandbrook. 2019. "Protecting Half of the Planet Could Directly Affect over One Billion People." *Nature Sustainability* 2 (12): 1094–96. https://doi.org/10.1038/s41893-019-0423-y.

United Nations (UN). 2003. *Agreement Under the United Nations Convention on the Law of the Sea on the Conservation and Sustainable Use of Marine Biological Diversity of Areas Beyond National Jurisdiction*. United Nations. https://www.un.org/bbnjagreement/sites/default/files/2024-08/Text%20of%20the%20Agreement%20in%20English.pdf.

Wilson, Edward O. 2016. *Half-Earth: Our Planet's Fight for Life*. Liveright Publishing Corporation.

Worsdell, Thomas, Kundan Kumar, James R. Allan, et al. 2020. *Rights-Based Conservation: The Path to Preserving Earth's Biological and Cultural Diversity?* Washington, DC: Rights and Resources Initiative, November 24. https://rightsandresources.org/wpcontent/uploads/2020/11/Final_Rights_Conservation_RRI.pdf.

A

Accountability

Róisín Burke, Anna Kasperovich, and Sara Ouasti

Accountability, broadly speaking, may arise for states, individuals, and non-state actors such as corporations when they breach a legal obligation through an action or omission. This notion implies that those with authority can be held responsible for their conduct. It can include being held to account before a court of law and/or having consequences in terms of a need to provide reparation to victim(s) of wrongdoing.

International environmental law consists of a system of mechanisms and measures that attempt to better ensure actors' compliance with their obligations as well as to provide avenues for redress for harm occurring (Voigt 2021). Nevertheless, regulatory frameworks in environmental **governance** are fragmented and weak, which means that outcomes are often poor in terms of mitigating environmental harm (Kramarz and Park 2016).

With regards to states and international organizations, a wrongful act must be attributed to them and needs to violate a substantive legal obligation under primary international law (e.g., treaty law, customary international law) (Henckaerts 2001). Under the secondary international rules on state responsibility, it can be triggered either through actions or omissions of States through their organs and agents (Bodansky and Crook 2002). Responsibility can also be shared across multiple actors. Regarding environmental harm, state responsibility, in most cases, requires acting with due diligence (e.g., take reasonable steps to prevent harm, ensure compliance, assess risks, regulate private entities) (Scovazzi 2001). However, a common threshold of what is required of States to meet this obligation is not defined, and due diligence remains an obligation of conduct. State responsibility only arises in case of an injury caused by a violation of this obligation (*Certain Activities Carried Out by Nicaragua 2015*, ¶101, 180).

The current environmental law framework also imposes certain restrictions on States in their exercise of rights granted by both *jus ad bello* and *jus in bellum* (*Legality of the Threat or Use of Nuclear Weapons* (1996, 30–33). Pursuant to the law of armed **conflict**, accountability for environmental damage may amount to a **war** crime under certain circumstances. For instance, pursuant to the *Additional Protocol I to the 1949 Geneva Conventions*, the **environment** can be considered a civilian object,

DOI: 10.4324/9781003604143-3

and it is prohibited to use "any methods or means of warfare that are intended or may be expected to cause ... widespread, long-term and severe damage" (Additional Protocol I, Article 55). This rule is considered customary (Henckaerts et al. 2005; Gieseken and Murphy 2023). Violations of **international humanitarian law** can give rise to both State responsibility and individual criminal responsibility. The latter is guaranteed by the *Rome Statute* establishing the International Criminal Court (Rome Statute, Article 8(2)(b)(iv)). War crimes are first prosecutable at a national level.

Currently, the codification of the crime of **ecocide**, applicable in **peace** and wartime, perceived as wanton acts causing severe, widespread, and long-term environmental damage, is being discussed by international legal scholars, not least given the critical links between ecosystems, climate **adaptation**, and **human security**. Damage to the environment often has long-term effects on humans and ecosystems. Some national jurisdictions already criminalize environmental destruction, imposing fines on individual and corporate perpetrators to contribute to the restoration of damaged ecosystems.

Regarding private actors, recent years have seen an increase in academic writing on corporate environmental accountability, including for trade in **conflict resources**, such as diamonds and other **critical minerals**, which often fuel instability. There is also increased concern about the role of green minerals as a conflict multiplier or driver (Bruch et al. 2019). Beyond national level regulation of corporations, there is increased discussion on corporate responsibility and soft law codifications, and the need for environmental impact assessments of production at all stages, including in dealings with companies' partners (Directive (EU) 2024/1760, arts. 7–8). For example, EU Directive 2024/1203 lists environmental offenses to be criminalized by Member States, setting a minimum standard for environmental protection (Article 3).

At a transnational level, in recognition of how the extractive industry can feed into conflicts, there is increased pressure for transparency, voluntary professional standards, compliance **monitoring** and **evaluation**, and due diligence. The Organisation for Economic Cooperation and Development, for example, has developed *Due Diligence Guidance on Responsible Supply Chain of Minerals from Conflict-Affected and High-Risk Areas* and the Extractive Transparency Initiative. Nevertheless, enforcement mechanisms and sanctions for non-compliance remain weak (Cusato 2021; Kramarz and Park 2016).

On a positive note, it is increasingly highlighted that businesses engaged in conflict-affected areas can play a role in **peacebuilding**, for instance by facilitating track II **mediation**, feeding economic development, following and promoting the rule of law, adopting anti-corruption and transparency policies, and community development. This may be logical for smart business and ethical reasons.

References

Bodansky, Daniel, and John R. Cook. 2002. "The ILC's State Responsibility Articles: Introduction and Overview." *American Journal of International Law* 96 (4): 773–78. https://doi.org/10.2307/3070677.

Bruch, Carl, David Jensen, Mikiyasu Nakayama, and Jon Unruh. 2019. "The Changing Nature of Conflict, Peacebuilding, and Environmental Cooperation." *Environmental Law Reporter* 49 (2): 10134–54.
Certain Activities Carried Out by Nicaragua in the Border Area (Costa Rica/Nicaragua) and Construction of a Road in Costa Rica along the San Juan River (Nicaragua/Costa Rica) (Judgments) [2015] ICJ Rep 667.
Cusato, Eliana. 2021. "Transnational Law and the Politics of Conflict Minerals Regulation: Construing the Extractive Industry as a 'Partner' for Peace." *Transnational Legal Theory* 12 (2): 269–93. https://doi.10.1080/20414005.2021.1967683.
Directive 2024/1203 of the European Parliament and of the Council of April 11, 2024, on the Protection of the Environment through Criminal Law and Replacing Directives 2008/99/EC and 2009/123/EC.
Directive 2024/1760 of the European Parliament and of the Council of June 13, 2024, on Corporate Sustainability Due Diligence.
Gieseken, Helen, and Vanessa Murphy. 2023. "The Protection of the Natural Environment under International Humanitarian Law: The ICRC's 2020 Guidelines" *International Review of the Red Cross* 105 (924): 1180–207. https://doi.org/10.1017/S1816383123000401.
Henckaerts, Jean-Marie. 2001. "Study on Customary International Humanitarian Law: A Contribution to the Understanding and Respect for the Rule of Law in Armed Conflict." *International Review of the Red Cross* 87 (857): 175–212. https://doi.org/10.1017/S181638310018124X.
Henckaerts, Jean-Marie, Louise Doswald-Beck, Carolin Alvermann, Knut Dörmann, and Baptiste Rolle. 2005. *Customary International Humanitarian Law*. Cambridge University Press.
Kramarz, Teresa, and Susan Park. 2016. "Accountability in Global Environmental Governance: A Meaningful Tool for Action?" *Global Environmental Politics* 16 (3): 1–21. https://doi.org/10.1162/GLEP_a_00349.
Legality of the Threat or Use of Nuclear Weapons (Advisory Opinion). [1996] ICJ Rep 226
Protocol on Environmental Protection to the Antarctic Treaty or Madrid Protocol, signed October 4, 1991, 2941 UNTS 3 (entered into force January 14, 1998).
Protocol Additional to the Geneva Conventions of August 12, 1949 and Relating to the Protection of Victims of International Armed Conflicts (Protocol I), adopted June 8, 1977, 1125 UNTS 3.
Scovazzi, Tullio. 2001. "State Responsibility for Environmental Harm." *Yearbook of International Environment Law* 12 (1): 43–67. http://dx.doi.org/10.1093/yiel/12.1.43.
Voigt, Christina. 2021. "International Environmental Responsibility and Liability." In *The Oxford Handbook of International Environmental Law,* edited by Lavanya Rajamani and Jacqueline Peel. Oxford University Press.

Adaptation

Ann J. Le

Adaptation refers to strategies to reduce vulnerability and foster growth opportunities across human and natural systems in response to actual or projected climatic conditions. Effective adaptation measures adjust to climate-related risks and exploit potential beneficial opportunities (e.g., longer growing seasons or increased rainfall in some regions) associated with climate change.

Adaptation responses are distinguished by the timing and intent of actors involved. Temporal classifications of adaptation include proactive and reactive adaptation, as well as incremental and developmental adaptation. Proactive adaptation occurs before **climate change** impacts are realized, while reactive adaptation occurs after they have been observed. Incremental adaptation occurs when improvements to **infrastructure** and capacities are initiated only after specific emerging climatic risks are identified. In contrast, developmental adaptation considers broad-reaching improvements to infrastructure and capacity necessary to respond to existing and emerging climatic risks (Zhao et al. 2018).

Adaptation measures may also be organized in terms of the entities and **stakeholders** involved. Such classifications of adaptation include autonomous and planned adaptation, as well as private and public adaptation. Autonomous adaptation constitutes adaptation triggered by ecological shifts in natural systems and welfare or market shifts in human systems, independent of any conscious response to predicted climatic scenarios. Planned adaptation, conversely, is characterized by intentional policy decisions to maintain a status quo or achieve a desired climatic state (International Panel on Climate Change [IPCC] 2023). Private adaptation is initiated and implemented by private entities—companies, households, and individuals—as opposed to public adaptation, which is initiated and implemented by governments across all levels (IPCC 2023).

There is increasing recognition of and advocacy for peacekeeping approaches and research that simultaneously address climate adaptation and **peace**. They highlight how the coordination of climate adaptation and **peacebuilding** between relevant actors across time and scales can lead to positive, mutually reinforcing outcomes that can create conditions for lasting peace through ensuring the longevity of climate adaptation projects (Eklöw and Krampe 2019; Hegazi et al. 2021; Matthew 2018; Barnett 2019; Diehl 2018). For example, preventive deployment and early warning systems constitute proactive adaptation measures that may inform how peacebuilding strategies can address climate change impacts. Other strategies—including nature-based adaptation, sustainable **agriculture**, and natural **resource governance**—constitute planned adaptation measures that can be embedded in post-conflict peacebuilding strategies to enhance capacity building and promote bilateral, regional, or global **cooperation** between **conflict** actors (Krampe et al. 2024).

More empirical evidence is needed to understand the intersections between climate change, peacebuilding, and adaptation responses. Such understanding presents a major research gap in configuring effective climate adaptation measures that are responsive to the needs, knowledge, and resources of the communities that will be affected by climate change (Matthew 2018). Early empirical evidence suggests that climate adaptation efforts not designed and implemented with **conflict sensitivity** may risk perpetuating historical injustices that can undermine peacebuilding efforts, creating or escalating additional conflict (Krampe et al. 2021; Borras et al. 2020).

References

Barnett, Jon. 2019. “Global Environmental Change I: Climate Resilient Peace?” *Progress in Human Geography* 43 (5): 927–36. https://doi.org/10.1177/0309132518798077.

Borras, Saturnino M., Jennifer C. Franco, and Zau Nam. 2020. “Climate Change and Land: Insights from Myanmar.” *World Development* 129: 104864. https://doi.org/10.1016/j.worlddev.2019.104864.

Diehl, Paul F. 2018. “Mainstreaming Climate Change Adaptation into Peace Missions.” In *Climate Change and the UN Security Council*, edited by Shirley V. Scott and Charlotte Ku. Edward Elgar Publishing.

Eklöw, Karolina, and Florian Krampe. 2019. *Climate-Related Security Risks and Peacebuilding in Somalia*. Stockholm International Peace Research Institute, October. https://www.sipri.org/publications/2019/sipri-policy-papers/climate-related-security-risks-and-peacebuilding-somalia.
Hegazi, Farah, Florian Krampe, and Elizabeth Smith. 2021. *Climate-Related Security Risks and Peacebuilding in Mali*. Stockholm International Peace Research Institute. https://www.sipri.org/publications/2021/sipri-policy-papers/climate-related-security-risks-and-peacebuilding-mali.
International Panel on Climate Change (IPCC). 2023. *Climate Change 2023: Synthesis Report. Contribution of Working Groups I, II and III to the Sixth Assessment Report of the Intergovernmental Panel on Climate Change*. Edited by Core Writing Team, Hoesung Lee, and José Romero. IPCC. https://doi.org/10.59327/IPCC/AR6-9789291691647.
Krampe, Florian, Dylan O'Driscoll, McKenzie Johnson, Dahlia Simangan, Farah Hegazi, and Cedric de Coning. 2024. "Climate Change and Peacebuilding: Sub-Themes of an Emerging Research Agenda." *International Affairs* 100 (3): 1111–30. https://doi.org/10.1093/ia/iiae057.
Krampe, Florian, Elizabeth S. Smith, and Mohammad Daud Hamidi. 2021. "Security Implications of Climate Development in Conflict-Affected States: Implications of Local-Level Effects of Rural Hydropower Development on Farmers in Herat." *Political Geography* (90): 102454. https://doi.org/10.1016/j.polgeo.2021.102454.
Matthew, Richard A. 2018. "Climate Change Adaptation and Peacebuilding." In *Routledge Handbook of Environmental Conflict and Peacebuilding*, edited by Ashok Swain and Joakim Öjendal. Routledge.
Zhao, Chunli, Yan Yan, Chenxing Wang, et al. 2018. "Adaptation and Mitigation for Combating Climate Change – From Single to Joint." *Ecosystem Health and Sustainability* 4 (4): 85–94. https://doi.org/10.1080/20964129.2018.1466632.

Adaptive Management

Amanda Woomer and Larry A. Swatuk

Adaptive management is a structured, iterative approach to decision-making under uncertainty, where policies or actions are treated as experiments, outcomes are monitored, and lessons learned are used to adjust future decisions and actions to improve both social and ecological resilience over time (Holling 1973; Rogers and Macfarlan 2020). Adaptive interventions are characterized as flexible, iterative, and evolving (O'Donnell 2016).

Adaptive management first emerged in the early 1970s to deal with emerging challenges to ecosystems, especially desertification and **biodiversity** loss, posed by narrowly focused human actions and interests (Allen and Garmestani 2015; Holling 1973, 1978) and the hierarchical, command-and-control style and certainty-driven approaches people developed to address these challenges (Holling and Meffe 1996). As the climate crisis has worsened, approaches to adaptive management and the pursuit of adaptive capacity across systems and scales have taken center stage in institutional and organizational settings (Allen and Garmestani 2015).

While adaptive management has been used for decades in natural resource management, it is a newer approach within international development and humanitarian work. Humanitarian and development interventions incorporate adaptive management to more effectively respond to the fluid, interconnected, and dynamic contexts in which they operate, contexts that are often not amenable to linear, fixed plans or **theories of change**. Implementing adaptive management entails testing, gathering evidence and feedback, and making course corrections as needed. Adaptive interventions incorporate practices such as rapid feedback, reflection, **learning**, and participatory decision-making. These interventions also require a higher appetite for risk and a learning-focused operational **environment**. Good **monitoring, evaluation,** and learning (MEL) practices can support adaptive management by prioritizing actionable information and implementing MEL practices that empower intervention **stakeholders** to make important decisions quickly (Social Solutions International 2016).

Environmental peacebuilding operates almost exclusively in dynamic, complex contexts characterized by multiple, interrelated issues involving natural systems, sociocultural systems, and political systems. By working across systems and scales, environmental peacebuilding activities must consider the interrelated challenges of ecological **adaptation** (e.g., ecosystem adjustments to climate variability, **land use** change, resource extraction) and social adaptation (where communities, institutions, and states adjust their practices, policies, and relationships to cope with ecological change and social stressors). In environmental peacebuilding, ecological and social adaptation may be mutually reinforcing and potentially conflicting. "Fortress **conservation**" is a common example of the latter. It reflects hierarchical, command-and-control, one-off approaches to resource conservation that often exacerbate the problem—like **land degradation** or species decline—which decision-makers ostensibly seek to ameliorate (Bocarejo and Ojeda 2016). In contrast, Hsiao (2022) uses Rwanda's Akagera National Park as a case study to show how "convivial conservation" as a form of adaptive management may result in effective ecological peacebuilding (Büscher and Fletcher 2019).

Adaptive management can support the development and implementation of relevant and effective environmental peacebuilding interventions using tools like **participatory** context mapping, community-led **conflict** monitoring, co-producing early warning systems with key stakeholders, regular pause and reflect activities in the intervention/engagement cycle, and the tracking of unintended consequences through deliberate and critical reflection including the use of focus groups and community engagement (Bruch and Woomer 2023).

References

Allen, Craig, and Ahjond Garmestani, eds. 2015. *Adaptive Management of Social-Ecological Systems*. Springer.

Bocarejo, Diana, and Diana Ojeda. 2016. "Violence and Conservation: Beyond Unintended Consequences and Unfortunate Incidences." *Geoforum* 69: 176–83. https://doi.org/10.1016/j.geoforum.2015.11.001.

Bruch, Carl, and Amanda Woomer. 2023. *Toolkit on Monitoring and Evaluation of Environmental Peacebuilding*. Environmental Law Institute, Environmental Peacebuilding Association, and United States Institute of Peace. https://m-and-e.environmentalpeacebuilding.org/toolkit/.

Büscher, B., and R. Fletcher. 2019. "Towards Convivial Conservation." *Conservation and Society* 173 (3): 283–96.

Holling, C. S. 1973. "Resilience and Stability of Ecology Systems." *Annual Review of Ecology, Evolution, and Systematics* 4: 1–23. https://www.jstor.org/stable/2096802.

Holling, C. S., ed. 1978. *Adaptive Environmental Assessment and Management*. International Series on Applied Systems Analysis.

Holling, C. S., and Gary K. Meffe. 1996. "Command and Control and the Pathology of Natural Resource Management." *Conservation Biology* 10 (2): 328–37. https://doi.org/10.1046/j.1523-1739.1996.10020328.x.

Hsiao, Elaine. 2022. "Conviviality in Disrupted Socionatural Landscapes: Ecological Peacebuilding around Akagera National Park." *Conservation and Society* 20 (2): 79–91. https://doi.org/10.4103/cs.cs_24_21.

O'Donnell, Michael. 2016. *Adaptive Management: What it Means for CSOs*. Bond, September. https://www.bond.org.uk/wp-content/uploads/2022/03/adaptive_management_-_what_it_means_for_csos_0.pdf.
Rogers, Patricia, and Alice Macfarlan. 2020. *What Is Adaptive Management and How Does it Work?* Monitoring and Evaluation for Adaptive Management Working Paper Series, Number 2, September 7. https://www.betterevaluation.org/tools-resources/what-adaptive-management-how-does-it-work-working-paper-2.
Social Solutions International (SSI) Inc. 2016. *Monitoring and Evaluation for Adaptive Management: 8 Promising Practices*. Bureau for Policy, Planning and Learning of the USAID.

Agriculture

Guido Trivellini, Bernd Unger, and Tom Cavanagh

Agriculture is the organized practice of cultivating soil, growing crops, and raising livestock to produce resources essential for human and human-dependent communities. It involves transforming natural ecosystems through human labor to generate food, materials, and other ecological and economic goods that sustain and enhance human life.

A critical distinction exists between extensive (smallholder) and intensive (industrial) models of agriculture. While extensive agriculture often emphasizes **sustainability**, drawing on biological methods, crop diversity, and natural soil fertility, intensive agriculture prioritizes high yields and profit maximization, typically relying on mechanization, synthetic fertilizers, pesticides, and monocultures, impacting ecosystems and **natural resources**. Thus, the type of agriculture practiced has a significant impact on landscape **conservation**, with agriculture as the primary driver of **land use** changes globally (Food and Agriculture Organization of the United Nations [FAO] 2021).

In addition, agriculture plays a foundational role in the political economy of both developed and developing societies, though the nature of that role differs. In developing societies, agriculture remains the largest employer (Cheong et al. 2013) and a significant contributor to gross domestic product. It plays a crucial role in ensuring **food security**, sustaining rural livelihoods, and maintaining social stability. Agriculture is often central to national development strategies, with governments leveraging agricultural policies to reduce poverty and promote nation-building. In contrast, developed societies rely on a much smaller agricultural workforce (Cheong et al. 2013) due to high levels of intensification and industrialization. Despite this, agriculture retains substantial political influence, particularly through subsidies and trade policies.

Agriculture and **conflict** are linked in a complex, mutually influencing relationship in which resource scarcity, land disputes, and food insecurity can be a cause and an effect of conflict (Homer-Dixon 1994). Le Billon (2001) highlights how the scarcity of fertile land can lead to competition and **violence**. Good **governance** and an equitable distribution of land are, therefore, key factors for **peace** consolidation (de Soysa et al. 1999; Wehrmann 2008).

The spectrum of conflicts over the use of agricultural land and agricultural products ranges from minor disputes to armed conflicts. Lund et al. (2006) and Peña-Ramos et al. (2022) point out that such resource-related conflicts are often misunderstood as religious or cultural disputes or are oversimplified, even though their causes are in fact economic and ecological in nature.

The 20th century was marked by the expansion and intensification of agriculture, often at the expense of ecological systems. In response to the environmental consequences of this model, contemporary agricultural policies are beginning to incorporate **indicators** of **biodiversity** to promote more sustainable practices. This shift reflects a growing recognition that achieving global food security must be balanced with the urgent need to conserve biodiversity, an increasingly central challenge of the 21st century.

There is a range of perspectives on the impact of **climate change** on conflicts, ranging from multiplying existing threats to catalyzing new conflicts, all of which contribute to increased vulnerability across scales (Tirrell et al. 2021). Gleditsch and Nordås (2009) align with these arguments but point out that strong institutions and trustworthy **cooperation** can mitigate these risks. This has resulted in a global push to improve governance and land rights in regions with weak institutions and **land tenure** systems, framed through the lens of **peacebuilding** and **conflict prevention**. Martin-Shields and Stojetz (2018) also emphasize that food security is crucial for long-term **peace** and stability in the affected regions and, ultimately, worldwide. Gill (2010), moreover, underscores that long-term solutions require political stability, land reforms, and investments in sustainable agriculture.

This growing recognition of the need to align food production with ecological sustainability is increasingly challenged by the intensifying conflict between agriculture and **biodiversity** conservation. Rising global demand for agricultural land remains a major source of tension, with agricultural expansion identified as the leading driver of biodiversity loss worldwide (FAO 2021). Broader global trends, including climate change, globalization, and shifting policy regimes, further exacerbate these pressures. At the local level, debates over **land use** strategies, such as land-sparing versus land-sharing, highlight the complexity of reconciling competing priorities. While various regulatory, market-based, and voluntary mechanisms have been developed to address these issues, they remain insufficiently effective in striking a durable balance between agricultural productivity and biodiversity conservation (Lécuyer et al. 2021).

References

Cheong, Danny, Marc Jansen, and Ralf Peters, eds. 2013. *Shared Harvests: Agriculture, Trade, and Employment.* International Labour Office and United Nations Conference on Trade and Development. https://www.ilo.org/sites/default/files/wcmsp5/groups/public/@ed_emp/documents/genericdocument/wcms_212868.pdf.

de Soysa, Indra, Nils Petter Gleditsch, Michael Gibson, and Margareta Sollenberg. 1999. "To Cultivate Peace: Agriculture in a World of Conflict." *Environmental Change and Security Project Report* 5: 15–25. Woodrow Wilson International Center for Scholars. https://www.wilsoncenter.org/publication/to-cultivate-peace-agriculture-world-conflict.

Food and Agriculture Organization of the United Nations (FAO). 2021. "COP26: Agricultural Expansion Drives Almost 90 Percent of Global Deforestation." *FAO News and Media*, June 11. https://www.fao.org/newsroom/detail/cop26-agricultural-expansion-drives-almost-90-percent-of-global-deforestation/en.

Gill, Peter. 2010. *Famine and Foreigners: Ethiopia Since Live Aid.* Oxford University Press.

Gleditsch, Nils P., and Ragnhild Nordås. 2009. "Climate Change and Conflict: A Critical Overview." *Die Friedens-Warte* 84 (2): 11–28. https://www.jstor.org/stable/23773880.

Homer-Dixon, Thomas F. 1994. "Environmental Scarcities and Violent Conflict: Evidence from Cases." *International Security* 19 (1): 5–40. https://www.jstor.org/stable/2539147.
Lécuyer, Lauriane, Didier Allard, Sophie Calla, et al. 2021. "Conflicts Between Agriculture and Biodiversity Conservation in Europe: Looking to the Future by Learning from the Past." *Advances in Ecological Research* 65: 3–56. https://www.sciencedirect.com/science/article/abs/pii/S0065250421000209.
Lund, Christian, Rie Odgaard, and Espen Sjaastad. 2006. *Land Rights and Land Conflicts in Africa: A Review of Issues and Experiences*. DIIS Report. Danish Institute for International Studies. https://pure.diis.dk/ws/files/68278/Land_rights_and_land_conflicts_in_Africa_a_review_of_issues_and_experiences.pdf.
Martin-Shields, Charles, and Wolfgang Stojetz. 2018. *Food Security and Conflict: Empirical Challenges and Future Opportunities for Research and Policy-Making on Food Security and Conflict*. FAO Agricultural Development Economics Working Paper 18-04. Food and Agricultural Organization. September. https://openknowledge.fao.org/server/api/core/bitstreams/bf424047-b87c-4469-bb08-72b2d8673307/content.
Peña-Ramos, José Antonio, Rafael José López-Bedmar, Francisco Javier Sastre, and Adrián Martínez-Martínez. 2022. "Water Conflicts in Sub-Saharan Africa." *Frontiers in Environmental Science* 10: 863903. https://doi.org/10.3389/fenvs.2022.863903.
Philippe Le Billon, "The political ecology of war: natural resources and armed conflicts," *Political Geography* 20: 5, 2001,pp. 561–84
Tirrell, Alex, Pablo Cordero, and David Crane. 2021. "Addressing the Environment-Peace-Nexus." *Peace Review* 33 (3): 370–78. https://doi.org/10.1080/10402659.2021.1953802.
Wehrmann, Babette. 2008. *Land Conflicts: A Practical Guide to Dealing with Land Disputes*. Deutsche Gesellschaft für Technische Zusammenarbeit (GTZ), February. https://globalprotectioncluster.org/sites/default/files/202304/Land_Conflicts_Practical_Guide_Dealing_Land_Disputes_2008_EN.pdf.

Agricultural–Pastoral Conflict

Leif Brottem, Alan Channer, and Fakunle John Aremu

Agricultural–pastoral conflicts, also known as agropastoral conflicts or farmer-herder conflicts, are disputes, often violent and sometimes fatal, between sedentary farmers on the one hand and nomadic or semi-nomadic herders of livestock on the other.

Agricultural–pastoral conflicts occur when mobile pastoralists, in search of pasture and water for their livestock, clash in one way or another with sedentary farmers cultivating crops (Brottem 2016; Channer 2016; Kräetli and Toulmin 2020). It is prevalent in parts of the world where seasonal grazing abounds, or used to abound—notably in the Sahel, parts of Central Asia, and some dryland areas of India, Pakistan, and Afghanistan (Baksh and van Steerbergen 2020). The causes of agropastoral **conflict** are multiple, including seasonal incompatibility, land rights, resource competition, **governance**, and factors related to identity or ethnicity.

Pastoral movements can be incompatible with **agriculture** when cropping seasons and especially harvest periods are out of sync with the presence of livestock herds. This incompatibility is linked to land rights that often overlap and are based on access agreements between farmers and herders. As populations grow and resource availability changes, these rights come under strain, and farmers and herders compete for access to specific areas and often clash over which group can legitimately use them (Maiangwa 2017).

Climate change is increasingly a dimension of these conflicts, as changing ecological conditions disrupt pastoral and agricultural calendars (Landis 2014). Institutions that regulate land access are often ill-equipped to handle these changing conditions, even as demand for resources increases. Governance systems often favor one group over the other due to historic factors or uneven political **power** in a country or jurisdiction. Historically, modern nation-states have favored sedentary groups over pastoral livestock herders due to biases against **pastoralism** as an unproductive or archaic system.

Since agriculture and pastoralism are frequently linked to specific ethnic or cultural groups, disputes over land and resources can become intertwined with identity-based ethno-religious and political conflicts (Raleigh 2010). For example, the Fulani/Peul—the dominant pastoralist group in West Africa and the largest pastoralist ethnic group in the world—are invariably Muslim, while some of the farming communities with which they clash are Christian or animist (Nwankwo 2020). This has been a factor in recurrent intercommunal **violence** in certain countries, notably Nigeria. Investment by wealthy and influential individuals in large mobile herds of cattle has further inflamed tensions in the region.

Agropastoral conflict is an important focus of **environmental peacebuilding** initiatives, not least since they have the potential to escalate into intercommunal violence and contribute to regional insurgencies (Tadesse et al. 2015). Furthermore, traditional systems of reciprocal benefit—for example, where cattle would graze stubble in return for fertilizing fields—provide a strong basis for environmental peacebuilding initiatives (Eke and Byrne 2022). A good example exists in Baluchistan Province, Pakistan, where pastoralists moving between the highlands and lowlands purposefully foster good relations and mutual interdependence with farmers in the lowlands (Baksh and van Steerbergen 2020).

Agropastoral peacebuilding initiatives generally focus on intergroup relations at the local level through dialogue, **confidence-building measures**, communication channels, early warning systems, and the establishment of resource use conventions (Eke and Byrne 2022). Religious and traditional peacemakers are well placed to mediate ethno-religious agropastoral conflicts using environmental peacebuilding approaches, for example, blessing joint tree-planting ceremonies and emphasizing the joint benefits of collaboration in land stewardship. Such approaches have been deployed to good effect by the Interfaith Mediation Centre in Kaduna and Plateau States, Nigeria (Channer 2016; Vanguard 2017).

In West Africa, farmers and herders have traditionally shared access to land through semi-formal conventions rather than highly codified laws. Recent **peacebuilding** efforts by Humanitarian Dialogue in Mali, where farmer-herder conflict is linked to a violent insurgency, have emphasized the traditional legitimacy of local conventions as an effective bottom-up initiative, as well as agreements on the shared management of and access to **natural resources**, such as grazing routes and water points.

References

Baksh, Allah, and Frank van Steerbergen. 2020. "Symbiosis: Pastoralists and Farmers in Balochistan." *The Water Channel*, April 21. https://thewaterchannel.tv/thewaterblog/symbiosis-pastoralists-and-farmers-in-balochistan/.

Brottem, Leif. 2016. "Environmental Change and Farmer-Herder Conflict in Agro-Pastoral West Africa." *Human Ecology* 44 (5): 547–63. https://doi.org/10.1007/s10745-016-9846-5.

Channer, Alan. 2016. "Trust Building and Mobile Pastoralism in Africa." In *Land Restoration: Reclaiming Landscapes for a Sustainable Suture*, edited by Ilan Chabay, Martin Frick, and Jennifer F. Helgeson. Elsevier.

Eke, Surulola, and Sean Byrne. 2022. "Bringing the Indigenous into Mainstream Peacemaking and Peacebuilding in Farmer-Herder Conflicts: Some Critical Reflections." In *Routledge Handbook of Peacebuilding and Ethnic Conflict*, edited by Jessica Senehi, Imani M. Scott, Sean Byrne, and Thomas G. Matyok. Routledge.

Kräetli, Saverio, and Camilla Toulmin. 2020. *Farmer-Herder Conflict in Sub-Saharan Africa: Rethinking the Phenomenon?* International Institute for Environment and Development, October. https://www.iied.org/sites/default/files/pdfs/2021-01/10208IIED.pdf.

Landis, Steven T. 2014. "Temperature Seasonality and Violent Conflict: The Inconsistencies of a Warming Planet." *Journal of Peace Research* 51 (5): 603–18. https://doi.org/10.1177/0022343314538275.

Maiangwa, Benjamin. 2017. "Conflicting Indigeneity and Farmer–Herder Conflicts in Postcolonial Africa." *Peace Review* 29 (3): 282–88. https://doi.org/10.1080/10402659.2017.1344527.

Nwankwo, Cletus F. 2020. "Essentializing Critical Geopolitics of the Farmers–Pastoralists Conflicts in West Africa." *GeoJournal* 85: 1291–308. https://link.springer.com/content/pdf/10.1007/s10708-019-10023-3.pdf.

Raleigh, Clionadh. 2010. "Political Marginalization, Climate Change and Conflict in African Sahel States." International Studies Review 12 (1): 69-86. https://www.jstor.org/stable/40730710.

Tadesse, Bamlaku, Fekadu Beyene, Workneh Kassa, and Richard Wentzell. 2015. "The Dynamics of (Agro) Pastoral Conflicts in Eastern Ethiopia." *Ethiopian Journal of the Social Sciences and Humanities* 11 (1): 29–60. https://ejol.aau.edu.et/index.php/EJOSSAH/article/view/6275/7471.

Vanguard. 2017. "Nigerian Pastor, Imam Win Intercultural Innovation Award at UN." November 30. https://www.vanguardngr.com/2017/11/nigerian-pastor-imam-win-intercultural-innovation-award-un/.

Anthropocene

Osama Abu-Libda and Simon Dalby

Anthropocene is an earth science term proposed to define the current epoch of Earth's geological history marked by humanity's profound effect on Earth's geology, climate, and ecosystems, far surpassing the effects of natural processes, underscoring disruptions in the planetary system and their fateful impact on social and environmental stability.

The term the Anthropocene, coined by Paul Crutzen in the early 2000s, describes the significant global impact of human activities fundamentally altering natural processes at the level of the planetary system (Crutzen 2002, 2006). It emphasizes that anthropogenic processes (urbanization, industrialization, **agriculture**, etc.) and the development of recent technologies are the driving forces shaping today's Earth systems (Zalasiewicz et al. 2019). The Anthropocene concept evolved with contributions from various scholars who identified important events and signs associated with its emergence. For example, the "Great Acceleration" developed in the mid-20th century (Dalby 2020) shows the dramatic increase in the magnitude and rate of human-induced change in Earth's system at various scales.

The Anthropocene epoch is marked by specific human impacts on Earth's geological and biological processes, demonstrating human interference in natural processes, and is identified through environmental signals and interconnected changes. These indicators include deforestation, resource overexploitation, pollution, and abnormal changes in atmospheric and climate systems, such as elevated carbon dioxide levels from industrial processes and fossil fuel burning. This leads to serious risks of habitat destruction, **biodiversity** loss, resource scarcity, global warming, and ocean acidification (Waring et al. 2024; Zalasiewicz et al. 2020). In addition, the emergence of novel materials, often synthetic and non-biodegradable, is a signature of the Anthropocene. These materials, such as plastics, are often persistent in the **environment**, disrupting biogeochemical cycles and posing challenges to ecosystems and human well-being (Dalby 2020). These changes result in rapid alterations and significant transformations in Earth's systems, thereby setting the Anthropocene apart from earlier geological periods characterized by nature-induced variations instead of human-induced influences (Jørgensen et al. 2019). Consequently, the Anthropocene presents a complex array of environmental issues that demand non-traditional solutions and innovative **governance** and policymaking approaches to mitigate adverse impacts, ensure system **sustainability**, and foster effective **resilience** (Bornemann 2021; Cassotta 2021).

The Anthropocene Earth's system changes also have critical implications for global **peace** and **security**, especially by intensifying resource conflicts, migration, and geopolitical disputes. With the escalation of **climate change**, the competition for scarce **natural resources** like water and arable land becomes increasingly evident, which could result in violent conflicts and social unrest (Spring 2021). Additionally, **environmental degradation** contributes to **displacement**, forcing populations to migrate (temporarily, such as to evacuate wildfires or permanently) and potentially straining host communities and igniting tensions over resource allocation (Martine and Alves 2019). These migratory pressures often amplify existing geopolitical frictions, as nations grapple with border security and integrating displaced persons (Hardt 2021). The same pressures are also contributing to animal and plant migration. Addressing and navigating these security challenges necessitates a comprehensive governance approach and adaptive policy frameworks that integrate international **cooperation**, scientific evidence, and **sustainable development** with **conflict** resolution strategies, thus fostering an environment where peace and security can be maintained despite the Anthropocene's turbulent changes.

References

Bornemann, Basil. 2021. "Environmental Governance in the Anthropocene: Challenges, Approaches, and Critical Perspectives." In *International Relations in the Anthropocene: New Agendas, New Agencies and New Approaches*, edited by David Chandler, Franziska Müller, and Delf Rothe. Palgrave Macmillan. https://doi.org/10.1007/978-3-030-53014-3_17.

Cassotta, Sandra. 2021. "The Development of Environmental Law within a Changing Environmental Governance Context: Towards a New Paradigm Shift in the Anthropocene Era." *Yearbook of International Environmental Law* 30 (1): 54–67. https://academic.oup.com/yielaw/article-abstract/30/1/54/6154437.

Crutzen, Paul J. 2002. "The 'Anthropocene.'" *Journal de Physique IV* 12 (10): 1–5. https://doi.org/10.1051/jp4:20020447.

Crutzen, Paul J. 2006. "The 'Anthropocene.'" In *Earth System Science in the Anthropocene*, edited by Eckart Ehlers and Thomas Krafft. Springer. https://doi.org/10.1007/3-540-26590-2_3.

Dalby, Simon. 2020. *Anthropocene Geopolitics: Globalization, Security, Sustainability*. University of Ottawa Press.

Hardt, Judith N. 2021. "Research Perspectives and Boundaries of Thought: Security, Peace, Conflict, and the Anthropocene." *Revista de Estudios en Seguridad Internacional* 7 (1): 11–28. https://shs.hal.science/halshs-03546723/.

Jørgensen, Peter S., Carl Folke, and Scott P. Carroll. 2019. "Evolution in the Anthropocene: Informing Governance and Policy." *Annual Review of Ecology, Evolution, and Systematics* 50 (1): 527–46. https://doi.org/10.1146/annurev-ecolsys-110218-024621.

Martine, George, and Jose Eustaguio Diniz Alves. 2019. "Disarray in Global Governance and Climate Change Chaos." *Brazilian Journal of Population Studies* 36: 1–30. https://doi.org/10.20947/S102-3098a0075.

Spring, Úrsula Oswald. 2021. "Decolonising Peace in the Anthropocene: Introduction Towards an Alternative Understanding of Peace and Security." In *Decolonising Conflicts, Security, Peace, Gender, Environment and Development in the Anthropocene*, edited by Úrsula Oswald Spring and Hans Günter Brauch. Springer. https://link.springer.com/chapter/10.1007/978-3-030-62316-6_15.

Waring, Timothy M., Zachary T. Wood, and Eörs Szathmáry. 2024. "Characteristic Processes of Human Evolution Caused the Anthropocene and May Obstruct Its Global Solutions." *Philosophical Transactions of the Royal Society B* 379 (1893): 20220259. https://royalsocietypublishing.org/doi/abs/10.1098/rstb.2022.0259.

Zalasiewicz, Jan, Colin Waters, and Mark Williams. 2020. "The Anthropocene." In *Geologic Timescale 2020*, edited by Felix M. Gradstein, James G. Ogg, Mark D. Schmitz, and Gabi M. Ogg. Elsevier. https://www.sciencedirect.com/science/article/pii/B9780128243602000310.

Zalasiewicz, Jan, Colin N. Waters, Mark Williams, and Colin P. Summerhayes, eds. 2019. *The Anthropocene as a Geological Time Unit: A Guide to the Scientific Evidence and Current Debate*. Cambridge University Press.

B

Backdraft

Geoffrey Dabelko, Lauren Herzer Risi, and Meaghan Parker

Backdraft is defined as the potential of climate responses—including mitigation, adaptation, and resilience efforts—to cause conflict.

The term, first published by the Wilson Center's **Environmental Change and Security Program** in *Backdraft: The Conflict Potential of Climate Mitigation and Adaptation* (Dabelko et al. 2013), expands the focus of climate-conflict research and practice to encompass the impact of efforts to address **climate change**.

Since 2007, climate change and **conflict** links have garnered significant attention at national and international levels, centered on the conflict-generating potential of direct climate change impacts (e.g., drought, extreme weather events) and immediate responses to those impacts (e.g., migration or state **fragility**). The concept of backdraft adds to the discussion of the knock-on effects, often unintended, generated by proactive, usually positive efforts to mitigate and adapt to direct and indirect impacts of climate change.

The backdraft concept was based in part on a deductive argument that the scale of changes deemed necessary in **mitigation** and **adaptation** proposals would likely impact existing conflicts or generate new ones. Examples include the following: (1) changing access to and use of forests as carbon sinks; (2) ramping up mining for the inputs needed to deploy **renewable energy** technologies at scale; (3) using land for large-scale solar **power** generation installations; and (4) increasing large-scale hydropower dams. These desirable components of the "green transition" create new winners and losers in ways likely to be contested and even conflictual (Dabelko et al. 2022).

Critics of the backdraft concept flagged its limited evidence base. They also argued that highlighting conflicts generated by climate responses could be used by fossil fuel supporters to oppose renewable energy projects. Subsequently, a wide-ranging research agenda emerged across geographies, scales, and sectors (Gilmore and Buhaug 2021). Scholars and practitioners conducted research that would enable actors to anticipate potential conflict and develop **conflict sensitivity** in their approaches to climate action. Research on a "**boomerang effect**" explored how negative effects of local climate adaptation could impact the state (Swatuk and Wirkus 2018). Both

DOI: 10.4324/9781003604143-4

bilateral and multilateral institutions supported the development of guidelines for a conflict-sensitive adaptation to climate change for use in environmental, development, and **conflict prevention** efforts (Tänzler et al. 2022).

Climate and **security** research and practice remain heavily weighted toward understanding how direct and indirect climate impacts can cause conflict. Yet backdraft and boomerang dynamics are increasingly reflected in leading intergovernmental and national climate and conflict assessments and programs. To avoid climate-related conflict, it is critical to understand the conflict potential of both the impacts of climate change and the efforts to mitigate and adapt to it.

References

Dabelko, Geoffrey D., Anniek Barnhoorn, Noah Bell, et al. 2022. *Navigating a Just and Peaceful Transition (Part 3)*. Stockholm International Peace Research Institute.

Dabelko, Geoffrey D., Lauren Herzer, Schuyler Null, Meaghan Parker, and Russell Sticklor, eds. 2013. "Backdraft: The Conflict Potential of Climate Change Adaptation and Mitigation." *Environmental Change and Security Program Report* 14 (2). Woodrow Wilson International Center for Scholars.

Gilmore, Elizabeth A., and Halvard Buhaug. 2021. "Climate Mitigation Policies and the Potential Pathways to Conflict: Outlining a Research Agenda." *WIREs Climate Change* 12 (5): e722. https://doi.org/10.1002/wcc.722.

Ide, Tobias. 2025. "Rethinking Climate Conflicts: The Role of Climate Action and Inaction." *World Development* 186: 106845. https://doi.org/10.1016/j.worlddev.2024.106845.

Swatuk, Larry A., and Lars Wirkus, eds. 2018. *Water, Climate Change, and the Boomerang Effect: Unintentional Consequences of Resource Insecurity*. Routledge.

Tänzler, Dennis, Nikolas Schere, and Adrien Detges. 2022. "Guidelines for Conflict-Sensitive Adaptation to Climate Change." Report for the German Environment Agency.

Basic Services

Ananya Peddibhotla

Basic services are provisioning services that are important for improving people's lives and fostering peace in society. Examples of basic services include delivery of waste management, water and sanitation, health care, and education.

In **environmental peacebuilding** literature, provision of basic services is seen as a vital dimension of post-conflict peacebuilding (Bruch et al. 2022; Tignino and Kebebew 2023). Re-establishment of basic services should be among the immediate priorities for **peacebuilding** interventions (Crawford et al. 2015). Particularly, the restoration and development of basic services can provide significant **peace dividends** by improving people's lives and showing the value of **peace** (and the cost of returning to **conflict**) (Bruch et al. 2022).

Importantly, the lack of basic services such as electricity and water has been linked to the main cause of violent conflicts in urban centers (Dresse et al. 2021). On the other hand, extreme conflicts such as **war** often obliterate **infrastructure**, which is necessary for the provision of basic services (Weinthal et al. 2011). In this sense, there is a logical connection between basic services and conflict, and hence, it becomes vital to consider how successful service delivery can foster peacebuilding efforts that can also, in turn, reduce the potential for conflicts.

An important argument in the literature (e.g., Lai 2024) concerns how welfare and basic services provision should be subject to universal provision as basic **human rights**, rather than as **reparations** or a measure useful for achieving peace. This is because basic services are a precondition for basic livelihood that constitutes peace, rather than being a vehicle for it. A report published by the United Nations Development Programme (UNDP) and United Nations Environment Programme (UNEP) (2013) further highlights how **infrastructure** improvements and the establishment of basic services are the foundation for economic growth in general. In this sense, basic services as part of fundamental human rights have a direct influence on sustaining other areas of an individual's life, such as livelihood and economic growth, and this interlinkage further makes provisioning of basic services a critical avenue for peacebuilding efforts to also reduce conflict-induced extreme poverty.

The literature also underscores the importance of good **governance** in provisioning basic services and fostering peace. Successful service delivery by state actors has a positive effect on the relationship between society and the state, and consequently on peace (Krampe et al. 2021). This can be an important criterion for peacebuilding as it improves state-society social contracts and creates more peaceful **accountability**. If done well, the provisioning of basic services can foster peacebuilding efforts and improve overall trust in society.

References

Bruch, Carl, David Jensen, and Monica Emma. 2022. "Defining Environmental Peacebuilding." In *Routledge Handbook of Environmental Security*, edited by Crystal Murphy, Evgenia Nizkorodov, and Richard Matthew. Routledge. https://doi.org/10.4324/9781315107592-2.

Crawford, Alec, Angie Dazé, Anne Hammill, Jo-Ellen Parry, and Alicia Natalia Zamudio. 2015. *Promoting Climate-Resilient Peacebuilding in Fragile States*. International Institute for Sustainable Development, April 8. https://www.iisd.org/library/promoting-climate-resilient-peacebuilding-fragile-states.

Dresse, Anaïs, Jonas Østergaard Nielsen, and Itay Fischhendler. 2021. "From Corporate Social Responsibility to Environmental Peacebuilding: The Case of Bauxite Mining in Guinea." *Resources Policy* 74: 102290. https://doi.org/10.1016/j.resourpol.2021.102290.

Krampe, Florian, Farah Hegazi, and Stacy D. VanDeveer. 2021. "Sustaining Peace Through Better Resource Governance: Three Potential Mechanisms for Environmental Peacebuilding." *World Development* 144: 105508. https://doi.org/10.1016/j.worlddev.2021.105508.

Lai, Daniela. 2024. "Everyday Political Economies of Peacebuilding." In *Routledge Handbook of Peacebuilding*, edited by Roger Mac Ginty. Routledge.

Tignino, Mara, and Tadesse Kebebew. 2023. "Access to Water and Peacebuilding." In *Research Handbook on International Law and Environmental Peacebuilding*, edited by Daniëlla Dam-de Jong and Britta Sjöstedt. Edward Elgar Publishing.

United Nations Environment Programme and United Nations Development Programme. 2013. *The Role of Natural Resources in Disarmament, Demobilization and Reintegration Addressing Risks and Seizing Opportunities*. United Nations Environment Programme.

Weinthal, Erika, Jess Troell, and Mikiyasu Nakayama. 2011. "Water and Post-Conflict Peacebuilding: Introduction." *Water International* 36 (2): 143–53. https://doi.org/10.1080/02508060.2011.561772.

Basin at Risk

Melissa McCracken and Emmanuel Valax

A basin at risk (BaR) is a conceptual framework with a range of methodological approaches for assessing the current and future risk for a river basin that is at risk of conflict over water resources between various actors in the basin over present and projected future periods, and the approach to assess these risks.

Within this concept and the related methodological approaches, **conflict** is defined broadly to include the range of interactions (e.g., verbal accusations, diplomatic tensions, and economic disputes up to violent conflict (Wolf et al. 2003)) between actors that perceive having mutually incompatible goals and values about the use, development, and/or protection of a basin's water resources. The status of a surface or groundwater basin as a BaR will change over time with respect to the hydrologic, social, political, and economic systems.

The term originated in the Basins at Risk (BaR) project conducted by researchers at Oregon State University beginning in the late 1990s and following the idea of Regions at Risk of international water conflict by Gleick (1993). The BaR project aimed to systematically assess conflict and **cooperation** within internationally shared freshwaters, identify indicators of future potential tensions, identify basins at risk, and understand potential mitigatory factors (Wolf et al. 2003). Subsequent scholars have built on this foundation, using the term to capture the prediction and discussion of basins prone to conflict or cooperation (Motta et al. 2025; Bernauer and Böhmelt 2014; Turton 2005), as well as predict conflict risk by 2050, where conflict risk may be determined by the absence of cooperation under a wide range of hydrological, climatic, and socioeconomic developments (De Bruin et al. 2023). Others recognized that this was not limited to internationally shared water resources and broadened the definition to subnational, using data to explore water scarcity effects on conflict risk (Bernauer et al. 2012).

The expansion of the concept "basin at risk" has continued, where a basin can be "at risk" from issues other than conflict, such as a recent example evaluating the water scarcity and climate sensitivity of a basin to assess its socio-ecological vulnerability (Jiang et al. 2025). The Transboundary Water Assessment Programme conceptualized how basins could be at risk to a range of issues beyond conflict by identifying five interconnected thematic groups of data and indicators–water quantity, water quality, ecosystems, **governance**, and socioeconomics–which may have second-order effects in contributing to or mitigating conflict potential (United Nations Environment Programme–DHI Centre on Water and Environment 2016). While the evolution of this term has nuanced our understanding of contributing and mitigating factors to conflict risk at multiple scales, it has also muddied the linkage between its usage and the risk of conflict at its core. Therefore, this definition recenters the term within its origins on the potential risk of conflict over water resources shared between different actors with competing goals and values at present and in the future.

To summarize, the term "basin at risk" can be viewed as a conceptual framework with a range of methodological approaches for assessing the current and future risk level. From the original BaR project, we know that the likelihood and intensity of conflict rise at the rate of change in the basin exceeding the institutional capacity to absorb the change (Wolf et al. 2003; Yoffe and Wolf 1999). This identifies two categorical variables that contribute to a Basin at Risk status: (1) the change or disturbance in a basin, such as political regime change, large **infrastructure**, or hydrologic change, and (2) the level of institutional capacity to handle the changes, such as treaties, institutions, policies, relationships, or technical capacity. Together, these factors contribute to or mitigate the potential for conflict within a basin and, therefore, whether and to what extent one can qualify it as a basin at risk. For the **environmental peacebuilding** field, the conceptual framework and methodological approaches that underpin this term inform situational awareness, design, implementation, and **monitoring** of interventions at the basin and sub-basin levels. Perhaps more importantly, it offers the means to identify mitigating and contributing factors, while inviting deeper analyzes of the potential sources of **resilience** that may be targeted for development.

References

Bernauer, Thomas, and Tobias Böhmelt. 2014. "Basins at Risk: Predicting International River Basin Conflict and Cooperation." *Global Environmental Politics* 14 (4): 116–38. https://doi.org/10.1162/GLEP_a_00260.

Bernauer, Thomas, Tobias Böhmelt, Halvard Buhaug, et al. 2012. "Water-Related Intrastate Conflict and Cooperation (WARICC): A New Event Dataset." *International Interactions* 38 (4): 529–45. https://doi.org/10.1080/03050629.2012.697428.

De Bruin, Sophie, Susanne Schmeier, Rens Van Beek, and Marijn Gulpen. 2023. "Projecting Conflict Risk in Transboundary River Basins by 2050 Following Different Ambition Scenarios." *International Journal of Water Resources Development* 40 (1): 7–32. https://doi.org/10.1080/07900627.2023.2184650.

Gleick, Peter H. 1993. "Water and Conflict: Fresh Water Resources and International Security." *International Security* 18 (1): 79–112. https://doi.org/10.2307/2539033.

Jiang, Ruijie, Hui Lu, Deliang Chen, et al. 2025. "Transboundary Conflict from Surface Water Scarcity under Climate Change." *Nature Communications* 16: 8166. https://doi.org/10.1038/s41467-025-63568-y.

Motta, Stew, Isabella Böck, Johanna Koehler, Aaron T. Wolf, and Philipp Pattberg. 2025. "The Financialization of Rivers: Clean Development Mechanism (CDM) Subsidized Hydropower in the Mekong Region's Basins at Risk." *Global Environmental Change* 90: 102962. https://doi.org/10.1016/j.gloenvcha.2024.102962.

Turton, Anthony. 2005. *A Critical Assessment of the Basins at Risk in the Southern African Hydropolitical Complex*. Helsinki University of Technology.

United Nations Environment Programme–DHI Centre on Water and Environment (UNEP-DHI). 2016. *Transboundary River Basins: Status and Trends. Volume 3: River Basins*. UNEP.

Wolf, Aaron T., Shira B. Yoffe, and Mark Giordano. 2003. "International Waters: Identifying Basins at Risk." *Water Policy* 5 (1): 29–60. https://doi.org/10.2166/wp.2003.0002.

Yoffe, Shira B., and Aaron T. Wolf. 1999. "Water, Conflict and Cooperation: Geographical Perspectives." Cambridge Review of International Affairs 12(2): 197-213

Benchmarking

Richard Matthew

Benchmarking refers to: (1) setting standards or targets, and (2) measuring performance against these standards or targets.

Benchmarking has been adopted by the United Nations and many other organizations as a practice that facilitates collaboration around identifying concrete goals in a wide range of contexts, improves coordination of state and non-state actors in pursuing these goals, promotes transparency, supports **monitoring** and **evaluation** systems, and strengthens **accountability** (United Nations 2010). In recent decades, the practice of benchmarking has expanded dramatically in the global arena. A database compiled by the University of Warwick, for example, identifies hundreds of global benchmarks, most of which have been created since 2000 (University of Warwick 2021).

Informed by the work of De Coning and Romita (2009) and many others, the United Nations prepared and released a landmark report in 2010—*Monitoring Peace Consolidation: United Nations Practitioners' Guide to Benchmarking* (United Nations 2010). The goal of this document was "to provide basic principles, guidelines and resources that will enable United Nations field presences to measure progress towards or regress away from **peace** consolidation" in the rapidly changing, high stakes, inadequately resourced, and often contested context of **peacebuilding** operations (United Nations 2010, 2). The guide provided a seven-step process for benchmarking: (1) prepare for benchmarking, (2) establish benchmarks, (3) establish a data collection system, (4) attribute **indicators** to the benchmarks, (5) aggregate and analyze data, (6) establish a reporting system, and (7) evaluate the process and adjust as needed.

This approach has been viewed positively by some analysts (Olin 2013; Caplan 2019; Bellamy 2020), while others have raised concerns about establishing targets without adequate contextual understanding and local input, in conditions that are highly fluid and unstable, to consolidate an outcome—peace—that may take a long time to achieve and may not lend itself to the sort of metrics for which data can be readily collected (Broome and Quirk 2015). Analysts and practitioners seem to be in broad agreement that benchmarking should be a part of the peacebuilding toolkit, but that its effectiveness depends on the quality of data available, continuous local input, adaptability, and a high degree of contextual sensitivity.

Environmental peacebuilding is well-positioned to draw upon UN peacebuilding guidance and on many environmentally focused benchmarking systems within and outside the United Nations (e.g., World Benchmarking Alliance 2022–24). While the general practice of setting goals and measuring performance against them has been a part of environmental peacebuilding since its inception in the 1990s, as the field has evolved (Ide et al. 2021), new opportunities for benchmarking have also emerged. These include (but are not limited to) developing benchmarks for fully assessing environmental damage, establishing priorities for reconstruction and mainstreaming these into national plans, building local and state capacity, developing trusted and effective systems for natural resource management, building **resilience** to **climate change** impacts, and identifying environmental factors linked to **conflict** and peace. The integration of AI and other **digital technologies** into environmental peacebuilding activities also has enormous potential to support benchmarking in this domain, especially through improving data collection and analysis, if the risks posed by AI that are currently being assessed in many different fora can be effectively and fairly managed.

References

Bellamy, Alex J. 2020. "How Much Peace is Enough? A Reflection on Richard Caplan's Measuring Peace." *Ethnopolitics* 19 (3): 1–4. https://doi.org/10.1080/17449057.2020.1759917.
Broome, André, and Joel Quirk. 2015. "Governing the World at a Distance: The Practice of Global Benchmarking." *Review of International Studies* 41 (5): 819–41. https://doi.org/10.1017/S0260210515000340.
Caplan, Richard. 2019. *Measuring Peace: Principles, Practices, and Politics.* Oxford University Press.
De Coning, Cedric, and Paul Romita. 2009. *Monitoring and Evaluation of Peace Operations.* International Peace Institute (IPI).
Ide, Tobias, Carl Bruch, Alexander Carius, et al. 2021. "The Past and Future(s) of Environmental Peacebuilding." *International Affairs* 97 (1): 1–16. https://doi.org/10.1093/ia/iiaa177.
Olin, Nathaniel. 2013. *Measuring Peacekeeping: A Review of the Security Council's Benchmarking Process for Peacekeeping Missions.* Social Science Research Council. https://ssrc-cdn1.s3.amazonaws.com/crmuploads/new_publication_3/%7BCEA59B0A-ECFA-E211-AFF4-001CC477EC84%7D.pdf.
United Nations (UN). 2010. *Monitoring Peace Consolidation: United Nations Practitioners' Guide to Benchmarking.* UN. https://www.un.org/peacebuilding/sites/www.un.org.peacebuilding/files/documents/monitoring_peace_consolidation.pdf.
University of Warwick. 2021. *Global Benchmarking Database.* University of Warwick. https://warwick.ac.uk/fac/soc/pais/research/csgr/benchmarking/database
World Benchmarking Alliance (WBA). 2022. *Nature Benchmark.* WBA. https://www.worldbenchmarkingalliance.org/publication/nature/.

Biodiversity

Rachel Weaver, Julia Gorricho, and Janet Edmond

Biodiversity, short for biological diversity, refers to the variety of life on Earth, encompassing genetic diversity, species diversity, and ecosystem diversity, including plants, animals, fungi, algae, and microorganisms (Convention on Biological Diversity 2022; Wilson 1988).

Biodiversity is crucial for ecosystem **resilience**, providing important services that sustain life (Romanelli et al. 2015; Wilson 1988). Biodiversity is a multidimensional term, often reflecting varying ecological, cultural, and economic values (Diaz and Malhi 2022). The concept of biodiversity gained prominence in scientific discourse and literature in the 1980s as global environmental challenges escalated (Diaz and Malhi 2022). Its relevance has expanded from strictly ecological contexts to include broader social, political, and economic fields, including **environmental peacebuilding**.

In environmental peacebuilding, biodiversity **conservation** and management foster environmental **sustainability** and **social cohesion** (Romanelli et al. 2015). The conservation of biodiversity can mitigate **environmental degradation** that can contribute to or exacerbate conflicts, especially in regions where **natural resources** are scarce or contested (Rist 2024). Biodiversity can provide **ecosystem services** such as water regulation, clean air, **food security**, and **disaster risk reduction**, all of which contribute to **peace** by enhancing community resilience (Romanelli et al. 2015). Collaborative biodiversity conservation and management efforts (e.g., transboundary **peace parks**) can serve as an opportunity for **cooperation**, which can, over time, build trust and collaboration (Ali 2007).

Conflict is a significant driver of biodiversity loss (Rist et al. 2024). The geographic overlap between biodiversity hotspots and conflict-affected regions is notable (Hanson 2009). **War**, civil unrest, and military activities can result in severe biodiversity loss from deforestation, **land degradation**, and overexploitation of resources in both terrestrial and marine ecosystems (Hanson et al. 2009; Machlis and Hanson 2008; Hanson 2018; Parsons et al. 2008).

Biodiversity loss itself can fuel and exacerbate conflict, particularly in areas where communities and livelihoods depend on **natural resources** and **ecosystem services** (Romanelli et al. 2015). When ecosystems degrade, communities lose access to essential resources like food and water, intensifying competition and increasing the likelihood of conflict (Romanelli et al. 2015). This creates a destructive feedback loop, where environmental degradation and conflict perpetuate one another (Klare 2001).

Despite the challenges of armed conflict, biodiversity conservation efforts can persist, benefiting rural communities (McNeely 2003) and maintaining essential ecosystem services in protected areas (Butsic et al. 2015; de Merode et al. 2007; Dávalos et al. 2011). Key strategies for protecting biodiversity in conflict zones include securing international funding (Hart et al. 1997; Stephenson and Newby 1997), supporting and training local staff (Gorricho and Schultze-Kraft 2021; Plumptre 2003), recognizing the role of local communities (de Merode et al. 2007; Hamilton et al. 2000), and fostering collaboration between conservation and humanitarian organizations (Milburn 2012). Post-conflict, there is an urgent need to re-establish conservation programs within broader **peacebuilding** initiatives that include local communities as key allies for long-term success.

The Convention on Biological Diversity (**CBD**) is a key international treaty promoting biodiversity conservation, sustainable use of resources, and fair and equitable sharing of benefits from genetic resources. While the CBD has shaped global biodiversity policy, national strategies, and international collaboration to address biodiversity loss, it has explored the link between biodiversity, peace, and insecurity. The Sixteenth Conference of Parties (COP16) in 2024 (Cali, Colombia) marks a significantly higher profile for the "Peace with Nature" theme to be highlighted within the CBD framework, urging **conflict sensitivity** within inclusive, rights-based approaches in biodiversity conservation.

As global biodiversity declines (Isbell et al. 2022), its conservation must be increasingly recognized as essential to peacebuilding initiatives, ensuring we safeguard the **environment** and lay stronger foundations for peaceful, resilient societies.

References

Ali, Saleem H. 2007. *Peace Parks: Conservation and Conflict Resolution*. MIT Press.

Butsic, Van, Matthias Baumann, Anja Shortland, Sarah Walker, and Tobias Kuemmerle. 2015. "Conservation and Conflict in the Democratic Republic of Congo: The Impacts of Warfare, Mining, and Protected Areas on Deforestation." *Biological Conservation* 191: 266–73. https://doi.org/10.1016/j.biocon.2015.06.037.

Convention on Biological Diversity. 2022. United Nations Treaty Series No. 1760. https://www.cbd.int/convention/articles/default.shtml?a=cbd-02.

Dávalos, Liliana M., Adriana C. Bejarano, Mark A. Hall, H. Leonardo Correa, Angelique Corthals, and Oscar J. Espejo. 2011. "Forests and Drugs: Coca-Driven Deforestation in Tropical

Biodiversity Hotspots." *Environmental Science and Technology* 45 (4): 1219–27. https://doi.org/10.1021/es102373d.
de Merode, Emmanuel, Kes Hillman Smith, Katherine Homewood, Richard Pettifor, Marcus Rowcliffe, and Guy Cowlishaw. 2007. "The Impact of Armed Conflict on Protected-Area Efficacy in Central Africa." *Biology Letters* 3 (3): 299–301. https:/doi.org/10.1098/rsbl.2007.0010.
Díaz, Sandra, and Yadvinder Malhi. 2022. "Biodiversity: Concepts, Patterns, Trends, and Perspectives." *Annual Review of Environment and Resources* 47: 31–63. https://doi.org/10.1146/annurev-environ-120120-054300.
Gorricho, Julia, and Markus Schultze-Kraft. 2021. "Wartime Protected Area Governance: The Case of Colombia's Alto Fragua Indiwasi National Park." *Third World Quarterly* 42 (10): 2160–77. https://doi.org10.1080/01436597.2021.1892482.
Hamilton, Alan, Andrew Cunningham, Dominic Byarugaba, and Fredrick Kayanja. 2000. "Conservation in a Region of Political Instability: Bwindi Impenetrable Forest, Uganda." *Conservation Biology* 14 (5): 1448–51. https://doi.org/10.1111/j.1523-1739.2000.99452.x.
Hanson, Thor, Thomas M. Brooks, Gustavo A. B. Da Fonseca, Thomas M. Brooks, Gustavo A. B. Da Fonseca, Michael Hoffmann, John F. Lamoreux, Gary Machlis, Cristina G. Mittermeier, Russell A. Mittermeier, and John D. Pilgrim. 2009. "Warfare in Biodiversity Hotspots." *Conservation Biology* 23 (3): 578–87. https://doi.org/10.1111/j.1523-1739.2009.01166.x.
Hanson, Thor. 2018. "Biodiversity Conservation and Armed Conflict: A Warfare Ecology Perspective." *Annals of the New York Academy of Sciences* 1429 (1): 50–65. https://doi.org/10.1111/nyas.13689.
Hart, Terese, John Hart, Cheryl Fimbel, et al. 1997. "Conservation and Civil Strife: Two Perspectives from Central Africa." *Conservation Biology* 11 (2): 308–14. https://www.jstor.org/stable/2387605.
Isbell, Forest, Patricia Balvanera, Akira S. Mori, et al. 2022. "Expert Perspectives on Global Biodiversity Loss and Its Drivers and Impacts on People." *Frontiers in Ecology and the Environment* 21 (2): 94–100. https://doi.org/10.1002/fee.2536.
Klare, Michael T. 2001. "The New Geography of Conflict." *Foreign Affairs* 80 (3): 49-61. https://www.foreignaffairs.com/articles/russia-fsu/2001-05-01/new-geography-conflict.
Machlis, Gary E., and Thor Hanson. 2008. "Warfare Ecology." *BioScience* 58 (8): 729–36. https://doi.org/10.1641/B580809.
McNeely, Jeffrey A. 2003. "Conserving Forest Biodiversity in Times of Violent Conflict." *Oryx* 37 (2): 142–52. https://doi.org/10.1017/S0030605303000334.
Milburn, Richard. 2012. "Mainstreaming the Environment into Postwar Recovery: The Case for 'Ecological Development'." *International Affairs* 88 (5): 1083–100. https://doi.org/10.1111/j.1468-2346.2012.01120.x.
Parsons, E. Chris M., Sarah J. Dolman, Andrew J. Wright, Naomi A. Rose, and Wil C. G. Burns. 2008. "Navy Sonar and Cetaceans: Just How Much Does the Gun Need to Smoke Before We Act?" *Marine Pollution Bulletin* 56 (7): 1248–57. https://doi.org/10.1016/j.marpolbul.2008.04.025.
Plumptre, Andrew J. 2003. "Lessons Learned from On-the-Ground Conservation in Rwanda and the Democratic Republic of the Congo." In *War and Tropical Forests*, edited by Steven Price. CRC Press.
Plumptre, Andrew, Deo Kujirakwinja, Adrian Treves, Isaiah Owiunji, and Helga Rainer. 2007. "Transboundary Conservation in the Greater Virunga Landscape: Its Importance for Landscape Species." *Biological Conservation* 134 (2): 279–87. https://doi.org/10.1016/j.biocon.2006.08.012.
Rist, Lucy, Albert Norström, and Cibele Queiroz. 2024. "Biodiversity, Peace, and Conflict: Understanding the Connections." *Current Opinion in Environmental Sustainability* 68: 101431. https://doi.org/10.1016/j.cosust.2024.101431.
Romanelli, Cristina, David Cooper, Diarmid Campbell-Lendrum, et al. 2015. *Connecting Global Priorities: Biodiversity and Human Health: A State of Knowledge Review*. United Nations Environment Programme, Convention on Biological Diversity, and World Health Organization.
Stephenson, Peter, and John E. Newby. 1997. "Conservation of the Okapi Wildlife Reserve, Zaïre." *Oryx* 31 (1): 49–58. https://doi.org/10.1046/j.1365-3008.1997.d01-3.x.
Wilson, E. O. 1988. *Biodiversity*. National Academy Press.

Blame Avoidance

Meryam El Bouhati and Rami Abi Ammar

Blame avoidance is the use of strategies by political actors and institutions to shift responsibility for issues like resource scarcity or environmental degradation, often by framing external factors—such as refugee populations or climate migration—as the cause. This allows governments and institutions to divert attention from their own shortcomings in resource management or environmental policies.

Blame avoidance has been primarily examined in political science in relation to how politicians seek to distance themselves from responsibility for their decisions to protect, maintain, or improve their chances of reelection (Weaver 1986). The literature has focused extensively on the strategies employed to evade blame, leading to a growing inventory of such strategies and their categorizations. Weaver (1986) identifies eight strategies that public officials can use to address blame-inducing pressures: (1) limit the agenda, (2) redefine the issue, (3) throw good money after bad, (4) pass the buck, (5) find a scapegoat, (6) jump on the bandwagon, (7) circle the wagons, and (8) "stop me before I kill again." Hood (2002) groups these strategies into three broad categories: agency strategies, presentational strategies, and policy strategies.

While the use of blame avoidance is evident across various forms of government (Heinkelmann-Wild et al. 2023), the opportunities to avoid blame differ significantly between them. In sectors such as climate and **security**, the impact of domestic policy decisions on worsening, rather than mitigating, climate effects and fostering **conflict** is often overlooked (Daoudy et al. 2022). Still, those policy decisions drive public officials, private, and international actors to adopt blame avoidance strategies in shaping policies related to the **environment** and security. Research suggests that delegating **governance** tasks to third parties can bolster a government's capacity to evade blame (Bach and Wegrich 2019; Hinterleitner 2020), with international organizations, including the European Union, playing a particularly helpful role in this regard (Schimmelfennig 2020).

These strategies are evident in discourses surrounding water scarcity and governance, particularly in **displacement** and humanitarian relief contexts. In such scenarios, political actors and governments employ scapegoating strategies that place emphasis on the impact of population pressure to evade blame for pre-existing shortages in water supply, weak **infrastructure**, or a decline in water quality. Refugees and migrant communities are prevalent scapegoats for blame avoidance. Hussein et al. (2020) identify the use of blame shifting in Lebanon and Jordan following the influx of Syrian refugees. In both countries, deliberate framing of resource depletion, contamination, and infrastructure pressures as issues stemming from refugee demand and inefficient consumption is used to downplay inefficiencies in governments' water allocation, management, and infrastructure maintenance capacities. Political and economic motivations underlie this form of blame avoidance, as it allows both Jordan and Lebanon to tap into capital flows from development funding and global investment in **mitigation**. In Lebanon, the impact of the refugee crisis is underscored by the need to mobilize international and humanitarian aid to support emergency response. In Jordan, these narratives seek to attract funding for expensive water megaprojects, like the Disi Water Conveyance Project (Hussein et al. 2020). Governments may use climate migration **securitization** as a tactic to shift the responsibility away from their lack of environmental regulations or **climate change** readiness (Weinthal et al. 2015). Authorities efficiently deflect blame by focusing on an external "threat" narrative rather than their own part in accelerating **environmental degradation** or neglecting to ameliorate climate impacts by portraying **climate migrants** as a national security threat. This strategy, which emphasizes security over **accountability** for climate challenges, is frequently used to defend strict immigration laws and call for international assistance (Bettini 2013; Hartmann 2010).

Blame avoidance allows governments to deflect accountability by attributing issues to external factors like refugee influxes, attracting funding while obscuring policy failures. These dynamics have significant implications for **environmental peacebuilding** through processes of **depoliticization** that obscure **power** relations and structural inequalities. By scapegoating migrants and refugees, governments securitize environmental challenges as threats, framing them as technical problems caused by external populations rather than addressing systemic governance failures. As Ide (2019) suggests, such approaches can reinforce existing power structures and marginalize vulnerable populations. Blame avoidance strategies thus obstruct participatory governance mechanisms, limiting marginalized communities' contributions to sustainable resource management solutions and contributing to the reproduction of inequality.

References

Bach, Tobias, and Kai Wegrich. 2019. "Blind Spots, Biased Attention, and the Politics of Non-Coordination." In *The Blind Spots of Public Bureaucracy and the Politics of Non-Coordination*, edited by Tobias Bach and Kai Wegrich. Palgrave Macmillan. https://doi.org/10.1007/978-3-319-76672-0_1.

Bettini, Giovanni. 2013. "Climate Barbarians at the Gate? A Critique of Apocalyptic Narratives on 'Climate Refugees.'" *Geoforum* 45: 63–72. https://doi.org/10.1016/j.geoforum.2012.09.009.

Daoudy, Marwa, Jeannie Sowers, and Erika Weinthal. 2022. "What Is Climate Security? Framing Risks around Water, Food, and Migration in the Middle East and North Africa." *WIREs Water* 9 (3): e1582. https://doi.org/10.1002/wat2.1582.

Hartmann, Betsy. 2010. "Rethinking Climate Refugees and Climate Conflict: Rhetoric, Reality and the Politics of Policy Discourse." *Journal of International Development* 22 (2): 233–46. https://doi.org/10.1002/jid.1676.

Heinkelmann-Wild, Tim, Bernhard Zangl, Berthold Rittberger, and Lisa Kriegmair. 2023. "Blame Shifting and Blame Obfuscation: The Blame Avoidance Effects of Delegation in the European Union." *European Journal of Political Research* 62 (1): 221–38. https://doi.org/10.1111/1475-6765.12503.

Hinterleitner, Markus. 2020. *Policy Controversies and Political Blame Games*. Cambridge University Press. https://doi.org/10.1017/9781108860116.

Hood, Christopher. 2002. "The Risk Game and the Blame Game." *Government and Opposition* 37 (1): 15–37. https://doi.org/10.1111/1477-7053.00085.

Hussein, Hussam, Alberto Natta, Abed Al Kareem Yehya, and Baha Hamadna. 2020. "Syrian Refugees, Water Scarcity, and Dynamic Policies: How Do the New Refugee Discourses Impact Water Governance Debates in Lebanon and Jordan?" *Water* 12 (2): 325. https://doi.org/10.3390/w12020325.

Ide, Tobias. 2019. "The Dark Side of Environmental Peacebuilding." *World Development* 127: 104777. https://doi.org/10.1016/j.worlddev.2019.104777.

Schimmelfennig, Frank. 2020. "Politicisation Management in the European Union." *Journal of European Public Policy* 27 (3): 342–61. https://doi.org/10.1080/13501763.2020.1712458.

Weaver, R. Kent. 1986. "The Politics of Blame Avoidance." *Journal of Public Policy* 6 (4): 371–98. https://doi.org/10.1017/S0143814X00004219.
Weinthal, Erika, Neda Zawahri, and Jeannie Sowers. 2015. "Securitizing Water, Climate, and Migration in Israel, Jordan, and Syria." *International Environmental Agreements: Politics, Law and Economics* 15 (3): 293–307. https://doi.org/10.1007/s10784-015-9279-4.

Boomerang Effect

Larry A. Swatuk, Lars Wirkus, Florian Krampe, Bejoy K. Thomas, and Luis Paulo Batista da Silva

The boomerang effect is "the emergence of largely unanticipated and unintended negative consequences of climate change adaptation and mitigation policies and programs on domestic non-state actors that result in negative feedback on the state" (Swatuk et al. 2019, 3, 2021).

> Put more comprehensively, deriving from state actor **climate change** adaptation or **mitigation** policies and programs, the implementation of these climate interventions (through state or state-authorized private actors) often has unanticipated and unintended negative social, political, economic, and ecological effects that impact local communities on various spatial and temporal scales (i.e., local-level side effects). These impacts, in turn, negatively feed back to the state on multiple levels (e.g., local, regional, national), at various scales (e.g., watershed, forest, landscape, ecosystem), with numerous impacts—i.e., boomerang effects—(e.g., political economic instability, social unrest, and **violence**), thus undermining **climate security**.
>
> This definition builds on related framings such as "**backdraft**" (Dabelko et al. 2013) and "**maladaptation**" (Barnett and O'Neill 2013). Whereas backdraft and maladaptation are often used as blanket concepts, the "boomerang effect" constitutes an attempt to refine the analytical ability to differentiate among complementary effects too often lumped together: i.e., (1) local level side effects and (2) state-level boomerang effects. The value of the boomerang effect extends beyond analytical refinement toward improved policymaking. If policymakers can grasp that climate actions often boomerang back upon them, then, in theory at least, there will be improved scope and political incentive for more nuanced climate action. In practice, however, states seem only too willing to bear the various costs of their actions as "the price of doing business" (Swatuk and Wirkus 2019). At the same time, evidence suggests that high consumption (HC) countries, through global **governance** mechanisms such as the Green Climate Fund (GCF) and **Global Environment Facility (GEF)**, seek to offset boomerang effects at the national level by supporting **adaptation** and mitigation activities in low consumption (LC) countries to help meet their NDCs (Nationally Determined Contributions) (Mirumachi et al. 2019; Achampong 2022; Tomlinson 2022). For example, financial support for REDD (Reducing Emissions from Deforestation and Forest Degradation in Developing Countries) and **REDD+** (i.e., additional forest-related activities that protect the climate) programs in Kenya and Vietnam (Kioko 2022; Maxwell et al. 2019) and **renewable energy** projects in Tanzania (Collins 2021) displace the possibility for both local level side effects and boomerang effects from HC to LC countries. For Krampe and colleagues (2021), in fragile, post-conflict settings, where local-level side effects of climate adaptation and mitigation projects

result in unequal outcomes, the potential for **sustainable peace** is questionable. The widespread occurrence of what Ide (2025, 6) calls "direct climate action conflicts related to adaptation" elevates the importance of **environmental peacebuilding** through **conflict**-sensitive planning, programming, and implementation.

References

Achampong, Leia. 2022. "How Lessons from Development Finance Can Strengthen Climate Finance." In *The Political Economy of Climate Finance: Lessons from International Development*, edited by Corrine M. Cash and Larry A. Swatuk. Palgrave Macmillan.

Barnett, Jon, and Saffron J. O'Neill. 2013. "Minimising the Risk of Maladaptation: A Framework for Analysis." In *Climate Adaptation Futures*, edited by Jean P. Palutikof, Sarah L. Boulter, Andrew J. Ash, et al. Wiley-Blackwell.

Collins, Tom. 2021. "Kenya's Maasai Fear Mass Displacement from Geothermal Plant." *African Business*. https://african.business/2021/03/energy-resources/kenyas-maasai-fear-geothermal-plant-development.

Dabelko, Geoffrey D., Lauren Herzer Risi, Schuyler Null, Meaghan Parker, and Russel Stricklor. 2013. *Backdraft: The Conflict Potential of Climate Change Adaptation and Mitigation*. Woodrow Wilson Center. https://www.wilsoncenter.org/documents/publication/backdraft.pdf.

Ide, Tobias. 2025. "Rethinking Climate Conflicts: The Role of Climate Action and Inaction." *World Development* 186: 106845. https://doi.org/10.1016/j.worlddev.2024.106845.

Kioko, Eric M. 2022. "Forest Crime in Africa: Actors, Markets and Complexities." In *African Futures,* edited by Clemens Greiner, Steven van Wolputte and Michael Bollig. Brill. https://library.oapen.org/handle/20.500.12657/54653.

Krampe, Florian, Elizabeth S. Smith, and Mohammad Daud Hamidi. 2021. "Security Implications of Climate Development in Conflict-Affected States: Implications of Local-Level Effects of Rural Hydropower Development on Farmers in Herat." *Political Geography* 90: 102454. https://doi.org/10.1016/j.polgeo.2021.102454.

Maxwell, Laura, Vidya Nair, Stephanie Soloman, and Larry A. Swatuk. 2019. "As Assessment of UN-REDD in Lam Dong Province, Vietnam." In *Water, Climate Change and the Boomerang Effect: Unintentional Consequences for Resource Insecurity,* edited by Larry Swatuk and Lars Wirkus. Routledge.

Mirumachi, Naho, Amiera Sawas, and Mark Workman. 2019. "Unveiling the Security Concerns of Low Carbon Development: Climate Security Analysis of the Undesirable and Unintended Effects of Mitigation and Adaptation." *Climate and Development* 12 (2): 97–109. https://doi.org/10.1080/17565529.2019.1604310.

Swatuk, Larry A., Bejoy K. Thomas, Larry Wirkus, Florian Krampe, and Luis Paulo Batista da Silva. 2021. "The 'Boomerang Effect': Insights for Improved Climate Action." *Climate and Development* 13 (1): 61–67. https://doi.org/10.1080/17565529.2020.1723470.

Swatuk, Larry A., and Lars Wirkus. 2019. *Water, Climate Change and the Boomerang Effect: Unintentional Consequences for Resource Insecurity*. Routledge.

Swatuk, Larry A., Lars Wirkus, Florian Krampe, Bejoy Thomas, and Luis Paulo Batista da Silva. 2019. "The Boomerang Effect: Overview and Implications for Climate Governance." In *Water, Climate Change and the Boomerang Effect: Unintentional Consequences for Resource Insecurity*, edited by Larry A. Swatuk and Lars Wirkus. Routledge.

Tomlinson, Brian. 2022. "International Climate Finance and Development Effectiveness." In *The Political Economy of Climate Finance: Lessons from International Development*, edited by Corrine Cash and Larry A. Swatuk. Palgrave Macmillan. https://doi.org/10.1007/978-3-031-12619-2_3.

C

Capacity Development

Annika Erickson-Pearson, Rebecca L. Farnum, and Faith Mutegi

Capacity development is the process of enhancing and expanding an actor's ability to understand and respond to issues that affect them. This ability, often referred to as capacity, can be strengthened through education, skill building, and increased access to resources, among other approaches.

In his turn-of-the-millennium book on *Development as Freedom*, Nobel Prize-winning economist Amartya Sen stated that true development "consists of the removal of various types of unfreedoms that leave people with little choice and little opportunity of exercising their reasoned agency" (Sen 1999, 12). Advocating for political freedoms, economic facilities, social opportunities, transparency guarantees, and protective **security**, Sen (1999, 10) posits that these freedoms "are not only the primary ends of development, but they are also among its principal means." His framework came to be known as the "capabilities approach" and was further developed by scholars such as Martha Nussbaum (2000). Sen's work is particularly notable for arguing in economic terms that ensuring that individuals, groups, communities, and nations have the necessary capabilities to shape their own destinies is both an inherent (i.e., of worth in and of itself) and extrinsic (e.g., economically productive) good.

Twenty-five years later, international institutions have come to engage more with agency, capacities, and **empowerment** as core elements of **positive peace** and **sustainable development**. Indicators such as the Human Development Index and the Gender Inequality Index have their origins in the capabilities approach; instruments such as the **United Nations Convention on Biological Diversity** (**CBD**), **United Nations Framework Convention on Climate Change** (**UNFCCC**), and the 2030 Agenda for Sustainable Development explicitly engage with capacity development as critical to their implementation and success.

The United Nations Development Group defines capacity development as "the process whereby people, organizations and society as a whole unleash, strengthen, create, adapt and maintain [the ability to manage their affairs successfully] over time" (United Nations Development Group 2017). Skills, knowledge, and resources are necessary for actors to achieve their goals and, on a collective level, build and sustain their desired systems. It's important to recognize that capacity development is

DOI: 10.4324/9781003604143-5

essential at *all* levels and for all actors, not merely those assumed to be less capable in mainstream approaches. Government officials, corporate leaders, and academic researchers are also in need of capacity development: they need to build their skills in participatory approaches, collaboration, and decolonial practices. Focusing solely on less privileged communities and the Global South reinforces a "deficit mentality" and reinforces "capacity building" as a form of neocolonialism (Mormina and Istratii 2021).

Capacities are often diminished in armed **conflict** contexts, as economic, political, and social systems are thrown into disarray and **governance** mechanisms are disrupted. Schools may be shut down; adults may be conscripted into military service rather than gaining professional skills in their desired vocation. On a larger scale, political leadership and other decision-makers are likely to put more time, attention, and resources into conflict logistics for short-term gains than community-centric institutions for long-term impact.

When it comes to **environmental peacebuilding** concerns, a vicious cycle may exist. Communities with more limited capacities are less likely to be able to successfully adapt to **climate change** shocks and **environmental degradation** and may resort to **violence** as a result. This violent conflict, in turn, further detracts from capacities and further damages the **environment**. Though this may paint a bleak picture, it highlights capacity development as a key entry point for **peacebuilding,** both as a preventive measure and a conflict disruptor.

References

Mormina, Maru, and Romina Istratii. 2021. "'Capacity for What? Capacity for Whom?' A Decolonial Deconstruction of Research Capacity Development Practices in the Global South and a Proposal for a Value-Centred Approach." *Welcome Open Research* 6: 129. https://doi.org/10.12688/wellcomeopenres.16850.1.

Nussbaum, Martha. 2000. *Women and Human Development: The Capabilities Approach.* Cambridge University Press.

Sen, Amartya. 1999. *Development as Freedom.* Knopf.

United Nations Development Group. 2017. *Capacity Development: UNDAF Companion Guidance.* United Nations Development Assistance Framework. https://unsdg.un.org/sites/default/files/UNDG-UNDAF-Companion-Pieces-8-Capacity-Development.pdf.

Cities

Silvia Danielak

Cities are dense population settlements with correspondingly dense infrastructure, housing, and economic activity in relation to their surrounding land.

Cities are important centers of economic, cultural, and intellectual production, often centralizing political **power** and institutions. Their very character constitutes a significant resource footprint globally. Internally, cities are marked by density and heterogeneity, thereby posing distinct opportunities and challenges for their **governance,** ranging from collaboration, innovation, and experimentation, communal wealth creation, to fragmentation, urban **violence**, and **conflict** (Amin and Thrift 2017).

Cities are the product of urbanization processes, which describe a growing agglomeration. Cities do not exist in a vacuum and maintain important connections and influence over periurban and rural geographies (Brenner 2014). Urbanization can be the consequence of **displacement** from armed conflict, environmental hazards, and **climate change**, even if cities are equally exposed to them, creating further cycles of (urban) displacement. What happens in cities is thus of importance beyond their cadastral boundaries.

Today, more than half of the world's population is urban (United Nations Department of Economic and Social Affairs n.d.). While conflict research has caught on to this new reality, fewer studies have explored the explicit links to **environmental peacebuilding**. Thus far, studies in urban settings have showcased the intersection of conflict with environmental themes, including in periurban areas and through the lens of protest, public service provision, or structural violence (Gizelis et al. 2021; Nagle 2013; Stockmans and Büscher 2017). Urban planning and urban political ecology have challenged the long-standing assumption that nature and the city are opposites. Instead, these fields (especially urban political ecology) emphasize how nature is embedded in urban environments and how sustainability is an integral part of city life (Braun 2005; Angelo and Wachsmuth 2020; Campbell 1996; Heynen 2014).

Cities are exposed to natural hazard-induced disasters just as they are subject to climate change, which they experience distinctly and intensely due to their compressed built **environment** and large population assembled on a relatively small territory. Environmental stressors impact urban governance and urban dwellers' livelihoods and thereby shape conflict dynamics, for example, in urban flooding, heat island effects, and toxic pollution (Danielak 2022; Yeeles 2015). In parallel, cities can also be significant polluters with low levels of air quality, high levels of gray, impermeable surfaces, and massive amounts of waste produced. This is all the truer during armed conflict, when cities are destroyed—and rebuilt—in the process, releasing waste, toxins, rubble, and carbon emissions (Weinthal and Sowers 2019; Neimark et al. 2024).

Even in the absence of **war**, efforts at greening cities are a physical, social, and political endeavor (Heynen et al. 2006; Angelo 2021). Urban interventions for climate **adaptation**, **mitigation**, and **disaster risk reduction** are crucial to ensure city dwellers' safety and well-being, but they carry their own risk for harm and conflict. Well-intended measures may put a disproportionate burden on low-income and otherwise marginalized groups in the city, for example, through "climate gentrification" (Immergluck and Balan 2018; Shokry et al. 2020; Chiesi and Forte 2022), especially when collaborative planning is cut short.

Positively, urban density also allows for carbon-neutral alternatives in the transport and energy sector, innovative circular economy and waste management, or urban **agriculture**. Cities need to build urban **resilience** in equitable ways to provide dignified, stable, and safe livelihoods for their residents (Lamb and Vale 2024). This is the case for cities seeking to build resilience in the face of disaster vulnerability (Sanderson et al. 2016), just as those experiencing protracted conflict, violence, and destruction (Muggah 2014; Wahba et al. 2021; Elfversson and Höglund 2023).

References

Amin, Ash, and Nigel Thrift. 2017. *Seeing Like a City*. John Wiley and Sons.
Angelo, Hillary. 2021. *How Green Became Good: Urbanized Nature and the Making of Cities and Citizens*. University of Chicago Press. https://press.uchicago.edu/ucp/books/book/chicago/H/bo61910401.html.
Angelo, Hillary, and David Wachsmuth. 2020. "Why Does Everyone Think Cities Can Save the Planet?" *Urban Studies* 57 (11): 2201–21. https://doi.org/10.1177/0042098020919081.
Braun, Bruce. 2005. "Environmental Issues: Writing a More-than-Human Urban Geography." *Progress in Human Geography* 29 (5): 635–50. https://doi.org/10.1191/0309132505ph574pr.
Brenner, Neil, ed. 2014. *Implosions/Explosions: Towards a Study of Planetary Urbanization*. JOVIS.
Campbell, Scott. 1996. "Green Cities, Growing Cities, Just Cities? Urban Planning and the Contradictions of Sustainable Development." *Journal of the American Planning Association* 62 (3): 296–312. https://doi.org/10.1080/01944369608975696.
Chiesi, Leonardo, and Giuseppina Forte. 2022. "Design for Climate Change in the Neoliberal Present: Gentrification, Ecocide, and the Loss of Urbanity in New York City." *Social Sciences* 11 (10): 451. https://doi.org/10.3390/socsci11100451.
Danielak, Silvia. 2022. "Risk, Vulnerability, and Pragmatic Inevitability: The Conflict–Disaster Nexus and Urban Governance in Johannesburg, South Africa." *Disasters* 46 (1): 271–95. https://doi.org/10.1111/disa.12461.
Elfversson, Emma, and Kristine Höglund. 2023. "Urban Growth, Resilience, and Violence." *Current Opinion in Environmental Sustainability* 64: 101356. https://doi.org/10.1016/j.cosust.2023.101356.
Gizelis, Theodora-Ismene, Steve Pickering, and Henrik Urdal. 2021. "Conflict on the Urban Fringe: Urbanization, Environmental Stress, and Urban Unrest in Africa." *Political Geography* 86: 102357. https://doi.org/10.1016/j.polgeo.2021.102357.
Heynen, Nik. 2014. "Urban Political Ecology I: The Urban Century." *Progress in Human Geography* 38 (4): 598–604. https://doi.org/10.1177/0309132513500443.
Heynen, Nik, Maria Kaika, and Erik Swyngedouw. 2006. "Urban Political Ecology: Politicizing the Production of Urban Natures." In *In the Nature of Cities: Urban Political Ecology and the Politics of Urban Metabolism*, edited by Nik Heynen, Maria Kaika, and Erik Swyngedouw. Routledge.
Immergluck, Dan, and Tharunya Balan. 2018. "Sustainable for Whom? Green Urban Development, Environmental Gentrification, and the Atlanta Beltline." *Urban Geography* 39 (4): 546–62. https://doi.org/10.1080/02723638.2017.1360041.
Lamb, Zachary B., and Lawrence J. Vale. 2024. *The Equitably Resilient City: Solidarities and Struggles in the Face of Climate Crisis*. MIT Press. https://bookshop.org/p/books/the-equitably-resilient-city-struggles-and-solidarities-in-the-face-of-climate-crisis-lawrencej-vale/21005869.
Muggah, Robert. 2014. "Deconstructing the Fragile City: Exploring Insecurity, Violence and Resilience." *Environment and Urbanization* 26 (2): 345–58. https://doi.org/10.1177/0956247814533627.
Nagle, John. 2013. "'Unity in Diversity': Non-Sectarian Social Movement Challenges to the Politics of Ethnic Antagonism in Violently Divided Cities." *International Journal of Urban and Regional Research* 37 (1): 78–92. https://doi.org/10.1111/j.1468-2427.2012.01156.x.
Neimark, Benjamin, Patrick Bigger, Frederick Otu-Larbi, and Reuben Larbi. 2024. *A Multitemporal Snapshot of Greenhouse Gas Emissions from the Israel-Gaza Conflict*. SSRN Scholarly Paper. Social Science Research Network. https://doi.org/10.2139/ssrn.4684768.
Sanderson, David, Jerold S. Kayden, and Julia Leis, eds. 2016. *Urban Disaster Resilience: New Dimensions from International Practice in the Built Environment*. Routledge.
Shokry, Galia, James J. T. Connolly, and Isabelle Anguelovski. 2020. "Understanding Climate Gentrification and Shifting Landscapes of Protection and Vulnerability in Green Resilient Philadelphia." *Urban Climate* 3: 100539. https://doi.org/10.1016/j.uclim.2019.100539.
Stockmans, Jep, and Karen Büscher. 2017. "A Spatial Reading of Urban Political-Religious Conflict: Contested Urban Landscapes in Addis Ababa, Ethiopia." *The Journal of Modern African Studies* 55 (1): 79–104. https://doi.org/10.1017/S0022278X1600077X.

United Nations Department of Economic and Social Affairs. n.d. *68 Percent of the World Population Projected to Line in Urban Areas by 2050*. United Nations. Accessed August 13, 2025. https://www.un.org/uk/desa/68-world-population-projected-live-urban-areas-2050-says-un.
Wahba, Sameh, Maitreyi Bordia Das, and Yuna Chun. 2021. "Building Back Stronger: Urban Resilience Through Post-Conflict Reconstruction." *Journal of International Affairs* 74 (1): 227–42. https://www.jstor.org/stable/27169781.
Weinthal, Erika, and Jeannie Sowers. 2019. "Targeting Infrastructure and Livelihoods in the West Bank and Gaza." *International Affairs* 95 (2): 319–40. https://doi.org/10.1093/ia/iiz015.
Yeeles, Adam. 2015. "Weathering Unrest: The Ecology of Urban Social Disturbances in Africa and Asia." *Journal of Peace Research* 52 (2): 158–70. https://www.jstor.org/stable/24557452.

Citizen Science

Karam F. Robeil and Nynke Schaap

Citizen science is a form of participatory research that involves lay citizens as active researchers, to varying degrees and in varying steps of the research process, in the generation or verification of knowledge through data collection, analysis, and/or dissemination. This knowledge is usually aimed at contributing to scientific research, public awareness, and/or policymaking.

The practice of citizen science has spread since the beginning of the 21st century as new technologies emerged, including **digital technologies** and the internet (Bonney et al. 2014; Hodgkinson et al. 2022). Its utility spans across a variety of fields, with environmental and ecological sciences being the most common (Fraisl et al. 2022). Although there is no consensus on the criteria for citizen science, its concept has evolved from using citizens as data collectors to engaging citizens as scientists (Lakshminarayanan 2007). Some scholars go further to argue that citizen science involves the public beyond the research process to engage them in the scientific debate (Schrögel and Kolleck 2019). However, in many contexts, science is not the main purpose of citizen science; rather, it is used as an evidence-based approach to achieve other goals related to public awareness or advocacy.

In fragile and conflict-affected settings (FCAS), citizen science has been a useful tool to gather data from areas that researchers often lack access to due to safety concerns and politically motivated restrictions (Weir et al. 2019). The increased recognition of climate, **environment**, and **conflict** linkages has opened new opportunities for citizen science. For example, citizen science can be carefully applied to help with measuring **toxic remnants of war** and industrial waste after conflicts, where toxic residues and environmental pollution pose a serious risk to local communities that weakened and distracted national authorities may be unwilling or unable to deal with (Garrity and Zwijnenburg 2016; Weir et al. 2019). In such settings, data is often scarce, and citizen science can help bridge this data gap (Garrity and Zwijnenburg 2016; Weir et al. 2019). As FCAS are often characterized by a lack of trust, citizen science as part of **environmental peacebuilding** can promote public trust in science, **governance**, and democracy (Gundersen et al. 2022). Furthermore, citizen science can flag issues that need attention, both toward communities and the government. Accordingly, it can go beyond the purpose of informing science and inform public awareness and policymaking as well. Many FCAS are also highly vulnerable to **climate change** impacts (Moran et al. 2018), including extreme weather events. Citizen science can be utilized in support of early warning, as part of environmental **monitoring** (Hicks et al. 2019). Yet, in some cases, government officials in FCAS may still be reluctant to trust or rely on data collected by citizens, and as such, significant efforts may be required for governments to utilize the data.

A major value of citizen science is that it includes local, situated knowledge and perspectives (Skarlatidou et al. 2024; Weir et al. 2019). Integrating **Indigenous knowledge** and perspectives of vulnerable communities in scientific research contributes positively to environmental **peacebuilding**. Additionally, citizen science has the potential to democratize science (Herzog and Lepenies 2022; Strasser et al. 2019), which contributes to enriching social capital (Conrad and Hilchey 2011). Citizen science is empowering for those involved, especially after they have experienced conflict, where communities have faced exclusion and skepticism about the motives of others (Garrity and Zwijnenburg 2016; Weir et al. 2019). It may promote public trust in science and, more generally, between different actors. Finally, another empowering factor is that citizen science "enables participants to engage in science-protected politics" (Skarlatidou et al. 2024). This gives participants political agency in a context of collapsed institutions and fragmented governance (Weir et al. 2019). In a post-conflict country like Iraq, for example, where civic space is shrinking and environmental activism is increasingly becoming a dangerous endeavor (Khalaf 2023), citizen science may offer a safer and more supported alternative.

References

Bonney, Rick, Jennifer L. Shirk, Tina B. Phillips, et al. 2014. "Next Steps for Citizen Science." *Science* 343 (6178): 1436–37. https://doi.org/10.1126/science.1251554.

Conrad, Cathy C., and Krista G. Hilchey. 2011. "A Review of Citizen Science and Community-Based Environmental Monitoring: Issues and Opportunities." *Environmental Monitoring and Assessment* 176: 273–91. https://doi.org/10.1007/s10661-010-1582-5.

Fraisl, Dilek, Gerid Hager, Baptiste Bedessem, et al. 2022. "Citizen Science in Environmental and Ecological Sciences." *Nature Reviews Methods Primers* 2: 64. https://doi.org/10.1038/s43586-022-00144-4.

Garrity, Andy, and Wim Zwijnenburg. 2016. "Using Citizen Science to Assess Environmental Damage in the Syrian Conflict." *Bellingcat*. https://www.bellingcat.com/resources/how-tos/2016/01/22/using-citizen-science-to-assess-environmental-damage-in-the-syrian-conflict.

Gundersen, Torbjørn, Donya Alinejad, Teresa Yolande Branch, et al. 2022. "A New Dark Age? Truth, Trust, and Environmental Science." *Annual Review of Environment and Resources* 47: 5–29. https://doi.org/10.1146/annurev-environ-120920-015909.

Herzog, Lisa, and Robert Lepenies. 2022. "Citizen Science in Deliberative Systems: Participation, Epistemic Injustice, and Civic Empowerment." *Minerva* 60 (4): 489–508. https://doi.org/10.1007/s11024-022-09467-8.

Hicks, Anna, Jenni Barclay, Jason Chilvers, et al. 2019. "Global Mapping of Citizen Science Projects for Disaster Risk Reduction." *Frontiers in Earth Science* 7: 226. https://doi.org/10.3389/feart.2019.00226.

Hodgkinson, Ian R., Sahar Mousavi, and Paul Hughes. 2022. "New Development: Citizen Science—Discovering (New) Solutions to Wicked Problems." *Public Money and Management* 42 (2): 133–36. https://doi.org/10.1080/09540962.2021.1967630.

Khalaf, Safaa. 2023. "Environmental Mobilization in Iraq: NGOs, Local Actors and the Challenge of Climate Change." *Arab Reform Initiative*. https://www.arab-reform.net/publication/environmental-mobilization-in-iraq-ngos-and-local-actors-and-the-challenge-of-climate-change.

Lakshminarayanan, Shyamal. 2007. "Citizens to do Science Versus Citizens as Scientists." *Ecology and Society* 12 (2): r2. http://www.ecologyandsociety.org/vol12/iss2/resp2.

Moran, Ashley, Joshua Busby, Clionadh Raleigh, et al. 2018. "The Intersection of Global Fragility and Climate Risks." *Agency for International Development*. https://www.strausscenter.org/wp-content/uploads/The-Intersection-of-Global-Fragility-and-Climate-Risks-2018.pdf.
Schrögel, Philipp, and Alma Kolleck. 2019. "The Many Faces of Participation in Science: Literature Review and Proposal for a Three-Dimensional Framework." *Science and Technology Studies* 32 (2): 77–99. https://doi.org/10.23987/sts.59519.
Skarlatidou, Artemis, Muki Haklay, Simon Hoyte, Michiel van Oudheusden, and Isabel J. Bishop. 2024. "How Can Bottom-Up Citizen Science Restore Public Trust in Environmental Governance and Sciences? Recommendations From Three Case Studies." *Environmental Science and Policy* 160: 103854. https://doi.org/10.1016/j.envsci.2024.103854.
Strasser, Bruno, Jérôme Baudry, Dana Mahr, Gabriela Sanchez, and Elise Tancoigne. 2019. "'Citizen Science'? Rethinking Science and Public Participation." *Science and Technology Studies* 32 (2): 52–76. https://doi.org/10.23987/sts.60425.
Weir, Doug, Dan McQuillan, and Robert A. Francis. 2019. "Civilian Science: The Potential of Participatory Environmental Monitoring in Areas Affected by Armed Conflicts." *Environmental Monitoring and Assessment* 191: 618. https://doi.org/10.1007/s10661-019-7773-9.

Climate Change

Maureen J. Purcell and Richard Matthew

Climate change is a human-driven destabilization of long-term global patterns of weather characterized by current and long-term widespread and rapid change to Earth's natural systems and cycles, and human systems.

This phenomenon is attributable to human activity, which directly or indirectly changes the atmosphere's global makeup. Climate change is separate from and in addition to natural climate variations over similar timeframes (United Nations 1992). Global impacts due to climate change include a to-date observed increase of more than 1 degree Celsius in global surface temperature above the 1850–1900 pre-industrial marker of historical stability; altered composition of Earth's atmosphere from carbon-based emissions and other greenhouse gases; rising global mean sea level of 0.20m from 1901–2018 and a current rate of sea level rise of 3.7 mm/yr; changes to severity, frequency, and predictability of extreme weather events such as droughts, heatwaves, heavy precipitation, tropical cyclones, and compound events; melting of glaciers and ice caps and thawing of permafrost; and increased **biodiversity** loss due to damage to coastal, ocean, terrestrial, freshwater, and cryospheric ecosystems (Intergovernmental Panel on Climate Change [IPCC] 2023).

Changes to predictability of livelihoods, **water security**, **food security**, health and well-being, physical **infrastructure**, and economic and social systems have been observed globally because of climate change (IPCC 2023). In this way, climate change directly and indirectly amplifies factors associated with **conflict**, namely economic security and state capacity, as well as increased competition for **natural resources** (Crawford and Church 2022; Matthew 2022). Furthermore, climate-driven extremes are leading to the **displacement** of people in most of the world (IPCC 2023), which can lead to strained or exacerbated strain on resources in host countries and can be a precursor to conflict (Crawford and Church 2022). Global **power** dynamics will very likely be altered because of the changing climate; for instance, some countries will gain access to new mineral wealth in a thawing Arctic, or critical resources like water and land will shift in economic and geopolitical value (Matthew 2022). Climate change is a **threat multiplier**, exacerbating and complicating issues of resource security, political, social, and economic stability, thereby risking **security** and **peacebuilding** efforts (Goodman and Baudu 2023).

References

Crawford, Alec, and Clare V. Church. 2022. "Climate Change as a Contributor to Conflict." In *Routledge Handbook of Environmental Security*, edited by Richard Matthew, Evgenia Nizkorodov, and Crystal Murphy. Routledge.

Goodman, Sherri, and Pauline Baudu. 2023. "Climate Change as a 'Threat Multiplier': History, Uses and Future of the Concept." *Center for Climate and Security Council on Strategic Risks Briefer* 38.https://councilonstrategicrisks.org/wp-content/uploads/2023/01/38-CCThreatMultiplier.pdf.

Intergovernmental Panel on Climate Change (IPCC). 2023. "Summary for Policymakers." In *Climate Change 2023: Synthesis Report. Contribution of Working Groups I, II and III to the Sixth Assessment Report of the Intergovernmental Panel on Climate Change*, edited by Core Writing Team, Hoesung Lee, and José Romero. IPCC. https://doi.org/10.59327/IPCC/AR6-9789291691647.001.

Matthew, Richard. 2022. "Integrating Climate Change Adaptation into Peacebuilding." In *Routledge Handbook of Environmental Security*, edited by Richard Matthew, Evgenia Nizkorodov, and Crystal Murphy. Routledge.

United Nations. 1992. "Article 1." In *United Nations Framework Convention on Climate Change*. https://treaties.un.org/doc/treaties/1994/03/ch_xxvii_07.pdf.

Climate Diplomacy

Aliakbar Shams and Atieh Khatibi

Climate diplomacy refers to using various diplomatic strategies, such as formal and informal meetings and negotiations, to formulate and implement climate policies and agreements, such as the Paris Agreement, and promote cooperation among states and other stakeholders to reduce greenhouse gas (GHG) emissions, adapt to climate change impacts, and support sustainable development (Locci and Jammoul 2023).

The roots of climate diplomacy can be traced at least to the First World Climate Conference held in Geneva in 1979, which was integral to early international **climate change** agreements, such as the **United Nations Framework Convention on Climate Change (UNFCCC)** and the Kyoto Protocol. International climate diplomacy is highly institutionalized, with the Conference of the Parties (COPs) negotiations as its foundation (Doussis and Dikaios 2025). Since 2015, the Paris Agreement has been the central focus of climate diplomacy (Tskipurishvili 2022).

While international climate diplomacy provides the general framework for global **cooperation**, regional organizations play a crucial role in translating commitments into concrete strategies. The European Commission, for instance, identifies four key dimensions of climate diplomacy at the political level: (1) committing to multilateralism in climate policy with a particular emphasis on implementing the Paris Agreement, (2) addressing the impacts of climate change on **peace** and **security**, (3) promoting accelerated actions at the national level and fostering increased global ambition, and (4) strengthening international climate cooperation (Council of the European Union 2018). Building on these regional approaches, at the national level, climate diplomacy can influence and inform foreign policy initiatives. Building on these regional approaches, at the national level, climate diplomacy can influence and inform foreign policy initiatives; for example, Germany's Climate Foreign Policy Strategy emphasizes the intersection of **climate security** and international **peace** efforts, acknowledging that climate-induced instability can intensify conflicts (German Federal Foreign Office 2023).

Climate diplomacy encompasses a range of **stakeholders**, including governmental and non-governmental actors, to formulate a new approach for the **participation** of voices from affected areas at the table (Mabey et al. 2013). Looking ahead, stakeholders could identify sectors affected by climate change, particularly fragile and conflict-sensitive ones, to ensure a coordinated climate change planning process incorporating peace and security. This new approach should consider the range of policies in the context of climate change, including development cooperation, **conflict prevention**, humanitarian aid, climate change adaptation, and GHG emissions **mitigation** (Tänzler and Carius 2013). Furthermore, by promoting international cooperation on these policies, climate diplomacy can help reduce the possibility of resource-based conflicts and promote regional stability through the development of regional or bilateral agreements. For example, in South Sudan, the Intergovernmental Authority on Development (IGAD) used regional diplomacy around a shared climate-**resilience** agenda for the Horn of Africa. Through IGAD's Drought Disaster Resilience and Sustainability Initiative (IDDRSI), member states, including South Sudan, agreed on common priorities, adopted Country Programming Papers, and launched cross-border resilience programs in pastoral borderlands. These diplomatic processes aim to lower **conflict** risks related to climate shocks and promote climate **adaptation** measures (IGAD 2019).

Climate diplomacy plays a critical role in **environmental peacebuilding** by addressing the security implications of climate change. Indeed, in the closing days of the Cold War, US President George H.W. Bush and Soviet Premier Mikhail Gorbachev identified climate change as an area of potential international cooperation, leading to the UN Framework Convention on Climate Change (George Washington University 1989). Climate change is increasingly recognized as a "**threat multiplier**," intensifying existing vulnerabilities and exacerbating sociopolitical and economic insecurities, posing risks to international peace and security (Rüttinger et al. 2015; Goodman 2024). Climate diplomacy serves as a bridge connecting climate action and **peacebuilding.** It helps to mitigate security risks through strategies such as building transformative pathways through low-carbon development and **renewable energy** expansion; ensuring conflict-sensitive climate policies; strengthening climate adaptation action, particularly in fragile areas; and promoting regional and international cooperation, particularly in managing shared and transboundary resources like water to prevent the possibility of tensions (Tänzler 2018).

References

Council of the European Union. 2018. *Council Conclusions on Climate Diplomacy: Promoting the External Dimension of Climate Action*. ST 6125/2018/INIT. https://data.consilium.europa.eu/doc/document/ST-6125-2018-INIT/en/pdf.

Doussis, Emmanuella, and George Dikaios. 2025. "Climate Diplomacy under the UNFCCC: Past, Present, and Future." In *Paraskevi Naskou-Perraki, ed. Contemporary Diplomatic and Consular Relations: Selected Aspects*. Springer. https://doi.org/10.1007/978-3-031-99243-8_10.

George Washington University (GWU). 1989. "Soviet Transcript of the Malta Summit, December 2–3, 1989." National Security Archive. https://nsarchive2.gwu.edu/NSAEBB/NSAEBB298/Document%2010.pdf.

German Federal Foreign Office. 2023. *Climate and Security: Strategy Paper on German Foreign Policy for Climate Change and Security*. https://www.auswaertiges-amt.de/resource/blob/2633116/a4e03e8283b9479559ef2dc3b741624a/kap-strategie-en-data.pdf.

Goodman, Sherri. 2024. *Threat Multiplier: Climate, Military Leadership, and the Fight for Global Security*. Island Press.

Intergovernmental Authority on Development (IGAD). 2019. *The IGAD Drought Disaster Resilience and Sustainability Initiative (IDDRSI): The IDDRSI Strategy (2019–2024)*. September. https://icpald.org/wp-content/uploads/2019/10/IDDRSI-STRATEGY.pdf.

Mabey, Nick, Liz Gallagher, and Camilla Born. 2013. *Understanding Climate Diplomacy: Building Diplomatic Capacity and Systems to Avoid Dangerous Climate Change*. E3G (Third Generation Environmentalism), October. https://cdkn.org/sites/default/files/files/E3G-Understanding-Climate-Diplomacy.pdf

Locci, Emanuela, and Hamza Jammoul, eds. 2023. *Climate Diplomacy: International and National Responses*. Dar Abaad.

Rüttinger, Lukas, Gerald Stang, Dan Smith, Dennis Tänzler, Janani Vivekananda, and Geoffrey Debelko. 2015. *A New Climate for Peace: Taking Action on Climate and Fragility Risks*. adelphi. https://climate-diplomacy.org/sites/default/files/2020-11/NewClimateForPeace_FullReport_small_0.pdf.

Tänzler, Dennis. 2018. "Climate Diplomacy and Peace." In *Routledge Handbook of Environmental Conflict and Peacebuilding*, edited by Ashok Swain and Joakim Öjendal. Routledge. https://doi.org/10.4324/9781315473772-21.

Tänzler, Dennis, and Alexander Carius. 2013. "Beyond International Climate Negotiations: Climate Diplomacy from a Foreign Policy Perspective." In *Climate Change: International Law and Global Governance. Volume II: Policy, Diplomacy and Governance in a Changing Environment*, edited by Oliver C. Ruppel, Christian Roschmann, and Katharina Ruppel-Schlichting. Nomos. https://doi.org/10.5771/9783845242774_259.

Tskipurishvili, Ana. 2022. *New Climate Diplomacy under the Paris Climate Agreement*. Geneva Global Policy Brief, No. 1/2022. February 10, 2022. https://www.meig.ch/ggpb-n1-2022-new-climate-diplomacy-under-the-paris-climate-agreement/.

Climate Migrant

Gabriela Nagle Alverio, Saeede Mokhtarzade, and Atieh Khatibi

A climate migrant is a person who relocates, either forcibly or voluntarily, and either temporarily or permanently, due to the direct or indirect impacts of climate change-related events or disasters.

As **climate change** intensifies, many individuals who are unable to adapt in place are forced to or choose to move out of harm's way (Vinke et al. 2020). Climate change is not typically the sole driver of migration; rather, it exacerbates pre-existing drivers, including social, political, and economic factors like **conflict**, weak **governance**, lack of livelihood-generating opportunities, social inequalities, and food insecurity (International Organization for Migration [IOM] 2024). Climate change leads to unique tipping points, leading to both an increase in migration and novel migration dynamics (Black et al. 2011). Furthermore, climate change acts as a **threat multiplier**, leading to increasing tensions and conflicts (Goodman 2021) and undermining the environmental **peace** consolidation process. In addition, the increased number of migrants could contribute to conflicts in receiving areas (Abel et al. 2019), particularly those struggling with resource scarcity (Reuveny 2007).

Projections for climate migration by the end of the 21st century vary widely, with no consensus on exact numbers (Schewel et al. 2024). Estimates vary from 250 million people to 1.2 billion people becoming climate migrants, temporarily or permanently, by 2050 (Baggerman 2022; Rigaud et al. 2018). Experts agree, however, that climate change will continue to increase migration rates, primarily internally within countries and regions (Rigaud et al. 2018). Despite the established connection between climate impacts and migration, there is no universally agreed-upon definition for a climate migrant, with the terms "environmental migrant" and "climate-induced migrant" often being used interchangeably (IOM 2021). Climate migration dynamics differ depending on the nature of climate impacts, which may be slow-onset (e.g., droughts) or rapid-onset events (e.g., hurricanes). Migration in response to these events can be short-term or permanent, involve internal **displacement** or cross-border movement, and may be voluntary or forced (IOM 2021). Additionally, some individuals may choose not to migrate in the face of climate impacts or may be unable to migrate even when they desire to, a phenomenon often referred to as climate immobility (Boas et al. 2022).

As much as possible with climate change-related risks, solutions targeted at climate migrants should aim to promote agency, allowing individuals to choose whether they stay or go (Vinke et al. 2020). To that end, climate **adaptation** measures, such as climate-resilient seeds or sea walls, can allow communities to remain in place in the face of climate impacts (Adger et al. 2005). For those who migrate internally, urban development policies that create affordable housing, protect labor rights, and provide social services can promote long-term well-being for migrants as well as economic development for the urban areas they migrate to (Nagle Alverio et al. 2024). Such measures may also reduce the potential for **conflict** to arise from competition over jobs, housing, or resources (Koubi 2019).

Planned relocation, the coordinated move of a community to a safer location, can also help preserve networks, livelihoods, and lifestyles, but only if those affected are involved in the decision-making process and have ample time and resources for the relocation process (Bower et al. 2023). In cases where individuals cross borders, domestic legal protections are needed, as international refugee law does not recognize climate change as a justification for asylum (Nishimura 2015). For instance, the Falepili Union Treaty provides a pathway for legal residency, occupation, and education in Australia for 280 Tuvaluans per year in response to climate-induced sea level rise affecting the Pacific Islands (Australian Government 2023). Ultimately, as climate change impacts worsen, they will continue to exacerbate pre-existing drivers of migration and prevent people from leaving unsafe situations. Climate migration calls for a comprehensive **environmental peacebuilding** approach that addresses both migration drivers and impacts; integrates climate adaptation with **conflict prevention**; and promotes **resource governance** that builds **resilience** and promotes harmony between host communities and climate migrants.

References

Abel, Guy J., Michael Brottrager, Jesus Crespo Cuaresma, and Raya Muttarak. 2019. "Climate, Conflict and Forced Migration." *Global Environmental Change* 54: 239–49. https://doi.org/10.1016/j.gloenvcha.2018.12.003.

Adger, W. Neil, Nigel W. Arnell, and Emma L. Tompkins. 2005. "Successful Adaptation to Climate Change Across Scales." *Global Environmental Change* 15 (2): 77–86. https://doi.org/10.1016/j.gloenvcha.2004.12.005.

Australian Government Department of Foreign Affairs and Trade. 2023. *Falepili Union Treaty*. Australian Government. https://www.dfat.gov.au/countries/tuvalu/treaty-text-falepili-union.

Baggerman, Jill. 2022. "Tethering to Human Rights in the Pushes and Pulls of Human Mobility." *Stimson Center*, March 3. https://www.newsecuritybeat.org/2022/03/tethering-human-rights-pushes-pulls-human-mobility/.

Black, Richard, W. Neil Adger, Nigel W. Arnell, Stefan Dercon, Andrew Geddes, and David Thomas. 2011. "The Effect of Environmental Change on Human Migration." *Global Environmental Change* 21 (1): S3–11. https://doi.org/10.1016/j.gloenvcha.2011.10.001.

Boas, Ingrid, Hanne Wiegel, Carol Farbotko, Jeroen Warner, and Mimi Sheller. 2022. "Climate Mobilities: Migration, Im/Mobilities and Mobility Regimes in a Changing Climate." *Journal of Ethnic and Migration Studies* 48 (14): 3365–79. https://doi.org/10.1080/1369183X.2022.2066264.

Bower, Erica R., Anvesh Badamikar, Gabrielle Wong-Parodi, and Christopher B. Field. 2023. "Enabling Pathways for Sustainable Livelihoods in Planned Relocation." *Nature Climate Change* 13 (9): 919–26. https://doi.org/10.1038/s41558-023-01753-x.

Goodman, Sherri. 2021. *Threat Multiplier: How Climate Change is Reshaping National Security and the World*. Potomac Books.

International Organization for Migration (IOM). 2021. *Institutional Strategy on Migration, Environment and Climate Change 2021–2030*. IOM. https://publications.iom.int/books/institutional-strategy-migration-environment-and-climate-change-2021-2030.

International Organization for Migration (IOM). 2024. "Climate Change, Food Insecurity and Human Mobility: Interlinkages, Evidence And Action." In *World Migration Report 2024*. IOM. https://worldmigrationreport.iom.int/what-we-do/world-migration-report-2024-chapter-7/climate-change-food-insecurity-compounding-and-direct-drivers-human-mobility.

Koubi, Vally. 2019. "Climate Change and Conflict." *Annual Review of Political Science* 22: 343–60. https://doi.org/10.1146/annurev-polisci-050317-070830.

Nagle Alverio, Gabriela, Jeannie Sowers, and Erika Weinthal. 2024. "Climate Change, Conflict, and Urban Migration." *Environment and Security* 3 (2): 134–44. https://doi.org/10.1177/27538796241259242.

Nishimura, Lauren. 2015. "'Climate Change Migrants': Impediments to a Protection Framework and the Need to Incorporate Migration into Climate Change Adaptation Strategies." *International Journal of Refugee Law* 27 (1): 107–34. https://doi.org/10.1093/ijrl/eev002.

Reuveny, Rafael. 2007. "Climate Change-Induced Migration and Violent Conflict." *Political Geography* 26 (6): 656–73. https://doi.org/10.1016/j.polgeo.2007.05.001.

Rigaud, Kanta Kumari, Alex de Sherbinin, Bryan Jones, et al. 2018. *Groundswell: Preparing for Internal Climate Migration*. World Bank. https://doi.org/10.1596/29461.

Schewel, Kerilyn, Sarah Dickerson, B. Madson, and Gabriela Nagle Alverio. 2024. "How Well Can We Predict Climate Migration? A Review of Forecasting Models." *Frontiers in Climate* 5. https://doi.org/10.3389/fclim.2023.1189125.

Vinke, Kira, Jonas Bergmann, Julia Blocher, Himani Upadhyay, and Roman Hoffman. 2020. "Migration as Adaptation?" *Migration Studies* 8 (4): 626–34. https://doi.org/10.1093/migration/mnaa029.

Climate-Related Security Risks

Sinéad Barry, Albert Martínez, and Alina Viehoff

Climate-related security risks are systemic, compounding risks that emerge through direct or indirect interactions between (1) climate stressors and (2) contextual factors (e.g., social, economic, demographic, health, political, and conflict).

Climate-related security risks are risks to the **peace** and **security** of individuals, communities, institutions, and states driven by climatic stressors. The framework for understanding interactions between **climate change** and **human security** systems originates in academic and policy developments in the late 1980s and early 1990s which began to link climate change and/or the natural **environment** with security foci, including through **food security**, migration, institutional **resilience**, interstate tensions, and violent **conflict**, etc. (Homer-Dixon 1993; Perelet 1994; United Nations Development Programme 1994, 24; Scheffran 1999). In the last 30 years, the **climate security** field has slowly developed, with researchers, policymakers, and practitioners increasingly acknowledging the interconnection between a changing climate and human security (Austin and Bruch 2000; Barnett 2003; European Commission 2008; Brauch et al. 2011; Rüttinger et al. 2015; Intergovernmental Panel on Climate Change 2023; Munayer et al. 2024). Some important policy developments include the launch of the UN **Climate Security Mechanism** in 2018, NATO's 2021 Climate Change and Security Action Plan, and the **African Union**'s 2022 Climate Change and Resilient Development Strategy and Action Plan.

Climate-related security risks can pose a direct risk to life and well-being through rapid onset events (e.g., flash floods) or through slow onset events (e.g., droughts). Most often, they occur along indirect pathways that emerge when climate stressors compound with contextual conditions, including social, economic, demographic, political, and conflict factors (Goodman and Baudu 2023; Sedová et al. 2024; Rüttinger et al. 2023). As a result, climate-related security risks are highly case-dependent and vary widely (Detges and Foong 2023). Climate-related security risks can multiply existing vulnerabilities of institutions, communities, or individuals. Fragile and conflict-affected regions and traditionally marginalized groups, including internally displaced persons, low-income groups, women, and **youth**, are particularly vulnerable to these risks (Daoudy et al. 2022; Detges and Foong 2023). In addition, pre-existing insecurity, such as in the context of fragile and conflict-affected states, reduces the resilience of individuals and groups to withstand the impacts of climate-related security risks. Pre-existing insecurity can be shaped by factors including high rates of socioeconomic inequality, social exclusion, or weak **governance**, etc. Resilience may be shaped by factors including financial resources, the presence of formal and informal support networks, and institutional strength. Climate-related security risks can thus trap vulnerable populations in cycles of climate-related insecurity (Buhaug and Uexkull 2021).

Climate-related security risks also include risks to peace and stability more broadly. These risks typically play out through intermediary factors. For example, reduced availability or quality of **natural resources** may undermine livelihoods, increase competition for natural resources, or shift social, political, or migration dynamics, leading to competition and conflict (Vivekananda 2014; Brzoska 2016; Koubi 2019). For example, **environmental degradation** and overexploitation have significantly reduced the availability of and access to water across Yemen, leading to conflict over water between farmers (European Institute of Peace 2024). In southern Iraq, climate impacts have forced many farmers to migrate to **cities** such as Basra, leading to tensions between long-term residents and new arrivals (Löble et al. 2024).

From an **environmental peacebuilding** perspective, as conflict-affected countries are often the most vulnerable to climate change, failing to consider the impact of climate-related security risks in peace processes or actions could risk undermining the effectiveness and **sustainability** of peace (Moran et al. 2018). On the other hand, addressing climate-related security risks is increasingly seen by peacebuilders and peacemakers as a useful entry point to mitigate and resolve conflicts (Johnson et al. 2021). Climate and environmental issues are frequently perceived as less partisan than other political topics, meaning that dialogues on issues such as climate vulnerability, environmental degradation, and access to natural resources are used as relatively neutral topics for **confidence-building measures** between groups in conflict, improving inclusivity of women and youth, and strengthening **peace agreements** to new climate realities (United Nations Department of Political and Peacebuilding Affairs 2022; Viehoff et al. 2024). Similarly, in post-conflict situations, integrating climate **adaptation**, environmental restoration, and natural resources management approaches can help prevent conflict relapse, foster collaboration between conflict groups, improve political leadership, and support the livelihoods of groups that depend on natural resources, contributing to resilience (United Nations Environmental Programme 2022). As a result, addressing climate-related security risks in peacemaking and **peacebuilding** can offer opportunities for more **sustainable peace** in climate-vulnerable and conflict-affected countries.

References

Austin, Jay, and Carl Bruch. 2000. *The Environmental Consequences of War: Legal, Economic, and Scientific Perspectives*. Cambridge University Press.

Barnett, Jon. 2003. "Security and Climate Change." *Global Environmental Change* 13 (1): 7–17. https://doi.org/10.1016/S0959-3780(02)00080-8.

Brauch, Hans Günter, Úrsula Oswald Spring, Czeslaw Mesjasz et al. 2011. *Coping with Global Environmental Change, Disasters and Security: Threats, Challenges, Vulnerabilities and Risks*. Springer. https://link.springer.com/content/pdf/10.1007/978-3-642-17776-7.pdf.

Brzoska, Michael, and Christiane Fröhlich. 2016. "Climate Change, Migration and Violent Conflict: Vulnerabilities, Pathways and Adaptation Strategies." *Migration and Development* 5 (2): 190–210. https://doi.org/10.1080/21632324.2015.1022973.

Buhaug, Halvard, and Nina von Uexkull. 2021. "Vicious Circles: Violence, Vulnerability, and Climate Change." *Annual Review of Environment and Resources* 46: 545–68. https://doi.org/10.1146/annurev-environ-012220-014708.

Conca, Ken, and Geoffrey D. Dabelko. 2024. "The International (In)security Order and the Climate-Conflict-Security Nexus." *Environment and Security* 2 (4): 501–24. https://doi.org/10.1177/27538796241287358.

Daoudy, Marwa, Jeannie Sowers, and Erika Weinthal. 2022. "What is Climate Security? Framing Risks around Water, Food, and Migration in the Middle East and North Africa." *Wiley Interdisciplinary Reviews: Water* 9 (3): e1582. https://doi.org/10.1002/wat2.1582.

Detges, Adrien, and Adrian Foong. 2023. "Context Matters: A Review of the Evidence of How Economic, Social, and Other Variables Influence the Relationship Between Climate and

Security." *adelphi*, June. https://weatheringrisk.org/sites/default/files/document/Context_Matters_Report.pdf

European Commission. 2008. *Climate Change and International Security: Paper from the High Representative and the European Commission to the European Council.* European Commission, March 14. https://www.consilium.europa.eu/uedocs/cms_data/docs/pressdata/en/reports/99387.pdf.

European Institute of Peace (EIP). 2024. *Environmental Pathways for Reconciliation in Yemen.* EIP, April. https://www.eip.org/environmental-pathways-for-reconciliation-in-yemen-consultation-report/.

Goodman, Sherri, and Pauline Baudu. 2023. *Climate Change as a "Threat Multiplier": History, Uses and Future of the Concept.* Center for Climate and Security Council on Strategic Risks, January 3. https://councilonstrategicrisks.org/CCThreatMultiplier.pdf.

Homer-Dixon, Thomas F. 1993. *Environmental Scarcity and Global Security.* Foreign Policy Association. https://files.eric.ed.gov/fulltext/ED384561.pdf.

Johnson, McKenzie F., Luz A. Rodríguez, and Manuela Quijano Hoyos. 2021. "Intrastate Environmental Peacebuilding: A Review of the Literature." *World Development* 137: 1–18. https://doi.org/10.1016/j.worlddev.2020.105150.

Koubi, Vally. 2019. "Climate Change and Conflict." *Annual Review of Political Science* 22 (1): 343–60. https://doi.org/10.1146/annurev-polisci-050317-070830.

Löble, Nike, Andreas Schädel, Ammar Zakri, Alexandra Steinkraus, and Sinéad Barry. 2024. "Bridging Divides Between IDPs and Host Communities in Iraq Through Dialogues on Climate Change." *Weathering Risk*, July 11. https://weatheringrisk.org/en/impact-note/bridging-divides-between-idps-and-host-communities-iraq-through-dialogues-climate.

Moran, Ashley, Joshua Busby, Clionadh Raleigh, et al. 2018. *The Intersection of Global Fragility and Climate Risks.* United States Agency for International Development. https://www.strausscenter.org/wp-content/uploads/The-Intersection-of-Global-Fragility-and-Climate-Risks-2018.pdf.

Munayer, Raquel, Sinéad Barry, and Beatrice Mosello. 2024. "The State of Climate and Security: An Overview." In *Climate Changes Security: Navigating Climate Change and Security Challenges in the OSCE Region*, edited by Hans Lampalzer and Gerald Hainzl. Republic of Austria, Federal Ministry of Defence.

Perelet, Renat. 1994. "The Environment as a Security Issue." In *The Environment: Towards a Sustainable Future*, edited by The Dutch Committee for Long-Term Environmental Policy. Springer.

Rüttinger, Lukas, Dan Smith, Gerald Stang, Dennis Tänzler, and Janani Vivekananda. 2015. *A New Climate for Peace: Taking Action on Climate and Fragility Risks.* adelphi, International Alert, Woodrow Wilson International Center for Scholars, and European Union Institute for Security Studies.

Rüttinger, Lukas, Janani Vivekananda, and Alexandra Steinkraus. 2023. *Weathering Risk Climate Security Risk Assessment Methodology–Guide and Tools.* adelphi.

Scheffran, Jürgen. 1999. "Environmental Conflict and Sustainable Development: A Conflict Model and its Application to Climate and Energy Policy." In *Environmental Change and Security: A European Perspective*, edited by Alexander Carius and Kurt M. Lietzmann. Springer.

Sedova, Barbora, Lisa Binder, Sidney Michelini, Marie Schellens, and Lukas Rüttinger. 2024. "A Review of Climate Security Risk Assessment Tools." *Environment and Security* 2 (1): 1–36. https://doi.org/10.1177/27538796241226996.

United Nations Department of Political and Peacebuilding Affairs (DPPA). 2022. *The Implications of Climate Change for Mediation and Peace Processes: DPPA Practice Note.* DPPA, September. https://peacemaker.un.org/dppapracticenotetheimplicationsofclimatechangeformediationandpeaceprocesses.pdf.

United Nations Development Programme (UNDP). 1994. *Human Development Report.* UNDP. https://hdr.undp.org/system/files/documents/hdr1994encompletenostats.pdf.

United Nations Environment Programme (UNEP). 2022. *Climate Change and Security Partnership Project–Final Report: March 2017–February 2022.* UNEP and European Union, September. https://wedocs.unep.org/20.500.11822/40549.

Viehoff, Alina, Laís Clemente Pereira, and Sinéad Barry. 2024. "Using Climate and Environmental Issues to Build Trust in Conflict-Affected Areas: Q&A with the Weathering Risk Peace Pillar."

Climate Diplomacy, September 10. https://climate-diplomacy.org/magazine/cooperation/using-climate-and-environmental-challenges-build-trust-conflict-affected-areas.
Vivekananda, Janani, Janpeter Schilling, and Dan Smith. 2014. "Climate Resilience in Fragile and Conflict-Affected Societies: Concepts and Approaches." *Development in Practice* 24 (4): 487–501. https://www.jstor.org/stable/24565406.

Climate Security

Grazia Pacillo, Emmanuel Valax, and Radhika Singh

Climate security encompasses the complex connections between climate change and security at multiple scales, from the individual to the international (Läderach et al. 2021; McDonald 2013; Buhaug and von Uexkull 2021a).

This concept highlights how climate-related risks can exacerbate the root causes of **fragility, conflict**, and **displacement**, further intensify vulnerabilities, and undermine **human security**, limiting the capacity to effectively respond to climate impacts.

It also underscores the need to integrate climate **resilience** into traditional development, **peace**, and **security** interventions. This term further recognizes the need to reduce risks of climate **maladaptation**, enhance the climate sensitivity of development, humanitarian, and peace operations, and leverage climate action as a tool for **conflict prevention** and **peacebuilding.**

In the past few decades, definitions of climate security have varied widely, reflecting different perspectives from national security to human security. The national security approach, largely influenced by the Homer–Dixon tradition and by the work of Goodman (Homer-Dixon 1999; Goodman 2024), tends to frame **climate change** as a "**threat multiplier**" to state stability and borders, often focusing on the risks of conflict, **migration**, and **resource scarcity** (CNA Military Advisory Board 2007; Alam et al. 2024). Critics of this approach argue that the "**securitization**" of climate issues can shift the discourse toward prioritizing militarized responses, emphasizing defense strategies that may prioritize state interests over collaborative solutions, potentially straining international **cooperation** (Abrahams 2019; Dunlap and Fairhead 2014). Such an approach also raises concerns about sovereignty and the use of military force in response to climate change, as it risks framing environmental challenges through a conflict-oriented lens (Goodman 2024; Duffy 2016; Rothe 2016; Hardt et al. 2024).

The human security perspective focuses on the well-being of individuals and communities, prioritizing their resilience to climate impacts and emphasizing cooperation, equity, and trust-building. It centers on the welfare of individuals and communities, scrutinizing how climate change intensifies vulnerabilities, conflicts, and disparities. It delineates a vicious cycle (Buhaug and von Uexkull 2021b; Zografos et al. 2014) in which climate change and variability, resource limitations, and **political ecology** issues escalate risks and sustain insecurity. Political ecology factors, such as **power** relations, resource access, **land tenure** and rights, and **governance** failures, serve as pivotal mediators that shape whether climate-induced stresses lead to significant human security issues. The effectiveness of governance and adaptive measures either softens or intensifies these effects, impacting how communities can effectively address climatic challenges (Pacillo et al. 2024).

Ensuring climate security remains a "wicked" problem due to the uncertainties surrounding the impacts of climate change on complex socio-ecological systems. The concept is notoriously difficult to operationalize at programmatic and project levels and warrants careful consideration when designing interventions in fragile and conflict-affected societies. This complexity challenges traditional science-policy models, complicating the efforts of any single actor to control or fully understand the nexus between climate and conflict (Pacillo et al. 2024; Beaumont and Coning 2022). The broad range of academic disciplines relevant to climate security, such as **environmental change** (Barnett et al. 2010; O'Brien and Barnett 2013; Daoudy 2021), armed conflict (Gleditsch 2012; Mach et al. 2019), **environmental peacebuilding** (Ide et al. 2021; Johnson et al. 2021; Loehr et al. 2022), political science (Tschakert 2012; Le Billon and Duffy 2018), **environmental justice** (Velicu and Kaika 2017), and **decolonization** (Rodríguez and Liz Inturias 2018; Vélez-Torres et al. 2022), further contributes to the complexity of the nexus.

Acknowledgment

The authors thank Cedric De Coning, Brittany Croll, Giulia Caroli, Leonardo Medina, Frans Schapendonk, and Adam Savelli for their review of this entry. This work was carried out with support from the CGIAR Climate Action and Food Frontiers and Security Science Programmes. We would like to thank all funders who supported this research through their contributions to the CGIAR Trust Fund: https://www.cgiar.org/funders/.

References

Abrahams, Daniel. 2019. "From Discourse to Policy: US Policy Communities' Perceptions of and Approaches to Climate Change and Security." *Conflict, Security and Development* 19 (4): 323–45. https://doi.org/10.1080/14678802.2019.1637080.

Alam, Ashraful, Hasanul Banna, Ahmed W. Alam, Md. Badrul U. Bhuiyan, and Nurul B. Mokhtar. 2024. "Climate Change and Geopolitical Conflicts: The Role of ESG Readiness." *Journal of Environmental Management* 353: 120284. https://doi.org/10.1016/j.jenvman.2024.120284.

Barnett, Jon, Richard A. Matthew, and Karen L. O'Brien. 2010. "Global Environmental Change and Human Security: An Introduction." In *Global Environmental Change and Human Security*, edited by Richard A. Matthew, Jon Barnett, and Karen L. O'Brien. MIT Press.

Beaumont, Peter, and Cedric de Coning. 2022. "Coping with Complexity: Toward Epistemological Pluralism in Climate–Conflict Scholarship." *International Studies Review* 24 (4): viac055. https://doi.org/10.1093/isr/viac055.

Buhaug, Halvard, and Nina von Uexkull. 2021a. "Security Implications of Climate Change: A Decade of Scientific Progress." *Journal of Peace Research* 58 (1): 3–17. https://doi.org/10.1177/0022343320984210.

Buhaug, Halvard, and Nina von Uexkull. 2021b. "Vicious Circles: Violence, Vulnerability, and Climate Change." *Annual Review of Environment and Resources* 46 (1): 545–68. https://doi.org/10.1146/annurev-environ-012220-014708.

CNA Military Advisory Board. 2007. *National Security and the Threat of Climate Change*. CNA, April 15. https://www.cna.org/reports/2007/national-security-and-the-threat-of-climate-change.

Daoudy, Marwa. 2021. "Rethinking the Climate–Conflict Nexus: A Human–Environmental Climate Security Approach." *Global Environmental Politics* 21 (3): 4–25. https://doi.org/10.1162/glep_a_00609.

Duffy, Rosaleen. 2016. "War, by Conservation." *Geoforum* 69: 238–48. https://doi.org/10.1016/j.geoforum.2015.09.014.

Dunlap, Alexander, and James Fairhead. 2014. "The Militarisation and Marketisation of Nature: An Alternative Lens to 'Climate-Conflict.'" *Geopolitics* 19 (4): 937–61. https://doi.org/10.1080/14650045.2014.964864.

Gleditsch, Nils Petter. 2012. "Whither the Weather? Climate Change and Conflict." *Journal of Peace Research* 49 (1): 3–9. https://doi.org/10.1177/0022343311431288.

Goodman, Sherri. 2024. *Threat Multiplier: Climate, Military Leadership, and the Fight for Global Security*. Island Press.

Hardt, Judith N., Dhanasree Jayaram, Cameron Harrington, et al. 2024. "The Challenges of the Increasing Institutionalization of Climate Security." *PLOS Climate* 3 (4): e0000402. https://doi.org/10.1371/journal.pclm.0000402.

Homer-Dixon, Thomas F. 1999. *Environment, Scarcity, and Violence*. Princeton University Press. https://doi.org/10.1515/9781400822997.

Ide, Tobias, Carl Bruch, Alexander Carius, et al. 2021. "The Past and Future(s) of Environmental Peacebuilding." *International Affairs* 97 (1): 1–16. https://doi.org/10.1093/ia/iiaa177.

Johnson, Michael F., Luis A. Rodríguez, and Marcela Quijano Hoyos. 2021. "Intrastate Environmental Peacebuilding: A Review of the Literature." *World Development* 137: 105150. https://doi.org/10.1016/j.worlddev.2020.105150.

Läderach, Peter, Grazia Pacillo, Philip Thornton, Diego Osorio, and Dan Smith. 2021. "Food Systems for Peace and Security in a Climate Crisis." *The Lancet Planetary Health* 5 (5): e249–50. https://doi.org/10.1016/S2542-5196(21)00056-5.

Le Billon, Philippe, and Rosaleen Duffy. 2018. "Conflict Ecologies: Connecting Political Ecology and Peace and Conflict Studies." *Journal of Political Ecology* 25 (1): 1–22. https://doi.org/10.2458/v25i1.22704.

Loehr, Karin, Helena Morales-Muñoz, Tania Rodriguez, et al. 2022. "Integrating the Concept of Peacebuilding in Sustainability Impact Assessment." *Environmental Impact Assessment Review* 95: 106803. https://doi.org/10.1016/j.eiar.2022.106803.

Mach, Katharine J., Caroline M. Kraan, W. Neil Adger, et al. 2019. "Climate as a Risk Factor for Armed Conflict." *Nature* 571 (7764): 193–97. https://doi.org/10.1038/s41586-019-1300-6.

McDonald, Matt. 2013. "Discourses of Climate Security." *Political Geography* 33: 42–51. https://doi.org/10.1016/j.polgeo.2013.01.002.

O'Brien, Karen, and Jon Barnett. 2013. "Global Environmental Change and Human Security." *Annual Review of Environment and Resources* 38 (1): 373–91. https://doi.org/10.1146/annurev-environ-032112-100655.

Pacillo, Grazia, Laura Medina, Thomas Liebig, et al. 2024. "Measuring the Climate Security Nexus: The Integrated Climate Security Framework." *PLOS Climate* 3 (10): e0000280. https://doi.org/10.1371/journal.pclm.0000280.

Rodríguez, Iokiñe, and Mirna Liz Inturias. 2018. "Conflict Transformation in Indigenous Peoples' Territories: Doing Environmental Justice with a 'Decolonial Turn'." *Development Studies Research* 5 (1): 90–105. https://doi.org/10.1080/21665095.2018.1486220.

Rothe, Delf. 2016. *Securitizing Global Warming: A Climate of Complexity*. Routledge. https://doi.org/10.4324/9781315677514.

Tschakert, Petra. 2012. "From Impacts to Embodied Experiences: Tracing Political Ecology in Climate Change Research." *Geografisk Tidsskrift–Danish Journal of Geography* 112 (2): 144–58. https://doi.org/10.1080/00167223.2012.741889.

Velicu, Irina, and Maria Kaika. 2017. "Undoing Environmental Justice: Re-Imagining Equality in the Roşia Montană Anti-Mining Movement." *Geoforum* 84: 305–15. https://doi.org/10.1016/j.geoforum.2015.10.012.

Vélez-Torres, Irene, Katherine Gough, Juan Larrea-Mejía, Giulia Piccolino, and Karol Ruette-Orihuela. 2022. "'Fests of Vests': The Politics of Participation in Neoliberal Peacebuilding in Colombia." *Antipode* 54 (2): 586–607. http://dx.doi.org/10.1111/anti.12785.

Zografos, Christos, Michael C. Goulden, and Giorgos Kallis. 2014. "Sources of Human Insecurity in the Face of Hydro-Climatic Change." *Global Environmental Change* 29: 327–36. https://doi.org/10.1016/j.gloenvcha.2013.11.002.

Co-benefits

Héctor Morales-Muñoz and Elise Doumergue

Co-benefits are "the positive effects that a policy or measure aimed at one objective might have on other objectives, thereby increasing the total benefits for society or the environment" (Intergovernmental Panel on Climate Change [IPCC] 2019).

The term "co-benefits" gained traction in the 1990s within environmental and climate policy, emphasizing the multiple positive outcomes of climate **mitigation** efforts, such as improved public health and energy efficiency. These ideas were formalized in frameworks like the 1992 Earth Summit and the Kyoto Protocol adopted in 1997, and they are now central to **IPCC** reports (Mayrhofer and Gupta 2016). Co-benefits challenge the traditional view that environmental **conservation** and socioeconomic development are mutually exclusive by demonstrating that initiatives can achieve both objectives simultaneously and promote policy coherence (Wong 2022).

The concept of co-benefits is integral to the field of **environmental peacebuilding,** which encompasses initiatives seeking to simultaneously address both environmental and **peacebuilding** objectives. Such projects are designed to mitigate **environmental degradation** while fostering **cooperation** between conflicting parties or communities, creating vertical and horizontal trust, and enhancing **social cohesion** (Hachmann et al. 2023). Co-benefits can include improved access to and **governance** of shared **natural resources**, climate **adaptation**, sustainable livelihoods, enhanced **food security** and social capital, and the reduction of **environment**-related **fragility** and **conflict** drivers.

In this context, co-benefits do not usually arise spontaneously—they require integrated and conflict-sensitive approaches that acknowledge the interconnectedness of environmental and social systems (Morales-Muñoz et al. 2022). For example, agroforestry initiatives aiming to absorb carbon in conflict-affected areas should include local **stakeholders,** such as farmers, victims of **conflict**, and former combatants, be implemented through negotiated conservation agreements, and acknowledge cultural preferences (Morales-Muñoz et al. 2023). To achieve co-benefits from conservation activities, coordinated interventions among environmental and peacebuilding actors must be implemented.

Environmental peacebuilding co-benefits can play a significant role in addressing the ongoing **polycrisis,** as they can strengthen individual and institutional **resilience** to conflict, climate, and economic shocks at various scales. Co-benefits have therefore become a cornerstone of advocacy efforts seeking to incentivize investments in fragile and conflict-affected settings (United Nations Development Programme 2021). Moving beyond the win-win strategy discourse will require practitioners to build the evidence base for such gains. This can be achieved by developing measurement frameworks that rely on a clear **theory of change** and locally relevant **indicators** of multidimensional outcomes.

References

Hachmann, Samyra, Katharina Löhr, Héctor Morales-Muñoz, Luca Eufemia, Stefan Sieber, and Michelle Bonatti. 2023. "Conceptualizing Community-based Environmental Peacebuilding in Cesar, Colombia." *Human Ecology* 51: 221–35. https://doi.org/10.1007/s10745-023-00399-9.

Intergovernmental Panel on Climate Change (IPCC). 2019. "Annex I: Glossary." https://www.ipcc.ch/site/assets/uploads/2019/11/11_Annex-I-Glossary.pdf.

Mayrhofer, Jan P., and Joyeeta Gupta. 2016. "The Science and Politics of Co-Benefits in Climate Policy." *Environmental Science and Policy* 57: 22–30. https://doi.org/10.1016/j.envsci.2015.11.005.

Morales-Muñoz, Héctor, Arwen Bailey, Katharina Löhr, et al. 2022. "Co-Benefits Through Coordination of Climate Action and Peacebuilding: A System Dynamics Model." *Journal of Peacebuilding and Development* 17 (3): 304–23. https://doi.org/10.1177/15423166221132149.

Morales-Muñoz, Héctor, Leigh Martens, Katharina Löhr, et al. 2023. "Integrating Climate Mitigation and Environmental Peacebuilding Objectives through Sustainable Land Use Systems: Theory of Change and Indicators." *PLOS Climate* 2 (5): e0000075. https://doi.org/10.1371/journal.pclm.0000075.

United Nations Development Programme (UNDP). 2021. *Climate Finance for Sustaining Peace: Making Climate Finance Work for Conflict-Affected and Fragile Contexts*. UNDP, December 14. https://www.undp.org/publications/climate-finance-sustaining-peace-making-climate-finance-work-conflict-affected-and-fragile-contexts.

Wong, Catherine. 2022. "Climate Finance and the Peace Dividend, Articulating the Co-benefits Argument." In *The Political Economy of Climate Finance: Lessons from International Development*, edited by Corrine Cash and Larry A. Swatuk. Palgrave Macmillan. https://doi.org/10.1007/978-3-031-12619-2_9.

Confidence-Building Measures

Saher Husain, Gülsena Çevirken, and Róisín Burke

Confidence-building measures (CBMs) are strategic instruments in conflict-affected areas, at any level (interpersonal to international), aimed at creating an environment conducive to dialogue, negotiations, de-escalation, trust-building, or violence prevention in peace processes. CBMs are designed to reduce fear and enhance transparency, promoting predictable patterns of behavior among conflicting parties through agreed protocols.

A CBM is an action that can be instigated throughout the **conflict** life cycle and can be military or non-military in nature. The goal is to work toward creating **sustainable peace** between parties at conflict, bringing about mutual benefits to the parties or communities affected, which is critical to trust-building. Thus, the main objectives of CBMs include preventing escalation of the conflict, initiating and deepening negotiations, and consolidating the process and its outcome (Mason and Siegfried 2013). Negotiators, decision-makers, and constituencies are among the different types of **stakeholders** that are involved in these CBMs (Mason and Siegfried 2013).

A common objective of CBMs is to reduce misunderstandings that could lead to unintended escalation. Measures may include information sharing, hotlines, and verification activities to monitor military or non-military activities and intentions. The 1975 Helsinki Final Act exemplified this approach, where East and West bloc countries agreed to notify each other in the event of major military exercises, restrict certain activities, and allow on-site inspections. Agreements of this nature were crucial during the Cold War, significantly reducing the risk of a surprise attack. CBMs may also include diplomatic actions, such as cultural exchange programs to promote mutual understanding. A seminal example was Anwar Sadat's 1977 visit to Jerusalem, a major CBM in diplomacy that led to the Camp David Accords and later a **peace** treaty between Egypt and Israel.

Environmental CBMs, such as environmental or natural resource concerns, can serve as an entry point for confidence-building. They may include collaborative planning, joint projects, and joint training in response to natural/man-made disasters, or **cooperation** on water **infrastructure** in a river basin that is shared by states or communities in conflict with each other or emerging from conflict. For example, the 1994 Treaty of Peace between Israel and Jordan can be seen as the result of an environmental CBM effort that was aimed at joint management of water resources. Environmental CBMs could also relate to bilateral or multilateral dialogues between parties in conflict. Another example is the water CBM within Georgia, involving cooperation between the Georgian government and South Ossetia (a breakaway territory within Georgia) regarding water infrastructure running along the Administrative Border Line (ABL) (between the breakaway region of South Ossetia in Georgia). The CBM, with cooperation under the Incident Prevention and Response Mechanism since 2009, relies on the assistance of the EU **monitoring** mission (EUMM) and a hotline to report issues, resulting in increased access to water for villages on both sides of the ABL (Geneva Water Hub 2019).

An example of CBMs in responding to natural disasters includes lending support to programs in situations after earthquakes between Turkey and Greece in 1999 (Mavroulis et al. 2022), and in situations after the 2004 tsunami between Indonesia and Aceh (Gaillard et al. 2008). This improved, somewhat, diplomatic relations between the two countries in conflict.

Environmental-related CBMs may create mutual interdependencies to encourage **resilience** and further cooperation. In practice, CBMs are most effective when accompanied by verification mechanisms, such as inspections or satellite monitoring, to ensure compliance. From the perspective of **environmental peacebuilding**, CBMs can help reduce suspicion, create opportunities for dialogue and engagement, and have a spillover effect into other issue areas. They may even restructure established **security** frameworks, allowing nations to redefine security needs over time and with greater focus on broader conceptions of security than the mere absence of armed conflict.

References

Gaillard, Jean-Christophe, Elsa Clavé, and Ilan Kelman. 2008. "Wave of Peace? Tsunami Disaster Diplomacy in Aceh, Indonesia." *Geoforum* 39 (1): 511–26. https://doi.org/10.1016/j.geoforum.2007.10.010.

Geneva Water Hub. 2019. "The Geneva List of Principles on the Protection of Water Infrastructure." August. https://www.genevawaterhub.org/sites/default/files/atoms/files/gva_list_of_principles_protection_water_infra_www.pdf.

Mason, Simon J. A., and Matthias Siegfried. 2013. "Confidence Building Measures (CBMs) in Peace Processes." In *Managing Peace Processes: Process Related Questions. A Handbook for AU Practitioners*. African Union and the Centre for Humanitarian Dialogue. https://peacemediation.ch/wp-content/uploads/2013/07/AU-Handbook_Confidence-Building-Measures-in-Peace-Processes.pdf.

Mavroulis, Spyridon, Makbule Ilgaç, M. Tunçağ, et al. 2022. "Emergency Response, Intervention, and Societal Recovery in Greece and Turkey after the 30th October 2020, MW = 7.0, Samos (Aegean Sea) Earthquake." *Bulletin of Earthquake Engineering* 20 (14): 7933–55. https://doi.org/10.1007/s10518-022-01317-y.

Conflict

Laura E. R. Peters and Bashar Alshawwa

Conflict is the real or perceived incompatibility of goals, interests, or values between two or more parties, where perceived incompatibilities may be as important as real ones (Kriesberg 2007; United Nations Department of Political Affairs [UNDPA] and United Nations Environment Programme [UNEP] 2015). A conflict can be violent or non-violent.

Conflict is a natural part of everyday social life (Lederach 1997) due to the inherent diversity within and across human societies. However, this does not mean that conflict naturally leads to **violence** and instability, underscoring the need to consider that conflict is not interchangeable with violence. Applied to the field of **environmental peacebuilding,** environmental conflicts do not naturally lead to violence, though there may be environmental dimensions of ongoing conflict and violence. Conflict and **cooperation** co-exist in societies spanning from armed conflict to peaceful contexts, both with the potential to be constructive or destructive (Zeitoun and Mirumachi 2008).

Conflicts differ from disputes. Disputes are short-term disagreements over specific, often negotiable issues, while conflicts are more complex, systemic, and long-term, often involving deeper-seated interests or values (Burton 1990). Environmental issues, including water, are rarely (if ever) the central source of armed conflict (Wolf 1998), though low-intensity water conflicts have increased since 2000 (Turgul et al. 2024). Disasters leading to conflict (or contributing to **peace**) depend strongly on contextual risk factors and human actions or inaction (Canavan and Ide 2024; Peters 2022). Similar trends may be seen for various environmental issues and concerns related to potentially distinct but mutually influencing environmental and political disputes and conflicts. Conflicts around resource scarcity, distribution and access, rights and ownership, meaning, and values have the potential to lead to violence when they are met with **governance** failures, social and economic inequities, **power** imbalances, or other political dynamics (UNDPA and UNEP 2015).

In most environmental **peacebuilding** literature, concerns focus on where non-violent and mutually agreeable mechanisms do not exist or break down. In such situations, adversaries may pursue conflict resolution through violence, using threats, intimidation, coercion, or physical force. In its most extreme forms, conflict can escalate into armed conflict. Such forms of destructive conflict have the tendency to become self-perpetuating, as escalating hostilities can amplify and generate grievances, identities become increasingly entrenched and adversarial, and parties become invested in the conflict, including some actors who reap benefits from its perpetuation (Kriesberg 2007). However, Mead (2022) emphasized that modern organized warfare, including tools of warfare, is a cultural construct that is avoidable and can be "uninvented" through such alternative social mechanisms for conflict.

Formal and informal mechanisms for non-violent conflict resolution that satisfy the needs of all involved parties offer opportunities to work through conflict without the use of any form of violence and arrive at mutually agreeable outcomes (Galtung 1996). "Constructive conflict" emerges when the parties involved engage in dialogue and negotiation in good faith, seeking solutions that address underlying interests and grievances. Constructive outcomes are more likely when conflicts are approached with the goal of creating equitable, inclusive solutions that improve societal relations (Kriesberg 2007). Environmental conflicts can paradoxically offer entry points for building peace through the **environment**, including where there may be a spillover from "low" to "high politics" (Dresse et al. 2019). Critical peace scholars have further challenged the dichotomy between conflict and peace, arguing that **learning** to live with conflict can be the basis for more peaceful relations. For example, the concept of "agonistic peace" has been put forward as a practice of transforming enemies into political adversaries, contesting conflicting interests through political rather than violent means as the foundation for political pluralism and deliberative democracy (Aggestam et al. 2015; Mouffe 2005), applied recently to the field of environmental peacebuilding (Peters et al. 2025).

References

Aggestam, Karin, Fabio Cristiano, and Lisa Strömbom. 2015. "Towards Agonistic Peacebuilding? Exploring the Antagonism–Agonism Nexus in the Middle East Peace Process." *Third World Quarterly* 36 (9): 1736–53. https://www.jstor.org/stable/24523147.

Burton, John W. 1990. *Conflict: Resolution and Provention*. St. Martin's Press.

Canavan, Chloe, and Tobias Ide. 2024. "Contention, Cooperation, and Context: A Systematic Review of Research on Disasters and Political Conflicts." *International Journal of Disaster Risk Reduction* 101 (1): 104558. https://doi.org/10.1016/j.ijdrr.2024.104558.

Dresse, Anaïs, Itay Fischhendler, Jonas Østergaard Nielsen, and Dimitrios Zikos. 2019. "Environmental Peacebuilding: Towards a Theoretical Framework." *Cooperation and Conflict* 54 (1): 99–119. https://doi.org/10.1177/0010836718808331.

Galtung, Johan. 1996. *Peace by Peaceful Means: Peace and Conflict, Development and Civilization*. Peace Research Institute Oslo.

Kriesberg, Louis. 2007. *Constructive Conflicts: From Escalation to Resolution*. Rowman and Littlefield.

Lederach, John Paul. 1997. *Sustainable Reconciliation in Divided Societies*. United States Institute of Peace.

Mead, Margaret. 2022. "Warfare Is Only an Invention—Not a Biological Necessity." In *Conflict After the Cold War*, edited by Richard K. Betts. Routledge. 6th ed.

Mouffe, Chantal. 2005. *On the Political*. Routledge.

Peters, Laura E. R. 2022. "Disasters as Ambivalent Multipliers: Influencing the Pathways from Disaster to Conflict Risk and Peace Potential Through Disaster Risk Reduction." *Journal of Peacebuilding and Development* 17 (2): 151–72. https://doi.org/10.1177/15423166221081516.

Peters, Laura E. R., Ken Conca, Aaron T. Wolf, Ari Lippi, and Jamon Van Den Hoek. 2025. "The Many Voices of Environmental Cooperation: A Relational Analysis of 30 years of Environmental Peacebuilding over Shared Waters in Israel, Jordan, and Palestine." *Environment and Security*. https://doi.org/10.1177/27538796251347820.

Turgul, Alexandra, Melissa McCracken, Susanne Schmeier, Zoe H. Rosenblum, Lynette de Silva, and Aaron T. Wolf. 2024. "Reflections on Transboundary Water Conflict and Cooperation Trends." *Water International* 49 (3–4): 274–88. https://doi.org/10.1080/02 508060.2024.2321727.

United Nations Department of Political Affairs (UNDPA) and United Nations Environment Programme (UNEP). 2015. *Natural Resources and Conflict: A Guide for Mediation Practitioners*. https://wedocs.unep.org/bitstream/handle/20.500.11822/9294/-Natural_resources_and_conflic .pdf?sequence=2&amp%3BisAllowed.

Wolf, Aaron T. 1998. "Conflict and Cooperation Along International Waterways." *Water Policy* 1 (2): 251–65. https://doi.org/10.1016/S1366-7017(98)00019-1.
Zeitoun, Mark, and Naho Mirumachi. 2008. "Transboundary Water Interaction I: Reconsidering Conflict and Cooperation." *International Environmental Agreements: Politics, Law and Economics* 8 (4): 297–316. https://doi.org/10.1007/s10784-008-9083-5.

Conflict Environmental Assessments

Hassan Partow

Conflict environmental assessments refer to evaluations conducted during or after armed conflicts to understand the scale and nature of environmental damage, the risks it poses to ecosystems and livelihoods, and the steps needed for environmental rehabilitation. These assessments serve as critical tools for guiding remedial action and informing broader short- to medium-term recovery efforts.

One of the most recognized frameworks in this field is the Post-Conflict Environmental Assessment (PCEA), spearheaded by the **United Nations Environment Programme** (United Nations Environment Programme [UNEP] 2022a). Designed to assess **environmental degradation** in the aftermath of **conflict**, PCEAs aim to determine the consequences on ecosystems, society, and **governance** capacity. However, since around 2015, the relevance of PCEAs has declined, primarily due to the protracted nature of many modern conflicts, where there is no clear ceasefire or **peace agreement** in sight. In response, new interim approaches have emerged to assess environmental damage as conflicts rage on. These include desk-based preliminary assessments that provide an initial snapshot of environmental impacts, relying on literature review, satellite imagery, and social media analysis to document environmental harm in active conflict zones (UNEP 2017; UNEP 2022b, UNEP 2024a).

Another key framework is the Recovery and Peacebuilding Assessment (RPBA), a joint initiative by the **World Bank**, European Union, and United Nations (Recovery and Peacebuilding Assessment n.d.). RPBAs take a multi-sectoral approach to evaluating conflict-related damages, losses, and recovery needs, with a strong emphasis on quantifying economic impact. The **environment** is one of the cross-cutting areas in a standard RPBA. However, like PCEAs, RPBAs have been largely discontinued in recent years due to the prolonged nature of contemporary conflicts. In their place, Rapid Damage Needs Assessments (RDNA) have become the preferred approach—conducted periodically to provide updated data and economic cost estimates for immediate recovery efforts (United Nations Development Programme [UNDP] et al. 2019, 2025).

The environmental impacts of conflicts are typically assessed at three levels. The first level is direct impacts, which result from military action during the conflict itself—from the destruction of **forests** and farmland to pollution from bombed industrial sites, military debris, and toxic waste. In some cases, warfare itself uses **weaponization of natural resources**, for instance, by intentionally setting oil wells ablaze, contaminating water supplies, or devastating entire ecosystems. The exploitation of **natural resources** by militant groups to finance conflict is another facet of direct impacts (SDG Academy and UNEP n.d.). The second level is indirect impacts, which emerge in the wake of conflict as communities struggle to survive. Displaced populations often overexploit natural resources, while illegal trade in minerals, timber, and wildlife flourishes in lawless environments. Even well-intended humanitarian and peacekeeping efforts can have unintended environmental consequences when large-scale operations lack **sustainability** measures (SDG Academy and UNEP n.d.). The third and last level is environmental governance impacts, which are often the most enduring and difficult to reverse. **War** weakens institutions, leaving environmental protection laws unenforced, **conservation** efforts abandoned, and resource management plans in disarray. Unlike direct pollution or deforestation, these governance failures extend beyond immediate conflict zones, creating lasting challenges for recovery and stability (SDG Academy and UNEP n.d.).

The geographic scale of an environmental assessment depends on the conflict itself—it may focus on a specific region, an entire country, or extend across borders. The methodologies used are equally diverse, combining historical data reviews, satellite **monitoring, stakeholder** consultations, field surveys, and environmental sampling to build a comprehensive picture of the situation and its consequences.

At its core, assessing the environmental impacts is not just about documenting damage—it is about providing practical recommendations and a roadmap for recovery. By identifying priority areas for action, these assessments seek to guide policy decisions, international aid programming, and longer-term rebuilding efforts that integrate environmental sustainability into recovery and reconstruction initiatives. Since the late 1990s, dozens of conflict environmental assessments have been conducted in regions such as Kosovo, Afghanistan, Iraq, Liberia, Sudan, the Democratic Republic of Congo, Ukraine, and the Gaza Strip. These assessments have played a key role in generating scientifically grounded evidence that has supported the mobilization of resources for more sustainable recovery programs, enhancing stability and reinforcing **peacebuilding** efforts.

References

Recovery and Peacebuilding Assessment (RPBA). n.d. "Home Page." Accessed August 26, 2025. https://www.recoveryandpeacebuilding.org/content/rpba/en/home.html.

SDG Academy and United Nations Environment Programme (UNEP). n.d. *Environmental Security and Sustaining Peace (Online Course)*. UN SDG Learn. Accessed August 26, 2025. https://www.unsdglearn.org/courses/environmental-security-and-sustaining-peace/.

United Nations Development Programme (UNDP), European Union, and World Bank. 2019. *Guidance for PDNA in Conflict Situations*. UNDP. https://www.undp.org/publications/pdna-guidance-integrating-conflict-sensitivity.

United Nations Development Programme (UNDP), World Bank Group, European Commission, and Government of Ukraine. 2025. *Ukraine – Fourth Rapid Damage and Needs Assessment (RDNA4): February 2022–December 2024*. UNDP. https://www.undp.org/ukraine/publications/ukraine-fourth-rapid-damage-and-needs-assessment-rdna4-february-2022-december-2024-english.

United Nations Environment Programme (UNEP). 2017. *Environmental Issues in Areas Retaken from ISIL: Mosul, Iraq*. UNEP. https://www.unep.org/resources/publication/environmental-issues-areas-retaken-isil-mosul-iraq-technical-note.

United Nations Environment Programme (UNEP). 2022a. *Timeline: 25 Years of Crisis Response*. UNEP. https://www.unep.org/explore-topics/disasters-conflicts/timeline-25-years-crisis-response.

United Nations Environment Programme (UNEP). 2022b. *The Environmental Impact of the Conflict in Ukraine: A Preliminary Review*. UNEP, October 14. https://www.unep.org/resources/report/environmental-impact-conflict-ukraine-preliminary-review.

Conflict Life Cycle

Mara Pusic and Carl Bruch

The "conflict life cycle" is a conceptual framework used to understand the temporal progression of conflicts from their emergence through escalation to violence, resolution, and recovery. It situates the various phases of conflict in relation to one another over time and provides a structure for analyzing how conflict dynamics evolve and how interventions can be integrated across the lifespan of a conflict (Ramsbotham et al. 2005).

The **conflict** life cycle has been developed primarily in the context of conflict and conflict resolution research (Pleşanu and Badulescu 2018). The term "conflict life cycle" avoids concerns associated with the "conflict cycle," which suggests that conflicts (including **environment**-related conflicts) necessarily will recur. While the literature indicates that many conflicts do recur within five or ten years, an even greater percentage do not recur (Suhrke and Samset 2007).

In the context of **environmental peacebuilding**, the conflict life cycle provides an overarching framework for understanding the roles that **environment** and **natural resources** play in the development of disputes, escalation to **violence**, the conduct of hostilities, the conclusion of the conflict, and recovery from conflict. See, for example, Figure 1.1 (Bruch et al. 2019):

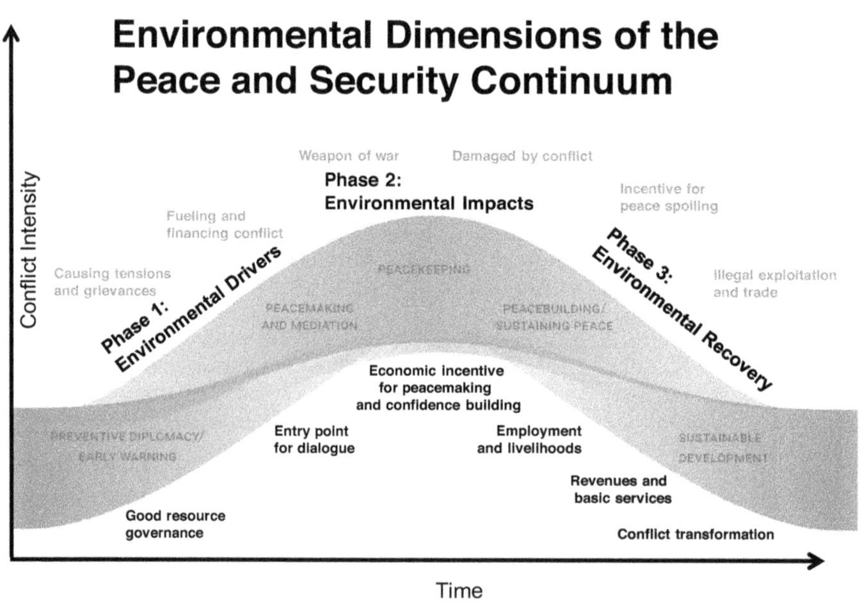

Figure 1.1 Environmental Dimensions of the Conflict Life Cycle

While not captured by Figure 1.1, failures in peacekeeping and post-conflict **peacebuilding** can lead to a relapse of conflict, rather than a transition to **sustainable development.**

A key aspect of the figure is that it links discrete environmental peacebuilding dynamics and interventions across time, objectives, and institutions. As a practical matter, many dynamics and interventions focus on a particular moment in time. Environmental peacebuilding seeks to integrate these dynamics and associated interventions in theory and practice, and the conflict life cycle helps to visualize these linkages. For example, selecting appropriate post-conflict environmental peacebuilding measures depends in large part on the causes of the conflict, the environmental impacts of conflict, the relevant environmental provisions of the **peace** agreement, and the potential presence of peace spoilers. Similarly, while organizations before and during conflict need to focus on their core mandate, it is important for them to consider how their choices and actions might set up—or impair—subsequent interventions by other institutions (Conca and Wallace 2009).

A second key aspect of this framework is that it includes both conflict risks and opportunities for peace. Environmental impacts caused by conflicts are associated with various risks to human health, livelihoods, and ecosystem services. Furthermore, natural resources and the environment can play a role in contributing to violence, due to conflicts over resource access, ownership, benefit sharing, and decision-making (Lujala 2009; Rustad and Binningsbø 2012). Addressing natural resources and the environment in the post-conflict peacebuilding process is vital to avoiding a relapse into conflict (United Nations Environment Programme 2009; Bruch et al. 2016).

In 2022, the United Nations **International Law Commission** adopted Principles on the **Protection of the Environment in Relation to Armed Conflicts (PERAC)** (United Nations 2022). Of note, these principles captured existing and emerging international law across the conflict life cycle (i.e., before, during, and after)—a substantial development over previous efforts that focused either on international law applicable during conflict or after conflict (Stahn et al. 2014).

References

Bruch, Carl, David Jensen, Mikiyasu Nakayama, and Jon Unruh. 2019. "The Changing Nature of Conflict, Peacebuilding, and Environmental Cooperation." *Environmental Law Reporter* 49 (2): 10134–54.

Bruch, Carl, Carroll Muffett, and Sandra S. Nichols. 2016. "Part 4: Local Institutions and Marginalized Populations." In *Governance, Natural Resources, and Post-Conflict Peacebuilding*, edited by Carl Bruch, Carroll Muffett, and Sandra S. Nichols. Earthscan.

Conca, Ken, and Jennifer Wallace. 2009. "Environment and Peacebuilding in War-Torn Societies: Lessons From the U.N. Environment Programme's Experience With Postconflict Assessment." *Global Governance* 15 (4): 485–504. https://www.jstor.org/stable/27800777.

Lujala, Päivi. 2009. "Deadly Combat Over Natural Resources: Gems, Petroleum, Drugs, and the Severity of Armed Civil Conflict." *Journal of Conflict Resolution* 53 (1): 50–71. https://doi.org/:10.1177/0022002708327644.

Pleşanu, Teodora, and Daniela Badulescu. 2018. *A New Approach to the Life Cycle of the Conflict*. Carol I National Defence University. https://proxy.lib.umich.edu/login?url=https://www.proquest.com/conference-papers-proceedings/new-approach-life-cycle-conflict/docview/2089263664/se-2.

Ramsbotham, Oliver, Tom Woodhouse, and Hugh Miall. 2005. *Contemporary Conflict Resolution: The Prevention, Management and Transformation of Deadly Conflicts*. Polity Press.
Rustad, Siri, Camilla Aas, and Helga Malmin Binningsbø. 2012. "A Price Worth Fighting For? Natural Resources and Conflict Recurrence." *Journal of Peace Research* 49 (4): 531–46. https://doi.org/10.1177/0022343312444942.
Stahn, Carsten, Jennifer S. Easterday, and Jens Iverson, eds. 2014. *Jus Post Bellum: Mapping the Normative Foundations*. Oxford University Press. https://doi.org/10.1093/acprof:oso/9780199685899.001.0001.
Suhrke, Astri, and Ingrid Samset. 2007. "What's in a Figure? Estimating Recurrence of Civil War." *International Peacekeeping* 14 (2): 195–203. https://doi.org/10.1080/13533310601150776.
United Nations. 2022. *Protection of the Environment in Relation to Armed Conflicts*. G/22/348/04. https://documents.un.org/doc/undoc/ltd/g22/348/04/pdf/g2234804.pdf.
United Nations Environment Programme (UNEP). 2009. *From Conflict to Peacebuilding: The Role of Natural Resources and the Environment*. https://www.iisd.org/system/files/publications/conflict_peacebuilding.pdf.

Conflict Prevention

Timothy Balag'kutu, Madeleine Loll, and Arabella Angeline Vallee

Conflict prevention is a proactive mechanism to contain tensions, settle, and/or avoid the degeneration of disputes into armed conflict, often involving intercessory mechanisms.

Sometimes referred to as "preventive diplomacy" (Lund 1995, 1999) or "crisis prevention" (Lund 2009), conflict prevention targets causes and triggers of **conflict** through operational and structural mechanisms, as well as transnational efforts in international contexts (Stewart 2003). *Operational Prevention* involves diplomatic, political, and **security** mechanisms such as **mediation**, negotiations, and international arms sanctions; actions aimed at fostering international **cooperation** and commitment to international law. *Structural Prevention* targets systemic factors, including social and economic inequality, identity-based marginalization, and political exclusion, to engender inclusive and equitable systems and prevent societal tensions through long-term initiatives. *Transnational actors* are often instrumental in conflict prevention, mainly employing mediation and negotiation to ease tensions, awareness creation, and education and advocacy for **peacebuilding**. Overall, conflict prevention facilitates peaceful dispute resolution and addresses underlying triggers of disagreement (Ackermann 2003; Wallensteen and Möller 2003; Carment and Schnabel 2003).

Conflict prevention can be used to prevent violence from occurring in a brewing conflict or can be employed post-conflict to prevent the resumption of violent conflict. Examples of conflict prevention can be found in earlier historical processes, including the post-Napoleonic Wars Congress of Vienna (1814–1815) and the post-WWII Marshall Plan (1945), but the formal, modern conceptualization was shaped by global **peace** policy platforms in the late-20th century. UN Secretary General Dag Hammarskjöld is credited with coining the term "preventive diplomacy" in 1960 to describe the need to contain Cold War proxy conflicts. In the 1990s, Secretary General Boutros Boutros-Ghali applied the concept to the discussion of the prevention of the conflict in post-Cold War Yugoslavia (Lund 2009), with Kofi Annan broadening the usage by advocating for a "culture of prevention" in 2001 and 2006 (Melander and Pigache 2007). These conceptualizations and applications informed and gave rise to conflict prevention as a scholarly and policy concept in peace and security.

Increasingly, intergovernmental and transnational organizations advance conflict prevention in security and **governance** agendas (Ackermann 2003; Igarape Institute 2018; Melander and Pigache 2007). The concept underpins much of the international system. It informed the League of Nations and remains the foundation of the UN system. For instance, it is reflected in the UN Charter's Chapters VI and VII that authorize actions by the Security Council, the Secretary-General, and the General Assembly. Conflict prevention is also embedded in norms like the Responsibility to Protect, programs like the Millennium Development Goals and Sustainable Development Goals, and reviews on peace operations, peacebuilding architecture, and global study on women, peace and security highlight prevention (Igarape Institute 2018). Conflict prevention also drives regional peace efforts, including the EU's European Programme for Prevention of Violent Conflict, as well as the **African Union**'s Constitutive Act, Peace and Security Architecture, and in its collaboration with regional economic communities.

Increasingly, conflict prevention combines operational and structural mechanisms into integrated approaches that utilize top-down policy measures to address immediate conflict while building long-term strategies in partnership with local actors to address structural issues (Cliffe and Steven 2017). The growing nexus of conflict, international security, and **climate change** necessitates prevention strategies that involve multiple actors and critical lenses across security, economics, environmental, social, and **human rights** spheres. Inclusive institutions and theoretical approaches, such as **Indigenous Knowledge** systems and **environmental peacebuilding** tools, are increasingly relevant to modern conflict prevention. Further, the combination of top-down policy interventions and bottom-up community efforts is necessary to encourage and build long-term commitment and partnership between local and international **stakeholders** to foster sustainable and inclusive peace. In addition to greater inclusion, integrated conflict prevention strategies reduce financial and social costs of **violence** and, through localized and community-led approaches, help address broader geopolitical, cultural, and historical conflict contexts.

At its core, a comprehensive conflict prevention framework must operate across multiple levels without imposing a prescriptive solution. Yet, despite its potential, the global recurrence of conflict suggests that conflict prevention remains underutilized and demonstrates a lack of international commitment to operationalizing its core principles. To shift this, a more robust, coordinated, and sustained global commitment to conflict prevention is urgently needed.

References

Ackermann, Alice. 2003. "The Idea and Practice of Conflict Prevention." *Journal of Peace Research* 40 (3): 339–47. https://www.jstor.org/stable/3648335.

Carment, David, and Albrecht Schnabel. 2003. "Introduction–Conflict Prevention: A Concept in Search of a Policy." In *Conflict Prevention: Path to Peace or Grand Illusion*, edited by David Carment and Albrecht Schnabel. United Nations University Press.

Cliffe, Sarah, and David Steven. 2017. *An Integrated Approach to Prevention: The Links Between Prevention, the 2030 Agenda, and Sustaining Peace*. NYU Center on International Cooperation,

December 19. https://cic.nyu.edu/resources/an-integrated-approach-to-prevention-the-links-between-prevention-the-2030-agenda-and-sustaining-peace/
Igarape Institute. 2018. "Defining Conflict Prevention Report Title." In *The Handbook of Conflict Prevention Report*. Igarpe Institute. https://www.jstor.org/stable/resrep20660
Lund, Michael S. 1995. "Underrating Preventive Diplomacy." *Foreign Affairs* 74 (4): 160–63. https://doi.org/10.2307/20047285.
Lund, Michael S. 1999. "Preventive Diplomacy for Macedonia, 1992–1997: Containment Becomes Nationbuilding." In *Preventive Diplomacy in the Post–Cold War World: Opportunities Missed, Opportunities Seized, and Lessons to Be Learned*, edited by Bruce Jentleson. Rowman and Littlefield.
Lund, Michael S. 2009. "Conflict Prevention: Theory in Pursuit of Policy and Practice." In *The SAGE Handbook of Conflict Resolution*, edited by Jacob Bercovitch, Victor Kremenyuk, and I. William Zartman. SAGE Publications.
Melander, Erik, and Claire Pigache. 2007. "Conflict Prevention: Concepts and Challenges." In *Konflikt Prävention: Zwischen Anspruch und Wirklichkeit*. Austrian National Defence Academy.
Stewart, Emma. 2003. "Conflict Prevention: Consensus or Confusion." *Peace, Conflict and Development* 3 (3). https://doi.org/10.7246/pcd.0301.
Wallensteen, Peter, and Frida Möller. 2003. *Conflict Prevention: Methodology for Knowing the Unknown*. Uppsala University.

Conflict Resources

Daniëlla Dam-de Jong and Atieh Khatibi

Conflict resources are natural resources that, when systematically extracted and traded in conflict settings, play a role in fueling, benefiting from, or leading to serious human rights abuses, violations of international humanitarian law, or crimes under international law.

The term "conflict diamonds" is outlined in the Kimberley Process Certification Scheme (KPCS), a voluntary agreement between States, civil society, and the diamond sector to combat the trade in rough diamonds financing armed conflicts. According to the KPCS (2013), conflict diamonds are "rough diamonds used by rebel movements or their allies to finance **conflict** aimed at undermining legitimate governments." This definition is narrow as it only focuses on armed groups. The NGO Global Witness has broadened the scope of the notion and extended it to other **natural resources**. It defines conflict resources as:

> Natural resources whose systematic exploitation and trade in a context of conflict contribute to, benefit from, or result in the commission of serious violations of **human rights**, violations of **international humanitarian law** or violations amounting to crimes under international law.
>
> (Global Witness 2006)

In addition to trade and exploitation, other factors, including control and taxation of natural resources, have been considered in a World Bank report as contributing to or benefiting from armed conflicts (Bannon and Collier 2003).

Many armed conflicts over the past three decades have been fueled or financed by conflict resources, such as diamonds (Angola, Sierra Leone, Côte d'Ivoire), timber (Cambodia, Liberia), or gold (Democratic Republic of the Congo, Central African

Republic, Colombia) (Le Billon 2014). Conflict resources began to attract attention in the 1990s, when, because of the end of the Cold War, internal armed conflicts became "self-financing" (Ballentine and Sherman 2003, 1–3). Conflict resources provided armed groups with a source of funding, replacing external superpower funding (Ballentine and Sherman 2003). In addition to providing funding opportunities, conflict resources have been found to prolong armed conflicts and to provide a disincentive for parties to a conflict to negotiate a **peace** settlement (Le Billon and Nicholls 2007; Bannon and Collier 2003). A contested issue is whether the presence of high-value natural resources has an impact on the outbreak of an armed conflict, either because it encourages rent-seeking behavior or because it creates opportunities for armed groups to take up arms (Ross 2004; Collier and Hoeffler 2004; Collier et al. 2007). The presence of high-value natural resources has, however, been associated with higher risks of conflict relapse, primarily if not adequately addressed in **peace agreements** (Webersik and Levy 2016; Mason et al. 2016).

Considering the negative contributions of conflict resources to **environmental peacebuilding**, breaking the nexus between natural resources and armed conflict has become a significant concern for the international community. The UN has been at the forefront of efforts to curb the trade in conflict resources, with the UN Security Council taking a leading role through sanctions (Aldinger et al. 2018; Dam-de Jong 2023). The UNSC has relied on market-based instruments such as the KPCS and OECD Guidance targeting the private sector (Dam-de Jong 2017a), while strengthening the mandate of UN peace operations to address conflict resources (Dam-de Jong 2017b). Furthermore, important economic powers such as the United States and the European Union have followed suit by adopting legislation to curb the trade in conflict resources. While such approaches have had positive effects, they have also been subject to criticism. Fundamental critiques are the failure of unilateral initiatives to engage with the impact on the population of conflict-affected states and, more generally, the one-sided focus of both UN sanctions and unilateral initiatives on armed groups (Okowa 2020; Dam-de Jong 2015). Considering these critiques and the challenges of curbing the trade in conflict resources more generally, it is crucial to continue **monitoring** and fine-tuning approaches to transform conflict resources into resources for peace.

References

Aldinger, Peter, Carl Bruch, and Sofia Yazykova. 2018. "Revisiting Securitization: An Empirical Analysis of Environment and Natural Resource Provisions in United Nations Security Council Resolutions, 1946–2016." In *Routledge Handbook of Environmental Conflict and Peacebuilding*, edited by Ashok Swain and Joakim Öjendal. Routledge.

Ballentine, Karen, and Jake Sherman. 2003. *The Political Economy of Armed Conflict: Beyond Greed and Grievance*. Lynne Rienner Publishers.

Bannon, Iain, and Paul Collier. 2003. *Natural Resources and Violent Conflict: Options and Actions*. World Bank.

Collier, Paul, and Anke Hoeffler. 2004. "Greed and Grievance in Civil War." *Oxford Economic Papers* 56: 563–95. https://www.jstor.org/stable/3488799.

Collier, Paul, Anke Hoeffler, and Dominic Rohner. 2007. "Beyond Greed and Grievance: Feasibility and Civil War." *Oxford Economic Papers* 61 (1): 1–27.

Dam-de Jong, Daniëlla. 2015. *International Law and Governance of Natural Resources in Conflict and Post-Conflict Situations*. Cambridge University Press.

Dam-de Jong, Daniëlla. 2017a. "UN Natural Resources Sanctions Regimes: Incorporating Market-Based Responses to Address Market-Driven Problems." In *Research Handbook on UN Sanctions and International Law*, edited by Larissa van den Herik. Edward Elgar Publishing.

Dam-de Jong, Daniëlla. 2017b. "Standard-Setting Practices for the Management of Natural Resources in Conflict-Torn States: Constitutive Elements of Jus Post Bellum." In *Environmental Protection and Transitions from Conflict to Peace: Clarifying Norms, Principles, and Practices*, edited by Carsten Stahn, Jens Iverson, and Jennifer Easterday. Oxford University Press.

Dam-de Jong, Daniëlla. 2023. "The Contribution of the UN Security Council to Environmental Peacebuilding." In *Research Handbook on International Law and Environmental Peacebuilding*, edited by Daniëlla Dam-de Jong and Britta Sjöstedt. Edward Elgar Publishing.

Global Witness. 2006. *The Sinews of War: Eliminating the Trade in Conflict Resources*. Global Witness.

Kimberly Process Certification Scheme. 2013. "Core Document." https://www.kpcivilsociety.org/wpcontent/uploads/2019/10/20131122_kpcs_core_document_with_amending_ads.pdf

Le Billon, Philippe. 2014. *Wars of Plunder: Conflicts, Profits and the Politics of Resources*. Oxford University Press.

Le Billon, Philippe, and Eric Nicholls. 2007. "Ending 'Resource Wars': Revenue Sharing, Economic Sanction or Military Intervention?" *International Peacekeeping* 14 (5): 613–32. https://doi.org/10.1080/13533310701753941.

Mason, Simon, Damiano Sguaitamatti, and María del Pilar Ramirez Gröbli. 2016. "Stepping Stones to Peace? Natural Resource Provisions in Peace Agreements." In *Governance, Natural Resources, and Post-Conflict Peacebuilding*, edited by Carl Bruch, Carroll Muffett, and Sandra S. Nichols. Earthscan.

Okowa, Phoebe. 2020. "The Pitfalls of Unilateral Legislation in International Law: Lessons From Conflict Minerals Legislation." *International and Comparative Law Quarterly* 69 (3): 685–701. https://doi.org/10.1017/S0020589320000238.

Ross, Michael. 2004. "What Do We Know About Natural Resources and Civil War?" *Journal of Peace Research*, 41 (3): 337–56. https://doi.org/10.1177/0022343304043773.

Webersik, Christian, and Marc Levy. 2016. "Reducing the Risk of Conflict Recurrence: The Relevance of Natural Resource Management." In *Governance, Natural Resources, and Post-Conflict Peacebuilding*, edited by Carl Bruch, Carroll Muffett, and Sandra S. Nichols. Earthscan.

Conflict Sensitivity

Amanda Woomer

Conflict sensitivity is the mechanism by which a person, intervention, or organization understands the potential for conflict in the context in which it is engaging and the interaction between itself and that context, and uses that information to avoid negative effects and maximize positive ones (Saferworld 2004).

The concept of conflict sensitivity emerged following the humanitarian and development community's response to the 1994 Rwandan genocide, which served to inadvertently exacerbate **conflict** and **violence** (Brown et al. 2009). Conflict sensitivity is a way to counteract this risk by developing a better understanding of conflicts (often through conflict analyses) and mitigating them to the extent possible. The term is often linked with Do No Harm, an approach put forth by the Local Capacities for Peace Project (LCPP) in 1993 to help aid workers engage without making a conflict worse (Anderson 1999). Principles underlying conflict sensitivity include **participation**, inclusivity, transparency, respect, **accountability**, partnership, complementarity, and timeliness (Saferworld 2004). It is important to note that conflict-sensitive work is not the same as **peacebuilding** work. Conflict sensitivity can be applied to all interventions in any context, not just those with a specific peacebuilding or conflict-related objective. Working in conflict-affected contexts requires conflict sensitivity even if not working to address conflict dynamics as within peacebuilding (Interpeace 2010).

In the field of **environmental peacebuilding**, conflict sensitivity is necessary for an intervention (and the **monitoring** and **evaluation** of that intervention) to be appropriately aware of and responsive to a conflict context (Bruch and Woomer 2023). Possible mechanisms include regular conflict analysis, early warning indicators, context indicators, indicators of **conflict sensitivity**, and the engagement of diverse teams that include conflict experts and people with lived experience in the contexts. Conflict sensitivity also requires a detailed consideration of how adherence to principles of participation and transparency may exacerbate conflict; environmental peacebuilding work must balance these considerations carefully.

References

Anderson, Mary B. 1999. *Do No Harm: How Aid Can Support Peace – Or War.* Lynne Rienner Publishers.

Brown, Sarah, Rachel Goldwyn, Hesta Groenewald, and Joan McGregor. 2009. "Conflict Sensitivity Consortium Benchmarking Paper." Conflict Sensitivity Consortium.

Bruch, Carl, and Amanda Woomer. 2023. *Toolkit on Monitoring and Evaluation of Environmental Peacebuilding.* Environmental Law Institute, Environmental Peacebuilding Association, and United States Institute of Peace. https://m-and-e.environmentalpeacebuilding.org/toolkit/.

Interpeace. 2010. *What Is Peacebuilding? Do No Harm, Conflict-Sensitivity and Peacebuilding.* https://www.interpeace.org/wpcontent/uploads/2010/08/2010_IP_What_Is_Peacebuilding_Do_No_Harm_Conflict_Sensitivity_And_Peacebuilding.pdf.

Saferworld. 2004. "An Introduction to Conflict-sensitive Approaches to Development, Humanitarian Assistance and Peacebuilding." In *Conflict-sensitive Approaches to Development, Humanitarian Assistance and Peacebuilding.* Saferworld. https://www.saferworld-global.org/downloads/pubdocs/chapter_1__265.pdf.

Conflict Transformation

Alexandra Turgul, Lynette de Silva, and Shahab Khalid

Conflict transformation is an approach or process to comprehend and address the underlying relationships of conflict dynamics, viewing conflict as an opportunity to foster long-term harmony and peace.

Developed by John Paul Lederach in the 1980s, conflict transformation emphasizes shifting the underlying structures, relationships, and dynamics that create and sustain **conflict** (Lederach 2014; Zeitoun et al. 2020). The conflict transformation approach aligns with the concept of **positive peace**, coined by John Galtung, seeking to understand **peace** not merely as conflict resolution to end direct **violence**, but as an opportunity to change relational tensions and create conditions fostering social justice and long-term peace (Galtung 1969; Lederach 2014). Crucial to understanding and correctly implementing conflict transformation is a holistic analysis of underlying and deeply rooted conflict dynamics. Adopting a conflict transformation lens entails viewing conflict as an opportunity to constructively address how conflict is natural to relationships; conflict presents an opportunity to better understand one another and the systems we create (Lederach 2014). In **environmental peacebuilding**, conflict could be triggered by competition over **natural resources, environmental degradation**, and/or ecological breakdown (Simangan et al. 2022). However, with a conflict transformation lens, practitioners can better see that the underlying need for change in the environmental, economic, political, or social relationships is the reason why the **conflict life cycle** occurs. When one or more institutions (government, social service, legal entity) cannot adequately adjust to new realities, it compounds the crisis (Wolf 2007). In environmental peacebuilding contexts, the intersectoral relationships behind the Water-Energy-Food-Ecosystems (WEFE) Nexus (also known as the **Water-Energy-Food Nexus**) are key to positive peace.

Conflict transformation is about elevating self-consciousness through cultivating connections in a process that extends beyond the present dispute to examine communication channels, relationship boundaries, and dynamics among parties over time and space (de Silva 2022; Lederach 2014). It transforms oppressive and adversarial relationships, especially in situations of **power** asymmetry, where one party dominates or marginalizes another (Jeong 2020). Much of the underlying issues in a conflict relate to conflicting parties that have fundamentally diverging perspectives of what their society (including its food, water, and environmental systems) should look like in the future (Connolly and Baggerman 2014). Power imbalances, negative othering between actors, and drastic political change can contribute to the breakdown of **environmental cooperation** and violent escalation of conflicts over scarce resources. In an environmental peacebuilding context, an understanding of the history in which a particular conflict is rooted, the power dynamics that underpin it, and the ways in which knowledge about the resource is constructed and framed can better ensure that conflict transformation leads to a more positive outcome for all.

For conflict transformation, collaboration requires the ability to build a foundation of trust and listen to the needs of others while working toward joint decision-making and capacity enhancement (Wolf 2007). Establishing trust is a gradual process and invites participants to focus on identifying the interests underpinning often entrenched positions. This can bring clarity on the history of the conflict and facilitate the identification of shared values and opportunities for consensus-building and relationship-building (de Silva 2022).

Recognizing that life is a constant unfolding of new possibilities, conflict transformation requires the nurturing of relationships focused on shared goals toward honest, authentic, and earnest **cooperation.** It resolves social and economic disparities, brings **environmental justice**, and mitigates ecological breakdown to work toward long-term solutions that promote positive peace.

Acknowledgment

This work was carried out with support from the CGIAR Science Programs on Climate Action and Food Frontiers and Security. We would like to thank all funders who supported this research through their contributions to the CGIAR Trust Fund: https://www.cgiar.org/funders/.

References

Connolly, Lesley, and Jill Baggerman. 2014. *Getting Coherence and Coordination Right: Principles for the Peacebuilding Policy Community*. The African Centre for the Constructive Resolution of Disputes, August. https://www.accord.org.za/publication/getting-coherence-and-coordination-right/

de Silva, Lynette. 2022. "Water Conflict Transformation." In *Resolving Water Conflicts Workbook*, edited by Lynette de Silva and Chris Maser. Taylor and Francis Group.

Galtung, Johan. 1969. "Violence, Peace, and Peace Research." *Journal of Peace Research* 6 (3): 167–91. https://doi.org/10.1177/002234336900600301.

Galtung, Johan. 1990. "Cultural Violence." *Journal of Peace Research* 27 (3): 291–305. https://journals.sagepub.com/doi/10.1177/0022343390027003005.

Jeong, Ho Won. 2020. "Conflict Transformation." In *Routledge Companion to Peace and Conflict Studies*, edited by Sean Byrne, Thomas Matyók, Imani Michelle Scott, and Jessica Senehi. Routledge.

Lederach, John P. 2014. *Little Book of Conflict Transformation: Clear Articulation of the Guiding Principles by a Pioneer in the Field*. Good Books.

Simangan, Dahlia, Chui Ying Lee, Ayyoob Sharifi, John Lee Candelaria, and Shinji Kaneko. 2022. "A Global Analysis of Interactions between Peace and Environmental Sustainability." *Earth System Governance* 14: 100152. https://doi.org/10.1016/j.esg.2022.100152.

Wolf, Aaron T. 2007. "Shared Waters: Conflict and Cooperation." *Annual Review of Environment and Resources* 32: 241–69. https://www.annualreviews.org/content/journals/10.1146/annurev.energy.32.041006.101434.

Zeitoun, Mark, Naho Mirumachi, Jeroen Warner, Matthew Kirkegaard, and Ana Cascão. 2020. "Analysis for Water Conflict Transformation." *Water International* 45 (4): 365–84. http://dx.doi.org/10.1080/02508060.2019.1607479.

Conservation

Jacobo Grajales, Rachel Weaver, and Julia Gorricho

Conservation is the preservation and protection of natural resources and ecosystems. It focuses on the protection of biodiversity, ecosystems, and natural habitats, with the aim of maintaining their ecological functions and resilience.

> Conservation encompasses a range of strategies, including protected areas, community-based conservation, and ecosystem-based management. Conservation should not be seen merely as a technical response or solution to the challenges of **biodiversity** loss, but rather as a set of political techniques shaped by historical trajectories of state formation and social group differentiations such as class, ethnicity, race, caste, and **gender**. In other words, conservation raises questions about the relationship between territories, social groups, and political **power**.
>
> **Conflict** can negatively impact conservation in many ways, including the destruction of habitat and the degradation or increased exploitation of **natural resources** (Hammill et al. 2009). Military operations can also affect fragile environments, as in the case of US bombing in Vietnam (Westing 1971) or aerial fumigation of coca leaf fields in Colombia (Johnson et al. 2024). Resettlement of displaced populations can create pressure on **land use**, leading governments or local authorities to use protected areas to relieve pressure on land access, as in the case of Rwanda (Huggins 2014). Conservation initiatives are also undermined by the decline or collapse of state institutions and loss of administrative capacity (Fischer 2004; Glew and Hudson 2007; Álvarez 2008; Loucks et al. 2009).

The exploitation of natural resources to finance wars by both state and non-state groups can pose significant challenges to conservation. These elements of **war** economies, often linked to global value chains (e.g., timber in Liberia, rare minerals in the DRC), challenge conservation initiatives (Le Billon 2012). Conservation policy is also being reshaped by militarization processes involving the use of military, paramilitary, and **security** actors in conservation efforts (Lunstrum 2014; Duffy et al. 2019). This reflects the long-term and complex relationship between state territorialization, conservation goals, and attempts to control and repress populations perceived as threatening protected ecosystems (Brockington 2002; Jacoby 2003). As a result, a range of new actors, including the military, private security companies, and surveillance and technology actors, are playing an increasingly important role in conservation (Massé and Lunstrum 2016).

Conservation can serve as a mechanism for **peacebuilding** by providing an opportunity to foster **cooperation**, improve relations between states and social groups, and build **peace** (Ali 2007; Conca and Dabelko 2002). At the same time, conservation has the potential to introduce new sources of conflict, restrict access to key livelihood resources, and lead to an unequal distribution of benefits. **Environmental peacebuilding** and conservation are also linked because many of the world's conflict zones are also areas of high biodiversity value (Hanson et al. 2009). This creates some specific challenges for conservation organizations working in conflict-affected areas, shaping the overall political context of conservation policy. In armed conflict contexts, conservation can be shaped and influenced by broader political objectives beyond biodiversity, such as state attempts to extend control into marginal areas, counterinsurgent strategies (Ybarra 2018), and the relationship between conflict-affected states and the donor community (Beevers 2019).

Within environmental peacebuilding, the strategy and rubric of conflict-sensitive conservation have gained prominence, especially among practitioners (Bruch et al. 2024; Ide et al. 2021; Hammill et al. 2008, 2009). Conflict-sensitive conservation integrates conservation programming and implementation with conflict analysis and conflict management strategies. Conflict-sensitive conservation actively considers the potential for environmental projects to influence, exacerbate, or mitigate conflict. It emphasizes understanding local social, economic, and political dynamics to prevent conservation initiatives from unintentionally igniting or exacerbating conflict (Bruch et al. 2024).

References

Ali, Saleem H (ed.). 2007. *Peace Parks: Conservation and Conflict Resolution*. MIT Press. https://doi.org/10.7551/mitpress/7433.001.0001

Álvarez, María D. 2008. "Forests in the Time of Violence." *Journal of Sustainable Forestry* 16 (3–4): 47–68. https://doi.org/10.1300/J091v16n03_03.

Beevers, Michael D. 2019. *Peacebuilding and Natural Resource Governance After Armed Conflict*. Palgrave Macmillan.

Brockington, Dan. 2002. *Fortress Conservation–The Preservation of the Mkomazi Game Reserve, Tanzania*. International African Institute, in Association with James Currey and Indiana University Press.

Bruch, Carl, Geeta Batra, Anupam Anand, Shehla Chowdhury, and Sierra Killian. 2024. *Conflict-Sensitive Conservation Lessons from the Global Environment Facility*. Routledge.

Conca, Ken, & Geoffrey Dabelko (eds.). 2002. *Environmental Peacemaking.* Johns Hopkins University Press and Woodrow Wilson Center Press.
Duffy, Rosaleen, Francis Massé, Emile Smidt, et al. 2019. "Why We Must Question the Militarisation of Conservation." *Biological Conservation* 232: 66–73. https://doi.org/10.1016/j.biocon.2019.01.013.
Fischer, Frauke. 2004. "Status of the Comoé National Park, Côte d'Ivoire, and the Effects of War." *Parks* 14 (1): 17–25. https://www.researchgate.net/profile/Andriy-Kovalchuk-3/post/Does-Armed-Conflict-have-any-impact-on Biodiversity/attachment/59d63bba79197b8077998bf6/AS%3A412318681845760%401475315842503/download/14_1.pdf#page=19.
Glew, Louise, and Malcom D. Hudson. 2007. "Gorillas in the Midst: the Impact of Armed Conflict on the Conservation of Protected Areas in Aub-Saharan Africa." *Oryx* 41 (2): 140–50. https://doi.org/10.1017/S0030605307001755.
Hanson, Thor, Thomas M. Brooks, Gustavo A. B. Da Fonseca, Michael Hoffmann, John F. Lamoreux, Gary Machlis, Cristina G. Mittermeier, Russell A. Mittermeier, and John D. Pilgrim. 2009. "Warfare in Biodiversity Hotspots." Conservation Biology 23 (3): 578--87. https://doi.org/10.1111/j.1523-1739.2009.01166.x.
Hammill, Anne, Alec Crawford, and Charles Besançon. 2008. *Gorillas in the Midst: Assessing the Peace and Conflict Impacts of International Gorilla Conservation Programme (IGCP) Activities.* International Institute for Sustainable Development.
Hammill, Anne, Alec Crawford, Robert Craig, Robert Malpas, and Richard Matthew. 2009. *Conflict-Sensitive Conservation Practitioners' Manual.* International Institute for Sustainable Development.
Huggins, Chris. 2014. "Land Grabbing and Land Tenure Security in Post-Genocide Rwanda." In *Losing Your Land: Dispossession in the Great Lakes*, edited by An Ansoms and Thea Hilhorst. Boydell and Brewer. https://doi.org/10.1017/9781782043416.008.
Ide, Tobias, Carl Bruch, Alexander Carius, et al. 2021. "The Past and Future(s) of Environmental Peacebuilding." *International Affairs* 97 (1): 1–16. https://doi.org/10.1093/ia/iiaa177.
Jacoby, Karl. 2003. *Crimes Against Nature – Squatters, Poachers, Thieves, and the Hidden History of American Conservation.* University of California Press.
Johnson, McKenzie F., Luz A. Rodríguez, and Manuela Quijano Hoyos. 2024. "Governing after FARC: Environmental Peacebuilding in Caquetá, Colombia." *The Journal of Peasant Studies* 52 (1): 129–56. https://doi.org/10.1080/03066150.2024.2326577.
Le Billon, Philippe. 2012. *Wars of Plunder: Conflicts, Profits and the Politics of Resources.* Columbia University Press.
Loucks, Colby, Michael B. Mascia, Andy Maxwell, etc. 2009. "Wildlife Decline in Cambodia, 1953-2005: Exploring the Legacy of Armed Conflict." *Conservation Letters* 2 (2): 82–92. https://doi.org/10.1111/j.1755-263X.2008.00044.x.
Lunstrum, Elizabeth. 2014. "Green Militarization: Anti-Poaching Efforts and the Spatial Contours of Kruger National Park." *Annals of the Association of American Geographers* 104 (4): 816–32. https://doi.org/10.1080/00045608.2014.912545.
Massé, Francis, and Elizabeth Lunstrum. 2016. "Accumulation by Securitization: Commercial Poaching, Neoliberal Conservation, and the Creation of New Wildlife Frontiers." *Geoforum* 69: 227–37. https://doi.org/10.1016/j.geoforum.2015.03.005.
Westing, Arthur H. 1971. "Ecological Effects of Military Defoliation on the Forests of South Vietnam." *BioScience* 21 (17): 893–98. https://doi.org/10.2307/1295667.
Ybarra, Megan. 2018. *Green Wars: Conservation and Decolonization in the Maya Forest.* University of California Press.

Cooperation

Héctor Morales Muñoz and Sinéad Barry

Cooperation is the process by which individuals, groups, or organizations work together toward mutual goals. Cooperation can happen in different contexts, such as within organizations or across various sectors. Cooperation involves reciprocity and mutual benefit, where each participant contributes without expecting immediate rewards (Amici and Bietti 2015).

In **environmental peacebuilding**, cooperation is typically viewed both as a mechanism and a result, emerging from natural resource management (NRM), climate **governance**, or other environmental protection initiatives through which **conflict** can be prevented, mitigated, or resolved. This materializes through various processes, for instance: (1) technical environmental challenges lead to cooperation between groups, seeing positive spillover effects, (2) common environmental threats spark recognition of shared interests, (3) joint NRM facilitates regular interactions, leading to a shared purpose, expectations, and values which over time contribute to building a community of practice, and (4) cooperation establishes a tradition of joint NRM, leading to "expectations for productive, long-term interactions" (Johnson et al. 2021, 5; Doughman and Dabelko 2002, 209; Mosello and Rüttinger 2020, 6).

In practice, cooperation over joint environmental and climatic threats is often viewed by peacebuilders as a strategic entry point, with a comparative advantage for trust-building between groups in or at risk of conflict (United Nations Environment Programme [UNEP] 2024a; adelphi n.d., 2015). This is often because environmental and climatic topics are often viewed as less partisan than other political issues. In conflict settings, groups that may not usually engage with one another on many issues may be willing to cooperate over the shared identification of environmental and climatic challenges or NRM, leading to increased intergroup trust, advancing to cooperation and, therefore, contributing to **peace** (Martínez et al. 2024). On the other hand, presenting cooperation as neutral or apolitical has risks, particularly in contexts of asymmetric **power** relations. For example, water-sharing cooperation efforts between Israel and Palestine have been viewed as obscuring the structural inequalities at the root of the resource dispute (Barron 2024; Selby et al. 2022; Selby 2003).

Evidently, the ways in which cooperation plays out in environmental peacebuilding vary by context and level. At the intrastate level, cooperation may involve joint community action to manage and sustain **natural resources**, such as water (Carius 2006; UNEP 2024b). Both at the community and national levels, cooperation has also been used as a starting point for larger reconciliation processes (e.g., post-tsunami Aceh in Indonesia, Colombia's national peace agreement) (Kelman 2012; Rüttinger et al. 2024). At the transboundary level, interstate cooperation over technical issues, sometimes referred to as science diplomacy, aims to strengthen relations between states by establishing spaces for engagement, norms for cooperation, reducing environmental destabilizers of peace, and increasing the number of interactions between states, as evidenced in the Bay of Bengal and the Korab Massif area, among others (Royal Society-AAAS 2010; Sugunan et al. 2024; Viehoff and Mosello 2024). **Transboundary cooperation** initiatives also pose risks to **human security**, particularly during vast **infrastructure** projects that do not adequately consider the effects of NRM on local communities, e.g., along the Mekong River (Ide 2020, 4). More recently, processes of ecosystem restoration at the transboundary level have been recognized as an entry point for international cooperation when implemented in a conflict-sensitive manner (United Nations Convention to Combat Desertification

2024). Furthermore, the IPCC's Fifth Report identified two reasons why international cooperation on **climate change** is essential: it is a global commons issue requiring coordinated global action, and cooperative solutions can enhance economic efficiency by addressing the diverse **mitigation** costs and opportunities across countries (Intergovernmental Panel on Climate Change 2014).

Finally, cooperation often depends on specific conditions to meet positive outcomes. First, mutual trust and transparency between **stakeholders** could be necessary to build confidence for collaborative efforts (Morales-Muñoz et al. 2023). In contrast, International Relations scholars such as Keohane argue that cooperation must align with principles of sovereignty and self-help to develop. For these scholars, cooperation may be challenging and frustrating, particularly in contexts of "international regimes," understood as "liberal international arrangements for trade and international finance as responses to the need for policy coordination created by interdependence" (Keohane 1986, 2005, 8). In that sense, for cooperation to foster peace, equitable resource-sharing frameworks addressing marginalization and benefit disparities are essential. Robust governance and institutional mechanisms must provide **accountability**, dispute resolution, and multi-level coordination (Morales-Muñoz et al. 2022). Finally, long-term commitments from state and non-state actors could be necessary to ensure sustained efforts toward common and essential environmental goals.

References

adelphi. 2015. "Environment, Conflict and Cooperation Factbook." https://climate-diplomacy.org/case-studies?filter[geolocation][center]=19.0933679,13.1100849&widget[map][zoom]=3.

adelphi. n.d. "The Weathering Risk Peace Pillar." Accessed October 21, 2024. https://weatheringrisk.org/en/peacepillar.

Amici, Federica, and Lucas Bietti. 2015. "Coordination, Collaboration and Cooperation: Interdisciplinary Perspectives." *Interaction Studies* 16 (3): 7–11. https://doi.org/10.1075/is.16.3.

Barron, Elsa. 2024. "On Dialogue and Beyond: Positive Environmental Peacebuilding in Palestine." *The Journal of Social Encounters* 8 (1): 21–28. https://doi.org/10.69755/2995-2212.1239.

Carius, Alexander. 2006. "Environmental Cooperation as an Instrument of Crisis Prevention and Peacebuilding: Conditions for Success and Constraints." *adelphi*, January. https://adelphi.de/system/files/mediathek/bilder/us_503_carius_environmental_peacemaking_06-07-02_1.pdf.

Doughman, Pamela, and Geoffrey Dabelko. 2002. "Water Cooperation in the U.S.-Mexico Border Region." In *Environmental Peacemaking*, edited by Ken Conca and Geoffrey D. Dabelko. Woodrow Wilson Center Press.

Ide, Tobias. 2020. "The Dark Side of Environmental Peacebuilding." *World Development* 127: 104777. https://doi.org/10.1016/j.worlddev.2019.104777.

Intergovernmental Panel on Climate Change (IPCC). 2014. *Climate Change 2014: Synthesis Report. Contribution of Working Groups I, II and III to the Fifth Assessment Report of the Intergovernmental Panel on Climate Change*. IPCC.

Johnson, McKenzie, Luz Rodríguez, and Manuela Quijano Hoyos. 2021. "Intrastate Environmental Peacebuilding: A Review of the Literature." *World Development* 137: 105150. https://doi.org/10.1016/j.worlddev.2020.105150.

Kelman, Ilan. 2012. *Disaster Diplomacy: How Disasters Affect Peace and Conflict*. Routledge.

Keohane, Robert O. 1986. "Reciprocity in International Relations." *International Organization* 40 (1): 1–27. https://www.jstor.org/stable/2706740.

Keohane, Robert O. 2005. *After Hegemony: Cooperation and Discord in the World Political Economy*. Princeton University Press.

Martínez, Albert, Suaad al-Salahi, Laura Geres, Sinéad Barry, and Alina Viehoff. 2024. "Voices From Yemen – How Environmental Dialogues can Contribute to Environmental Resilience and Peace." *Weathering Risk*, October 7. https://weatheringrisk.org/en/impact-note/voices-yemen-how-environmental-dialogues-can-contribute-environmental-resilience-and?back_url=/en/impact&hmac=Ba5CSzYtwcbrPHB4dTfQA9uqYsPRvIGfJyjvEiutZvo.

Morales-Muñoz, Héctor, Arwen Bailey, Katharine Löhr, et al. 2022. "Co-benefits Through Coordination of Climate Action and Peacebuilding: A System Dynamics Model." *Journal of Peacebuilding and Development* 17 (3): 304–23. https://doi.org/10.1177/15423166221132149.

Morales-Muñoz, Héctor, Leigh Martens, Katharina Löhr, et al. 2023. "Integrating Climate Mitigation and Environmental Peacebuilding Objectives through Sustainable Land Use Systems: Theory of Change and Indicators." *PLOS Climate* 2 (5): e0000075. https://doi.org/10.1371/journal.pclm.0000075.

Mosello, Beatrice, and Lukas Rüttinger. 2020. "Linking Adaptation and Peacebuilding–Lessons Learned and the Way Forward." *adelphi*, January 22. https://climate-diplomacy.org/magazine/conflict/linking-adaptation-and-peacebuilding-lessons-learned-and-way-forward.

Royal Society-AAAS. 2010. *New Frontiers in Science Diplomacy: Navigating the Changing Balance of Power*. The Royal Society.

Rüttinger, Lukas, Lucas Destrijcker, Héctor Morales-Muñoz, Adrien Foong, Jakob Gomolka, Lisa Binder, et al. 2024. "Africa Climate Security Risk Assessment." *adelphi*, August. https://weatheringrisk.org/sites/default/files/document/240924_ACRA_Full_Report.pdf.

Selby, Jan. 2003. "Dressing up Domination as 'Cooperation': The Case of Israeli-Palestinian Water Relations." *Review of International Studies* 29 (1): 121–38. https://www.jstor.org/stable/20097837.

Selby, Jan, Gabrielle Daoust, and Clemens Hoffmann. 2022. "Peace." In *Divided Environments: An International Political Ecology of Climate Change, Water and Security*, edited by Jan Selby, Gabrielle Daoust, and Clemens Hoffmann. Cambridge University Press.

Sugunan, Sreejith, Nicholas Chu, Sebastian Kratzer, Lina Hillert, Sinéad Barry, and Alexandra Steinkraus. 2024. "Voices from the Bay–Making a Fishing Ban Work in the Bay of Bengal." *Weathering Risk. Effective*, August 8. https://weatheringrisk.org/en/impact-note/voices-bay-making-fishing-ban-work-bay bengal?back_url=/en/impact&hmac=Ba5CSzYtwcbrPHB4dTfQA9uqYsPRvIGfJyjvEiutZvo.

United Nations Convention to Combat Desertification (UNCCD). 2024. *Ground for Peace: Land Restoration for International Peace and Security*. United Nations Convention to Combat Desertification.

United Nations Environment Programme (UNEP). 2024a. *Environmental Cooperation and Peacebuilding*. UNEP, October 21. https://www.unep.org/topics/disasters-and-conflicts/environment-security/environmental-cooperation-and-peacebuilding.

United Nations Environment Programme (UNEP). 2024b. *Lessons Learned from the Implementation of the Wadi El Ku Catchment Management Project (Phase 2): Avenues for Climate Security and Environmental Peacebuilding Programming*. UNEP.

Viehoff, Alina, and Beatrice Mosello. 2024. "Climate Change and Security in South-Eastern Europe: Strengthening Joint Co-operation in the Shar/Sara Mountains and Korab Massif Area." In *Climate Changes Security Navigating Climate Change and Security Challenges in the OSCE Region*, edited by Hans Lampalzer and Gerald Hainzl. Federal Ministry of Defence, Republic of Austria.

Critical Minerals

Stacy D. VanDeveer, Erik Post, and Tracy Hart

Critical minerals are a subset of minerals considered crucial to firms, economic growth, and national security—especially as related to the energy and high-technology sectors—whose supply chains include vulnerabilities/risks to disruption.

The EU and many states maintain different lists of "critical" materials. Local and transnational **governance** challenges and risks are frequently associated with supply disruption at mining sites and in processing and distribution supply chains (Carr-Wilson et al. 2024). Cobalt and lithium, both necessary for **renewable energy** and the high-tech sectors, receive substantial policymaking and scholarly attention. Supply risks are frequently associated with environmental/ecological harm, political instability, civil and interstate **violence**, and geopolitical risks stemming from highly asymmetric geographic concentration of mining (such as for cobalt and lithium), which have a high concentration for extraction in South America and the Democratic Republic of the Congo and undergo minerals processing in China (VanDeveer et al. 2023). Policymaking and scholarly attention to governance have been focused on efforts to increase financial transparency, corporate social responsibility, and reduce risks to international firms and state **security**, in the context of the global need to decarbonize energy production (Auld et al. 2018; VanDeveer et al. 2023).

Geographers and other scholars studying the criticality of materials note that this is a dynamic category that emerges in relation to **power** structures, which means asking questions about for whom and why certain materials are deemed critical. This is illustrated by the historical association of critical materials such as nitrate, coal, iron, and later oil among others, with industrialization and militarization. From such perspectives, the framing of critical materials as necessary to addressing **climate change** signals the emergence of a drive for control over the extraction of these materials. Critical scholarship warns that such security framings enable the obfuscation of violence and the concentration of economic and military power in the name of environmentalism (Post and Le Billon 2024).

Governments and industries can support the transformation of illicit flows of minerals into responsible trade models through strengthening natural **resource governance**, transparency, industry-led traceability initiatives, and due diligence mechanisms. Increased in-region engagement and heightened due diligence by midstream and downstream companies related to their supply chains can help enable more consistent progress toward addressing risks and contributing to **peace** (U.S. Department of State Bureau of African Affairs 2024). The ability of communities and civil society to act as watchdogs, through the rapid and widespread dissemination of information on mining activities, continues to grow and underscores the importance for companies to maintain their social license to operate and minimize the reputational risks associated with **conflict** minerals (Church and Crawford 2018). Legitimizing and strengthening local and traditional structures for expressing and resolving grievances is one of the most important avenues to empower the voices of impacted communities (Ahmadnia et al. 2022). Sustainable resource management, responsible sourcing, and the use of resources and associated rents for **basic services** provide important **conflict prevention, mitigation**, and resolution opportunities. Finally, restoration of landscapes affected by critical minerals extraction and efforts to move to post-extractive alternative futures based on more circular and sustainable modes of production and consumption can create conflict prevention, mitigation, and resolution opportunities. **Peacebuilding** efforts must aim to prevent, mitigate, and resolve the aggravating effects of the extraction of critical minerals on conflict dynamics while harnessing the positive opportunities.

References

Ahmadnia, Shaadee, Agathe Marie Christien, Phoebe Spencer, Tracy Hart, and Caio Cesar de Araujo Barbosa. 2022. *Defueling Conflict: Environment and Natural Resource Management as a Pathway to Peace*. The World Bank.

Auld, Graham, Michele M. Betsill, and Stacy D. VanDeveer. 2018. "Transnational Governance along the Minerals Life Cycle." *Annual Review of Energy and Resources* 43 (1): 425–53. https://doi.org/10.1146/annurev-environ-102017-030223.

Carr-Wilson, Savannah, Subhrendu N. Pattanayak, and Erika Weinthal. 2024. "Critical Minerals in the Energy Transition: A Systematic Review of the Environmental, Social, and Governance Risks and Opportunities." *Energy Research and Social Science* 116: 1–14. https://doi.org/10.1016/j.erss.2024.103672.

Church, Clare, and Alex Crawford, 2018. *Green Conflict Minerals: The Fuels of Conflict in the Transition to a Low-Carbon Economy*. Institutional Institute for Sustainable Development. https://www.iisd.org/story/green-conflict-minerals/.

Post, Erik, and Philippe Le Billon. 2024. "The 'Green War': Geopolitical Metabolism and Green Extractivisms." *Geopolitics* 30 (2): 760–800. https://doi.org/10.1080/14650045.2024.2385411.

U.S. Department of State, Bureau of African Affairs. 2024. *Statement of Concern Related to Certain Minerals Supply Chains from Rwanda and Eastern Democratic Republic of the Congo Contributing to the Ongoing Conflict*. U.S. Department of State. https://www.state.gov/statement-of-concern-related-to-certain-minerals-supply-chains-from-rwanda-and-eastern-democratic-republic-of-the-congo-contributing-to-the-ongoing-conflict/.

VanDeveer, Stacy D., Hyeyoon Park, Yixian Sun, and Michele M. Betsill. 2023. "Extractive Industries and Mineral Resources: Turbulence All Around." In *Global Environmental Politics in a Turbulent Era,* edited by Peter Dauvergne and Leah Shipton. Edward Elgar Publishing.

D

Decolonization

Theresa (Isa) Arriola, Megan Baker, and Sarah Njeri

Decolonization is the simultaneous process of dismantling colonial systems and restoring Indigenous sovereignty, which was and continues to be undermined by colonization. Whether classic colonialism or settler colonialism, decolonization aims to remake the world that colonialism fundamentally disrupted.

Colonialism, which sought to extract from lands and people for the benefit of an outside group, required a destruction of Indigenous lifeways to claim and exploit lands, labor, and resources. Indigenous sovereignty is the ability to make social, political, and economic decisions without colonial interference, which includes but is not limited to colonial institutions such as state governments, private entities, or public corporations that seek to colonize, displace, restrict, or denigrate Indigenous Peoples and territories for sociopolitical or economic gain.

Settler colonialism is focused on eliminating Indigenous peoples to claim their lands and create a new settler society (Wolfe 2006). Under such colonial and imperial systems, all aspects of society—such as politics, the economy, social structures, and ideologies—are shaped and reshaped. Most critically, colonialism transforms relations between peoples, lands, and the non-human, commodifying and creating hierarchies that facilitate the exploitation of those placed in lower tiers of the social hierarchy. Without an ideological **infrastructure** premised upon the subjection of populations, settlers and colonizers would not be able to undermine Indigenous connections to the land for the purpose of controlling and settling those lands. Rejecting colonial formations and returning to Indigenous modes of relations and **governance** is paramount to decolonization.

Colonial **violence** is often linked to environmental violence against Indigenous Peoples because it disrupts their relationships with the land (Dhillon 2022; Whyte 2018). This form of harm often occurs as what Rob Nixon (2011) terms "slow violence," a gradual, invisible form of violence, accumulating over time and disproportionately impacting Indigenous communities. **Environmental degradation** often includes destroying the environment or making it inhospitable to growing Indigenous food crops or forcing a population to rely on unhealthy, processed foods that lead to health problems over generations. For instance, the growing militarization of the Pacific

DOI: 10.4324/9781003604143-6

Islands to prepare for **war** has led to continued contamination and military testing that disrupts Indigenous Pacific Islanders' relationships to their lands. Likewise, the contamination of waterways affecting Māori communities in Aotearoa (New Zealand) illustrates how environmental and colonial harms are deeply entwined.

Decolonization is a process that must be carried out with agency by Indigenous Peoples themselves and supported by non-Indigenous Peoples and institutions to restore and continuously cultivate Indigenous sovereignty. In 2016, the Standing Rock movement demonstrated how Indigenous peoples foregrounded their sovereignty and way of life to prevent further degradation of the water and lands for all people (Estes 2019).

While the concept of sovereignty has been robustly critiqued, Indigenous sovereignty is distinct from its Western conceptualization and formulation. Audra Simpson (2014) has shown ethnographically how sovereignty can be nested, existing within one another, underscoring the difference and distinctiveness of Indigenous sovereignty. Meanwhile, Paul Nadasdy has argued that "initial efforts to theorize sovereignty went hand-in-hand with broader efforts to delegitimize the political systems of colonized people" (Nadasdy 2003). Thus, any attempt to theorize Indigenous sovereignty has always been an inherently fraught process, which brings with it a slew of cultural "entailments" (Nadasdy 2003) that have historically worked to undermine Indigenous governance. Indigenous sovereignty, when rooted in Indigenous epistemologies and philosophies, often rejects the commodification of land and people and offers an alternative to the world that has since been shaped by colonization. Recentering Indigenous sovereignty is integral to the practice of decolonization.

Decolonization seeks to rectify past colonial harms and imagine healthier futures by recentering and respecting Indigenous ways of being and reorganizing society according to Indigenous laws and philosophies that predate settlers and colonizers. Indigenous agency and decision-making should lead all aspects of life, including social, political, economic, and psychological. It works toward reinvigorating Indigenous lifeways, especially their sovereignty and relationships with their lands. It would stop the exploitation of land, a hallmark feature of colonialism. Decolonization is inherently violent, as Frantz Fanon has noted, since it requires a total dismantling of the systemic colonialism that has permeated all aspects of our lives, including the lands we live on, our psychological well-being, our worldviews, and our spiritual lives (Fanon 1961). Without decolonization, **environmental peacemaking** is not possible.

References

Dhillon, Jaskiran, ed. 2022. *Indigenous Resurgence: Decolonization and Movements for Environmental Justice*. Berghahn Books.

Estes, Nick. 2019. *Our History Is the Future: Standing Rock Versus the Dakota Access Pipeline, and the Long Tradition of Indigenous Resistance*. Verso.

Fanon, Frantz. 1961. *The Wretched of the Earth*. François Maspero.

Nadasdy, Paul. 2003. "Reevaluating the Co-Management Success Story." *Arctic* 56 (4): 367–80. https://doi.org/10.14430/arctic634.

Nixon, Rob. 2011. *Slow Violence and the Environmentalism of the Poor*. Harvard University Press.

Simpson, Audra. 2014. *Mohawk Interruptus: Political Life Across the Borders of Settler States*. Duke University Press.

Whyte, Kyle. 2018. "Settler Colonialism, Ecology, and Environmental Injustice." *Environment and Society* 9 (1): 125–44. https://doi.org/10.3167/ares.2018.090109.
Wolfe, Patrick. 2006. "Settler Colonialism and the Elimination of the Native." *Journal of Genocide Research* 8 (4): 387–409. https://doi.org/10.1080/14623520601056240.

Depoliticization

Ambreen Tour Ben-Shmuel, Radhika Singh, and Meryam El Bouhati

Depoliticization is the process of framing environmental issues and solutions as neutral, primarily technical, or administrative, while minimizing or excluding their political, social, and historical dimensions. In environmental peacebuilding, this approach can both facilitate initial cooperation and potentially undermine long-term conflict transformation.

Depoliticization is a concept emerging from international relations and political science that has gained increasing attention since the 1990s and early 2000s (Wood 2016). In the **environmental peacebuilding** field, it has multiple dimensions and applications.

In some instances, parties involved in environmental peacebuilding strategically employ depoliticization to enable **cooperation** and build trust, focusing on mutual educational, scientific, and technical efforts without addressing political disputes. This approach, which focuses on shared interests rather than political differences, can enable cooperation while reducing resistance toward these efforts (Ide and Tubi 2019). At times, this strategy is employed purposefully, and other times, it is used unintentionally by practitioners and organizations, creating entry points for joint problem-solving and environmental engagement.

However, depoliticization has important limitations. Actors using this strategy during **peacebuilding** processes often overlook asymmetrical **power** dynamics and neglect underlying structural inequalities contributing to **environmental degradation** and social and political **conflict** (Ide 2020). For instance, *discursively*, water scarcity might be framed as a purely technical rather than a political problem; *normatively*, universal **conservation** standards may take precedence over local sociopolitical contexts; or *technocratically*, scientific assessments may be prioritized while excluding traditional ecological knowledge or **Indigenous Knowledge** (Petiteville 2016). This approach can sideline discussions around grievances, feelings of injustice, and causes of underlying vulnerabilities (Aggestam et al. 2013). As a result, depoliticization can marginalize certain communities and generate local resistance to cooperative efforts, potentially undermining the long-term success of environmental peacebuilding initiatives (Ben-Shmuel and Halle 2023).

Additionally, the involvement of private actors and NGOs in peacebuilding has deepened the technocratic focus, resulting in what some scholars describe as a "depoliticized peace" where political processes are seen as too complex or contentious to address (Aggestam and Sundell 2016). Funding dynamics further reinforce this technocratic orientation. In conflict settings, local organizations often face pressure to adopt apolitical strategies to secure resources from external actors. As a result, discussions of underlying disputes may be discouraged, with an emphasis placed instead on technical solutions that sidestep historical or contemporary political responsibilities.

The concept of depoliticization in environmental peacebuilding remains contested, highlighting the field's struggle to address both immediate environmental concerns and deeper sociopolitical issues. While some view it as a necessary pragmatic approach in highly charged contexts, others argue for more explicitly political and **justice**-oriented frameworks. For practitioners, this suggests the need for adaptive approaches, leveraging technical aspects of cooperation while creating space for addressing underlying political issues. The varying impacts of depoliticization in different contexts remain an important area for further research and practice.

Acknowledgment

This work was supported by the Truman Institute and the Azrieli Foundation.

References

Aggestam, Karin, and Annika Sundell-Eklund. 2013. "Situating Water in Peacebuilding: Revisiting the Middle East Peace Process." *Water International* 39 (1): 10–22. https://doi.org/10.1080/02508060.2013.848313.

Aggestam, Karin, and Annika Sundell-Eklund. 2016. "Depoliticizing Water Conflict: Functional Peacebuilding in the Red Sea–Dead Sea Water Conveyance Project." *Hydrological Sciences Journal* 61 (7): 1302–12. https://doi.org/10.1080/02626667.2014.999778.

Ben-Shmuel, Ambreen, and Silja Halle. 2023. "Beyond Greenwashing: Prioritizing Environmental Justice in Conflict-Affected Settings." *Environment and Security* 1 (3–4): 209–18. https://doi.org/10.1177/27538796231186697.

Ide, Tobias. 2020. "The Dark Side of Environmental Peacebuilding." *World Development* 127: 104777. https://doi.org/10.1016/j.worlddev.2019.104777.

Ide, Tobias, and Amit Tubi. 2019. "Education and Environmental Peacebuilding: Insights from Three Projects in Israel and Palestine." *Annals of the American Association of Geographers* 110 (1): 1–17. https://doi.org/10.1080/24694452.2019.1613954.

Petiteville, Franck. 2016. "Les Organisations Internationales Dépolitisent-elles les Relations Internationales?" (Do International Organizations Depoliticize International Relations?) *Gouvernement et Action Publique* 5 (3): 113–29. https://doi.org/10.3917/gap.163.0113.

Wood, Matthew. 2016. "Politicisation, Depoliticisation and Anti-Politics: Towards a Multilevel Research Agenda." *Political Studies Review* 14 (4): 521–33. https://doi.org/10.1111/1478-9302.12074.

Diagnostic Framework

Larry A. Swatuk and Rachel Weaver

A diagnostic framework is a structured approach or system used to identify, analyze, and understand problems, issues, or conditions within a specific context. It provides a set of criteria, methods, and tools to help diagnose the root causes of problems, assess the situation, and guide decision-making for potential solutions.

Diagnostic frameworks are used across the social, physical, health, and management sciences. A diagnostic framework generally combines theoretical and empirical elements, often developing new concepts to more accurately frame the cause-and-effect relationship among variables, thereby heightening the likelihood of actions resulting in desired outcomes. Additionally, diagnostic frameworks can be adapted to different scales, from local to global contexts.

Diagnostic frameworks have diverse applications in environmental contexts, including in environmental peacebuilding, addressing challenges such as land-use **conflict**, **climate change adaptation**, and resource management (Moser and Ekstrom 2010; Magyera and Genskow 2013; Andersson et al. 2023). In the field of environmental peacebuilding, a great deal of attention is given to transboundary (meaning interstate) water **governance** and management (Conca and Dabelko 2002). Among the many frameworks and models is the **Global Environment Facility** (GEF) International Water Program's *Transboundary Diagnostic Analysis*. As described on the **Integrated Water Resources Management** (IWRM) Action Hub website,

> The TDA process allows to break down complex transboundary issues into smaller components for action. The main objectives of TDA are to: (1) Identify and prioritise the transboundary problems and challenges; (2) collect and analyze information on environmental impact and socio-economic dimension of each problem; (3) assess immediate, underlying and root causes of each problem, suggesting areas suffering from or under threat of **environmental degradation**.
>
> (Integrated Water Resources Management Action Hub n.d.)

A similar approach is seen in the *Toolkit on Monitoring and Evaluation of Environmental Peacebuilding*. The toolkit is shaped around an "intervention cycle," providing guidance on designing, monitoring, evaluating, and **learning** from environmental peacebuilding efforts across diverse contexts (Bruch and Woomer 2023). The toolkit integrates key components such as **conflict sensitivity**, participation and inclusion, and gender to ensure environmental peacebuilding initiatives protect the communities most vulnerable to climate change.

References

Andersson, Krister, Christina Boyes, and Kenneth Richards. 2023. "Diagnosing Environmental Problems and Their Possible Policy Responses: A Tool for Assessing Initial Options." *Elementa: Science of the Anthropocene* 11 (1): 00137. https://doi.org/10.1525/elementa.2022.00137.

Bruch, Carl, and Amanda Woomer. 2023. *Toolkit on Monitoring and Evaluation of Environmental Peacebuilding*. Environmental Law Institute, Environmental Peacebuilding Association, and United States Institute of Peace. https://m-and-e.environmentalpeacebuilding.org/toolkit/.

Conca, Ken, and Geoffrey D. Dabelko, eds. 2002. *Environmental Peacemaking*. Johns Hopkins University Press and Woodrow Wilson Center Press.

Ide, Tobias, Carl Bruch, Alexander Carius, et al. 2021. "The Past and Future(s) of Environmental Peacebuilding." *International Affairs* 97 (1): 1–16. https://doi.org/10.1093/ia/iiaa177.

Integrated Water Resources Management (IWRM) Action Hub. n.d. "Transboundary Diagnostic Analysis." https://iwrmactionhub.org/learn/iwrm-tools/transboundary-diagnostic-analysis.

Magyera, Kyle, and Kenneth Genskow. 2013. "Toward Integrated Wetland Conservation: A Diagnostic Framework." *Journal of Environmental Planning and Management* 56 (1): 121–39. https://doi.org/10.1080/09640568.2012.654631.

Moser, Susanne C., and Julia A. Ekstrom. 2010. "A Framework to Diagnose Barriers to Climate Change Adaptation." *Proceedings of the National Academy of Sciences* 107 (51): 22026–31. https://doi.org/10.1073/pnas.1007887107.

Digital Technologies

Ann J. Le

Digital technologies include all electronic tools used to create, find, organize, and apply available data to communicate, disseminate, analyze, and predict unknown information. They refer to both hardware and software—including devices, mobile phone-based and computer applications, as well as social media platforms (Hirblinger 2020; Mediation Support Unit and Centre for Humanitarian Dialogue 2019).

The term digital technologies is increasingly used within the **peacebuilding** and peacemaking community in place of "information and communication technologies," with the latter additionally encompassing some nondigital technologies. A momentum to further the potential use of digital technologies has taken shape within peacebuilding strategies to monitor, (re)construct, and maintain the **environment** and natural resources to foster inclusive decision-making and collaboration to facilitate **sustainable development** and **peace** in contested spaces (UNEP 2024). This narrative toward "techno-solutionism" posits digital technologies as promising tools to enable opportunities for equitable and sustainable outcomes within the peacebuilding process—including **infrastructure** reconstruction, economic development agendas, and environmental assessment and programming—to tackle intertwined challenges associated with **climate change** and violent conflicts around the globe. Proponents of digital technologies, for example, highlight the potential of spatial data infrastructures (SDIs) to offer broad access to objective environmental data— recognized as a step toward democratizing knowledge access, ensuring equitable resource management, and establishing **land tenure** rights (Döring et al. 2024; Hordiichuk et al. 2024). Supporters of digital technologies additionally point to the potential for modeling and spatial visualizations to shape data-driven and transparent decision-making in environmental management and peacebuilding processes; for example, hydrological modeling of rivers shared by countries in **conflict** as entry points for conversation and **mediation** (Döring et al. 2024). These sentiments presently shape the priorities of the United Nations, which characterizes technological solutions and inclusion as necessary to many of its Sustainable Development Goals and the broader peacebuilding agenda (Sætra 2024; Mikulaschek and Esberg 2021).

Conversely, the use of digital technologies has spurred heightened concerns over the potential escalation of surveillance, control, and persecution of vulnerable populations across the globe. Scholars, activists, and development practitioners have called attention to the potential ethical implications of applying digital technologies—most notably, the consequences of increased digital visibility for civilian groups in unstable peace contexts, such as forcibly displaced populations facing imminent risk of discrimination and persecution (Pelletier et al. 2024; Jacobsen and Landau 2003). Critics additionally point to the challenges of applying digital technologies in politically fragile or unstable contexts, citing barriers posed by low acceptance of such tools by conflicting parties and limits to data accessibility, accuracy, reliability, and data source discrepancies (Rwandarugali and Ngetar 2022). In addition to the **security** implications and challenges associated with digital technologies, critics of techno-solutionist sentiments question the motivations of Western states—whose monopoly over technological development has placed the Global North within the driver's seat in configuring imported and/or extractive solutions within the Global South (Sætra 2024). In line with these criticisms are concerns that the West's trickling-down approach to disseminating technological advances cannot be realized equitably but instead heightens the risk of reinforcing existing **power** hierarchies and exacerbating political repression (Sætra 2024; Verweijen and Hoffmann 2024).

References

Döring, Stefan, Kyungmee Kim, and Ashok Swain. 2024. "Integrating Socio-Hydrology, and Peace and Conflict Research." *Journal of Hydrology* 633: 131000. https://doi.org/10.1016/j.jhydrol.2024.131000.

Esberg, Jane, and Christoph Mikulaschek. 2021. *Digital Technologies, Peace and Security: Challenges and Opportunities for United Nations Peace Operations*. United Nations Department of Peace Operations. https://peacekeeping.un.org/sites/default/files/esberg_and_mikulaschek_-_conflict_peace_and_digital_technologies_-_v3_210825.pdf.

Hirblinger, Andreas T. 2020. *Digital Inclusion in Mediated Peace Processes: How Technology Can Enhance Participation*. United States Institute of Peace. https://www.usip.org/publications/2020/09/digital-inclusion-mediated-peace-processes-how-technology-can-enhance.

Hordiichuk, Valerii, Petro Snitsarenko, Nina Andriianova, Bohdan Molodetskyi, and Yevhenii Kapran. 2024. "Advanced Information Technologies (C4ISR) in the Peacekeeping Field Service. Advanced." *Advanced Information Systems* 8 (4): 93–102. https://doi.org/10.20998/2522-9052.2024.4.12.

Jacobsen, Karen, and Loren B. Landau. 2003. "The Dual Imperative in Refugee Research: Some Methodological and Ethical Considerations in Social Science Research on Forced Migration." *Disasters* 27 (3): 185–206. https://doi.org/10.1111/1467-7717.00228.

Mediation Support Unit and Centre for Humanitarian Dialogue. 2019. *Digital Technologies and Mediation in Armed Conflict*. United Nations Department of Political and Peacebuilding Affairs (DPPA) and Centre for Humanitarian Dialogue (HD). https://www.hdcentre.org/wp-content/uploads/2019/07/Digital-technologies-and-mediation-in-armed-conflict.pdf.

Pelletier, Alexandre, Laddawan Kaewkitipong, and Matthieu J. Guitton. 2024. "Using Technology to Study Refugee, Conflict-Affected, and Hard-to-Reach Populations: Methodological and Ethical Considerations." *Computers in Human Behavior* 152: 108053. https://doi.org/10.1016/j.chb.2023.108053.

Rwandarugali, Stanislas, and Njoya Ngeta. 2022. "The Application of Geographical Information Systems to Armed Violent Conflict Resolution and Peacebuilding: A Literature Review." *South African Journal of Geomatics* 11 (2): 234–46. https://doi.org/10.4314/sajg.v11i2.5.

Sætra, Henrik Skaug, and Evan Selinger. 2024. "Technological Remedies for Social Problems: Defining and Demarcating Techno-Fixes and Techno-Solutionism." *Science and Engineering Ethics* 30 (60). https://doi.org/10.1007/s11948-024-00524-x.

United Nations Environment Programme (UNEP) and the Environmental Peacebuilding Association. 2024. *Digital Technologies for Environmental Peacebuilding: Horizon Scanning of Opportunities and Risks*. Nairobi. https://tech.environmentalpeacebuilding.org/dtep.

Verweijen, Judith, and Kasper Hoffmann. 2024. "Dangerous Environments: Environmental Peacebuilding's Technomoral Imaginary and Its Power-Knowledge Effects." *Ecology and Society* 29 (3). https://doi.org/10.5751/ES-15391-290323.

Disaster Diplomacy

Laura E. R. Peters, Ilan Kelman, and Vincent Otieno

Disaster diplomacy is a field of study examining how and why disaster-related activities—before, during, and after disasters—may or may not affect peace and conflict (Kelman

2012). Disasters are highly disruptive situations at any scale, often requiring outside support for coping (Kelman 2020).

Disaster diplomacy relies on the distinction between hazards and disasters. Hazards are natural occurrences, though human decisions and activities may increase the negative impact of any hazard. Disasters occur when a hazard causes negative consequences for people, livelihoods, or ecosystems. Therefore, the field of disaster diplomacy avoids the phrase "natural disaster," since disasters are caused by human decisions rather than by nature, even if the initial hazard involves environmental phenomena such as rainfall, earthquakes, or viruses. In other instances, disasters are entirely societal or technological, such as nuclear power plant failures, wars, and riots. The field of disaster diplomacy concludes that disasters and disaster-related actions are unlikely to spur the creation of new, lasting **peace** processes, but they may be leveraged to temporarily pause **violence** and provide opportunities to support formal and informal peace processes already underway (Kelman 2012). Such actions are not confined to the post-disaster space, but may occur at any stage of disaster prevention, planning, preparation, preparedness, **mitigation**, management, relief, response, recovery, and reconstruction. Furthermore, these actions can be undertaken through actors in both formal and informal channels, including the interaction between those channels (Duda and Kelman 2023).

Disaster diplomacy emphasizes the central importance of taking intentional actions to maximize **peacebuilding** potential through environmental, political, and social initiatives, such as addressing disaster inequities and engaging in mutual aid (Peters 2022). This approach is termed "active disaster diplomacy," in contrast to "passive disaster diplomacy," in which situations are permitted to unfold regarding peace and **conflict**, rather than trying to pursue preferred pathways and avert undesired ones. Many kinds of approaches—whether active or passive, formal or informal—are employed in disaster diplomacy, including different types of approaches being used simultaneously. These findings align with the central argument of the field of **environmental peacebuilding** that managing environmental issues can be done through approaches and pathways that actively contribute to **conflict prevention**, resolution, and transformation (Ide et al. 2021).

The connections between disasters and various forms of violent conflict have been a topic of concern in academia, especially since the 1990s. The 2000s saw the emergence of interest in connections between disasters and peace, as experiences began to amass of disasters defying disaster-conflict logics (Peters and Kelman 2020). Disaster diplomacy coalesced formally as a term and a field in 2000 (Kelman and Koukis 2000), although there were earlier uses of the phrase and earlier examinations of the idea (Quarantelli and Dynes 1976). A recent systematic review of empirical literature on disaster-conflict linkages found that nearly a quarter of the studies reviewed argued that disasters decrease conflict risks (Canavan and Ide 2024). In a qualitative comparative analysis of 36 case studies, Ide (2023) found that a quarter of the cases experienced armed conflict de-escalation related to a major disaster.

The field of disaster diplomacy, since its conception, has featured case studies with counterexamples on the assumed relationship between disasters and conflict. Several studies argued the need for the warming of political relations as a precondition for effective disaster-related **cooperation** and coordination. One example is between the governments of the United States and Cuba during hurricane season (Glantz 2000). More recently, some evidence emerged that cooperation over disasters in Svalbard impacted broader international relations between Russia and Norway (Kelman et al. 2020). These explorations have considered various forms of conflict and peace, from the individual level to international relations. These findings expose limitations in understanding the impacts of disaster diplomacy, including with ill-defined and poorly delineated words and concepts with peace and conflict (i.e., peace-focused fields including terms like cooperation and rapprochement, as compared to conflict-focused fields using terms like **war** and terrorism and not wishing to communicate or have formal ties).

The 2004 Indian Ocean Tsunami tested disaster diplomacy and its various forms. Studies focused on the disaster catalyzing immediate ceasefires in both Aceh, Indonesia, and Sri Lanka following decades of civil war in each place, but then leading to divergent longer-term outcomes, with a comprehensive peace agreement in Aceh and renewed violence in Sri Lanka (e.g., Scanlon 2006; Le Billon and Waizenegger 2007; Enia 2008; Gaillard et al. 2008; Klitzsch 2014).

Overall, disaster diplomacy research reveals the potential for disasters to open windows of opportunity for peacebuilding, though peace outcomes are not inevitable. This highlights the need for more work to develop approaches that integrate **disaster risk reduction (DRR)** and disaster response with peacebuilding efforts over the long term, rather than solely and reactively mitigating the factors that contribute to conflict and instability (Peters 2024).

References

Canavan, Chloe, and Tobias Ide. 2024. "Contention, Cooperation, and Context: A Systematic Review of Research on Disasters and Political Conflicts." *International Journal of Disaster Risk Reduction* 108: 104558. https://doi.org/10.1016/j.ijdrr.2024.104558.

Duda, Patrizia I., and Ilan Kelman. 2023. "Informal Disaster Diplomacy." *Societies* 13 (1): 8. https://doi.org/10.3390/soc13010008.

Enia, Jason S. 2008. "Peace in its Wake? The 2004 Tsunami and Internal Conflict in Indonesia and Sri Lanka." *Journal of Public and International Affairs* 19 (1): 7–27. https://jpia.princeton.edu/document/384.

Gaillard, Jean-Christophe, Elsa Clavé, and Ilan Kelman. 2008. "Wave of Peace? Tsunami Disaster Diplomacy in Aceh, Indonesia." *Geoforum* 39 (1): 511–26. https://doi.org/10.1016/j.geoforum.2007.10.010.

Glantz, Michael H. 2000. "Climate-Related Disaster Diplomacy: A US–Cuban Case Study." *Cambridge Review of International Affairs* 14 (1): 233–53. https://doi.org/10.1080/09557570008400340.

Ide, Tobias. 2023. *Catastrophes, Confrontations, and Constraints: How Disasters Shape the Dynamics of Armed Conflicts*. The MIT Press.

Ide, Tobias, Carl Bruch, Alexander Carius, et al. 2021. "The Past and Future(s) of Environmental Peacebuilding." *International Affairs* 97 (1): 1–16. https://doi.org/10.1093/ia/iiaa177.

Kelman, Ilan. 2012. *Disaster Diplomacy: How Disasters Affect Peace and Conflict*. Routledge. https://doi.org/10.4324/9780203806210.

Kelman, Ilan. 2020. *Disaster by Choice: How Our Actions Turn Natural Hazards into Catastrophes*. Oxford University Press.

Kelman, Ilan, and Theodore Koukis, eds. 2000. "Introduction." Special section in: Charlotte Lindberg Clausen (ed.), *Cambridge Review of International Affairs* 14 (1): 214. https://doi.org/10.1080/09557570008400338.

Kelman, Ilan, Are Kristoffer Sydnes, Patrizia Isabelle Duda, Elena Nikitina, and Christian Webersik. 2020. "Norway-Russia Disaster Diplomacy for Svalbard." *Safety Science* 130: 104896. https://doi.org/10.1016/j.ssci.2020.104896.

Klitzsch, Nicole. 2014. "Disaster Politics or Disaster of Politics? Post-Tsunami Conflict Transformation in Sri Lanka and Aceh, Indonesia." *Cooperation and Conflict* 49 (4): 554–73. http://dx.doi.org/10.1177/0010836714545692.

Le Billon, Philippe, and Arno Waizenegger. 2007. "Peace in the Wake of Disaster? Secessionist Conflicts and the 2004 Indian Ocean Tsunami." *Transactions of the Institute of British Geographers* 32 (3): 411–27. https://doi.org/10.1111/j.1475-5661.2007.00257.x.

Peters, Laura E. R. 2022. "Disasters as Ambivalent Multipliers: Influencing the Pathways from Disaster to Conflict Risk and Peace Potential Through Disaster Risk Reduction." *Journal of Peacebuilding and Development* 17 (2): 151–72. https://doi.org/10.1177/15423166221081516.

Peters, Laura E. R. 2024. "The Peace Imperative for the Sendai Framework for Disaster Risk Reduction." *International Journal of Disaster Risk Science* 16: 9–15. https://doi.org/10.1007/s13753-024-00596-0.

Peters, Laura E. R., and Ilan Kelman. 2020. "Critiquing and Joining Intersections of Disaster, Conflict, and Peace research." *International Journal of Disaster Risk Science* 11: 555–67. https://doi.org/10.1007/s13753-020-00289-4.

Quarantelli, Enrico L., and Russell R. Dynes. 1976. "Community Conflict: Its Absence and Presence in Natural Disasters." *Mass Emergencies* 1: 139–52. https://cir.nii.ac.jp/crid/1370568470456863872.

Scanlon, Joseph. 2006. "Dealing with the Tsunami Dead: Unprecedented International Co-operation." *The Australian Journal of Emergency Management* 21 (2): 57–61. https://search.informit.org/doi/10.3316/informit.158211966166541.

Disaster Risk Reduction

Chloe Canavan and Laura E. R. Peters

Disaster risk reduction (DRR) is a systematic approach to prevent and reduce new or pre-existing disaster risks while mitigating residual risks, contributing to resilience, and laying the foundation for sustainable development (United Nations Office for Disaster Risk Reduction [UNDRR] 2016).

At the core of disaster risk reduction (DRR) is the understanding that disasters are not natural but result from hazards interacting with long-term and dynamic vulnerability processes (Wisner et al. 2004). DRR seeks to systematically address all components of the disaster risk equation, including hazard **mitigation**, exposure reduction, vulnerability reduction, and capacity building. The current global policy framework, the *Sendai Framework for Disaster Risk Reduction 2015-2030*, and its predecessor, *The Hyogo Framework for Action 2005-2015: Building the Resilience of Nations and Communities to Disasters*, have provided the foundation for global DRR efforts to prevent and mitigate disasters rather than focus solely on response. Global DRR planning and action have emphasized risk-informed **governance**, community **participation**, partnerships, and the integration of climate **adaptation** in disaster planning (UNDDR 2005, 2015; Wisner et al. 2012).

The field of DRR has long engaged with concepts of **conflict** and **violence** relevant to **environmental peacebuilding**. Much research has focused on how the occurrence of disasters contributes to the causes and dynamics of conflict (Ide 2023). Others have pointed to a mutually influencing relationship between disaster and conflict risks, aligning with DRR's recent focus on complex, compound, and cascading risks.

For example, the influential Pressure and Release Model by Wisner et al. (2004) articulated that armed conflict and violence contribute to vulnerability and risk. Building on this, recent research has demonstrated that conflict and violence impact all components of disaster risk (Peters 2021), making disasters more likely to occur in conflict-affected contexts (Caso et al. 2023). Global DRR policy and practice have increasingly embraced the need to adapt strategies for DRR in contexts of **fragility**, conflict, and violence despite their absence from the Sendai Framework (Peters et al. 2019a, 2019b). Other advancements include certain types of conflict and violence as hazards in the International Science Council-UNDRR Hazard Information Profiles (HIPs) (UNDRR and International Science Council 2021). The commonly used INFORM risk index includes conflict in the human dimensions of hazard and exposure (Joint Research Centre of the European Commission 2024).

The field of DRR has not engaged as explicitly with issues around **peace** and **peacebuilding** (Peters 2024). However, **disaster diplomacy**—the study of how disaster-related actions influence relations between and within countries—provides a critical lens to evaluate DRR's role in fostering **cooperation** or exacerbating tensions. Scholars have identified the linkages between **sustainability** and peace (Virji et al. 2019), with DRR playing a key role in this complex relationship (Peters and Peters 2021).

DRR presents pathways to environmental peacebuilding by engaging the interconnectedness between disasters, environmental management, conflict resolution, and peacebuilding. For example, environmental management that mitigates disaster risks can be integrated into peacebuilding and conflict resolution processes (Ide 2017), and peacebuilding efforts between communities can be pursued through DRR activities (Peters 2022). These connections become especially relevant in post-conflict and disaster-affected regions, where environmental resources and disaster **resilience** can shape recovery pathways (Peters et al. 2019b; Peters 2021). In fragile areas, DRR initiatives can reduce the likelihood of resource-driven conflict by fostering cooperation over shared vulnerabilities in the context of water scarcity or flood risks (Döring 2020; Ide and Detge 2018). Possible pathways for DRR to contribute to environmental peacebuilding include **conflict sensitivity** in DRR strategies, cooperative resource management and sharing, and inclusive governance, capacity building, and resilience (Ide et al. 2021; Peters 2022; Sändig et al. 2024).

References

Caso, Nicolás, Dorothea Hilhorst, and Rodrigo Mena. 2023. "The Contribution of Armed Conflict to Vulnerability to Disaster: Empirical Evidence From 1989 to 2018." *International Journal of Disaster Risk Reduction* 95: 103881. https://doi.org/10.1016/j.ijdrr.2023.103881.

Döring, Stefan. 2020. "From Bullets to Boreholes: A Disaggregated Analysis of Domestic Water Cooperation in Drought-prone Regions." *Global Environmental Change* 65: 102147. https://doi.org/10.1016/j.gloenvcha.2020.102147.

Ide, Tobias. 2017. "Space, Discourse and Environmental Peacebuilding." *Third World Quarterly* 38 (3): 544–62. https://doi.org/10.1080/01436597.2016.1199261.

Ide, Tobias. 2023. *Catastrophes, Confrontations, and Constraints: How Disasters Shape the Dynamics of Armed Conflicts*. The MIT Press. https://doi.org/10.7551/mitpress/14970.001.0001.

Ide, Tobias, Carl Bruch, Alexander Carius, et al. 2021. "The Past and Future(s) of Environmental Peacebuilding." *International Affairs* 97 (1): 1–16. https://doi.org/10.1093/ia/iiaa177.

Ide, Tobias, and Alexander Detges. 2018. "International Water Cooperation and Environmental Peacemaking." *Global Environmental Politics* 18 (4): 63–84. https://doi.org/10.1162/glep_a_00478.

Joint Research Centre of the European Commission. 2024. *Inform Report 2024–10 Years of Inform–Shared Evidence for Managing Crises and Disasters*. Publications Office of the European Union. https://data.europa.eu/doi/10.2760/555548.

Peters, Katie, Katherine Holloway, and Laura E. R. Peters. 2019a. *Disaster Risk Reduction in Conflict Contexts: An Agenda for Action*. The German Federal Ministry of Economic Cooperation and Development, Deutsche Gesellschaft für Internationale Zusammenarbeit, and the Overseas Development Institute, September. https://cdn.odi.org/media/documents/12910.pdf.

Peters, Katie, and Laura E. R. Peters. 2021. "Terra Incognita: The Contribution of Disaster Risk Reduction in Unpacking the Sustainability–Peace Nexus." *Sustainability Science* 16 (4): 1173–84. https://doi.org/10.1007/s11625-021-00944-9.

Peters, Katie, Laura E. R. Peters, John Twigg, and Colin Walch. 2019b. *Disaster Risk Reduction Strategies: Navigating Conflict Contexts*. German Federal Ministry of Economic Cooperation and Development (BMZ), Deutsche Gesellschaft für Internationale Zusammenarbeit (GIZ) and the Overseas Development Institute (ODI), May. https://media.odi.org/documents/12690.pdf.

Peters, Laura E. R. 2021. "Beyond Disaster Vulnerabilities: An Empirical Investigation of the Causal Pathways Linking Conflict to Disaster Risks." *International Journal of Disaster Risk Reduction* 55. https://doi.org/10.1016/j.ijdrr.2021.102092.

Peters, Laura E. R. 2022. "Disasters as Ambivalent Multipliers: Influencing the Pathways from Disaster to Conflict Risk and Peace Potential Through Disaster Risk Reduction." *Journal of Peacebuilding and Development* 17 (2): 151–72. https://doi.org/10.1177/15423166221081516.

Peters, Laura E. R. 2024. "The Peace Imperative for the Sendai Framework for Disaster Risk Reduction." *International Journal of Disaster Risk Science* 16: 9–19. https://doi.org/10.1007/s13753-024-00596-0.

Sändig, Jan, Natalia Dalmer, Tobias Ide, and Anselm Vogler. 2024. "From Climate Conflicts to Environmental Peacebuilding: Exploring Local Dimensions." *Environment and Security* 2 (1): 3–20. https://doi.org/10.1177/27538796241231090.

United Nations Office for Disaster Risk Reduction (UNDRR). 2005. *Hyogo Framework for Action 2005–2015: Building the Resilience of Nations and Communities to Disasters*. https://www.unisdr.org/files/1037_hyogoframeworkforactionenglish.pdf.

United Nations Office for Disaster Risk Reduction (UNDRR). 2016. *Report of the Open-ended Intergovernmental Expert Working Group on Indicators and Terminology Relating to Disaster Risk Reduction*. https://www.preventionweb.net/files/50683_oiewgreportenglish.pdf.

United Nations Office for Disaster Risk Reduction (UNDRR). 2015. *Sendai Framework for Disaster Risk Reduction 2015–2030*. https://www.undrr.org/publication/sendai-framework-disaster-risk-reduction-2015-2030.

United Nations Office for Disaster Risk Reduction (UNDRR) and International Science Council (ISC). 2021. *Hazard Information Profiles: Supplement to UNDRR-ISC Hazard Definition and Classification Review: Technical Report*. UNDRR. https://www.undrr.org/media/73913/download?startDownload=20250117.

Virji, Hassan, Ayyoob Sharifi, Shinji Kaneko, and Dahlia Simangan. 2019. "The Sustainability–Peace Nexus in the Context of Global Change." *Sustainability Science* 14 (6): 1467–8. https://doi.org/10.1007/s11625-019-00737-1.

Wisner, Ben, Piers Blaikie, Terry Cannon, and Ian Davis. 2004. *At Risk: Natural Hazards*. Routledge.

Wisner, Ben, J. C. Gaillard, and Ilan Kelman. 2012. *The Routledge Handbook of Hazards and Disaster Risk Reduction*. Routledge.

Displacement

Adam Savelli and Richard Matthew

Displacement is the movement of persons who have been forced or obliged to flee or to leave their homes or places of habitual residence, in particular because of or to avoid the effects of armed conflict, situations of generalized violence, violations of human rights, or natural or human-made disasters (International Organization for Migration 2019). It may also refer to the involuntary or forceful movement of other species (Hsiao et al. 2024).

Individuals forced to cross an international border "owing to well-founded fear of being persecuted for reasons of race, religion, nationality, membership of a particular social group or political opinion" and unable to return to their country of origin are considered refugees under international law (United Nations High Commissioner for Refugees [UNHCR] n.d.). Notably, people who cross an international border to avoid the negative impacts of climate or environmental hazards, a growing number (Rigaud et al. 2021), are not included in this legal definition, and thus, are ineligible for legal protections afforded to refugees (Behrman and Kent 2018; Melde 2014). Displacement can be either internal or international. Globally, most displacement occurs within national borders: as of mid-2024, approximately 122.6 million individuals were displaced worldwide, including 68.3 million internally displaced persons (IDPs), 37.9 million refugees, 8 million asylum seekers, and 5.8 million others in need of international protection (UNHCR 2024). Between 2013 and 2023, 400 million individuals were displaced internally (260 million or 65% due to disasters, and 141 million or 35% due to **conflict** and **violence**) (Internal Displacement Monitoring Centre Data Portal n.d.). While many of those displaced internally eventually return to their homes, the average time spent in displacement is growing more protracted: 20 years for refugees and 10 years for IDPs (European Commission n.d.).

Many living in displacement face legal precarity, socioeconomic marginalization, and exposure to conflict and environmental hazards. Despite international law, many refugees struggle to safely access international destinations or legal protections once there. Global North destinations such as the United States and Europe have taken steps to "externalize" their borders by incentivizing the transit countries that asylum seekers travel through to stop them while *en route* (Stock and Schultz 2019). As a result of this, as well as simple geographic proximity to origin areas, most refugees and internally displaced persons are hosted by states in the Global South that possess limited resources to support well-being. Typically, few opportunities are available for displaced populations to engage in formal employment, own land, or access public services. This restricts their ability to exercise agency and constrains overall economic opportunity, which decreases their ability to withstand future shocks.

While displacement rarely leads directly to conflict—and often leads to collaboration between host and displaced communities—it can inflame tensions related to social, political, economic, or environmental challenges (Raleigh et al. 2008; Burrows and Kinney 2016; Morales-Muñoz et al. 2020). For example, competition over economic resources or public services in destination areas can lead to grievances that may evolve into verbal or physical attacks, particularly in destination areas with a distinct ethnic identity or pre-existing resource scarcity (Savelli et al. 2023). State or private development initiatives can also force community members from their homes and lead to conflict, as land customarily belonging to communities is seized for other uses (Tan 2020).

Although displacement normally refers to physical displacement from a geographic area, the concept of in-situ displacement is gaining attention as community members who remain in place become dislocated from the social, economic, and ecological systems they have customarily enjoyed access to and relied on (Halstead 2020).

In terms of **environmental peacebuilding**, there are many opportunities to address the challenges of displacement in ways that build **resilience**. For example, greater collaboration and information sharing among humanitarian, migration, and **conservation** communities is important, though difficult to achieve under conditions of scarce resources and narrowly focused mandates. Improving natural resource management and climate resilience across the displacement arc is essential to support those who are trapped in places where conflict and environmental stress intersect, those in transit searching for settlement opportunities who need resources to survive, and those who add (or are perceived to be adding) pressure on scarce resources at a destination point (Biella-Battista 2019; Hsiao et al. 2024). Additionally, bolstering international recognition of environmental displacement alone and in conjunction with other variables is needed to provide a legal basis for assistance (Berchin et al. 2017; Chazalnoel and Ionesco 2018; McClain et al. 2022; Wood 2020). And finally, better predictive and modeling tools are needed so environmental peacebuilding can deploy assets efficiently.

References

Behrman, Simon, and Avidan Kent, eds. 2018. *Climate Refugees: Beyond the Legal Impasse?* Routledge. https://doi.org/10.4324/9781315109619.

Berchin, Issa I., Isabela B. Valduga, Jéssica Garcia, and José B. Salgueirinho, and Osorio de Andrade Guerra. 2017. "Climate Change and Forced Migrations: An Effort Towards Recognizing Climate Refugees." *Geoforum* 84: 147–50. https://doi.org/10.1016/j.geoforum.2017.06.022.

Biella-Battista, Roberto, Georg Bolits, Hanna Mark, and Andrea Salvini. 2019. *Reintegration Handbook–Practical Guidance on the Design, Implementation and Monitoring of Reintegration Assistance*. International Organization for Migration (IOM). https://publications.iom.int/system/files/pdf/iom_reintegration_handbook.pdf.

Burrows, Kate, and Kinney, Patrick. 2016. "Exploring the Climate Change, Migration and Conflict Nexus." *International Journal of Environmental Research and Public Health* 13 (4): 443. https://dx.doi.org/10.3390%2Fijerph13040443.

Chazalnoel, Mariam Traore, and Dina Ionesco. 2018. "Advancing the Global Governance of Climate Migration through the United Nations Framework Convention on Climate Change and the Global Compact on Migration Perspectives from the International Organization for Migration." In *Climate Refugees: Global, Local and Critical Approaches*, edited by Simon Behrman and Avidan Kent. Cambridge University Press. https://doi.org/10.4324/9781315109619.

European Commission. n.d. *Forced Displacement*. Civil Protection and Humanitarian Aid Operations. Accessed October 9, 2025. https://civilprotectionhumanitarianaid.ec.europa.eu/what/humanitarian-aid/forced-displacement_en#:~:text=Forced%20displacement%20is%20no%20longer,10%20years%20for%20most%20IDPs.

Halstead, Huw. 2020. "Reclaiming the Land: Belonging, Landscape, and In Situ Displacement on the Plain of Karditsa (Greece)." *History and Anthropology* 31 (5): 643–68. https://doi.org/10.1080/02757206.2019.1696325.

Hsiao, Elaine, Richard Matthew, Philippe Le Billon, and Galeo Saintz, eds. 2024. *Planet on the Move – Reimagining Conservation at the Intersection of Migration, Environmental Change, and Conflict*. International Union for Conservation of Nature.

Internal Displacement Monitoring Centre (IDMC) Data Portal. n.d. *Displacement Data*. IDMC. Accessed October 9, 2025. https://www.internal-displacement.org/database/displacement-data/.

International Organization for Migration (IOM). 2019. *Glossary on Migration*. IOM. https://publications.iom.int/system/files/pdf/iml_34_glossary.pdf.

McClain, Shanna N., Carl Bruch, Erin Daly, et al. 2022. "Migration With Dignity: A Legal and Policy Framework." *Journal of Disaster Research* 17 (3): 292–300. https://doi.org/10.20965/jdr.2022.p0292.

Melde, Susanne. 2014. *Migration, Environment and Climate Change: Evidence for Policy (MECLEP): Glossary*. International Organization for Migration, July. https://www.iom.int/sites/g/files/tmzbdl486/files/migrated_files/Country/docs/MECLEP-Glossary-EN.pdf.

Morales-Muñoz, Héctor, Srijna Jha, Michelle Bonatti, Henryk Alff, Sabine Kurtenbach, and Stefan Sieber. 2020. "Exploring Connections—Environmental Change, Food Security and Violence as Drivers of Migration—A Critical Review of Research." *Sustainability* 12 (14): 5702. https://doi.org/10.3390/su12145702.

Raleigh, Clionadh, Lisa Jordan, and Idean Salehyan. 2008. *Assessing the Impact of Climate Change on Migration and Conflict*. World Bank, January. https://environmentalmigration.iom.int/sites/g/files/tmzbdl1411/files/documents/2023-10/assessing_the_impact_of_climate_change_on_migratio.pdf.

Rigaud, Kanta, Alex de Sherbinin, Bryan Jones, Anna Taeko Casals Fernandez, and Susana Adamo. 2021. *Groundswell Africa: A Deep Dive into Internal Climate Migration in the Lake Victoria Basin Countries*. World Bank. http://hlrn.org/img/violation/Groundswell-Africa-A-Deep-Dive-Into-Internal-Climate-Migration-in-Uganda.pdf.

Savelli, Adam, Frans Schapendonk, Tanaya Dutta Gupta, Grazia Pacillo, and Peter Läderach. 2023. "Climate Change, Mobility and Violent Conflict: A Typology of Interlinked Pathways." *International Development Planning Review* 45 (4): 403–36. https://doi.org/10.3828/idpr.2023.2.

Stock, Inka, Ayşen Üstübici, and Susanne U. Schultz. 2019. "Externalization at Work: Responses to Migration Policies from the Global South." *Comparative Migration Studies* 7 (48): 1–9. https://doi.org/10.1186/s40878-019-0157-z.

Tan, Yan. 2020. "Development-Induced Displacement and Resettlement: An Overview of Issues and Interventions." In *Routledge Handbook of Migration and Development*, edited by Tanja Bastia and Ronald Skeldon. Routledge.

United Nations High Commissioner for Refugees (UNHCR). n.d. *The 1951 Refugee Convention and 1967 Protocol Relating to the Status of Refugees*. UNHCR. https://www.unhcr.org/media/1951-refugee-convention-and-1967-protocol-relating-status-refugees.

United Nations High Commissioner for Refugees (UNHCR). 2024. *Refugee Data Finder*. UNHCR. Accessed June 12, 2025. https://www.unhcr.org/refugee-statistics.

Wood, Tamara. 2020. "New Pact Paves Way for Innovative Solutions to Disaster and Climate Change Displacement in Africa." *Platform on Disaster Displacement*, March 2. https:// disasterdisplacement.org/staff-member/new-pact-paves-way-for-innovative-solutions-to-disaster-and-climate-change-displacement-in-africa.

E

Earth Observation

Ariel Ozdemir, Marie Schellens, and Emnet Negash

Earth observation is the collection and analysis of data about the Earth's physical, chemical, and biological systems through remote sensing technologies.

Earth observation (EO) uses satellites, aerial sensors on drones or aircraft, and ground-based observations to monitor and assess various aspects of the Earth's **environment** and resources. EO analyzes metrics such as land surface temperature, vegetation indices, and changes in land cover over time. It derives those metrics from capturing and processing electromagnetic radiation, such as visible light, infrared, or microwave signals, reflected or emitted from the Earth's surface. These data points are then processed and analyzed to understand various physical characteristics of the land, water, or atmosphere (Kramer 2002; Awange and Kiema 2019). Beginning with aerial imaging in the early 20th century and rudimentary radar systems to observe military activities during the First and Second World Wars, EO now integrates Geographic Information Systems (GIS), machine **learning**, and ground-truthing methods to identify and analyze environmental patterns and changes (European Space Agency n.d.; Read and Torrado 2009). While often used interchangeably with the term "**remote sensing**," the former refers specifically to the study of the Earth, while the latter refers to the technical approach of studying an object, area, or phenomenon, including but not limited to the Earth.

EO allows for the **monitoring** of **natural resources**, including water bodies, vegetation, and **forests**. It is often applied in agricultural monitoring, **biodiversity** studies, disaster management, urban planning, climate research, **conservation**, water resource management, and environmental compliance (Committee on Earth Observation Satellites 2023). In **conflict**-affected regions, EO allows for the monitoring of hard-to-reach and dangerous zones. Since the 1990s, there have been various applications of EO to conflict contexts (Witmer 2015). In the past decade, increased technical capacities related to more and newer satellites, better data processing, and open-source sharing of EO data have attracted an increasing number of environmental and conflict researchers to this research technique (Kaplan et al. 2022) and to conflict-related environmental damages and degradation. Recent studies include forest disturbance in Tigray and Colombia (Murillo-Sandoval et al. 2020; Negash et al. 2023; Schulte to Bühne et al. 2024), environmental exposure in refugee settlements (Owen

DOI: 10.4324/9781003604143-7

et al. 2023), land abandonment in Syria (Eklund et al. 2022), the detection of craters in Ukraine (Duncan et al. 2023), and the catastrophic damage in Gaza (Holail et al. 2024; Yin et al. 2025). As such, EO provides insights into environmental, resource, and climate-related conflict instigators toward **conflict prevention**. It can assess and monitor environmental conditions to provide humanitarian aid in an environmentally sensitive, sustainable way. EO can also support post-conflict **environmental peacebuilding**, including reconstruction and recovery, environmental **governance** strategies, holding perpetrators of conflict-related environmental damages **accountable**, and supporting restorative **justice** practices (Rapach et al. 2024).

Two key limitations of EO in conflict-affected contexts are (1) the lack of options for ground-truthing interpretations of EO data and (2) the lack of ethical considerations around the potentially harmful impacts of EO data generation, analysis, and publication on already vulnerable communities and ecosystems. The complexities and sensitivities of conflict and environmental contexts may unintentionally expose sensitive information that can be exploited by military, criminal, commercial, armed, or governmental actors, resulting in severe ecological damage and intensified human suffering (PAX forthcoming).

References

Awange, Joseph, and John Kiema. 2019. "Fundamentals of Remote Sensing." In *Environmental Geoinformatics: Extreme Hydro-Climatic and Food Security Challenges: Exploiting the Big Data*. Springer. https://doi.org/10.1007/978-3-030-03017-9_7.

Committee on Earth Observation Satellites (CEOS). 2023. *Earth Observation Handbook: Space Data for the Global Stocktake*. European Space Agency. https://eohandbook.com/gst/.

Duncan, Eric C., Sergii Skakun, Ankit Kariryaa, and Alexander V. Prishchepov. 2023. "Detection and Mapping of Artillery Craters with Very High Spatial Resolution Satellite Imagery and Deep Learning." *Science of Remote Sensing* 7: 100092. https://doi.org/10.1016/j.srs.2023.100092.

Eklund, Lina, Ole Magnus Theisen, Matthias Baumann, Andreas Forø Tollefsen, Tobias Kuemmerle, and Jonas Østergaard Nielsen. 2022. "Societal Drought Vulnerability and the Syrian Climate-Conflict Nexus are Better Explained by Agriculture than Meteorology." *Communications Earth and Environment* 3 (1): 85. https://doi.org/10.1038/s43247-022-00405-w.

European Space Agency (ESA). n.d. *History of Earth Observation*. ESA Eduspace, Accessed October 26. https://www.esa.int/SPECIALS/Eduspace_EN/SEM1NP3Z2OF_0.html.

Holail, Shimaa, Tamer Saleh, Xiongwu Xiao, et al. 2024. "Time-Series Satellite Remote Sensing Reveals Gradually Increasing War Damage in the Gaza Strip." *National Science Review* 11 (9): nwae304. https://doi.org/10.1093/nsr/nwae304.

Kaplan, Gordana, Tatjana Rashid, Mateo Gasparovic, Andrea Pietrelli, and Vincenzo Ferrara. 2022. "Monitoring War-Generated Environmental Security using Remote Sensing: A Review." *Land Degradation and Development* 33 (10): 1513–26. https://doi.org/10.1002/ldr.4249.

Kramer, Herbert J. 2002. *Observation of the Earth and Its Environment: Survey of Missions and Sensors*. Springer. https://doi.org/10.1007/978-3-642-56294-5.

Murillo-Sandoval, Paulo J., Kristina Van Dexter, Jamon Van Den Hoek, David Wrathall, and Robert Kennedy. 2020. "The End of Gunpoint Conservation: Forest Disturbance after the Colombian Peace Agreement." *Environmental Research Letters* 15 (3): 034033. https://dx.doi.org/10.1088/1748-9326/ab6ae3.

Negash, Emnet, Emiru Birhane, Aster Gebrekirstos, et al. 2023. "Remote Sensing Reveals How Armed Conflict Regressed Woody Vegetation Cover and Ecosystem Restoration Efforts in Tigray (Ethiopia)." *Science of Remote Sensing* 8: 100108. https://doi.org/10.1016/j.srs.2023.100108.

Owen, Michael, Andrew Kruczkiewicz, and Jamon Van Den Hoek. 2023. "Indexing Climatic and Environmental Exposure of Refugee Camps with a Case Study in East Africa." *Scientific Reports* 13 (1): 7533. https://doi.org/10.1038/s41598-023-31140-7.

PAX. Forthcoming. "Ethics in Action: Remote Sensing the Environment in Conflict Contexts."
Rapach, Seonaid, Annalisa Riccardi, and Rhonda Wheate. 2024. "Earth Observation Technology's Alignment with OHCHR Indicators for Strengthening Human Rights Breach Investigations and Adjudication." *Science and Justice* 64 (6): 710–27. https://doi.org/10.1016/j.scijus.2024.09.006.
Read, Jane M., and Marla J. Torrado. 2009. "Remote Sensing." In *International Encyclopedia of Human Geography*, edited by Rob Kitchin and Nigel Thrift. Elsevier. https://doi.org/10.1016/B978-008044910-4.00508-3.
Schulte to Bühne, Henrike, Eoghan Darbyshire, Teklehaymanot G. Weldemichel, Jan Nyssen, and Doug Weir. 2024. "Conflict-related Environmental Degradation Threatens the Success of Landscape Recovery in Some Areas in Tigray (Ethiopia)." *Ecology and Society* 29 (3): 20. https://doi.org/10.5751/ES-15138-290320.
Witmer, Frank. 2015. "Remote Sensing of Violent Conflict: Eyes from Above." *International Journal of Remote Sensing* 36 (9): 2326–52. https://doi.org/10.1080/01431161.2015.1035412.
Yin, He, Lina Eklund, Dimah Habash, Mazin B. Qumsiyeh, and Jamon Van Den Hoek. 2025. "Evaluating War-induced Damage to Agricultural Land in the Gaza Strip Since October 2023 Using PlanetScope and SkySat Imagery." *Science of Remote Sensing* 11: 100199. https://doi.org/10.1016/j.srs.2025.100199.

Ecocide

Stavros Evdokimos Pantazopoulos, Nataliia Hendel, and Zerrin Savaşan

Ecocide is a crime that generally refers to acts resulting in severe damage to the environment or parts thereof, often with long-term or irreversible impacts on the ecosystem.

The term "ecocide" has its roots in the early 1970s, when it was used to describe the environmental devastation caused by Agent Orange in the Vietnam War (Zierler 2011). Swedish Prime Minister Olof Palme first used the term in 1972 at the UN Conference on the Human Environment to condemn large-scale environmental harm. Definitions vary, but most emphasize human acts that result in severe, widespread, or long-term environmental damage, often with a reckless disregard for ecological health.

Efforts to criminalize ecocide as an international offense date back to the drafting of the International Criminal Court (ICC) Statute. Initial drafts included ecocide as a crime under ICC jurisdiction (article 26), but it was removed before adoption (ILC 1991, 107). The ICC Statute eventually included an **environment**-specific **war** crime in its Article 8(2)(b)(iv) applicable in international armed conflicts, which sets a high, cumulative threshold ("[i]ntentionally launching an attack in the knowledge that such attack will cause … widespread, long-term and severe damage to the natural environment, which would be clearly excessive in relation to the concrete and direct overall military advantage anticipated"). In 2016, the ICC's Office of the Prosecutor (OTP) indicated a stronger interest in prosecuting cases involving environmental harm, as evidenced by its Policy Paper on Case Selection and Prioritization (ICC OTP 2016, para. 41). Besides the crime of ecocide, it is noteworthy that international crimes under the ICC Statute carry the potential to encompass environmental harm (Gillett 2022), and the OTP launched a public consultation on a new policy initiative to advance **accountability** for **environmental crimes** under the Rome Statute in February 2024.

The modern push to define and criminalize ecocide began in 2010 when environmental lawyer Polly Higgins proposed its inclusion in the ICC Statute to the **International Law Commission** (Higgins et al. 2013). Recently, efforts have intensified to formally establish ecocide as a fifth international crime alongside genocide, crimes against humanity, war crimes, and aggression. The Stop Ecocide Foundation has led this initiative, with a 2021 proposal from an Independent Expert Panel (IEP) that defines ecocide as "unlawful or wanton acts committed with knowledge that there is a substantial likelihood of severe and either widespread or long-term damage to the environment" (Stop Ecocide Foundation 2021). This definition targets large-scale environmental destruction, intentional or reckless, and is currently under consideration by ICC member states.

Domestically, ecocide has been criminalized in various countries. The Commonwealth of Independent States (CIS), a regional intergovernmental organization, adopted a Model Criminal Code in 1996 that includes ecocide in Article 108. All post-Soviet states have criminalized the crime of ecocide in their national legislation (Russia (1996), Kazakhstan (1997), Kyrgyzstan (1997), Tajikistan (1998), Georgia (1999), Belarus (1999), Ukraine (2001), Moldova (2002), and Armenia (2003)) (EcocideLaw n.d.). For example, Article 441 of Ukraine's Criminal Code defines ecocide as the "mass destruction of flora and fauna, pollution of air or water, or other actions causing an environmental disaster," punishable by 8 to 15 years of imprisonment (Verkhovna Rada of Ukraine 2001). Since 2022, Ukraine has opened 14 ecocide cases under this article in response to environmental harm caused by Russian agents (Office of the Prosecutor General of Ukraine 2024a, 2024b).

Countries such as Vietnam, Ecuador, France, Brazil, Belgium, Peru, Italy, Mexico, and the European Union have also criminalized ecocide domestically (EcocideLaw n.d.). The ICC's potential adoption of ecocide as a standalone crime would signify a monumental step in international law, providing a legal basis for prosecuting those responsible for large-scale environmental damage, like existing categories of international crimes. The evolving recognition of ecocide as an international crime holds significant implications for **environmental peacebuilding**, insofar as it introduces a legal framework for addressing severe environmental harm, facilitates transitional **justice** mechanisms, and contributes to the reconstruction of ecologically and socially resilient post-conflict societies.

References

Ecocidelaw. n.d. "Ecocide / Serious Environmental Crimes in National Jurisdictions." Accessed August 1, 2025. https://ecocidelaw.com/existing-ecocide-laws/.

Gillett, Matthew. 2022. *Prosecuting Environmental Harm Before the International Criminal Court*. Cambridge University Press.

Higgins, Polly, Damien Short, and Nigel South. 2013. "Protecting the Planet: A Proposal for a Law of Ecocide." *Crime, Law, and Social Change* 59 (3): 251–66. https://doi.org/10.1007/s10611-013-9413-6.

International Law Commission. 1991. "Report of the International Law Commission on the Work of Its Forty-Third Session, 29 April – 19 July 1991." A/46/10. https://legal.un.org/ilc/documentation/english/reports/a_46_10.pdf.

Office of the Prosecutor General of Ukraine. 2024a. "The Shelling of a Nuclear Facility in Kharkiv – a Russian Colonel General and Four of His Subordinates Will Be Tried." https://www.gp.gov

.ua/ua/posts/obstrili-yadernoyi-ustanovki-u-xarkovi-suditimut-rosiiskogo-general-polkovnika-ta-cotiryox-iogo-pidleglix.
Office of the Prosecutor General of Ukraine. 2024b. "Three Years of the SEP: Despite Russian Armed Aggression and Martial Law in the Country, Prosecutors Counteract Environmental Crimes at All Levels." https://www.gp.gov.ua/ua/posts/tri-roki-sep-popri-rosiisku-zbroinu-agresiyu-i-vojennii-stan-u-derzavi-prokurori-protidiyut-zlocinam-proti-dovkillya-na-vsix-rivnyax.
Office of the Prosecutor, International Criminal Court (OTP ICC). 2016. "Policy Paper on Case Selection and Prioritization." https://www.icc-cpi.int/sites/default/files/itemsDocuments/20160915_OTP-Policy_Case-Selection_Eng.pdf.
Stop Ecocide Foundation. 2021. "Independent Expert Panel for the Legal Definition of Ecocide: Commentary and Core Text." June. https://www.stopecocide.earth/legal-definition.
Verkhovna Rada of Ukraine. 2001. "The Criminal Code of Ukraine." https://zakon.rada.gov.ua/laws/show/en/2341-14#top.
Zierler, David. 2011. *The Invention of Ecocide: Agent Orange, Vietnam, and the Scientists Who Changed the Way We Think About the Environment*. University of Georgia Press.

Ecological Distribution Conflicts

Richard Marcantonio

Ecological distribution conflicts are a collective manifestation of discontent that detonates when people organize themselves to denounce situations regarding both the unequal distribution of environmental benefits and the unequal distribution of environmental costs (Martinez-Alier and O'Connor 1996).

Ecological distribution conflicts can be instigated by a range of sources but are most often associated with pollution or environmental risks from industrial production, resource extraction, or environmental service deprivation. Ecological distribution conflicts arise out of environmental injustice: the unequal distribution of environmental risks, the benefits of producing that risk, and the inability to participate in processes impacting either. The term emerged from the scholarship of ecological economics and **political ecology**, first conceptualized by Joan Martinez-Alier and Martin O'Connor (1996). The concept of ecological distribution conflicts developed alongside growing awareness of **environmental justice**, pioneered by Robert Bullard (1993), and consciousness of how **environmental degradation** corresponds with unequal distribution of environmental risk in a globalized world. Ecological distribution conflicts can happen at different points in commodity chains or the life cycle of a product, from being extracted to its use to being discarded as waste. The proliferating rates of resource extraction, production, and consumption—increasing "social metabolism" (Martinez-Alier 2017)—across time and space have resulted in a rise in ecological distribution conflicts globally.

Causes of the inequality of benefits and risks in ecological distribution conflicts can vary (Navas et al. 2018). Lack of enforcement of environmental regulations by the national government institutions could give rise to unequal risks. The unequal distribution of the benefits of a project by local elites may initiate a conflict event. In other instances, the relationships between a foreign company and its local hosts can foment the conditions for a **conflict** to arise. Similarly, such conflicts can arise from acute instances of impacts—such as after a major toxic tailings release into a river from a mining company, harming downstream users—or more chronic issues of benefit deprivation or risk infringement. What might begin as a peaceful protest or process of political discourse can devolve into **violence**, constituting ecological distribution conflicts. Sources of ecological distribution conflicts have shifted over time as new commodities are sought, extracted, moved, used, and disposed of, and as markets and production chains shift (Martinez-Alier 2022). However, the underlying unequal political economies and ecologies are the same. For example, growing concerns around "green economy conflicts" and "transition mineral conflicts" are contributing to ecological distribution conflicts as communities where these resources are located vie for their rights, sometimes violently (Scott and Smith 2017). The Environmental Justice Atlas (https://ejatlas.org/) is the primary database tracking ecological distribution conflict case studies (Martinez-Alier 2021).

References

Bullard, Robert D. 1993. *Confronting Environmental Racism: Voices from the Grassroots*. South End Press.

Martinez-Alier, Joan. 2017. "A Guide to Environmental Justice Movements and the Language of Ecological Distribution Conflicts." In *Routledge Handbook of Ecological Economics,* edited by Clive L. Spash. Routledge.

Martinez-Alier, Joan. 2021. "Mapping Ecological Distribution Conflicts: The EJ Atlas." *The Extractive Industries and Society* 8 (4): 100883. https://doi.org/10.1016/j.exis.2021.02.003.

Martinez-Alier, Joan. 2022. "Circularity, Entropy, Ecological Conflicts and LFFU." *Local Environment* 27 (10–11): 1182–207. https://doi.org/10.1080/13549839.2021.1983795.

Martinez-Alier, Joan, and Martin O'Connor. 1996. "Ecological and Economic Distribution Conflicts." In *Getting Down to Earth: Practical Applications of Ecological Economics*, edited by Robert Costanza, Olman Segura, and Juan Martinez-Alier, 153–83. Island Press.

Navas, Grettel, Sara Mingorria, and Bernardo Aguilar-González. 2018. "Violence in Environmental Conflicts: The Need for a Multidimensional Approach." *Sustainability Science* 13 (3): 649–60. https://doi.org/10.1007/s11625-018-0551-8.

Scott, Dayna, and Adrian Smith. 2017. "'Sacrifice Zones' in the Green Energy Economy: Toward an Environmental Justice Framework." *McGill Law Journal* 62 (3): 861. https://lawjournal.mcgill.ca/article/sacrifice-zones-in-the-green-energy-economy-toward-an-environmental-justice-framework/.

Ecological Peace

Elaine Hsiao

Ecological peace centers a relationship of harmony and mutual care between humans and the rest of nature, which includes non-human species and life-sustaining systems, such as microbiomes, ecosystems, biodiversity, and habitats.

Ecological **peace** derives its conceptual foundations from various sources. These include Johann Galtung's concept of "nature peace" or interspecies **cooperation** (Galtung 1996, 31); fundamental principles of ecology (e.g., ecological bounds); and post-human and more-than-human theories that decenter the human species and recognize humans are assemblages of non-human life (Seaman 2007; Mitchell 2023), For example, Arne Naess' deep ecology or other biocentric ontologies (Naess 1988; Sessions 1987; Ibrahim Suryo 2025) and decolonial cosmovisions that see humans as a part of nature and in relationship to all other life forms and life-sustaining systems (Topa and Narvaez 2022; Whyte and Coolsaet 2020). Ecological peace differentiates itself from concepts of **environmental security** and conventional forms of **environmental peacebuilding** in its **stakeholders,** core objectives, and approaches. While environmental security and environmental **peacebuilding** often focus their attention on the natural **environment** for the purposes of enhancing the well-being, relationships, and state of humans, ecological peace has a broader multi-species and more-than-human purview, which is polycentric.

Ecological peace does not fit neatly into human theories of negative and **positive peace** (Hsiao and Le Billon 2021) because it is a more-than-human or post-human concept of peace. More-than-human theory reminds us that no species can survive in isolation and that all life relies on an assemblage of species to exist (Abram 1996; Whatmore 2002; Maller 2018, 2021). This assemblage also extends to what are typically perceived as non-living entities in the Western worldview, such as water, minerals, and air. While the more-than-human reminds us we exist in community with other life, it arguably still centers humanism; alternatively, critical posthumanism "rejects [the] species hierarchy and human exceptionalism" of anthropocentrism and 'Western Humanist ideal of 'Man' as the allegedly universal measure of all things" (Braidotti 2018, 339). In the same way, ecological peace is a call to restore coexistence within the complex web of life systems and other-than-human networks in ways that decenter peace exclusively between humans and human modes of peacebuilding.

Ecological peacebuilding is visible in calls to recognize the fundamental rights of other-than-human entities, such as rivers, mountains, or all of Mother Nature. Likewise, it is prevalent in the restoration of kinship and recognition of the fundamental interdependency of all life on Earth, including Indigenous land return and rematriation initiatives (Nash 1989; Boyd 2017; Hsiao 2012; Porter et al. 2020; Liboiron 2021; Nobiss 2023). As noted in *The Third Force*, Indigenous **conflict** resolution often has ecological peace at its core, seeking to right imbalances in the natural and spirit worlds caused by human rifts through the integration of story, ceremony, and **restorative justice** in its peace processes (Tuso and Flaherty 2016). Indigenous scholar Kyle Whyte writes of "collective continuance," an Indigenous concept of socio-ecological resilience based in systems of reciprocities or responsibilities that reflect the essential qualities of "consent, diplomacy, trust and redundancy" to form systems of governance rooted in seasonal round, caring for human and non-human life in adaptive and interdependent ways (Whyte 2018). Other Indigenous scholars, such as Robin Wall Kimmerer, offer different iterations of Original Instructions, which provide guidance on how to be a human in right relationship with non-human peoples (Wall Kimmerer 2015; Nelson 2008). In these ways, ecological peacebuilding moves beyond utilizing the environment to restore relations between humans or to right human relationships to do less harm to the environment.

References

Abram, David. 1996. *The Spell of the Sensuous: Perception and Language in a More-than-Human World*. Vintage Books.

Boyd, David R. 2017. *The Rights of Nature: A Legal Revolution That Could Save the World*. ECW Press.

Braidotti, Rosi. 2018. "Posthuman Critical Theory." In *Posthuman Glossary*, edited by Rosi Braidotti and Maria Hlavajova. Bloomsbury Academic.

Galtung, Johan. 1996. *Peace by Peaceful Means: Peace and Conflict, Development and Civilization*. International Peace Research Institute.
Hsiao, Elaine C. 2012. "Whanganui River Agreement: Indigenous Rights and Rights of Nature." *Environmental Policy and Law* 42 (6): 371–75. https://www.researchgate.net/publication/235671679_Whanganui_River_Agreement_-_Indigenous_Rights_and_Rights_of_Nature.
Hsiao, Elaine C., and Philippe Le Billon. 2021. "Connecting Peaces: TBCAs and the Integration of International, Social, and Ecological Peace." *International Journal on World Peace* XXXVII (4): 21–54. https://www.jstor.org/stable/27115983.
Ibrahim Suryo, Kevin. 2025. "Biocentrism and Its Role in Shaping Conservation Approaches in Protected Areas." *Journal of Critical Ecology* 2 (1). https://doi.org/10.61511/jcreco.v2i1.1740.
Liboiron, Max. 2021. *Pollution Is Colonialism*. Duke University Press.
Maller, Cecily. 2018. *Healthy Urban Environments: More-Than-Human Theories*. Routledge Studies in Environment and Health. Routledge.
Maller, Cecily. 2021. "Re-Orienting Nature-Based Solutions with More-than-Human Thinking." *Cities* 113 (June): 103155. https://doi.org/10.1016/j.cities.2021.103155.
Mitchell, Audra. 2023. "Nonhuman, More-Than-Human, and Post-Human International Relations and International Studies." In *Oxford Research Encyclopedia of International Studies*, edited by Nalanda Roy. Oxford University Press.
Naess, Arne. 1988. "The Deep Ecology Movement." In *Problems of International Justice*, edited by Steven Luper-foy. Routledge.
Nash, Roderick Frazier. 1989. *The Rights of Nature: A History of Environmental Ethics*. University of Wisconsin Press.
Nelson, Melissa K, ed. 2008. *Original Instructions: Indigenous Teachings for a Sustainable Future*. Bear & Company.
Nobiss, Sikowis. 2023. "Why 'ReMatriate' Is a More Inclusive Term for Returning Land to Indigenous Peoples." *News. PERSPECTIVES*, October 3. https://www.reckon.news/justice/2023/10/perspectives-why-rematriate-is-a-more-inclusive-term-for-returning-land-to-indigenous-peoples.html.
Porter, Libby, Julia Hurst, and Tina Grandinetti. 2020. "The Politics of Greening Unceded Lands in the Settler City." *Australian Geographer* 51 (2): 221–38. https://doi.org/10.1080/00049182.2020.1740388.
Seaman, Myra J. 2007. "Becoming More (than) Human: Affective Posthumanisms, Past and Future." *Journal of Narrative Theory* 37 (2): 246–75. https://doi.org/10.1353/jnt.2008.0002.
Sessions, George. 1987. "The Deep Ecology Movement: A Review." *Environmental Review* 11 (2): 105–25. https://doi.org/10.2307/3984023.
Topa, Wahinkpe, and Darcia Narvaez. 2022. *Restoring the Kinship Worldview: Indigenous Voices Introduce 28 Precepts for Rebalancing Life on Planet Earth*. North Atlantic Books.
Tuso, Hamdesa, and Maureen P. Flaherty. 2016. *Creating the Third Force: Indigenous Processes of Peacemaking*. Lexington Books.
Wall Kimmerer, Robin. 2015. *Braiding Sweetgrass: Indigenous Wisdom, Scientific Knowledge, and the Teachings of Plants*. Milkweed Editions.
Whatmore, Sarah. 2002. *Hybrid Geographies: Natures Cultures Spaces*. SAGE.
Whyte, Kyle. 2018. "Settler Colonialism, Ecology, and Environmental Injustice." *Environment and Society* 9: 125–44. https://doi.org/10.3167/ares.2018.090109.
Whyte, Kyle, and Brendan Coolsaet. 2020. "Indigenous Environmental Justice: Anti-Colonial Action Through Kinship." In *Environmental Justice: Key Issues*, 1st ed. Key Issues in Environment and Sustainability. Routledge. https://www.academia.edu/43803837/Indigenous_Environmental_Justice_Anti_Colonial_Action_through_Kinship_for_EJ_primer_?auto=download.

Ecosystem Services

Andres Suarez, Daniel E. Orenstein, and Esther C. Chinweuba

Ecosystem services (ES) represent nature's benefits to human societies.

Examples of ES include material benefits such as raw materials and food, as well as non-material benefits like spiritual or cultural values. Additionally, they provide regulating services such as climate control or water regulation (Pascual et al. 2017). However, the understanding and valuation of ES are not universal; they are shaped by historical **power** relations, social inequalities, and diverse cultural values. The value placed on these services by different demographic groups and individuals within the same ecosystem often depends on their needs, socioeconomic status, and relationship with the land. For instance, less affluent populations frequently rely on various ecosystems for their livelihoods and thus tend to place a higher value on the material benefits they provide, such as food, water, or medicinal plants. Conversely, communities that are more affluent may prioritize the aesthetic or ecological benefits of **biodiversity**, such as preserving landscapes or wildlife for recreational purposes. This creates a situation where there is a discrepancy between the valuation languages used by different groups (Gómez-Baggethun and Martín-López 2015).

These priority differences highlight how power imbalances influence ecosystem management (Arias-Arevalo et al. 2023). Groups that are more affluent frequently dominate the discourse, shaping policies that protect ecosystems for their cultural and ecological value, but at times their dominance excludes the more immediate survival needs of marginalized communities. This disparity often gives rise to **conflict**, as **conservation** goals clash with the basic provisioning needs of less privileged groups.

When these dynamics are considered in the context of **peacebuilding**, the picture becomes even more complex. ES are often a source of contention in areas affected by conflict, where access to land, water, and nature's other benefits is frequently disputed. In numerous instances, the uneven distribution of ecosystem benefits serves to intensify pre-existing social cleavages, giving rise to grievances that can result in **violence** (Benjaminsen and Svarstad 2021). In post-conflict settings, the restoration of equitable access to ES can serve as a pivotal step in the process of rebuilding trust and fostering reconciliation. For example, in Liberia, ES has been used to foster peacebuilding by creating legally regulated frameworks for resource management, addressing grievances over unregulated exploitation that had fueled conflict. By promoting transparent **governance** and equitable resource distribution in forestry and mining, these reforms helped rebuild trust among communities, transforming ecosystems from sources of contention into tools for economic stability and collaboration in the post-war period (Beevers 2015).

Moreover, the direct consequences of armed conflict endanger ES, destroying landscapes, polluting water sources, and displacing communities. Nevertheless, beyond these immediate threats, conflicts emerge when decisions regarding resource management are made without considering the diverse needs and priorities of all **stakeholders** or all relevant groups and individuals. A peacebuilding approach to ES must address both the physical restoration of ecosystems as a source of ES and work toward inclusive management that reflects the varying values, access capabilities, asymmetries, and dependencies of different social groups, mainly those more vulnerable.

In this line, ES are not merely environmental functions; rather, they represent a contested space of social, economic, and political struggle. The implementation of sustainable and just management practices requires the acknowledgment of the disparate ways in which different groups prioritize the benefits of nature.

References

Arias-Arévalo, Pablo, Erika Lazos-Chavero, Ana S. Monroy-Sais, et al. 2023. "The Role of Power in Leveraging the Diverse Values of Nature for Transformative Change." *Current Opinion in Environmental Sustainability* 64: 101352. https://doi.org/10.1016/j.cosust.2023.101352.

Beevers, Michael D. 2015. "Peace Resources? Governing Liberia's Forests in the Aftermath of Conflict." *International Peacekeeping* 22 (1): 26–42. https://doi.org/10.1080/13533312.2014.992568.

Benjaminsen, Tor A., and Hanne Svarstad. 2021. *Political Ecology: A Critical Engagement with Global Environmental Issues*. Springer Nature.

Gómez-Baggethun, Erik, and Berta Martín-López. 2015. "Ecological Economics Perspectives on Ecosystem Services Valuation." In *Handbook of Ecological Economics*, 260–82. Edward Elgar Publishing. https://doi.org/10.4337/9781783471416.00015.

Pascual, Unai, Patricia Balvanera, Sandra Díaz, et al. 2017. "Understanding the Diversity of Values of 'Nature's Contributions to People': Insights from the IPBES Assessment of Europe and Central Asia." *Sustainability Science* 12 (5): 676–90. https://doi.org/10.1007/s11625-017-0445-9.

Ecoterrorism

Jeremiah Ogonda Asaka, Elizabeth Chalecki, and Shannon O'Lear

Ecoterrorism is a type of terrorism perpetrated by pro-environment, non-state actors against governments and corporations engaged in environmentally destructive activities, particularly those in the extractive industry.

Ecoterrorism is distinct from **environmental terrorism**, which is a form of terrorism perpetrated by non-state actors against the **environment** for political reasons (Chalecki 2024; O'Lear 2003). As a portmanteau of the prefix "eco"—short for ecology—and terrorism, and considering its pro-environment stance, ecoterrorism has roots in both environmentalism and terrorism. However, unlike **environmental peacebuilding** (which is also pro-environment), ecoterrorism is specifically rooted in radical environmentalism. Thus, ecoterrorism can be viewed as a manifestation of radical environmentalism along the same lines as deep ecology.

Ecoterrorists employ violent acts or tactics—such as arson, "monkeywrenching" (Brown 2023, 363), bombing, and tree spiking—to force those engaged in environmentally destructive activities to stop further development. In essence, ecoterrorism relies on ecotage—sabotage for ecological reasons—to achieve its primary goal of environmental protection.

In comparison, environmental peacebuilding is premised on the assumption that environmental protection and **cooperation** can serve a critical role in **conflict prevention** as well as bringing about both negative and **positive peace** in post-conflict contexts (Black et al. 2022; Ide 2019; Balag'kutu et al. 2018). This typically happens through positive spillover effects or mitigated **conflict** risks (Krampe et al. 2021; Krampe 2017). In this regard, environmental peacebuilding is anti-**violence**.

Ecoterrorists and environmental peacebuilders can have differences over issues such as the extraction of resources or the development of land. For instance, environmental peacebuilders tend to work inside normal channels of the international system, such as the United Nations, whereas ecoterrorists feel they are forced to work outside normal channels to achieve their environmental protection goal(s). Moreover, practitioners engaged in environmental peacebuilding tend to be at odds with ecoterrorism's embrace of violent tactics as a necessary strategy for achieving pro-environment goals.

Thus, the ecoterrorism approach is contradictory to (and potentially complicates) the environmental peacebuilding cause, even though both aim to achieve environmental protection. Environmental peacebuilding focused on conflict risk **mitigation** at the intrastate level holds some promise as an alternative to ecoterrorism, but only if a balance between anthropocentrism and ecocentrism that appeals to all types of environmental activists can be reached.

In addition, **climate change** has been shown to intersect with most factors that have implications for **peace** and **security** (IPCC 2023). Therefore, it is important to consider what climate change may mean for the future of ecoterrorism. For instance, it is likely that some radical environmental groups could be motivated by what they consider to be government inaction on the climate crisis, and resort to ecotage and other violent acts to draw attention to the need for immediate climate action.

Examples of non-state actors that have been accused of engaging in ecoterrorism by the US Federal Bureau of Investigation (FBI) include Earth Liberation Front, Earth First!, Sea Shepherd Conservation Society, and several individuals. Other environmental groups, such as Greenpeace, have also been unfairly targeted by the FBI in the past (ABC News 2010), highlighting the trouble with the terrorism tag (Buell 2009).

References

ABC News. 2010. "FBI Spied on PETA, Greenpeace, Anti-War Activists." *ABC News*, September 20. https://abcnews.go.com/News/Blotter/fbi-spied-peta-greenpeace-anti-war-activists/story?id=11682844.

Balag'kutu, Tim, Jeremiah O. Asaka, Linda Holcombe, Jason McSparren, and Stacy VanDeveer. 2018. "Environmental Conflicts and Peacebuilding in Africa: Connecting Resources, Issues and Ongoing Initiatives." In *Routledge Handbook of Environmental Conflict and Peacemaking*, edited by Ashok Swain and Joakim Öjndela. Routledge.

Black, Richard, Joshua Busby, Geoffrey Dabelko, et al. 2022. *Environment of Peace: Security in a New Era of Risk*. Stockholm International Peace Research Institute (SIPRI), May. https://doi.org/10.55163/LCLS7037.

Brown, Joseph M. 2023. "Civil Disobedience, Sabotage and Violence in US Environmental Activism." In *The Oxford Handbook of Comparative Environmental Politics*, edited by Jeannie Sowers, Stacy D. VanDeveer, and Erika Weinthal. Oxford University Press.

Buell, Lawrence. 2009. "What Is Called Ecoterrorism." *Gramma: Journal of Theory and Criticism* 16: 153–66. http://www.enl.auth.gr/gramma/.

Chalecki, Elizabeth. 2024. "Environmental Terrorism Twenty Years On." *Global Environmental Politics* 24 (1): 1–9. https://doi.org/10.1162/glep_a_00728.

Ide, Tobias. 2019. "The Impact of Environmental Cooperation on Peacemaking: Definitions, Mechanisms, and Empirical Evidence." *International Studies Review* 21 (3): 327–46. https://doi.org/10.1093/isr/viy014.

Intergovernmental Panel on Climate Change. 2023. "Summary for Policymakers." In *Climate Change 2022 – Impacts, Adaptation and Vulnerability: Working Group II Contribution to*

the Sixth Assessment Report of the Intergovernmental Panel on Climate Change. Cambridge University Press.
Krampe, Florian. 2017. "Toward Sustainable Peace: A New Research Agenda for Post-Conflict Natural Resource Management." *Global Environmental Politics* 17 (4): 1–8. https://doi.org/10.1162/GLEP_a_00431.
Krampe, Florian, Farah Hegazi, and Stacy D. VanDeveer. 2021. "Sustaining Peace through Better Resource Governance: Three Potential Mechanisms for Environmental Peacebuilding." *World Development* 144: 105508. https://doi.org/10.1016/j.worlddev.2021.105508.
O'Lear, Shannon. 2003. "Environmental Terrorism: A Critique." *Geopolitics* 8 (3): 127–50. https://doi.org/10.1080/14650040412331307742.

Ecotheology

Aili Winstanley Channer, Alan Channer, and Daimon Mkwandawire

Ecotheology refers to religious teachings on the nature of humanity's relationship with the rest of the universe and with the divine, including the moral implications of this relationship.

Etymologically, the term can be understood as the study (*logos*) of the household (*oikos*) of the universe in connection with God (*theos*). This relationship is at the core of religious and spiritual traditions. It is manifest, for example, in the Maori concept of *whanaungatanga*, meaning a threefold kinship or "right relationship" to the Supreme Being (*ki te Atua*), to the people (*ki nga tangata*), and to the land (*ki te whenua*) (Cadigan 2001, 125). Ecotheology emerged across diverse traditions as a branch of theology in the 1970s because of growing awareness of the scale of ecological destruction caused by human activity, and the term gained currency in the 1990s. Drawing on earlier traditions of ethics, cosmology, and theological anthropology, ecotheology is rooted both in the unique endemic beliefs of each faith and in shared human values (Conradie 2023).

Ecotheology offers ethical guidelines to prevent damage to creation, highlighting faith-based values such as stewardship, compassion, **justice**, and moderation. The 2024 Islamic Covenant for the Earth, Al-Mizan, emphasizes the need for justice and **accountability** to maintain cosmic equilibrium (*mizan*) (Llewellyn and Khalid 2024). In 2001, Pope John Paul II called for an "ecological conversion" (John Paul II 2001, 179) where Christians would treat living sustainably as a spiritual practice. The appeal was later reiterated by Pope Francis (Francis 2015, 159).

Ecotheologies emphasize the effects of ecological destruction on the most vulnerable and marginalized in society; the papal encyclical *Laudato Si* urges listening to "the cry of the earth and the poor," a phrase also used in liberation theology (Kerber 2018, 630). Among the diverse ecotheologies of the Dharmic religions, the principle of radical nonviolence (*ahimsa*) toward all living beings is a common thread, providing a strong foundation for both ecology and **peacebuilding** (Chapple 2016).

Furthermore, ecotheologians have diagnosed the ecological crisis as a spiritual crisis, seeing consumerism and extractivism as resulting from secularization, the desacralization of the earth, or the idolization of material wealth (Francis 2015; Llewellyn and Khalid 2024). Ecotheology can also involve a critique of religious complicity in ecological destruction and a consequent reevaluation of doctrines and scriptures in the light of ecological awareness (Tucker and Grim 2016).

World religions are increasingly acting as a united front against ecological destruction, which creates opportunities for interfaith dialogue, peacebuilding, and collaborative initiatives in environmental protection and natural **resource governance** (Mayer 2023). Religious leaders have come together to issue joint declarations on the ecological crises, starting with the 1986 Assisi Declarations from representatives of Buddhism, Christianity, Hinduism, Islam, and Judaism. Faith is also increasingly influential in international bodies such as the **United Nations Environment Programme** and in international climate negotiations (Noll 2024).

Ecotheology encourages the role of faith communities in environmental protection and peacebuilding at the grassroots level. This involves community-based initiatives that address both **conflict** and **environmental degradation**, such as joint **conservation** efforts in contested territories and faith-based advocacy for policies that promote ecological welfare and **peace**. For instance, in Kenya, grassroots initiatives to conserve and restore mangrove **forests** have forged collaboration between Muslims and Christians with a history of communal suspicion (Channer 2022; Channer and Kobia 2025). In Nigeria, "green sermons" by imams and pastors have helped diffuse ethno-religious conflict between pastoralists and farmers (Channer 2015). In Colombia, the Catholic Church supports resistance to mining by mediating between the government and communities, providing affected communities with pastoral support, and leading local peace initiatives that "promote the collective construction of new relationships with the land and **natural resources**" (Polanía-Reyes and Henao 2022, 39–40).

References

Cadigan, Tui. 2001. "Land Ideologies that Inform a Contextual Maori Theology of Land." *Ecotheology: Journal of Religion, Nature and the Environment* 6 (1/2): 123. https://doi.org/10.1558/ecotheology.v6i2.123.

Channer, Alan. 2022. *Regreening Africa: Engaging Faith-Based Organisations in Land Restoration Across Africa—Insights and Opportunities*. World Agroforestry.

Channer, Alan. 2015. "Mobile Pastoralism and Trustbuilding in Africa." In *Land Restoration: Reclaiming Landscapes for a Sustainable Future*, edited by Ilan Chabay, Martin Frick, and Jennifer Helgeson. Elsevier. https://doi.org/10.1016/B978-0-12-801231-4.00012-4.

Channer, Alan, and Nkatha Kobia. 2025. "Faith in Mangrove Restoration." *OikoDiplomatique*, September. https://www.oikodiplomatique.org/report-2024-faith-in-mangrove-restoration.html.

Chapple, Christopher Key. 2016. "India." In *Routledge Handbook of Religion and Ecology*, edited by Willis Jenkins, Mary Evelyn Tucker, and John Grim. Taylor & Francis Group.

Conradie, Ernst. 2023. "Ecotheology." In *St Andrews Encyclopaedia of Theology*, edited by Brendan N. Wolfe. https://www.saet.ac.uk/Christianity/Ecotheology.

Francis, Pope. 2015. "Laudato Si': On Care for Our Common Home: Encyclical Letter." *The Vatican*, May 24. https://www.vatican.va/content/francesco/en/encyclicals/documents/papa-francesco_20150524_enciclica-laudato-si.html.

John Paul II, Pope. 2001. Catechesis (26 January). Insegnamenti di Giovanni Paolo II. *Insegnamenti* 41 (1): 179.

Kerber, Guillerme. 2018. "Latin American and Ecumenical Insights in *Laudato Si'*." *The Ecumenical Review* 70 (4): 627–36. https://doi.org/10.1111/erev.12389.

Llewellyn, Othman, and Fazlun Khalid. 2024. *Al-Mizan: A Covenant for the Earth*. United Nations Environment Programme. https://wedocs.unep.org/20.500.11822/44949.

Mayer, Farhana. 2023. *An Introduction to Qur'anic Ecology and Resonances with Laudato Si'*. Laudato Si' Research Institute and Randeree Charitable Trust.
Noll, Kristian. 2024. "The Religious Rhetoric of COP29." *LSE Religion and Global Society Blog*, December 10. https://blogs.lse.ac.uk/religionglobalsociety/2024/12/the-religious-rhetoric-of-cop29/.
Polanía-Reyes, Sandra, and Héctor Fabio Henao. 2022. "The Church in Colombia and Extractives: Pastoral Accompaniment Using an Eco-Theology of Peace." In *Catholic Peacebuilding and Mining: Integral Peace, Development, and Ecology*, edited by Caesar A. Montevecchio and Gerard F. Powers. Routledge.
Tucker, Mary Evelyn, and John Grim. 2016. "The Movement of Religion and Ecology." In *Routledge Handbook of Religion and Ecology*, edited by Willis Jenkins, Mary Evelyn Tucker, and John Grim. Taylor and Francis Group.

Energy Security

Kendra Dupuy, Abdullah Fahimi, and Larissa Jane Houston

Energy security is the ability of a country's citizens to access energy supplies on a consistent, regular basis. Among the many definitions of energy security, there is a consensus that it entails energy supplies being economically affordable, reliable, accessible, and available for citizens, and that the energy source(s) are socially acceptable (Sovacool 2011; Ang et al. 2015).

Energy supply in many countries still depends on traditional **fossil fuels** (oil, gas, and coal), though **renewable energy** sources like wind, solar, and hydropower are increasing. Yet, the variable nature of renewable energy sources—that is, their dependence on environmental conditions to produce energy, such as sufficiently windy or sunny conditions—poses challenges for achieving energy **security** in states without such renewable energy sources or where electricity grids are lacking (International Energy Agency [IEA] 2024).

The diversity and complexity of factors threatening energy security have grown in the past decades. These threats include, but are not limited to, **climate change**, competition over resources (including critical raw materials, **conflict**, and natural disasters), unreliable energy reserves, excessive reliance on a single source or supplier, manipulation by energy suppliers, and cyber-attacks on energy **infrastructure** such as electricity grids (European Academies Science Advisory Council 2025). Measures to mitigate risks to energy security and ensure uninterrupted energy supply include transitioning to renewable sources, diversifying energy sources, enhancing energy efficiency, improving the **resilience** of energy systems (such as grid flexibility and market integration), and investing in research and development for potential new technologies and uses (IEA 2024).

Energy infrastructure and supplies can be primary causes of armed conflict, an objective (or target) of conflict, or a means in conflict through weaponizing energy supplies. First, countries' quest for energy security by controlling supplies or energy infrastructure can trigger violent conflict, such as when Iraq invaded Kuwait in 1990 to gain control over oil fields (Sowers et al. 2017). Second, restrictions on access to energy by energy-producing states, whether done to drive up market prices or to conserve supply for domestic purposes, can lead to the outbreak of interstate conflict where energy supplies run critically low for energy-consuming states (Andrews-Speed et al. 2014). For instance, Egypt has threatened multiple times to resort to armed **violence** to prevent Ethiopia from implementing the Grand Renaissance Ethiopian Dam (GERD), Africa's largest dam and key to alleviating energy poverty in Ethiopia, but which Egypt and Sudan consider a national security threat with its estimated impacts on water supply (Pemunta et al. 2021). Third, large-scale energy projects (including in renewable sectors) can trigger conflict at local levels, through: (1) the loss of resources like land and associated livelihoods, (2) unfair project planning processes, (3) environmental damage, and (4) worsened inequality and vulnerability (Sovacool 2021; Månsson 2015; Avila 2018; Brunet et al. 2022). Finally, disrupting energy supplies can be a conflict objective. Attacking energy sources via physical or cyberattacks can serve as a way for armed groups to pursue conflict-related objectives, for instance, by weaponizing energy through reducing the supply of a resource to the enemy, as has occurred in the Russia-Ukraine **war** (Månsson 2014; La Belle 2023). In this way, actors can create energy insecurity and, in turn, political instability in other states.

Energy security can help to build **peace**. While low-intensity, local-level conflict is not uncommon in the hydropower sector, simultaneous interstate **cooperation** is more common than interstate war over shared water resources (Sovacool and Walter 2019; Del Bene et al. 2018; Bernhauer and Bohmelt 2020; Wolf 2007). Energy is key to economic development, which is key to peace, especially during the post-conflict period (Collier and Hoeffler 2004; Lujala and Rustad 2012). Cooperative management of energy supplies and the use of conflict-sensitive approaches to energy project development can help to reduce conflict potential and foster **peacebuilding** by better protecting vulnerable people and communities (Sawas et al. 2018).

References

Andrews-Speed, Philip, Coby van der Linde, and Kimon Keramidas. 2014. "Conflict and Cooperation over Access to Energy: Implications for a Low-Carbon Future." *Futures* 58: 103–14. https://doi.org/10.1016/j.futures.2013.12.007.

Ang, Bend Wah, W. L. Choong, and Ada Tsan Sheng Ng. 2015. "Energy Security: Definitions, Dimensions and Indexes." *Renewable and Sustainable Energy Reviews* 42: 1077–93. https://doi.org/10.1016/j.rser.2014.10.064.

Avila, Sofia. 2018. "Environmental Justice and the Expanding Geography of Wind Power Conflicts." *Sustainability* 13: 599–616. https://link.springer.com/article/10.1007/s11625-018-0547-4.

Bernhauer, Thomas, and Tobias Böhmelt. 2020. "International Conflict and Cooperation over Freshwater Resources." *Nature Sustainability* 3 (5): 350–6. https://www.nature.com/articles/s41893-020-0479-.

Brunet, Carole, Michel A. Bouchard, Pierre Baptiste, Oumarou Savadogo, Youba Sokona, and Nicholas Merveille. 2022. "Will Solar Energy Escape the Natural 'Resource Curse'?" *Energy Strategy Reviews* 44: 101010. https://doi.org/10.1016/j.esr.2022.101010.

Collier, Paul, and Anke Hoeffler. 2004. "Greed and Grievance in Civil War." *Oxford Economic Papers* 56: 563–95. https://www.jstor.org/stable/3488799.

Del Bene, Daniela, et al. 2018. "More Dams, More Violence? A Global Analysis on Resistances and Repression Around Conflictive Dams Through Co-Produced Knowledge." *Sustainability Science* 13: 617–33. https://doi.org/10.1007/s11625-018-0558-1.

European Academies Science Advisory Council (EASAC). 2025. *Security of Sustainable Energy Supplies* (EASAC Policy Report 47). EASAC. https://easac.eu/fileadmin/user_upload/EASAC_SoSES_final_310325.pdf.

International Energy Agency (IEA). 2024. *World Energy Outlook 2024*. https://www.iea.org/reports/world-energy-outlook-2024.
La Belle, Michael Carnegie. 2023. "Energy as a Weapon of War: Lessons from 50 Years of Energy Interdependence." *Global Policy* 14 (3): 531–47. https://doi.org/10.1111/1758-5899.13235.
Lujala, Païvi, and Siri Aas Rustad, eds. 2012. *High Value Natural Resources and Post-Conflict Peacebuilding*. Routledge.
Månsson, André. 2014. "Energy, Conflict, and War: Towards a Conceptual Framework." *Energy Research and Social Science* 4: 106–16. https://www.sciencedirect.com/science/article/abs/pii/S2214629614001170.
Månsson, André. 2015. "A Resource Curse for Renewables? Conflict and Cooperation in the Renewable Energy Sector." *Energy Research and Social Science* 10: 1–9. https://www.sciencedirect.com/science/article/abs/pii/S2214629615000791.
Pemunta, Ngambouk Vitalis, Ngo Valery Ngo, Choumbou Raoul Fani Djomo, et al. 2021. "The Grand Ethiopian Renaissance Dam, Egyptian National Security, and Human and Food Security in the Nile River Basin." *Cogent Social Sciences* 7 (1). https://doi.org/10.1080/23311886.2021.1875598.
Sawas, Amiera, Mark Workman, and Naho Mirumachi. 2018. "Climate Change, Low-Carbon Transitions and Security." Grantham Institute, March. https://www.imperial.ac.uk/media/imperial-college/grantham-institute/public/publications/briefing-papers/Climate-change,-low-carbon-transitions-and-security-WEB.pdf.
Sovacool, Benjamin K. 2011. "Introduction: Defining, Measuring and Exploring Energy Security." In *The Routledge Handbook of Energy Security*, edited by Benjamin K. Sovacool. Routledge.
Sovacool, Benjamin K. 2021. "Who are the Victims of Low-Carbon Transitions? Towards a Political Ecology of Climate Change Mitigation." *Energy Research and Social Science* 73. https://doi.org/10.1016/j.erss.2021.101916.
Sovacool, Benjamin K., and Götz Walter. 2019. "Internationalizing the Political Economy of Hydroelectricity: Security, Development and Sustainability in Hydropower States." *Review of International Political Economy* 26 (1): 49–79. https://www.tandfonline.com/doi/full/10.1080/09692290.2018.1511449.
Sowers, Jennifer, Erika Weinthal, Neda Zawahri. 2017. "Targeting Environmental Infrastructures, International Law, and Civilians in the New Middle Eastern Wars." *Journal of Peace Research* 48 (5): 410–30. https://doi.org/10.1177/0967010617716615.
Wolf, Aaron. 2007. "Shared Waters: Conflict and Cooperation." *Annual Review of Environment and Resources* 32: 241–69. https://doi.org/10.1146/annurev.energy.32.041006.101434.

Environment

Rachel Weaver and Carl Bruch

The environment comprises all external conditions that influence the life, development, and survival of living organisms (United Nations Environment Programme [UNEP] 2009a). The environment refers to both natural and anthropogenic systems, including the biophysical components of ecosystems, such as water, air, and soil, as well as human-built infrastructure and interactions within these systems.

> The environment can be perceived through different lenses. From an anthropocentric perspective, the environment is valued for its utility to human needs and use, emphasizing the extraction and use of resources, economic benefits, and **ecosystem services** (Donnelly and Bishop 2007; Casas and Burgess 2012). Conversely, from an ecocentric or inherent values perspective, ecosystems, natural processes, and **biodiversity** are recognized as having intrinsic value, independent of their contribution to human needs and economic use (Casas and Burgess 2012).

The environment extends beyond wilderness and rural areas to include urban areas, where human development and natural systems interact, affecting both ecological functions and public health (Marselle et al. 2021). Urban areas can be referred to as the "built environment" and are composed of human-built and shaped landscapes, interventions, and **infrastructure** such as **cities**, buildings, and transportation systems (Ghosh 2020). Infrastructure in the built environment facilitates the management and distribution of environmental resources, including water, energy, and sanitation systems (Fry et al. 2023). These systems govern access to **natural resources** and can shape environmental **sustainability** (Isopescu and Ungureanu 2023). The built environment can directly and indirectly degrade the natural environment through pollution, resource depletion, and greenhouse gas emissions, which in turn impact ecological processes, human health, and community well-being (US Environmental Protection Agency [EPA] 2020). The built environment is considered a major driver of **climate change** (Mahmoud 2020). Development patterns in the built environment exacerbate climate change impacts, intensifying natural disasters and threatening both human and environmental health (Fry et al. 2023; Mahmoud 2020).

In contrast, the natural environment includes ecosystems, habitats, climate, and natural resources (US EPA 2020). It comprises both renewable resources (e.g., timber, water, fisheries, and agricultural products) and non-renewable resources (e.g., oil, gas, and minerals). Although a resource may be classified as renewable, unsustainable management or use can deplete its availability (Bellinger 2004), making sustainable resource management critical to **environmental peacebuilding** efforts (Krampe et al. 2021).

In the context of environmental peacebuilding, the environment holds central significance, influencing both **conflict** dynamics and **peacebuilding** processes. It is therefore a potential source of both conflict and **cooperation** (Bruch et al. 2022; Conca and Beevers 2018). Environmental factors and social dynamics play a dual role, both constructive and destructive, across the **conflict life cycle**, affecting peacebuilding processes and recovery efforts (Bruch et al. 2019).

Renewable resources, non-renewable resources, and ecosystems possess distinct characteristics. They may be more localized or geographically dispersed, have substantially varying economic values, be essential to **food security**, hold differing cultural significance, and exhibit temporal variability in their availability (e.g., changes in available water resources). These different characteristics influence how specific environmental elements shape conflict causes, conflict dynamics, peacemaking, and post-conflict recovery efforts (Bruch et al. 2019). Accordingly, some environmental elements are particularly amenable to cooperation (such as water, biodiversity, and **disaster risk reduction**), while other elements more frequently drive contestations and conflict (such as minerals and territory boundaries). Mismanagement of environmental resources can exacerbate tensions, while cooperation over shared environmental challenges and threats can help establish trust, promote peaceful resolutions and post-conflict recovery, and create institutions and frameworks that facilitate rapprochement (Ide 2021; UNEP 2009b).

References

Bellinger, Edward G. 2024. *Water: Our Sustainable and Unsustainable Use*. Wiley.
Bruch, Carl, David Jensen, and Monica Emma. 2022. "Defining Environmental Peacebuilding." In *Routledge Handbook of Environmental Security*, edited by Crystal Murphy, Evgenia Nizkorodov, and Richard Matthew. Routledge. https://doi.org/10.4324/9781315107592-2.
Bruch, Carl, David Jensen, Mikiyasu Nakayama, and John Unruh. 2019. "The Changing Nature of Conflict, Peacebuilding, and Environmental Cooperation." *Environmental Law Reporter* 49 (2): 10134–54.
Casas, A. Brian, and Richard A. Burgess II. 2012. "The Practical Importance of Philosophical Inquiry for Environmental Professionals: A Look at the Intrinsic/Instrumental Value Debate." *Environmental Practice: Journal of the National Association of Environmental Professionals* 14 (3): 184–89. https://doi.org/10.1017/S146604661200018X.
Conca, Ken, and Michael D. Beevers. 2018. "Environmental Pathways to Peace." In *Routledge Handbook of Environmental Conflict and Peacebuilding*, edited by Ashok Swain and Joakim Öjendal. Routledge.
Donnelly, Bebhinn, and Patrick Bishop. 2007. "Natural Law and Ecocentrism." *Journal of Environmental Law* 19 (1): 89–101. https://doi.org/10.1093/jel/eql039.
Fry, Meridith M., Eric K. Chu, Jayajit Chakraborty, et al. 2023. "Built Environment, Urban Systems, and Cities." *Fifth National Climate Assessment*. https://nca2023.globalchange.gov/chapter/12/.
Ghosh, Mainak. 2020. "Built Environment in Response to the Ecology, Design, and Perception of the Global South." In *Perception, Design and Ecology of the Built Environment: A Focus on the Global South*, edited by Mainak Ghosh. Springer Geography. https://doi.org/10.1007/978-3-030-25879-5_1.
Ide, Tobias. 2021. *Environmental Peacebuilding*. Bogotá Instituto CAPAZ, March. https://www.instituto-capaz.org/wp-content/uploads/2021/03/DT-2-2021-Ingles-V3.pdf.
Isopescu, Dorina Nicolina, and Alexandra Ungureanu. 2023. "Sustainable Development of the Built Environment and Territorial Systematization." In *Circular Economy Implementation for Sustainability in the Built Environment*, edited by Nicoleta Cobirzan, Radu Muntean, and Raluca-Andreea Felseghi. IGI Global.
Krampe, Florian, Farah Hegazi, and Stacy D. VanDeveer. 2021. "Sustaining Peace through Better Resource Governance: Three Potential Mechanisms for Environmental Peacebuilding." *World Development* 144: 105508. https://doi.org/10.1016/j.worlddev.2021.105508.
Mahmoud, Hussam. 2020. "Barriers to Gauging Built Environment Climate Vulnerability." *Nature Climate Change* 10 (6): 482–85. https://doi.org/10.1038/s41558-020-0742-z.
Marselle, Melissa R., Sarah J. Lindley, Penny A. Cook, and Aletta Bonn. 2021. "Biodiversity and Health in the Urban Environment." *Current Environmental Health Reports* 8 (2): 146–56. https://doi.org/10.1007/s40572-021-00313-9.
United Nations Environment Programme (UNEP). 2009a. *From Conflict to Peacebuilding: The Role of Natural Resources and the Environment*. UNEP.
United Nations Environment Programme (UNEP). 2009b. *Integrating Environment in Post-conflict Needs Assessments*. UNEP, Disasters and Conflicts Programme.

Environmental Change

Galeo Saintz

Environmental change refers to significant alterations in the environment. It can be caused by complex and multifaceted phenomena, including both natural processes and human activities. It encompasses a wide range of phenomena, including climate change, pollution, deforestation, and biodiversity loss. These changes can occur on various scales and can have profound impacts on ecosystems and human societies.

> Environmental change describes a "systemic, related cluster" of physical changes to the natural **environment**, including an accelerating alteration to the climate, such as "**biodiversity** loss, ocean acidification, fertile soil loss, freshwater depletion and contamination, ... compounded by disruption to global elemental cycles" (Butler and McFarlane 2018, 453). Environmental change derives from both natural and human processes, where natural systems "transform the sun's energy into matter and cause changes by cycling materials through geological, biological, oceanic, and atmospheric processes" and human systems "transform materials and energy into products and services to meet human needs and aspirations" (EEA 1995). While human processes are thought to have historically contributed to environmental change in "relatively small" ways, human-induced environmental change now alters the flows of material and energy at "unprecedented scales" (EEA 1995). Physical environmental change is "accompanied by, and will also precipitate, great social changes" such as changes to food systems and human health (Butler and McFarlane 2018, 453).
>
> Vulnerability to environmental change is "socially differentiated across **gender**, class, race, and age" (Barnett 2009, 555). Poor and marginalized populations "tend to be more vulnerable to environmental change" for several reasons, including tendencies to be more heavily dependent on at-risk resources and **ecosystem services** and a greater likelihood to live in areas affected by **environmental degradation** (Barnett 2009, 555). Recognizing the unequal vulnerabilities and profound social impacts of environmental change highlights the imperative for **environmental peacebuilding** approaches to foster equity, **resilience**, and **cooperation** in the face of shared ecological challenges.

Acknowledgment

This entry has been adapted from Bruch and Woomer (2023).

References

Barnett, Jon. 2009. "Environmental Security." In *International Encyclopedia of Human Geography*, edited by Rob Kitchin and Nigel Thrift. Elsevier. https://doi.org/10.1016/B978-008044910-4.00774-4.

Bruch, Carl, and Amanda Woomer. 2023. *Toolkit on Monitoring and Evaluation of Environmental Peacebuilding*. Environmental Law Institute, Environmental Peacebuilding Association, and United States Institute of Peace. https://m-and-e.environmentalpeacebuilding.org/toolkit/.

Butler, Colin D., and Rosemary Anne McFarlane. 2018. "Climate Change, Food Security, and Population Health in the Anthropocene." In *Encyclopedia of the Anthropocene*, no. 2, 453–59. Elsevier. https://doi.org/10.1016/B978-0-12-409548-9.09745-1.

European Environment Agency (EEA). 1995. "Europe's Environment: The Dobříš Assessment. Chapter 2. Environmental Changes and Human Development." *EEA Task Force. Effective,* May 20. https://www.eea.europa.eu/en/analysis/publications/92-826-5409-5.

Environmental Cooperation

David Lehrer, Rina Kedem, and Kendra Dupuy

Environmental cooperation is the peaceful collaboration of individuals and groups to address shared environmental challenges and opportunities, thereby achieving a larger social benefit from collectively used environmental resources. It usually entails long-term processes of repeated interactions to govern natural resources (Ide 2018).

Environmental resources that are used by many individuals and groups, such as water, land, or **forests**, are known in economic terms as "commons." Because these types of resources are accessed by and benefit multiple users, managing or governing them raises collective action issues. Collective action problems in managing commonly used environmental resources arise when individuals are asked to take actions that are costly to themselves but benefit the whole group, a situation that incentivizes individuals to free-ride rather than contribute to group welfare (Taylor 1987). A classic negative consequence of free-riding in an environmental context is the "tragedy of the commons," wherein the lack of clear property rights of a resource (and the inability to create clear rules for resource use) incentivizes overuse of a resource. However, as observed by Ostrom (1990), groups of people and communities are capable of building functioning institutions and mechanisms to manage environmental resources in a cooperative manner, thus helping groups to avoid the collapse of a natural resource due to overuse (Ostrom 1990).

Environmental **cooperation** takes place at different **governance** levels: between and across local communities, subnational units, and between states. Transboundary Environmental Cooperation (TEC) most often refers to working across national borders to address environmental issues affecting more than one country/sovereign entity. But such **transboundary cooperation** can also take place across local communities. There is a growing literature within **environmental peacebuilding** on TEC and its potential peace-enhancing effects in post-conflict settings and post-conflict natural resource management. Tackling transboundary environmental issues can be a means to enhance collaboration between parties in **conflict** (including local communities) as well as a goal. Conca (2001) argues that environmental cooperation can contribute to international **peace** by fostering interdependence between states, establishing new norms, deepening transnational civil society, and enhancing government capacity. For instance, transboundary water resources such as large rivers are often peacefully managed through interstate cooperation, including through international organizations and interstate treaties (Sovacool and Walter 2019; Del Bene et al. 2018; Bernhauer and Bohmelt 2020). The same can occur across local communities that share resources, including through formal institutions like treaties and management bodies, or through informal rules and understandings (Ostrom 1990).

Environmental cooperation can be case- and resource-specific, targeting a particular shared hazard or natural resource. It can also address multifaceted challenges such as **climate change** or **renewable energy**. Environmental cooperation can be relevant in all conflict stages: before, during, and after conflict, and can include top-down policies and bottom-up initiatives. It can build peace by building trust and shared identities between groups, and by creating and strengthening institutions that facilitate reciprocity (Ide 2018; Maas et al. 2012).

An example of TEC is an agreement reached in July 2022 between the Palestinian Water Authority (PWA) and the Israeli Civil Administration (ICA) to issue a building permit for a 30-kilometer-long treated wastewater trunk line that would bring agricultural water from the Al Bireh wastewater treatment plant in the Ramallah area of Palestine to the Jericho Valley for use by Palestinian farmers. After years of disagreement on the route of the pipeline, two NGOs, the Arava Institute for Environmental Studies (IL) and Damour for Community Development (PA), jointly facilitated a process that led to an agreement between the PWA and the ICA. The two NGOs managed to move the sides from a situation of no direct communication to integrated teamwork based on common Palestinian/Israeli concerns for the health of the aquifers and ecosystems in the area and the scarcity of water for **agriculture**.

Local communities and actors within regions of a country can cooperate over shared resources like aquifers, **conservation** areas, **forests**, wetlands, and wildlife. Such cooperation can contribute to reducing tension and **violence** between communities that has been sparked by resource scarcity. For instance, in Ghana, herders and farmers have created informal agreements to cooperatively manage grazing and water rights, reducing conflict between the groups (Bukari et al. 2018).

Thus, environmental cooperation can be a critical aspect of promoting sustainable solutions, reinforcing intergroup trust, building confidence and solidarity, mitigating tensions, and preventing violent conflict.

References

Bernhauer, Thomas, and Tobias Böhmelt. 2020. "International Conflict and Cooperation Over Freshwater Resources." *Nature Sustainability* 3 (5): 350–56. https://doi.org/10.1038/s41893-020-0479-8.

Bukari, Kaderi Noagah, Papa Sow, and Jürgen Scheffran. 2018. "Cooperation and Co-Existence Between Farmers and Herders in the Midst of Violent Farmer-Herder Conflicts in Ghana." *African Studies Review* 61 (2): 78–102. https://doi.org/10.1017/asr.2017.124.

Conca, Ken. 2001. "Environmental Cooperation and International Peace." In *Environmental Conflict*, edited by Paul F. Diehl and Nils Petter Gleditsch. Routledge. https://doi.org/10.4324/9780429500794.

Ide, Tobias. 2018. "The Impact of Environmental Cooperation on Peacemaking: Definitions, Mechanisms, and Empirical Evidence." *International Studies Review* 21 (3): 327–46. https://doi.org/10.1093/isr/viy014.

Maas, Achim, Alexander Carius, and Anja Wittich. 2012. "From Conflict to Cooperation? Environmental Cooperation as a Tool for Peace-Building." In *Environmental Security,* edited by Rita Floyd and Richard Matthew. Routledge. https://doi.org/10.4324/9780203108635.

Ostrom, Elinor. 1990. *Governing the Commons: The Evolution of Institutions for Collective Action*. Cambridge University Press.

Sovacool, Benjamin K., and Götz Walter. 2019. "Internationalizing the Political Economy of Hydroelectricity: Security, Development and Sustainability in Hydropower States." *Review of International Political Economy* 26 (1): 49–79. https://www.tandfonline.com/doi/full/10.1080/09692290.2018.1511449

Taylor, Michael. 1987. "Introduction: The Problem of Collective Action." Chapter 1 in *The Possibility of Cooperation*. University of Cambridge.

Environmental Crimes

Zerrin Savaşan and Ekrem Yılmaz

Environmental crimes refer to illegal acts that directly harm the environment, typically violating national or international environmental laws. These acts often involve serious crimes and transnational organized crimes (UNEP-INTERPOL 2016).

Since there is no commonly accepted definition, different analyzes produce varying interpretations based on applicable principles, institutions, procedures, and implementation methods (Savaşan 2023). Broadly, it encompasses any act that breaches environmental law and causes environmental damage, regardless of whether it is human-caused or otherwise (Clifford 1998; Situ and Emmons 2000), effectively creating "a collective term" for all environmentally criminal acts (Zagaris 2015, 262). In some frameworks, it is considered a subclass of environmentally harmful behavior, with **ecocide** at the apex (Krott 2021). Furthermore, environmental harm can extend to damage resulting from activities not yet criminalized (Benton 2006; Megret 2010).

Environmental crimes include activities such as illegal logging, wildlife trafficking, illegal fishing, illicit waste dumping, and the illegal trade in ozone-depleting substances. These crimes are often transnational and organized, posing significant threats to ecosystems, **biodiversity**, and **sustainable development** (UN 1975, 1987, 1992). The debate around environmental crime also incorporates atrocity crimes, i.e., genocide, crimes against humanity, **war** crimes, and crimes of aggression (extended to include ethnic cleansing via the 2005 World Summit Outcome Document (paras. 138–139) (UN 2005), as provided under the Rome Statute of the International Criminal Court (ICC 1998). They are also involved in the environmental crime debate via a special court and penalty mechanism (UN 2014). A Policy Paper published by the Office of the Prosecutor signaled the possibility of considering environmental factors in future cases, hinting at a special court and penalty mechanism (Office of the Prosecutor, International Criminal Court 2016). Environmental crimes harm the **environment** and negatively affect all stages of an environmental-related **peace** process or **conflict life cycle.** Crucially, these crimes can disrupt social transformation and economic **adaptation** processes vital for both peacekeeping and **peacebuilding.** Issues like natural resource protection and joint environmental management, also critical for peacebuilding, can be severely affected (Savaşan and Duman 2021). A key reason for this disruption is that environmental crimes often serve as an important funding source for non-state armed groups through activities such as the illegal use or trade of **natural resources** and wildlife trafficking. For instance, coal exports leading to deforestation in Somalia, oil revenues in South Sudan, and illegal mining and mineral and wildlife trafficking in the Central African Republic and the Democratic Republic of the Congo have all helped sustain armed groups (Caparini 2022).

The UN, along with its related bodies and peacekeeping missions, actively works to address such crimes. For example, through Resolution 2036 (United Nations Security Council [UNSC] 2012), the **UNSC** banned charcoal exports from Somalia, which were financing armed groups and contributing to deforestation (Gridneff 2018). Similarly, in 2011, the UN Interim Security Force in Abyei (UNISFA) was mandated to provide

security for oil **infrastructure** in the Abyei Area in Sudan (UNSC 2011). The UNSC further adopted provisions related to the security and restriction of mineral trade in the Central African Republic through Resolutions 2127 and 2134 (UNSC 2013, 2014). Additionally, the UN Stabilization Mission in the Democratic Republic of the Congo (MONUSCO), in collaboration with INTERPOL's Environmental Crime Programme, supported efforts to implement international agreements to combat wildlife crime (Nellemann et al. 2012).

References

Benton, Ted. 2006. "Rights and Justice on a Shared Planet: More Rights or New Relations?" In *Green Criminology*, edited by Nigel South. Routledge. https://doi.org/10.4324/9781315093390.

Caparini, Marina. 2022. *Organized Environmental Crimes: Why it Matters for Peace Operations.* Stockholm International Peace Research Institute (SIPRI), May 12. https://www.sipri.org/commentary/topical-backgrounder/2022/organized-environmental-crime-why-it-matters-peace-operations#:~:text=Relevance%20for%20peace%20operations,groups%20and%20sustain%20armed%20conflict.

Clifford, Mary. 1998. *Environmental Crime: Enforcement, Policy, Social and Responsibility.* Aspen Publishers.

Gridneff, Ilya. 2018. "Burning Somalia's Future: The Illegal Charcoal Trade Between the Horn of Africa and the Gulf." In *Environmental Politics in the Middle East: Local Struggles, Global Connections*, edited by Harry Verhoeven. Oxford University Press. https://doi.org/10.1093/oso/9780190916688.003.0006.

International Criminal Court. 1998. *Rome Statute of the International Criminal Court.* July 17.

Krott, David. 2021. "The Ecocide-International Environmental Crime Nexus: When Can an International Environmental Crime be Called 'Ecocide'?" *International Law Blog*, June 25. https://internationallaw.blog/2021/06/25/the-ecocide-international-environmental-crime-nexus-when-can-an-international-environmental-crime-be-called-ecocide/.

Megret, Frederic. 2010. "The Challenge of an International Environmental Criminal Law." *SSRN Electronic Journal*: 1–22. https://dx.doi.org/10.2139/ssrn.1583610.

Nellemann, Christian, Rune Henriksen, Arnold Kreilhuber, et al., eds. 2016. *The Rise of Environmental Crime: A Growing Threat to Natural Resources Peace, Development and Security—A UNEP-INTERPOL Rapid Response Assessment.* United Nations Environment Programme (UNEP) and the International Criminal Police Organization (INTERPOL).

Office of the Prosecutor, International Criminal Court. 2016. "Policy Paper on Case Selection and Prioritization." https://www.icc-cpi.int/sites/default/files/itemsDocuments/20160915_OTP-Policy_Case-Selection_Eng.pdf.

Savaşan, Zerin. 2023. "Rethinking of Ecocrime with the R2P Debate." In *The Limits of Responsibility to Protect*, edited by Vasilka Sancin and Masa Kovic Dine. University of Ljubljana.

Savaşan, Zerin, and Fırat Duman. 2021. "Çevresel Barış, Çatışma ve İşbirliği Denklemi." In *Çevre Hukuku ve Politikaları, Kavramlar, Teoriler ve Tartışmalar*, edited by Zerrin Savaşan, Çağlar Söker, and Fırat Harun Yılmaz. Seçkin Yayıncılık.

Situ, Yingyi, and David Emmons. 2000. *Environmental Crime: The Criminal Justice System's Role in Protecting the Environment.* Sage Publications. https://doi.org/10.4135/9781452232256.

United Nations (UN). 1975. *Convention on International Trade in Endangered Species of Wild Fauna and Flora (CITES).*

United Nations (UN). 1987. *Montreal Protocol on Substances that Deplete the Ozone Layer.*

United Nations (UN). 1992. *Basel Convention on the Control of Transboundary Movements of Hazardous Wastes and their Disposal.*

United Nations (UN). 2005. *World Summit Outcome Document, A/RES/60/1.*

United Nations (UN). 2014. *Framework of Analysis for Atrocity Crimes: A Tool for Prevention.*

United Nations Security Council. 2011. S/RES/1990.

United Nations Security Council. 2012. S/RES/2036.

United Nations Security Council. 2013. S /RES/2127.

United Nations Security Council. 2014. S/RES/2134.

Zagaris, Bruce. 2015. "International Environmental Crime." In *International White-Collar Crime*, edited by Bruce Zagaris. Cambridge University Press.

Environmental Defenders

Philippe Le Billon, Veronica Limeberry, and Päivi Lujala

Environmental defenders are individuals, groups, or communities acting, often in the form of physical resistance, protest, or policy change, to protect environments against threats of extraction, pollution, or loss.

The United Nations (UN) defines environmental **human rights** defenders as "individuals and groups who, in their personal or professional capacity and in a peaceful manner, strive to protect and promote human rights relating to the **environment**, including water, air, land, flora and fauna" (UN General Assembly 2016). The term environmental and land defenders encompasses a wide range of people resisting projects threatening their environment, most often against extractive activities such as industrial mining, logging, and plantations. It includes peasants and Indigenous peoples seeking to (re)claim their lands, along with long-established rural communities facing "land grabs" by agro-industrial corporations or other industries.

The term encompasses people who may have very different understandings of their relationship with the environment and the land they seek to protect (Menton and Le Billon 2021). Many rural communities, for example, seek to ensure access to land for the sake of securing agrarian livelihoods. In doing so, they come to defend ways of life that can involve forest clearings to create farmland and assert de facto land rights in ways that may seem environmentally destructive and legally tenuous (Ghazoul and Kleinschroth 2018), even if such processes can (re)create forested and highly biodiverse landscapes (Hecht and Cockburn 2010). In contrast, park wardens seek to protect species and habitats, often through the militarized enforcement of human exclusion rules—called "fortress **conservation**"—that have evicted rural communities, undermined traditional livelihoods, and historically "re-wilded" the environment in the name of protecting game and **biodiversity** (Duffy 2016). Western conservation approaches that enclose and displace Indigenous Peoples and Local Communities (IPLCs) can themselves be a threat to these peoples and their environments (Agrawal and Redford 2009). Overall, the concept of environmental and land defenders is best applied to people involved in struggles for their community, territory, livelihoods, and healthy environments.

Defenders play an important role in **environmental peacebuilding** as they expose environmental and human rights abuses, struggle against colonial and extractivist frameworks promoting harmful projects, and seek to unsettle and decolonize **power** relations (Duffy and Le Billon 2018). As such, they tend to actively "repoliticize" environmental conflicts and demand changes to uneven power relations. Although environmental defenders mobilize through mostly peaceful forms of resistance, such as litigation and protests, they are frequently criminalized and repressed by corporate and state actors, with Indigenous Peoples being disproportionately affected (Le Billon and Lujala 2020). At least 2,106 environmental and land defenders were killed between 2012 and 2023, many in countries affected by high levels of **violence** or armed conflict such as Brazil, Colombia, Guatemala, Honduras, Mexico, Peru, and the Philippines (Global Witness 2024).

Environmental defenders often seek to mobilize the emancipatory potential of conflicts, help align environmental peacebuilding with broader **environmental justice** agendas, and promote **positive peace** and environmental **sustainability** alternatives beyond status quo options to merely "resolve" environmental conflicts (Davis et al. 2023; Le Billon and Duffy 2018). Environmental peacebuilding initiatives should thus actively seek to advance and protect the rights of environmental defenders (Khanna and Le Billon 2021). This includes addressing the structural causes of environmental injustices, including the intersectional dimensions of environmental defenders' struggles (Yoshida et al. 2021).

References

Agrawal, Arun, and Kent Redford. 2009. "Conservation and Displacement: An Overview." *Conservation and Society* 7 (1): 1. https://doi.org/10.4103/0972-4923.54790.
Davis, Katy, Laura E. Davis, Jamon Van Den Hoek, and Ken Conca. 2023. "Power in Environmental Peacebuilding." *World Development Sustainability* 3: 100110. https://doi.org/10.1016/j.wds.2023.100110.
Duffy, Rosaleen. 2016. "War, by Conservation." *Geoforum* 69: 238–48. https://doi.org/10.1016/j.geoforum.2015.09.014.
Ghazoul, Jaboury, and Fritz Kleinschroth. 2018. "A Global Perspective Is Needed to Protect Environmental Defenders." *Nature Ecology and Evolution* 2 (9): 1340–42. https://doi.org/10.1038/s41559-018-0640-1.
Global Witness. 2024. "Land and Environmental Defenders." https://www.globalwitness.org/en/campaigns/environmental-activists/numbers-lethal-attacks-against-defenders-2012/.
Hecht, Susanna B., and Alexander Cockburn. 2010. *The Fate of the Forest: Developers, Destroyers, and Defenders of the Amazon*. University of Chicago Press.
Khanna, Shivangi, and Philippe Le Billon. 2021. "Protecting and Supporting Defenders: A Review of Policies for Environmental and Land Defenders." *Policy Matters* 22 (3): 90–118. https://portals.iucn.org/library/sites/library/files/documents/Policy-Matters-Issue-22-vol3.pdf.
Le Billon, Philippe, and Rosaleen V. Duffy. 2018. "Conflict Ecologies: Connecting Political Ecology and Peace and Conflict Studies." *Journal of Political Ecology* 25 (1): 239–60. https://doi.org/10.2458/v25i1.22704.
Le Billon, Philippe, and Päivi Lujala. 2020. "Environmental and Land Defenders: Global Patterns and Determinants of Repression." *Global Environmental Change* 65: 102163. https://doi.org/10.1016/j.gloenvcha.2020.102163.
Menton, Mary, and Philippe Le Billon (eds.). 2021. *Environmental Defenders: Deadly Struggles for Life and Territory*. Routledge
United Nations General Assembly. 2016. *A/71/281*.
Yoshida, Keina, Hannah Bond, and Helen Kezie-Nwoha. 2021. *Defending the Future: Gender, Conflict, and Environmental Peace*. London School of Economics. https://eprints.lse.ac.uk/110308/1/Yoshida_defending_the_future_published.pdf.

Environmental Degradation

Richard Marcantonio

Environmental degradation is the deterioration of the environment through depletion of resources (such as quality of air, water, and soil), the destruction of ecosystems, habitat destruction, the extinction of wildlife, and pollution (Johnson et al. 1997).

Environmental degradation, along with its sources, can come in many forms—especially in **conflict**-affected areas. Environmental mediums that can be degraded include soil, fresh water or marine environments, air, wildlife, and other biotic species and abiotic materials. Each of these resources can be degraded in varied ways. For example, soil can be contaminated by unexploded ordinance, adding toxic materials to the soil composition. Soils can also be degraded because of nutrient depletion from overuse or erosion, affecting soil composition. Similarly, every environmental medium can be degraded either through the addition of materials that are disruptive to, or the subtraction of materials that are conducive to, ecosystem stability and functioning. Environmental degradation is often measured in terms of human health risks or environmental services. However, it can also be measured using other **indicators** of environmental quality, such as **biodiversity** (EPI 2022). Today, the primary driver of environmental degradation is human activities (Steffen et al. 2011). Some of these activities are more visibly destructive, including warfare and strip mining. Others are less visible, such as the carbon emissions from vehicle exhaust that contribute to global **climate change**. In global measures, such as the planetary boundaries framework, the Earth system has been broadly degraded, with assessments finding several key areas of the Earth system functioning are far outside "safe" parameters (Rockström et al. 2023). The most impactful form of environmental degradation on human health is air pollution, which results in an estimated seven million early deaths annually, shaving off about 2.9 years of global life expectancy (Lelieveld et al. 2020; Rentschler and Leonova 2023).

The degradation of the **environment** is often regulated at multiple levels, given the often-local impacts of environmentally damaging activities, even when they have potentially additional national, regional, and international impacts. In general, toxic pollution emissions tend to degrade the environmental quality of the areas immediately surrounding their emission sources and decrease in effect at further distances through dilution. In contrast, pollutants like greenhouse gas emissions impact the entire Earth system regardless of where they are emitted. The varied sources and types of pollutants make environmental and natural resource management schemes complex, as are the preferences for those causing degradation or being impacted by it. The transboundary nature of many pollution types, but especially air and water pollution, requires the coordination of numerous jurisdictional authorities to be effectively addressed. In **environmental peacebuilding**, environmental degradation can pose a potential risk of conflict or unfold because of conflict (Navas et al. 2018). Environmental degradation, or the risk of it, can also provide an opportunity to foster **cooperation** (Marcantonio 2022). It is thus important to assess the risks and possibilities of environmental degradation, such as through an environmental impact assessment (Glasson and Therivel 2019), in any **peacebuilding** programming.

References

Environmental Performance Index (EPI). 2022. *Environmental Performance Index 2022*. Yale Center for Environmental Law and Policy. https://epi.yale.edu/downloads/epi2022report06062022.pdf.

Glasson, John, and Riki Therivel. 2019. *Introduction To Environmental Impact Assessment*. 5th ed. Routledge. https://doi.org/10.4324/9780429470738.

Johnson, David L., Stanley H. Ambrose, Thomas J. Bassett, et al. 1997. "Meanings of Environmental Terms." *Journal of Environmental Quality* 26 (3): 581–89. https://doi.org/10.2134/jeq1997.00472425002600030002x.
Lelieveld, Jos, Andrea Pozzer, Ulrich Pöschl, Mohammed Fnais, Andy Haines, and Thomas Münzel. 2020. "Loss of Life Expectancy from Air Pollution Compared to Other Risk Factors: A Worldwide Perspective." *Cardiovascular Research* 116 (11): 1910–17. https://doi.org/10.1093/cvr/cvaa025.
Marcantonio, Richard. 2022. "Toxic Diplomacy Through Environmental Management: A Necessary Next Step for Environmental Peacebuilding." *World Development Perspectives* 28: 100471. https://doi.org/10.1016/j.wdp.2022.100471.
Navas, Grettel, Sara Mingorria, and Bernardo Aguilar-González. 2018. "Violence in Environmental Conflicts: The Need for a Multidimensional Approach." *Sustainability Science* 13 (3): 649–60. https://doi.org/10.1007/s11625-018-0551-8.
Rentschler, Jun, and Nadezda Leonova. 2023. "Global Air Pollution Exposure and Poverty." *Nature Communications* 14 (1): 4432. https://doi.org/10.1038/s41467-023-39797-4.
Rockström, Johan, Joyeeta Gupta, Dahe Qin, et al. 2023. "Safe and Just Earth System Boundaries." *Nature*: 1–10. https://doi.org/10.1038/s41586-023-06083-8.
Steffen, Will, Åsa Persson, Lisa Deutsch, et al. 2011. "The Anthropocene: From Global Change to Planetary Stewardship." *AMBIO* 40 (7): 739–61. https://doi.org/10.1007/s13280-011-0185-x.

Environmental Justice

Veronica Limeberry and Laurel Levin

Environmental justice (EJ) is both a social movement and a conceptual framework that seeks to address the inequitable distribution of environmental contaminants and resources. EJ emerged from Black and Indigenous civil rights activism, which exposed how environmental externalities, distribution of environmental burdens, inequitable policy, and unequal distribution of benefits have collectively reinforced existing structural inequalities along racial, ethnic, and other socioeconomic lines.

A commonly used definition of EJ from the US Environmental Protection Agency (US EPA) calls for "the fair treatment and meaningful involvement of all people regardless of race, color, culture, national origin, income, and educational levels with respect to the development, implementation, and enforcement of protective environmental laws, regulations, and policies" (U.S. EPA 2020). EJ scholarship and activism address the unequal distribution of environmental harms, often in communities already overburdened by pollution and toxins. Furthermore, EJ also examines and advocates for equitable environmental decision-making and resource reallocation.

EJ and **peacebuilding** are interwoven. At their core are principles of humanity, dignity, and fairness regarding **human security** and livelihoods. Both fields deal with the distribution of resource scarcity and abundance, and associated harms and privileges (Conca and Wallace 2009; Kojola and Pellow 2021; Martinez-Alier 2001). The United Nations (1987) report *Our Common Future* identified environmental problems as a cause of violent **conflict**. In these cases, conflicts may arise over access—or the inability to access—to land, water, and **critical minerals**.

The EJ movement emerged in the United States from the nexus of the civil rights movement and the environmental movement, where communities of color traced the correlation between public health disparities and proximity to pollution, to the legacy of colonialism and civic injustice (Kojola and Pellow 2021; Ranganathan

2017; Bullard and Johnson 2000, Bullard 2008). Early scholars, including Dr Robert Bullard in his seminal book *Dumping in Dixie* (1994), chronicled how race was the greatest **indicator** of one's exposure to toxic substances (Bullard 1994). In 1991, Bullard and other civil rights organizers came together for the inaugural multinational People of Color Environmental Leadership Summit. There, they developed the 13 Principles of Environmental Justice (1991), which include calls for colonial reconciliation and Indigenous sovereignty, political self-determination, opposition to military occupation, and compensation for environmental harm. To date, EJ focuses primarily on: (1) recognition of EJ communities' meaningful **participation** in environmental decision-making, (2) fair distribution of environmental burdens and benefits, and (3) how these impact individual and communal capabilities and functioning (Schlosberg 2009).

Since the early 1990s, EJ has proliferated across academic disciplines and organizing contexts. EJ has "developed as a concrete and material challenge to the dominant neoliberal discourse of ecological modernization" (Debbané and Keil 2004). Similarly, in **political ecology**—particularly urban political ecology—EJ "helps to understand links between globalization, urbanization, and nature" (Debbané and Keil 2004, 210). Since then, scholar-activism has expanded the concept of EJ to consider how it is a form of structural **violence**. In *The Environment as Freedom: A Decolonial Reimagining,* Ranganathan writes how "federally-enforced segregation, the criminalization of poverty, and neoliberal austerity [...] the lack of decent schools, poor food and nutrition, high maternal mortality, and the steady erosion of intergenerational wealth" (Ranganathan 2017, 4) are all examples of the "slow violence" of environmental injustice (Nixon 2011).

Understanding the connection between environmental injustice and human security is essential for effective **conflict prevention**, **mediation**, and **mitigation**, along with peacebuilding efforts (Dibaba 2020). EJ underpins analysis and commitment to understanding and engaging with social, environmental, and structural inequalities that perpetuate racial and ethnic disparities. Thus, making EJ a central concept in policymaking can help address the resource inequalities, social impacts of environmental harms, and inequitable outcomes of peacebuilding. Securing EJ is also a practice of advancing "positive peace." **Positive peace**, as defined in peacebuilding, pertains to not just the absence of violence but also ensuring people have the resources to thrive (Dibaba 2020). Hence, EJ is fundamental to both policy and theoretical frameworks of **environmental peacebuilding**, particularly related to vulnerable and oppressed groups.

References

Bullard, Robert D. 1994. *Dumping in Dixie: Race, Class, and Environmental Quality*. Westview Press.

Bullard, Robert D. 2008. "Toxic Wastes and Race at Twenty: Why Race Still Matters After All of These Years." *Environmental Law* 3: 371. https://www.jstor.org/stable/43267204.

Bullard, Robert D., and Glenn S. Johnson. 2000. "Environmental Justice: Grassroots Activism and Its Impact on Public Policy Decision Making." *Journal of Social Issues* 56 (3): 555-78. https://doi.org/10.1111/0022-4537.00184.

Conca, Ken, and Jennifer Wallace. 2009. "Environment and Peacebuilding in War-Torn Societies: Lessons From the U.N. Environment Programme's Experience With Postconflict Assessment." *Global Governance* 15 (4): 485–504. https://www.jstor.org/stable/27800777.
Debbané, Amanda, and Roger Keil. 2004. "Multiple Disconnections: Environmental Justice and Urban Water in Canada and South Africa." *Space and Polity* 8 (2): 209–25. https://doi.org/10.1080/1356257042000273968.
Dibaba, Ayantu. 2020. "Environmental Justice and Peacebuilding Practices in Ethiopia." *International Journal of Ecosystem* 10 (1). https://www.researchgate.net/publication/343390775_Environmental_Justice_and_Peacebuilding_Practices_in_Ethiopia.
First National People of Color Environmental Leadership Summit. 1991. *Principles of Environmental Justice*. Energy Justic Network. https://www.ejnet.org/ej/principles.pdf.
Kojola, Erik, and David N. Pellow. 2021. "New Directions in Environmental Justice Studies: Examining the State and Violence." *Environmental Politics* 30 (1–2): 100–18. https://doi.org/10.1080/09644016.2020.1836898.
Martinez-Alier, Joan. 2001. "Environmental Conflicts, Environmental Justice, and Valuation." Working paper, UHE/UAB. Universitat Autònoma de Barcelona.
Nixon, Rob. 2011. *Slow Violence and the Environmentalism of the Poor*. Harvard University Press. https://doi.org/10.2307/j.ctt2jbsgw.
Ranganathan, Malini. 2017. "The Environment as Freedom: A Decolonial Reimagining." In *Social Science Research Council Items*. African American Intellectual History Society, June 24. https://www.aaihs.org/the-environment-as-freedom-a-decolonial-reimagining/.
Schlolsberg, David. 2009. *Defining Environmental Justice: Theories, Movements, and Nature*. Oxford University Press.
U.S. Environmental Protection Agency (EPA) Office of Policy. 2020. "EJ 2020 Glossary." *Data and Tools*. https://19january2021snapshot.epa.gov/environmentaljustice/ej-2020-glossary.

Environmental Migration

Galeo Saintz and Alessandra Vaselli

Environmental migration is:

> the movement of persons or groups of persons who, predominantly for reasons of sudden or progressive changes in the environment that adversely affect their lives or living conditions, are forced to leave their places of habitual residence, or choose to do so, either temporarily or permanently, and who move within or outside their country of origin or habitual residence.
>
> (International Organization for Migration 2019)

Environmental—or **environment**-related—migration and **conflict** interact through complex and locally specific variables that intersect broader social, economic, and political dynamics across geographic and institutional scales (Swain 1996; Suhrke 1997; Freeman 2017; Adger et al. 2018; Savelli et al. 2023). The intersection of environmental migration and conflict could occur during or after migration, and the types of conflict could be those within or across national borders (i.e., conflicts between states, a government against a non-state group, or intergroup) (Swain 1996). Interstate conflicts can be triggered or intensified through transboundary environmental migration when the receiving state's policies clash with the origin state's mobility regulations and law enforcement measures (Swain and Öjendal 2018). Unsupported migration driven by **environmental degradation** and environment-related **displacement** can further undermine the livelihoods and **human security** of people on the move and in destination areas, exacerbating social tensions and contributing to conflict. Overcrowded and underserved destination areas can experience social discontent and trigger organized anti-government opposition, along with inter- and intra-group conflicts (Freeman 2017; Koubi et al. 2021). These types of conflicts are often grounded in resource competition, social inequalities, structural **violence**, marginalization and exploitation, and pre-existing cultural tensions, sometimes exacerbated by national narratives (Wiederkehr et al. 2022; Savelli et al. 2023; Sändig et al. 2024). The **environmental peacebuilding** literature underscores the potential to address these conflicts through effective environmental management and **governance**. Reducing environmental degradation, supporting livelihoods, and enhancing national and international **disaster risk reduction (DRR)** initiatives are expected to mitigate the drivers of involuntary environmental migration in origin areas and reduce resource- and disaster-related conflicts in destination areas (Brzoska 2019; He et al. 2024).

The concept of environmental migration also broadly addresses the critical intersection of **environmental change** and mobility challenges and opportunities, with significant implications for **environmental justice**, humanitarian aid, development, **peace**, and conflict. Since gaining prominence in policy and academic discourses, the term environmental migration has been subject to ongoing debate, particularly regarding the social implications of how the relationship between environmental change and human mobility is conceptualized (Boas and Wiegel 2021). Critics caution that the labels "environmental refugees" or "**climate migrants**" risk reinforcing environmental determinism and obscuring the complex interplay of political, social, and economic factors contributing to their movement (Oliver-Smith 2012; Boas and Wiegel 2021). In contrast, there is a large consensus that environmental change, including **climate change**-induced variabilities, does not directly cause migration. Scholars emphasize that the relationship between environmental change and migration is mediated by a range of factors that intersect with individual preferences and broader societal structures. The specific mechanisms through which environmental change influences migration, as well as the extent of its impact on migration decisions, remain debated (Warner 2010; Oliver-Smith 2012; Boas and Wiegel 2021; Adger et al. 2021; McLeman et al. 2021; Han et al. 2024). In this context, efforts to quantify and model environmental migration encounter significant methodological and theoretical challenges in capturing the complexity of multi-scalar connections between environmental change and migration, as well as the diversity of migration forms and decision-making processes (Obokata et al. 2014; Hoffmann et al. 2021).

Understanding the various locally specific dimensions of environmental migration at the intersection of environmental change, human mobility, and social stability emphasizes the role of environmental peacebuilding approaches that tackle environmental governance, social equity, and **conflict sensitivity** to foster inclusive peace in both origin and destination communities.

References

Adger, W. Neil, Ricardo Safra De Campos, and Colette Mortreux. 2018. "Mobility, Displacement and Migration, and Their Interactions with Vulnerability and Adaptation to Environmental Risks." In *Routledge Handbook of Environmental Displacement and Migration*, edited by

Robert McLeman and François Gemenne. Routledge. https://doi.org/10.4324/9781315638843-3.
Adger, W. Neil, Ricardo Safra De Campos, Tasneem Siddiqui, et al. 2021. "Human Security of Urban Migrant Populations Affected by Length of Residence and Environmental Hazards." *Journal of Peace Research* 58 (1): 50–66. https://doi.org/10.1177/0022343320973717.
Boas, Ingrid, and Hanne Wiegel. 2021. "Environmental Change and Migration." In *Introduction to International Migration*, edited by Jeannette Money and Sarah P. Lockhart. Routledge. https://doi.org/10.4324/9781003167631-18.
Brzoska, Michael. 2019. "Understanding the Disaster–Migration–Violent Conflict Nexus in a Warming World: The Importance of International Policy Interventions." *Social Sciences* 8 (6): 167. https://doi.org/10.3390/socsci8060167.
Freeman, Laura. 2017. "Environmental Change, Migration, and Conflict in Africa: A Critical Examination of the Interconnections." *The Journal of Environment and Development* 26 (4): 351–74. https://doi.org/10.1177/1070496517727325.
Han, Qing, Rupesh Kumar, and Amit Kumar. 2024. "Climate Change and Human Migration: Perspectives for Environmentally Sustainable Societies." *Journal of Geochemical Exploration* 256: 107352. https://doi.org/10.1016/j.gexplo.2023.107352.
He, Linlang, Elizabeth Kreske, Stephanie J. Nawyn, et al. 2024. "Interventions Addressing Conflict in Communities Hosting Climate-Influenced Migrants: Literature Review." *Environment and Security* 2 (1): 145–74. https://doi.org/10.1177/27538796231207919.
Hoffmann, Roman, Barbora Šedová, and Kira Vinke. 2021. "Improving the Evidence Base: A Methodological Review of the Quantitative Climate Migration Literature." *Global Environmental Change* 71: 102367. https://doi.org/10.1016/j.gloenvcha.2021.102367.
International Organization for Migration (IOM). 2019. *Glossary on Migration*. https://publications.iom.int/system/files/pdf/iml_34_glossary.pdf.
Koubi, Vally, Quynh Nguyen, Gabriele Spilker, and Tobias Böhmelt. 2021. "Environmental Migrants and Social-Movement Participation." *Journal of Peace Research* 58 (1): 18–32. https://doi.org/10.1177/0022343320972153.
McLeman, Robert, David Wrathall, Elisabeth Gilmore, Philip Thornton, Helen Adams, and François Gemenne. 2021. "Conceptual Framing to Link Climate Risk Assessments and Climate-Migration Scholarship." *Climatic Change* 165 (1–2): 24. https://doi.org/10.1007/s10584-021-03056-6.
Obokata, Reiko, Luisa Veronis, and Robert McLeman. 2014. "Empirical Research on International Environmental Migration: A Systematic Review." *Population and Environment* 36 (1): 111–35. https://doi.org/10.1007/s11111-014-0210-7.
Oliver-Smith, Anthony. 2012. "Debating Environmental Migration: Society, Nature and Population Displacement in Climate Change." *Journal of International Development* 24 (8): 1058–70. https://doi.org/10.1002/jid.2887.
Sändig, Jan, Natalia Dalmer, Tobias Ide, and Anselm Vogler. 2024. "From Climate Conflicts to Environmental Peacebuilding: Exploring Local Dimensions." *Environment and Security* 2 (1): 3–20. https://doi.org/10.1177/27538796241231090.
Savelli, Adam, Frans Schapendonk, Tanaya Dutta Gupta, Grazia Pacillo, and Peter Läderach. 2023. "Climate Change, Mobility and Violent Conflict: A Typology of Interlinked Pathways." *International Development Planning Review* 45 (4): 1–34. https://doi.org/10.3828/idpr.2023.2.
Suhrke, Astri. 1997. "Environmental Degradation, Migration, and the Potential for Violent Conflict." In *Conflict and the Environment*, edited by Nils Petter Gleditsch. Springer Netherlands. https://doi.org/10.1007/978-94-015-8947-5_16.
Swain, Ashok. 1996. "Environmental Migration and Conflict Dynamics: Focus on Developing Regions." *Third World Quarterly* 17 (5): 959–74. https://doi.org/10.1080/01436599615209.
Swain, Ashok, and Joakim Öjendal, eds. 2018. *Routledge Handbook of Environmental Conflict and Peacebuilding*. Routledge.
Warner, Koko. 2010. "Global Environmental Change and Migration: Governance Challenges." *Global Environmental Change* 20 (3): 402–13. https://doi.org/10.1016/j.gloenvcha.2009.12.001.
Wiederkehr, Charlotte, Tobias Ide, Ralf Seppelt, and Kathleen Hermans. 2022. "It's All About Politics: Migration and Resource Conflicts in the Global South." *World Development* 157: 105938. https://doi.org/10.1016/j.worlddev.2022.105938.

Environmental Peacebuilding

Tareq Abu Hamed, Carl Bruch, and Mara Pusic

Environmental peacebuilding comprises approaches and methodologies in which shared natural resource management, climate change interventions, and environmental threat-related initiatives—whether in programming or scholarship—are applied to prevent, mitigate, resolve, or recover from conflicts and support peace (Ide et al. 2021). It also includes the peace, conflict, and security dimensions of environmental programming, such as conflict-sensitive conservation (Woomer 2018).

Environmental peacebuilding is neither a distinct school of thought nor a concrete set of activities, but a broad umbrella term used by academics, practitioners, and decision-makers to describe the relationships and pathways that emerge at the nexus of **environment**, **conflict**, and **peace** (Dresse et al. 2016). Environmental peacebuilding is sometimes described as a meta-framework encompassing a range of **theories of change**, activities, and practices at the intersection of environment, conflict, and peace (Balinskaia 2024).

Environmental peacebuilding may involve **cooperation** over natural resource management, **disaster risk reduction**, and potentially climate **adaptation** (Conca and Dabelko 2002; Pieternel de Bruin 2022; Reiling and Brady 2015). Other framings of environmental peacebuilding emphasize the role of environment and **natural resources** in supporting specific peacebuilding objectives around **security**, livelihoods, economic recovery, **basic services**, and good **governance** (Bruch et al. 2016). Other framings focus on conflict-sensitive **conservation** (Ajroud et al. 2017).

A common element of environmental peacebuilding is the transboundary nature of environmental issues and the resulting cooperation that emerges from within ecosystem borders rather than politico-territorial borders (Dresse et al. 2018). In addition to using shared natural resources to prevent conflict between neighboring countries or communities, environmental peacebuilding also strengthens and sustains **peace agreements** and reduces the existential threat of **climate change**, even between countries that do not share borders (Sommer and Fassbender 2024).

Environmental diplomacy (including **climate diplomacy**) is a form of environmental peacebuilding in which diplomacy resolves disputes and prevents conflict regarding shared natural resources (Kurbalija 2020). In this context, the environment can be a diplomatic tool that plays a critical role in initiating and strengthening diplomatic relations that lead to addressing global environmental challenges and promoting **sustainable development** (Feder 2024).

Increasingly, environmental peacebuilding focuses on qualities such as **gender**, race, indigeneity, age, disability, and religion (Davis et al. 2023), especially how these characteristics relate to equality, **power**, and **environmental justice**. Environmental peacebuilding can take place at different scales within and between countries or communities. Environmental peacebuilding efforts to prevent, mitigate, resolve, and recover from conflicts and support peace include a diverse range of actors from the national government, local authorities, private sector, civil society, and intergovernmental organizations.

References

Ajroud, Brittany, Nathalie Al-Zyoud, Lydia Cardona, Janet Edmond, Danny Pavitt, and Amanda Woomer. 2017. *Environmental Peacebuilding: Training Manual.* Conservation International.

Balinskaia, Aleksandra. 2024. "Interstate Environmental Peacebuilding: A Systematic Review." *Environment and Security.* https://doi.org/10.1177/27538796241287366.

Bruch, Carl, Carroll Muffett, and Sandra Nichols, eds. 2016. *Governance, Natural Resources, and Post-Conflict Peacebuilding.* Routledge. https://doi.org/10.4324/9780203109793.

Conca, Ken, and Geoffrey Dabelko, eds. 2002. *Environmental Peacemaking.* Johns Hopkins University Press and Woodrow Wilson Center Press.

Davis, Katy, Laura E. R. Peters, Jamon Van Den Hoek, and Ken Conca. 2023. "Power in Environmental Peacebuilding." *World Development Sustainability* 3. https://doi.org/10.1016/j.wds.2023.100110.

Dresse, Anais, Itay Fischhendler, Jonas O. Nielsen, and Dimitrios Zikos. 2018. "Environmental Peacebuilding: Towards a Theoretical Framework." *Cooperation and Conflict* 54 (1): 99–119. https://doi.org/10.1177/0010836718808331.

Dresse, Anaïs, Jonas Østergaard Nielsen, and Dimitrios Zikos. 2016. "Moving Beyond Natural Resources as a Source of Conflict: Exploring the Human-Environment Nexus of Environmental Peacebuilding." Thesis Discussion Paper No. 2016-2. Humboldt-Universität.

Feder, Toni. 2024. "Q&A: Tareq Abu Hamed Champions Environmental Cooperation in the Middle East." *Physics Today* 77 (10): 26–28. https://doi.org/10.1063/pt.jqye.ljeb.

Ide, Tobias, Carl Bruch, Alexander Carius, et al. 2021. "The Past and Future(s) of Environmental Peacebuilding." *International Affairs* 97 (1): 1–16. https://doi.org/10.1093/ia/iiaa177.

Kurbalija, Jovan. 2020. "Environmental Diplomacy." *Blog*, December 11. https://www.diplomacy.edu/topics/environmental-diplomacy/.

Pieternel de Bruin, Sophie. 2022. "Security Dimensions of Climate Change Adaptation." In *Research Handbook on Climate Change Adaptation Law*, edited by Jonathan Verschuuren. Edward Elgar Publishing.

Reiling, Kirby, and Cynthia Brady. 2015. *Climate Change and Conflict: An Annex to the USAID Climate-Resilient Development Framework.* USAID.

Sommer, Udi, and Francesca Fassbender. 2024. "Environmental Peacebuilding: Moving Beyond Resolving Violence-Ridden Conflicts to Sustaining Peace." *World Development* 178: 106555. https://doi.org/10.1016/j.worlddev.2024.

Woomer, Amanda. 2018. "Conflict Sensitivity and Conservation: Evaluating Design, Implementation, and Practice." PhD diss., Kennesaw State University. https://digitalcommons.kennesaw.edu/incmdoc_etd/18/.

Environmental Peacemaking

Larry A. Swatuk

Environmental peacemaking is a process by which social conflict is prevented, reduced, or eliminated through environmental cooperation.

> Environmental peacemaking is the practice of using shared environmental interests—such as the stewardship of transboundary **natural resources** or joint **conservation** initiatives—to build trust, manage **conflict**, and foster **cooperation** between communities or nations. In essence, it reframes environmental challenges as opportunities for dialogue and **peace** rather than as sources of contention. Such cooperation can vary widely in form and scale, involving creative coalitions of community, nonprofit, private sector, and state actors.

Environmental peacemaking's typical **theory of change** builds on the idea that there are two general, often complementary pathways for positive change: changing the strategic climate and strengthening post-Westphalian **governance**. The first path involves transforming the more immediate problems of mistrust, uncertainty, suspicion, divergent interests, and short time horizons that typically accompany many conflictual situations. The second path views peace not simply as stable interstate relations but, more broadly, as a shared collective identity within which violent conflict becomes inconceivable (Conca 2002, 10). While the first focuses heavily on functional, practical, and discursive approaches to moving beyond conflict, the second is primarily institutionalist in nature. Institutionalists in this instance may be formal (e.g., the creation of transboundary river basin organizations) or informal, operating at a variety of social and socio-ecological scales: village, stream, country, river, region, and/or watershed.

Environmental peacemaking is a foundational component of **environmental peacebuilding**—it serves as an early-stage or entry-point strategy within the broader **peacebuilding** process by leveraging shared environmental interests to initiate cooperation and trust among divided parties. Put differently, environmental peacemaking is a tactical approach that contributes directly to environmental peacebuilding, which is the strategic, long-term process of embedding **environmental cooperation** into **sustainable peace**. Prominent inter- and intrastate areas of engagement include transboundary water management (Becker et al. 2007; Earle et al. 2010), **peace parks** and protected areas (Ali 2007; Duffy 2001; Swatuk 2022), and joint climate and environmental initiatives (Conca and Dabelko 2002; Sändig et al. 2024).

There are several ways in which environmental peacemaking contributes to environmental peacebuilding: (1) initiating dialogue in conflict-affected contexts, as environmental peacemaking creates neutral ground for communication (e.g., communities or states that distrust each other may still cooperate over shared watersheds or climate risks), (2) building trust and cooperative institutions, as repeated interaction over shared environmental issues fosters interpersonal and institutional trust, which is critical for more comprehensive post-conflict reconciliation and state-building, (3) shifting perceptions from zero-sum to shared benefits, as emphasizing mutual gains from environmental cooperation (e.g., shared water management, **renewable energy** projects) can reframe relationships from adversarial to collaborative—an essential step in sustainable peacebuilding, (4) supporting long-term **sustainable development**, as environmental peacemaking often introduces joint resource management or conservation practices that resolve immediate tensions and align with the goals of long-term environmental peacebuilding—such as **resilience**, equity, and ecological health, and (5) addressing root causes of conflict, as many conflicts have environmental roots (e.g., competition over land or water) and peacemaking efforts directly target these sources to help ensure that peacebuilding is not just reactive but is preventive.

References

Ali, Saleem H., ed. 2007. *Peace Parks, Conservation and Conflict Resolution*. MIT Press.

Becker, Gert, Jeroen Aerts, and Dave Huitema. 2007. "Transboundary Flood Management in the Rhine Basin: Challenges for Improved Cooperation." *Water Science and Technology* 56 (4): 125–35. https://doi.org/10.2166/wst.2007.544.

Conca, Ken. 2002. "The Case for Environmental Peacemaking." In *Environmental Peacemaking*, edited by Ken Conca and Geoffrey D. Dabelko. Johns Hopkins University Press and Woodrow Wilson Center Press.
Conca, Ken, and Geoffrey D. Dabelko, eds. 2002. *Environmental Peacemaking*. Johns Hopkins University Press and Woodrow Wilson Center Press.
Duffy, Rosaleen. 2001. "Peace Parks: The Paradox of Globalization." *Geopolitics* 6 (2): 1–26. https://doi.org/10.1080/14650040108407715.
Earle, Anton, Anders Jägerskog, and Joakim Öjendal. 2010. *Transboundary Water Management: Principles and Practice*. Routledge. https://doi.org/10.4324/9781849776585.
Sändig, Jan, Natalia Dalmer, Tobias Ide, and Anselm Vogler. 2024. "From Climate Conflicts to Environmental Peacebuilding: Exploring Local Dimensions." *Environment and Security* 2 (1). https://doi.org/10.1177/27538796241231090.
Swatuk, Larry A. 2022. "Peace Parks in Theory and Practice: Confronting the Elephant in the Room." In *Routledge Handbook of Environment Security*, edited by Richard Matthew, Evgenia Nizkorodov, and Crystal Murphy. Routledge.

Environmental Security

Stephen Lonergan and Aliakbar Shams

Environmental security is a contested term in part because it combines two ambiguous terms: environment and security. For some, environmental security relates to how human activities, including military activities, affect the natural environment, whether the impacts are caused by accident, mismanagement, or design. For others, it pertains to how environmental change might be a threat to national security or human security, including if environmental change causes conflict (Glenn et al. 2012 ; Barnett 2009).

The term environmental security initially arose in the context of the US military's use of defoliants in Southeast Asia during the Vietnam War. These activities focused international attention on the intentional and unintentional environmental damages caused by **war**. Two treaties soon addressed the issue: the Additional Protocol I to the 1949 Geneva Convention on the Protection of Victims of International Armed Conflicts (1977) and the Convention on the Prohibition of Military or Any Other Hostile Use of Environmental Modification Techniques (1977). Both treaties focus on the protection or **security** of the natural **environment** during international armed conflict.

At roughly the same time, a different use of the term environmental security arose from research and discussion on how **environmental degradation** and resource depletion affect human development and, over time, contribute to political and economic instability or **conflict** (Swain 1993; Dalby 1994; Matthew 1995). While there was previous research on the relationship between **environment**/resources and conflict dating back to the 1950s, a broader interest in non-conventional security threats, coupled with the realization that the world was facing significant global environmental issues, resulted in a coupling of the two terms (Barnett, 2009). Water stress, deforestation, declining fish stocks, drought, and pollution, along with slow-onset events like **climate change**, all have the potential to affect human and national security. Although finding evidence of a direct causal link between environment and security is problematic, research has shown that **environmental change** is related to insecurity through conditions of inequality, institutional weakening, and impoverishment

(Lonergan 2001). The term environmental security has been adopted by both national and international agencies, including various UN agencies (United Nations Development Programme 1994), the U.S. Department of State (2001), and the North Atlantic Treaty Organization (NATO) (2024).

The term "environmental security" is not without criticism. The term "environmental" is ill-defined and encompasses a vast array of problems, from industrial accidents to climate change. It is difficult, if not impossible, to relate these risks to violent conflict, let alone to a broader concept of security. Similarly, while redefining security to include non-conventional threats like environmental change might be a useful concept, nonmilitary threats make the role of militaries and their mandates almost untenable.

Despite these concerns, there is a close relationship between environmental security, however ill-defined this might be, and **environmental peacebuilding.** If environmental change is seen as a threat to human or national security, focusing on **environmental cooperation** and sustainable resource management is essential to mitigating these threats or adapting to them. If environmental change is influenced by efforts to promote human or military security, environmental cooperation and sustainable resource management are also crucial. In either scenario, the two concepts are inextricably linked.

References

Barnett, Jon. 2009. "Environmental Security." In *The International Encyclopedia of Human Geography*, 553–557. Elsevier. https://doi.org/10.1016/B978-008044910-4.00774-4.

Dalby, Simon. 1994. "The Politics of Environmental Security." In *Green Security or Militarized Environment*, edited by Jyrki Käkönen. Dartmouth Publishing Company.

Glenn, Jerome C., Theodore J. Gordon and Renat Perelet, 2012. Chapter 9: Environmental Security. In: Jerome C. Glenn, Theodore J. Gordon and Elizabeth Florescu, eds. 2012 State of the Future. Washington,DC: The Millennium Project. 102pp.

Lonergan, Steven. 2001. "Environmental Security." In *The International Encyclopedia of Global Change*, edited by Andrew S. Goudie and David J. Cuff. Oxford University Press.

Matthew, Richard. 1995. "Environmental Security and Conflict: An Overview of the Current Debate." *National Security Studies Quarterly* 1 (1): 1–10.

North Atlantic Treaty Organization (NATO). 2024. "Environment, Climate Change, and Security." *Effective*, July 18. https://www.nato.int/cps/en/natohq/topics_91048.htm.

Swain, Ashok. 1993. "Environment and Conflict: Analyzing the Developing World." Department of Peace and Conflict Research, Uppsala University. https://doi.org/10.13140/2.1.1676.2889.

United Nations Development Programme (UNDP). 1994. *The Human Development Report.* Oxford University Press.

U.S. Department of State. 2001. *Environmental Security Threat Report.* https://2001-2009.state.gov/p/eur/rls/rpt/2001/c4647.htm.

Environmental Terrorism

Shannon O'Lear, Elizabeth Chalecki, and Jeremiah Ogonda Asaka

Environmental terrorism is the targeting of environmental features such as commodity resources, infrastructure, ecosystem services, and other resources that are recognized as having some kind of value with the intention of communicating a political message or evoking fear.

Environmental resources and terrorism intersect in different ways. Two main categories of intersection include **ecoterrorism**, which are acts committed from a pro-environment stance, and environmental terrorism (Brown 2021), in which environmental features are tools or targets of a political message (Chalecki 2002, 2024).

Definitions of terrorism are extremely contested, and there is no universally agreed-upon definition. By some standards, terrorism includes intentional acts or threats of action that have targeted symbolic **violence**. Threats or actions of terrorism are outside of international laws of either **war** or peacetime (Schwartz 1998). Terrorism is an act of deliberate communication of a group's political objective through acts or threats of indiscriminate violence against a symbolic target or audience (Hoffman 2017).

A geographically nuanced approach to terrorism can be useful in understanding instances of environmental terrorism. Environmental terrorism can happen during wartime or **peace**time. International war involves sovereign states, but acts of environmental terrorism can happen at other spatial scales, such as within a state or when that state is not at war with other states. Avoiding general abstractions and distinctions such as "nonstate violence"—instead inquiring into how state-level policies, actions, and inactions can contribute to forms of tension and inequity—may also motivate violence involving environmental features (O'Lear 2003).

Environmental terrorism may intersect with **environmental peacebuilding** in different ways. Top-down strategies of **peacebuilding** can limit, rather than open, conversations about what **peace** means (Hammett and Marshall 2017), and environmental peacebuilding recognizes the importance of involving actors and efforts at multiple, interconnected spatial scales of activity. Environmental peacebuilding may be unlikely to curb the risk of terrorism—environmental or otherwise—depending on how tied a particular organization is to local funding, conditions, or recruitment potential (Hall 2024).

Since acts of environmental terrorism would focus on communicating through violence a political message or invoking fear, protecting, conserving, or reallocating benefits from environmental features would likely hold little interest for potential environmental terrorists—unless the benefits or value of environmental features, integrated with peaceful efforts, would address that organization's particular political concerns.

How might **climate change** influence environmental terrorism? An analysis of academic research on climate change, peace, and **conflict** indicates an increasing volume of work on these topics since 1990, including increased attention to risks such as terrorism (Sharifi et al. 2021). One collection of work focused on climate change and terrorism considers how climate change could exacerbate different kinds of tension associated with mass human migration, vulnerable agricultural systems and economies, and stressed water **infrastructure** in ways that incite acts of terrorism (Silke and Morrison 2022). Future research on how climate change is connected to terrorism could be usefully focused on context-specific **mitigation**, reduction of social vulnerability, and building **resilience** toward effective policy interventions at multiple spatial scales and in different geographic regions (Asaka 2021).

References

Asaka, Jeremiah O. 2021. "Climate Change-Terrorism Nexus? A Preliminary Review/Analysis of the Literature." *Perspectives on Terrorism* 15 (1): 81–92. https://www.jstor.org/stable/26984799.
Brown, Joseph M. 2021. "Civil Disobedience, Sabotage, and Violence in US Environmental Activism." In *Oxford Handbook of Comparative Environmental Politics*, edited by Jeannie Sowers, Stacy D. VanDeveer, and Erika Weinthal. Oxford University Press.
Chalecki, Elizabeth L. 2002. "A New Vigilance: Identifying and Reducing the Risks of Environmental Terrorism." *Global Environmental Politics* 2 (1): 46–64. https://doi.org/10.1162/152638002317261463.
Chalecki, Elizabeth L. 2024. "Environmental Terrorism Twenty Years On." *Global Environmental Politics* 24 (1): 1–9. https://doi.org/10.1162/glep_a_00728.
Hall, Jake. 2024. "Can Environmental Peacebuilding Tools Effectively Combat Terrorism in the Democratic Republic of the Congo? A Case Study of the Islamic State's Central Africa Province." *Cornell International Affairs Review* 17 (1): 9–62. https://cornelliar.org/issues.
Hammett, Daniel, and David Marshall. 2017. "Building Peaceful Citizens? Nation-Building in Divided Societies." *Space and Polity* 21 (2): 129–43. https://doi.org/10.1080/13562576.2017.1330383.
Hoffman, Bruce. 2017. *Inside Terrorism*. 3rd ed. Columbia University Press.
O'Lear, Shannon. 2003. "Environmental Terrorism: A Critique." *Geopolitics* 8 (3): 127–50. https://doi.org/10.1080/14650040412331307742.
Schwartz, Daniel M. 1998. "Environmental Terrorism: Analyzing the Concept." *Journal of Peace Research* 35 (4): 483–96. https://doi.org/10.1177/0022343398035004005.
Sharifi, Ayyoob, Dahlia Simangan, and Shinji Kaneko. 2021. "Three Decades of Research on Climate Change and Peace: A Bibliometrics Analysis." *Sustainability Science* 16: 1079–95. https://doi.org/10.1007/s11625-020-00853-3.
Silke, Andrew, and John Morrison. 2022. "Gathering Storm: An Introduction to the Special Issue on Climate Change and Terrorism." *Terrorism and Political Violence* 34 (5): 883–93. https://www.tandfonline.com/doi/pdf/10.1080/09546553.2022.2069444.

Environmental Violence

Richard Marcantonio

Environmental violence is the direct and indirect harm experienced by humans due to toxic and non-toxic pollutants put into a local—and concurrently the global—ecosystem through human activities and processes (Marcantonio 2022a).

> Environmental **violence** specifies and centers excess human-produced pollution as a violent environmental health hazard (Marcantonio and Fuentes 2023). Pollution is in excess when human flourishing has been maximized, and its production meets, or is not required to meet, a human need and instead causes more externalized human suffering than it serves to prevent. This definition recognizes that environmental violence arises as part of dynamic anthropogenic socio-ecological processes. In this process, ex-ante conditions and choices result in a material vector that causes damage to human health and flourishing, with potentially ex-post cascading and cumulative effects. These material vectors constitute a violent act because they are the result of intentional, excessive human consumption of resources and energy, and other activities, well beyond what is needed to maximize flourishing, all of which could lead to environmental hazards that directly and knowingly degrade human health. Seemingly paradoxically, many such practices and material consumption are in fact most often associated with the promotion of the human condition; however, they

ultimately become harmful to these ends if undertaken excessively. The persistence in causes and consequences of environmental violence is facilitated, at local to global scales, by the immense inequality undergirding it, particularly the inverse relationship between contributions and exposure. In general, the people most responsible for driving and producing environmental violence are the least exposed and impacted, and vice versa. For example, poverty is strongly correlated with toxic air pollution exposure, with 80% of the 7.3 billion people annually exposed to unsafe air pollution living in low- and middle-income countries and in low-income areas, despite generally contributing the least to its production (Marcantonio and Field 2025; Rentschler and Leonova 2023).

In **environment**, **conflict**, and **peace** research and practice, environment-conflict investigations and programming have predominantly focused on the environment as a contributing factor to direct violence. Environmental violence, progenerated by the concept of **environmental justice**, focuses on human-produced pollution as a material vector of violence—akin to but different from direct violence due to the difference of "intent" in its production (Marcantonio et al. 2024). Conceived of in this way, environmental violence is the largest source of early human mortality today (Pinho-Gomes et al. 2023; Fuller et al. 2022). Previous uses of environmental violence as a concept included additional factors such as violence against nature, the political economy of environmental issues impacting access to and distribution of resources, and other factors (Lee 2016; Narchi 2015). These more expansive definitions include factors important to understanding the broader range of impacts of human-produced pollution and/or human-environment interactions. However, due to issues of specificity, they make consistent, uniform operationalization more difficult while also shifting away from a primary focus on human health and flourishing outcomes, potentially reducing its analytical **power** (Marcantonio and Fuentes 2023). Environmental violence is closely related to Rob Nixon's (2011) "slow violence" and Kevin O'Brien's (2017) "climate violence," among others. Environmental vectors of violence, like toxic pollution, are nascent but increasingly recognized as a form of violence in peace and conflict studies (Marcantonio 2022b, 2024). Accounting for the acute and chronic varieties of environmental violence that are more and less visible, and their unequal geographic distribution even within communities, is critical for forging comprehensive approaches to **environmental peacebuilding**.

References

Fuller, Richard, Philip J. Landrigan, Kalpana Balakrishnan, et al. 2022. "Pollution and Health: A Progress Update." *The Lancet Planetary Health* S2542–5196 (22): 13. https://doi.org/10.1016/S2542-5196(22)00090-0.

Lee, Bandy X. 2016. "Causes and Cures VIII: Environmental Violence." *Aggression and Violent Behavior* 30: 105–9. https://doi.org/10.1016/j.avb.2016.07.004.

Marcantonio, Richard. 2022a. *Environmental Violence: In the Earth System and the Human Niche*. Cambridge University Press. https://www.cambridge.org/core/books/environmental-violence/30437FC2F540C830A23D9B0ECD93B560.

Marcantonio, Richard. 2022b. "Toxic Diplomacy through Environmental Management: A Necessary Next Step for Environmental Peacebuilding." *World Development Perspectives* 28: 100471. https://doi.org/10.1016/j.wdp.2022.100471.

Marcantonio, Richard. 2024. "The Glaring Gaps: Environmental Violence and Peace Research and Practice." *Peacebuilding* 2: 1–23. https://doi.org/10.1080/21647259.2024.2355765.

Marcantonio, Richard, and Sean Field. 2025. "Environmental Vulnerability and Conflict Occurrence Are Tightly Related." *Nature Communications Earth and Environment* 6 (1): 1–10. https://doi.org/10.1038/s43247-025-02300-6.
Marcantonio, Richard, and Agustín Fuentes. 2023. "Environmental Violence: A Tool for Planetary Health Research." *The Lancet Planetary Health* 7 (10): e859–67. https://doi.org/10.1016/S2542-5196(23)00190-0.
Marcantonio, Richard, John Paul Lederach, and Agustín Fuentes, eds. 2024. *Exploring Environmental Violence: Perspectives, Experience, Expression, and Engagement*. Cambridge: Cambridge University Press. https://www.cambridge.org/core/books/exploring-environmental-violence/exploring-environmentalviolence/8548BE5DB9E8D7F8241FBC09F406973C.
Narchi, Nemer E. 2015. "Environmental Violence in Mexico: A Conceptual Introduction." *Latin American Perspectives* 42 (5): 5–18. https://doi.org/10.1177/0094582X15579909.
Nixon, Rob. 2011. *Slow Violence and the Environmentalism of the Poor*. Cambridge: Harvard University Press.
O'Brien, Kevin J. 2017. *The Violence of Climate Change: Lessons of Resistance from Nonviolent Activists*. Georgetown University Press.
Pinho-Gomes, Ana-Catarina, Eleanor Roaf, and Gary Fuller et al. 2023. "Air Pollution and Climate Change." *The Lancet Planetary Health* 7 (9): e727–28. https://doi.org/10.1016/S2542-5196(23)00189-4.
Rentschler, Jun, and Nadezda Leonova. 2023. "Global Air Pollution Exposure and Poverty." *Nature Communications* 14 (1): 4432. https://doi.org/10.1038/s41467-023-39797-4.

Empowerment

Annika Erickson-Pearson, Rebecca L. Farnum, and Emmanuel Sindikubwabo

Empowerment refers to the expansion of agency and capacity, potentially including the redistribution of resources or changes in resource access. Empowerment is a process focusing on creating and enhancing opportunities for individuals and groups to have more influence over their surroundings and power over their lives.

A wide variety of fields, including psychology, sociology, and business, engage with ideas of empowerment. The concept entered international development conversations in the 1990s and has since become standard in poverty reduction initiatives and participatory methods (Calvès 2009).

In the early days of its contemporary usage, empowerment emphasized expanding actors' awareness and appreciation for the resources and influence they do hold, rather than focusing on feelings of relative powerlessness. Alice Walker, author of *The Color Purple*, contends that the "most common way people give up their **power** is by thinking they don't have any." Her assertion echoes Paulo Freire's *Pedagogy of the Oppressed*. Freire was an anti-colonial Brazilian theorist whose goal of "critical consciousness" for subjugated peoples has greatly influenced empowerment theory and action. Building from Freire's work, empowerment is often associated with marginalized groups. In the United States during the 1970s, empowerment gained popularity in sociopolitical arenas, including feminism, the Black Power movement, and the disability rights movement.

As the idea of empowerment gained popularity—being adopted and adapted by a greater number of fields, thinkers, and practitioners—it shifted from centering on a recognition of anti-oppression consciousness toward highlighting a material expansion of power, skills, and resources. Today, the term has become a catch-all for processes that expand individuals' and groups' ability to control what happens to them and their communities.

Conversely, disempowerment is the act of diminishing an individual's or group's influence, whether by actively removing forms of power and authority or by making them feel ineffective or unimportant. Disempowerment can be the result of social interactions or institutional policies, and can lead to feelings of helplessness, particularly for individuals and communities already experiencing marginalization.

Peacebuilding programs, international development agencies, and anti-poverty initiatives frequently engage with subfields of empowerment, such as: (1) economic empowerment, which focuses on supporting individuals to gain the skills and training they need to support themselves and their dependents, (2) social empowerment, which highlights individual agency amidst social relationships, encouraging actors to collectively change the institutions and cultural norms that exclude or constrain certain persons and communities, (3) political empowerment, which seeks representative equity in governmental institutions (e.g., gender parity in a senate) and greater engagement in decision-making processes, and (4) legal empowerment, which works to ensure that the rights of all individuals—especially those from strategically undervalued or historically disadvantaged groups—are protected, and that those individuals can use the law in appropriate ways to further their goals.

While relatively little **environmental peacebuilding** scholarship explicitly engages with empowerment (Davis et al. 2023), it is a regular tool for practitioners of environmental peacebuilding and sustainable development, who posit that empowered communities are better able to contribute to **positive peace** through **participatory processes** in **conflict** resolution and community-based natural resource management. Narratives about climate, peace, and **security** often address women's empowerment as a core issue (United Nations Environment Programme et al. 2020).

Critiques of empowerment as used by mainstream development agencies include the term's overuse (i.e. that its many meanings render the term meaningless); the simplified, individualized, and Westernized understanding of power that empowerment projects often aim for; and the cooptation of the term from a politicized, anti-colonial transformation of power dynamics to a neoliberal and instrumental tool that reinforces existing structures while proclaiming change.

References

Calvès, Anne-Emmanuèle. 2009. "Empowerment: The History of a Key Concept in Contemporary Development Discourse." *Revue Tiers Monde* 200 (4): 735–49. https://shs.cairn.info/article/E_RTM_200_0735.

Davis, Katy, Laura E. R. Peters, Jamon van Der Hoek, and Ken Conca. 2023. "Power in Environmental Peacebuilding." *World Development Sustainability* 3: 100110. https://doi.org/10.1016/j.wds.2023.100110.

United Nations Environment Programme, UN Women, UNDP and UNDPPA/PBSO. 2020. *Gender, Climate and Security: Sustaining Inclusive Peace on the Frontlines of Climate Change*. UN Women. https://www.unwomen.org/en/digital-library/publications/2020/06/gender-climate-and-security.

Evaluation

Leonardo Medina and Juha Uitto

Evaluation is the systematic and objective assessment of a planned, ongoing, or completed intervention's effectiveness, efficiency, relevance, coherence, and impact to determine its success in achieving intended goals and to inform future improvements. Evaluation also refers to the process of determining the worth or significance of an intervention. Evaluation should provide credible and useful information enabling the incorporation of lessons learned into decision-making processes (Organisation for Economic Co-operation and Development 2022).

Evaluations can be formative (conducted during the implementation of an intervention), summative (including outcome or impact evaluation), or prospective (assessing the likely outcomes of proposed interventions) (Morra Imas and Rist 2009).

In the **environmental peacebuilding** field, evaluation is a systematic process to assess the effectiveness, relevance, efficiency, and **sustainability** of interventions aimed at reducing **conflict** and promoting **peace** through environmental management and **conservation** (Bruch and Woomer 2023). Evaluation provides insights into whether activities—such as **conflict sensitivity** in natural resource management, community-based environmental programs, and cross-border **cooperation** on ecological issues—are achieving their intended **peacebuilding** and environmental goals.

This process begins with the establishment of clear **indicators** and criteria based on the intervention's objectives, such as improved community cohesion, reduced competition over resources, enhanced local **governance**, and strengthened trust among conflicting parties (Organisation for Economic Co-operation and Development 2019). Evaluators gather both qualitative and quantitative data through methods like surveys, interviews, participatory assessments, and environmental impact analyses to gauge whether expected outcomes are being met. These assessments consider the short- and long-term effects of interventions, looking at whether they have addressed root causes of conflict (like resource scarcity or exclusionary governance) and contributed to **sustainable peacebuilding** processes. They also pay attention to unanticipated or unintended consequences of interventions.

Evaluation is essential for supporting adaptive approaches to environmental interventions, especially in conflict-affected settings where circumstances are often volatile and outcomes unpredictable. In these contexts, the complexity of social, political, and ecological dynamics requires that interventions be flexible and responsive to emerging conditions, as static or rigid plans can quickly become ineffective or counterproductive.

By systematically assessing an intervention's performance, evaluation helps practitioners understand not only what works and why, but also what unintended effects may arise. For instance, a reforestation project in a conflict zone could inadvertently intensify local tensions if it does not account for existing land disputes or differing community needs, as has been documented in conflict-affected Colombia (Hein et al. 2020). Regular evaluation allows project teams to identify such issues early and adjust—whether by altering implementation strategies, engaging new **stakeholders**, or shifting resources. This process, often termed **adaptive management** or adaptive governance (Sharma-Wallace et al. 2018), relies on frequent feedback loops, where evaluation findings are actively used to refine the intervention, aligning it more closely with on-the-ground realities.

In conflict-affected settings, evaluations often focus on conflict sensitivity, meaning evaluators examine if interventions have unintentionally exacerbated tensions or created new conflicts, as well as whether they have built capacities for peaceful cooperation (Tänzler et al. 2022). Evaluators look for signs that environmental projects are fostering inclusion, building local capacities for peaceful resource management, and reinforcing the legitimacy of local governance structures. By carefully **monitoring** indicators of both peacebuilding and environmental health, evaluations can pinpoint adjustments needed to align interventions with evolving conflict dynamics or **power** shifts, as shown by Krampe and Gignoux (2018) in their assessment of post-conflict water service provision in East Timor.

By highlighting successes, failures, and unintended outcomes, evaluation serves as a **learning** tool for practitioners, policymakers, and communities, strengthening future peacebuilding efforts in environmentally fragile or conflict-affected settings.

Acknowledgment

This work was carried out with support from the CGIAR Initiative on Climate Resilience, ClimBeR, and the CGIAR Initiative on Fragility, Conflict, and Migration. We would like to thank all funders who supported this research through their contributions to the CGIAR Trust Fund: https://www.cgiar.org/funders/.

References

Bruch, Carl, and Amanda Woomer. 2023. *Toolkit on Monitoring and Evaluation of Environmental Peacebuilding*. Environmental Law Institute, Environmental Peacebuilding Association, and United States Institute of Peace. https://m-and-e.environmentalpeacebuilding.org/toolkit/.

Hein, Jonas, Carlos Del Cairo, Daniel Ortiz Gallego, et al. 2020. "A Political Ecology of Green Territorialization: Frontier Expansion and Conservation in the Colombian Amazon." *Erde* 151 (1): 37–57. https://doi.org/10.12854/erde-2020-456.

Krampe, Florian, and Suzanne Gignoux. 2018. "Water Service Provision and Peacebuilding in East Timor: Exploring the Socioecological Determinants for Sustaining Peace." *Journal of Intervention and Statebuilding* 12 (2): 185–207. https://doi.org/10.1080/17502977.2018.1466945.

Morra Imas, Linda G., and Ray C. Rist. 2009. *The Road to Results: Designing and Conducting Effective Development Evaluations*. The World Bank.

Organisation for Economic Co-operation and Development (OECD). 2019. *Better Criteria for Better Evaluation: Revised Evaluation Criteria Definitions and Principles for Use*. OECD.

Organisation for Economic Co-operation and Development (OECD). 2022. *Glossary of Key Terms in Evaluation and Results-Based Management*. 2nd ed. OECD.

Sharma-Wallace, Lisa, Sandra J. Velarde, and Anita Wreford. 2018. "Adaptive Governance Good Practice: Show Me the Evidence!" *Journal of Environmental Management* 222: 174–84. https://doi.org/10.1016/j.jenvman.2018.05.067.

Tänzler, Dennis, Nikolas Scherer, and Adrien Detges. 2022. *Guidelines for Conflict-Sensitive Adaptation to Climate Change*. Umweltbundesamt (UBA).

Extractive Resources

Erik Post and Tracy Hart

Extractive resources are substances or compounds extracted from the natural environment for use as inputs for production or manufacturing because of their properties (United Nations Environment Programme [UNEP] and International Resource Panel [IRP] 2024).

Extractive resources generally refer to non-renewable materials procured through mining. However, biomass resources obtained through agricultural, aquaculture, and wood harvest can also be extractive resources when harnessed at rates that exceed regeneration capacities (Post 2023). The annual rate of resource extraction has increased markedly from 30.9 billion tons in 1970 to over 100 billion tons due to rapid technological maturation, economic throughput, population growth, per capita income growth, and urbanization. According to the IRP, of all resources extracted globally in 2020, 48% were non-metallic minerals, 26% biomass, 16% **fossil fuels**, and 10% metal ores (UNEP and IRP 2024).

Resource extraction can result in the complete transformation of ecosystems, accompanied by near or complete loss of **biodiversity** and ecosystem functions and services. The extraction of biomass is the dominant driver of land-related and freshwater biodiversity losses as well as water stress (Intergovernmental Science-Policy Platform on Biodiversity and Ecosystem Services 2018). Mining of ores and fossil fuels is globally less relevant for global land-related biodiversity impacts (<3% of total global impacts), although it can generate locally significant environmental impacts through highly toxic wastes that can cause heavy metal pollution. Together, biomass cultivation, mineral and fossil resource extraction, and processing for materials, fuels, and food accounted for more than 55% of greenhouse gas emissions in 2022 (UNEP and IRP 2024). These trends have been accelerating globally, indicating that there is no decoupling of economic activity from environmental impacts (UNEP and IRP 2024).

The extraction of resources also transforms social and environmental relations that sustain communities. Communities near extraction sites can face severe impacts, such as livelihood loss and land dispossession, environmental pollution, threats to their knowledge systems, racial- and **gender-based violence**, as well as intimidation and even assassinations. Local communities are often excluded from, or insufficiently included in, decision-making processes related to extraction. Indigenous Peoples bear a disproportionate share, as they tend to steward lands where extraction is seeking to expand and intensify (Scheidel et al. 2023). Communities are more likely to resist extractive projects when they perceive a threat to their health or livelihoods (Conde 2017).

This can cause conflicts with governments and corporations, as a lack of trust in these entities becomes coupled with perceived deficient compensation for grievances suffered. Governments and companies are often viewed with suspicion because the economic rents that extraction can generate often create perverse incentives to distort **governance** processes, capture political systems, and foment corruption. The inflows of capital and economic specialization associated with extraction can also cause economic imbalances, such as the "Dutch Disease," which refers to how currency exchange rate appreciation due to natural resource exports stymies the economic development of other economic sectors. The observation that countries well-endowed with **natural resources** can generate significant rents but fail to improve on many economic, governance, and well-being **indicators** has prompted economists to speak of a **resource curse** (Humphreys et al. 2007). Governments can become highly dependent on extraction to fund public expenditures and placate elites with rents, risking an "extractive imperative" that prioritizes extraction over other objectives (Arsel et al. 2016).

Extractive resources can influence **conflict** dynamics (Lujala and Rustad 2012). Yet, extractive resources are at the basis of the built **environment** and globalized food systems, providing livelihoods to vast numbers of people. Improving working conditions and remuneration of people employed in extractive work would contribute to well-being, address global inequalities, and foster **peacebuilding**. Moreover, sustainable resource management, responsible sourcing, and the use of resources and associated rents for **basic services** provide important **conflict prevention, mitigation**, and resolution opportunities. Finally, restoration of landscapes affected by extraction, as well as efforts to move to post-extractive alternative futures based on more circular and sustainable modes of production and consumption, can also create conflict prevention, mitigation, and resolution opportunities. Peacebuilding efforts must aim to prevent, mitigate, and resolve the aggravating effects of resource extraction on conflict dynamics while harnessing the positive opportunities that resource extraction and the associated rents provide.

References

Arsel, Murat, Barbara Hogenboom, and Lorenzo Pellegrini. 2016. "The Extractive Imperative and the Boom in Environmental Conflicts at the End of the Progressive Cycle in Latin America." *The Extractive Industries and Society* 3 (4): 877–79. https://doi.org/10.1016/j.exis.2016.10.013.

Conde, Marta. 2017. "Resistance to Mining. A Review." *Ecological Economics* 132: 80–90. https://doi.org/10.1016/j.ecolecon.2016.08.025.

Humphreys, Macartan, Jeffrey Sachs, and Joseph E. Stiglitz. 2007. *Escaping the Resource Curse*. Columbia University Press.

Intergovernmental Science-Policy Platform on Biodiversity and Ecosystem Services (IPBES). 2018. *The IPBES Assessment Report on Land Degradation and Restoration*. Edited by Luca Montanarella, Robert Scholes, and Anastasia Brainich. IPBES Secretariat.

Lujala, Päivi, and Siri Aas Rustad. 2012. *High-Value Natural Resources and Post-Conflict Peacebuilding*. Routledge.

Post, Erik. 2023. "Expanding Extractivisms: Extractivisms as Modes of Extraction Sustaining Imperial Modes of Living." *International Development Policy | Revue Internationale de Politique de Développement* (16). https://doi.org/10.4000/poldev.5376.

Scheidel, Arnim, Álvaro Fernández-Llamazares, Anju Helen Bara, et al. 2023. "Global Impacts of Extractive and Industrial Development Projects on Indigenous Peoples' Lifeways, Lands, and Rights." *Science Advances* 9 (23): eade9557. https://doi.org/10.1126/sciadv.ade9557.

United Nations Environment Programme (UNEP) and International Resource Panel. 2024. *Global Resources Outlook 2024: Bend the Trend*. UNEP.

F

Faith-Based Environmental Peacebuilding

Joram Tarusarira and Elsa Barron

Faith-based environmental peacebuilding involves third-party interventions by faith-based actors—such as individuals, communities, or organizations—to address conflicts and violence connected to environmental resources. It also includes faith-based values and collaboration to address the root causes and impacts of climate and environmental crises that would otherwise exacerbate violence and injustice.

Faith-based actors have become crucial in addressing **violence** and **conflict**, including where environmental resources are at stake. These actors draw upon religious, traditional, and cultural values, beliefs, practices, creeds, and symbolic elements (e.g., texts, images, principles, rituals, myths, vocabulary, and narratives) to promote **peace** and foster **social cohesion**. Faith-based actors are motivated and inspired to do peace work by their spiritual and religious traditions, principles, and values (Bercovitch and Kadayifci-Orellana 2009). In cases where the parties to a conflict are religious or make decisions based on faith or religious resources, it increases the actors' credibility, legitimacy, trust, moral standing, and **power** of persuasion (Bercovitch and Kadayifci-Orellana 2009), as well as the willingness of the parties to the conflict to move toward resolving and transforming conflicts.

Three principles frequently guide faith actors' engagement with **environmental peacebuilding:** (1) relationship to **environment** and community, (2) dialogue and collaboration, and (3) healing through **justice** and reconciliation (Barron et al. 2022). These principles emphasize the significance of rootedness in place and long-standing community relationships, shared values about the environment within and across faith traditions, and an emphasis on moral progress and social and environmental healing as relatively more important than technological advancement.

In most parts of Africa, faith-based actors are often perceived as advocates of moral values such as justice for all, forgiveness, harmony, and human dignity (Tarusarira 2016; Johnston and Cox 2003; Bercovitch and Kadayifci-Orellana 2009). As a result, they are more likely to be perceived as balanced, trustworthy, and committed. They are highly regarded, and their opinions are generally valued. They are thought to be better able to understand their communities' physical and emotional needs because they engage with communities for long periods, developing a deeper understanding

DOI: 10.4324/9781003604143-8

of the communities. This immersion in communities is necessary for **peacebuilding** because **conflict transformation** is not an event but a process that starts long before a conflict ensues. The spiritual and religious motivation behind faith-based actors helps them understand communities where faith or religion has explanatory and decision-making power and when resources at the heart of conflicts are intertwined with beliefs, core values, identity, and territorial integrity (Smock 2001; Tarusarira 2022). They understand the language and register of the communities (Tarusarira 2016).

Faith-based actors tend to frame environmental conflicts and injustice not only as signs of geophysical phenomena but as spiritual phenomena (Pope Francis 2015), which resonate with the parties to the conflict. A case in point is Pope Francis's 2015 encyclical *Laudato Si: On Care for Our Common Home*, in which he challenges hyper-technical approaches that uphold the "myth of progress" and believes that "environmental problems will not be solved simply by applying new technologies." As a source of value commitment, this encyclical has influenced climate actions and **mediation** in environmentally related local conflicts, contributing to faith-based environmental peacebuilding. The Islamic perception of water as a gift from God, that water should be free or at least very low cost and the assumption that people have certain rights o low-cost water or are entitled to free water (Larson 2011), motivates **cooperation** and resource sharing, challenging selfish tendencies. Finally, spiritualities of Indigenous peoples and **Indigenous knowledge** often emphasize humanity's interconnectedness with non-human life and environments and invoke a deep reverence for land. This commitment to ecological well-being is at the forefront of **conservation** efforts around the world and operates intersectionally with peace and justice movements.

Acknowledgment

This work was carried out with support from the CGIAR Initiative on Climate Resilience, ClimBeR, and the CGIAR Initiative on Fragility, Conflict, and Migration. We would like to thank all funders who supported this research through their contributions to the CGIAR Trust Fund: https://www.cgiar.org/funders/.

References

Barron, Elsa, Huda Alkaff, Elyse Baden, Katie Chustak, and Matthieu Guiller. 2022. "Three Pillars for Faith's Engagement in Environmental Peacebuilding: The Transformative Potential of Faith and Spirituality in Relationship-Building, Dialogue, and Healing." *Ecosystem for Peace*. https://www.ecosystemforpeace.org/compendium/inclusion-1

Bercovitch, Jacob, and S. Ayse Kadayifci-Orellana. 2009. "Religion and Mediation: The Role of Faith-Based Actors in International Conflict Resolution." *International Negotiation* 14 (1): 175–204. https://doi.org/10.1163/157180609X406562.

Francis, Pope. 2015. *Laudato Si: On Care for Our Common Home*. Our Sunday Visitor Inc.

Johnston, Douglas, and Brian Cox. 2003. "Faith-Based Diplomacy and Preventive Engagement." In *Faith-Based Diplomacy: Trumping Realpolitik*, edited by Douglas Johnston. Oxford University Press.

Larson, Rhett B. 2011. "Holy Water and Human Rights: Indigenous Peoples' Religious-Rights Claims to Water Resources." *Arizona Journal of Environmental Law and Policy* 2: 83–5. http://hdl.handle.net/10150/675128. .

Smock, David. 2001. *Faith-Based NGOs and International Peacebuilding*. United States Institute of Peace. https://www.usip.org/publications/2001/10/faith-based-ngos-and-international-peacebuilding.
Tarusarira, Joram. 2022. "Religious Environmental Sensemaking in Climate-induced Conflicts." *Religions* 13 (3): 204. https://doi.org/10.3390/rel13030204.
Tarusarira, Joram. 2016. *Reconciliation and Religio-Political Non-Conformism in Zimbabwe*. Routledge.

Farmer-Managed Natural Regeneration

Alan Channer and Irene Ojuok

Farmer-managed natural regeneration (FMNR) is an agroforestry approach, typically based on Indigenous knowledge, where farmers systematically regrow indigenous trees and shrubs from stumps, sprouting root systems, and seeds.

FMNR has been described as generating "the largest environmental transformation in the Sahel and perhaps in the whole of Africa" (Hertsgaard 2011). It was brought to the world's attention by agronomist Tony Rinaudo in Niger. He noticed that protecting autochthonous stumps and saplings from cutting, grazing, and fire, in conjunction with selective pruning, could bring about more rapid and more sustainable restoration than planting trees in drought-prone contexts. FMNR builds on and systematizes Indigenous systems, knowledge, and practices, leveraging locally available resources (Rinaudo et al. 2019; Torquebiau 2024). It often finds due expression in vernacular languages. For example, in Tanzania, FMNR is called Kisiki hai (Swahili), meaning "living stump," and Tintuug lebge tii–"shrubs nurtured into trees" in northern Ghana.

FMNR can improve livelihoods by reducing soil erosion, enhancing soil fertility, and providing firewood, fodder, shade, fruits, and medicinal products. In addition, trees regenerated by FMNR sequester carbon and can improve the micro-climate (Weston et al. 2015). FMNR is particularly well-suited to agro-silvo pastoral systems in drylands (Abasse et al. 2023), where tree planting is challenging and where more land is available for extensive agroforestry than in higher-potential systems. Over 5 million hectares of agroforestry parkland have been restored by FMNR in Niger, with significant successes also in dryland contexts in other parts of Africa, Central America, and South Asia.

FMNR can present an opportunity to foster ecological and social restoration, and the synergies between them, concurrently, and, as such, can be a form of **environmental peacebuilding**. When undertaken on public or communal lands, it tends to depend on, and at the same time helps to forge, community agreements on (1) selecting areas for regeneration and preferred species, (2) thinning out, selecting, and pruning certain trees, (3) protecting those trees from livestock and inappropriate cutting, and (4) sustainable harvesting. So, while FMNR can strengthen environmental **resilience** and improve livelihoods, it can also strengthen community collaboration and cohesion.

FMNR has potential in situations of farmer-pastoralist **conflict** (Rinaudo 2019), bringing the conflict actors into collaboration to safeguard indigenous trees, as these can both provide fodder and shade for livestock and innumerable benefits to farmers.

There is growing evidence, albeit still circumstantial, that the return of vegetation in drylands is correlated with a reduction in insecurity and conflict and that FMNR can catalyze a virtuous cycle leading to environmental and social recovery in these contexts (Channer 2015).

Indeed, FMNR is part of an approach that could be termed "green **security**." On the one hand, FMNR can build community cohesion and reduce conflict over resources. On the other hand, it is also more likely to succeed in conflict-affected contexts when community-building dynamics are respected, conflict risks are understood, and dispute resolution mechanisms are integrated.

FMNR has significant potential for the **mitigation** of **climate change** through carbon sequestration. The drive to offset carbon by the private sector in the Global North may incentivize the uptake of FMNR by rural communities in the Global South, but it also comes with risks associated with the pressures and potential distortions of exogenous investments. **Peacebuilding**, community development, and **human rights** actors have insights and advice that can help make carbon programs involving FMNR more successful (Ojuok and Channer 2025). FMNR can also be empowering for women (Ojuok and Ndayizigiye 2021)—for example, when trees are regenerated close to homesteads, women have less distance to walk to collect firewood, saving time for other activities. Additionally, the shade from FMNR creates pleasant gathering spaces, and increased fodder supply can boost the productivity of sheep and goats managed by women, particularly when men migrate with cows for pasture during droughts.

In conclusion, FMNR, as a community-based environmental peacebuilding approach, has great potential to foster synergies between environmental restoration, **human security**, and climate resilience, particularly in drylands. Action research is required to understand and amplify these dynamics.

References

Abasse, Tougiani, Moussa Massaoudou, Rabiou Habou, Soumana Idrissa, and Dan Guimbo Iro. 2023. "Farmer Managed Natural Regeneration in Niger: The State of Knowledge." *Tropenbos International*. https://doi.org/10.55515/BYIZ5081.

Channer, Alan. 2015. "Mobile Pastoralism and Trustbuilding in Africa." In *Land Restoration: Reclaiming Landscapes for a Sustainable Future*, edited by Chbay Ilan, Martin Frick, and Jennifer Helgeson. Elsevier. https://doi.org/10.1016/B978-0-12-801231-4.00012-4.

Hertsgaard, Mark. 2011. *Hot: Living through the Next Fifty Years*. Island Press.

Ojuok, Irene, and Alan Channer. 2025. "Carbon Finance, Communities and Conflict: the Quest for Peace-positive Land Restoration in Africa." *Geneva Policy Outlook 2025*. https://www.genevapolicyoutlook.ch/carbon-finance-communities-and-conflict-the-quest-for-peace-positive-land-restoration-in-africa/.

Ojuok, Irene, and Tharcisse Ndayizigiye. 2021. "Women Participation in Farmer Managed Natural Regeneration for Climate Resilience: Laisamis, Marsabit County, Kenya." In *African Handbook of Climate Change Adaptation*, edited by Walter Leal Filho, Nicholas Oguge, Desalyegn Ayal, Lydia Adeleke, and Izael da Silva. Springer International Publishing. https://doi.org/10.1007/978-3-030-45106-6_152.

Rinaudo, Tony. 2019. "The Role of FMNR in Conflict Reduction." Global EverGreening Alliance. https://www.evergreening.org/the-role-of-fmnr-in-conflict-reduction/.

Rinaudo, Tony, Alice Muller, and Mary Morris. 2019. "Farmer Managed Natural Regeneration (FMNR) Manual. World Vision Australia." https://fmnrnetworkuganda.org/wp-content/uploads/2021/08/FMNR-Field-Manual_DIGITAL_FA-min.pdf.

Torquebiau, Edmund, ed. 2024. *Agroforestry at Work*. Tropical Forest Issues 62. Tropenbos International. https://doi.org/10.55515/OEQC4236.

Weston, Peter, Reaksmey Hong, Carolyn Kaboré, and Christian A. Kull. 2015. "Farmer-Managed Natural Regeneration Enhances Rural Livelihoods in Dryland West Africa." *Environmental Management* 55 (6): 1402–17. https://doi.org/10.1007/s00267-015-0469-1.

Food Security

Jennifer Clapp and Anastasia Papadopoulos

As defined by the Food and Agriculture Organization of the United Nations (FAO) at the 1996 World Food Summit, "Food security exists when all people, at all times, have physical, social and economic access to sufficient, safe and nutritious food that meets their dietary needs and food preferences for an active and healthy life" (FAO 2001).

Global food insecurity has worsened since 2015. As of 2024, approximately 700 million people around the world were chronically undernourished, and 2.3 billion people experienced moderate or severe food insecurity (FAO et al. 2025). Some of the highest prevalence of hunger occurs in **conflict**-affected regions.

In subsequent years, scholars and policymakers have identified four pillars of food security based on the FAO definition of the term: availability, access, utilization, and stability (FAO 2006; Webb et al. 2006). More recently, agency and **sustainability** have been added to this list to widen our understanding of food security to incorporate six interrelated dimensions (High Level Panel of Experts on Food Security and Nutrition [HLPE-FSN] 2020; Clapp et al. 2022), which are increasingly acknowledged in both scholarly and policy settings (Clapp et al. 2025). Availability refers to whether there is an adequate supply of food that can meet the dietary needs of populations in terms of both quantity and quality. Access to food is associated with people's ability to acquire food that meets their needs, such as market conditions, employment, and physical proximity. Food access has different implications for different groups, such as the impoverished or elderly populations. Utilization captures whether people can prepare and reap the nutritional benefits of the food they can acquire to meet their physiological needs (e.g., clean water, sanitary cooking conditions, cooking fuel) and carry out their sociocultural practices. Stability refers to the ability to access food on an ongoing basis, including in the event of sudden shocks or crises (e.g., economic crisis, conflicts, seasonal food insecurity). Stability has become particularly important in recent years, given ongoing crises that have impacted global food supply and accessibility (e.g., the war in Ukraine or the COVID-19 pandemic). Agency refers to the capacity of individuals or groups to make their own independent choices about how they engage with food systems. Having agency may refer to decisions around what they eat, the foods they grow and sell, how they process and distribute food, how they engage in food system policies, and **governance** debates (HLPE-FSN 2020). Strong governance structures built into sociopolitical systems are required to ensure that agency is upheld. Sustainability encompasses food system practices that contribute to the long-term regeneration of natural, social, and economic systems while safeguarding the food needs of future generations (HLPE-FSN 2020). Sustainability is a critical dimension of food security, acknowledging the complex relationships between ecosystems, **livelihoods**, nutrition, and the importance of long-term **resilience** (HLPE-FSN 2019).

In the absence of **peace**, each of these dimensions may be challenged, and not surprisingly, the highest levels of hunger occur in **conflict**-affected contexts (United Nations World Food Programme [WFP] and FAO 2025). Food production is severely hampered by conflict, for instance, when producers or market traders face danger from working in their fields or transporting goods, as observed in Gaza and Ukraine. Conflict disrupts livelihoods, making access especially challenging, as has become clear in Sudan, where nearly 25 million people face acute hunger (WFP n.d.). Conflict also severely disrupts markets, making them more volatile. Access to clean water and meeting cultural preferences cannot be taken for granted in such contexts. People's agency is undermined in conflict situations when governance structures fail, and sustainability is frequently compromised because agricultural landscapes are often affected by toxins from explosive weapons used in **war**.

References

Clapp, Jennifer, William G. Moseley, Paola Termine, and Barbara Burlingame. 2022. "The Case for a Six-Dimensional Food Security Framework." *Food Policy* 106: 102164. https://doi.org/10.1016/j.foodpol.2021.102164.

Clapp, Jennifer, William G. Moseley, Paola Termine, and Barbara Burlingame. 2025. "Multi-scalar Policy Uptake of the Six-Dimensional Food Security Framework." *Food Policy* 137: 102936. https://doi.org/10.1016/j.foodpol.2025.102936.

Food and Agriculture Organization (FAO). 2001. *The State of Food Insecurity in the World 2001*. http://www.fao.org/docrep/003/y1500e/y1500e00.htm.

Food and Agriculture Organization (FAO). 2006. *Food Security*. http://www.fao.org/forestry/13128–0e6f36f27e0091055bec28ebe830f46b3.pdf.

Food and Agriculture Organization (FAO), International Fund for Agricultural Development (IFAD), United Nations Children's Fund (UNICEF), United Nations World Food Programme (WFP) and the World Health Organization (WHO). 2025. *The State of Food Security and Nutrition in the World 2025 – Addressing High Food Price Inflation for Food Security and Nutrition*. FAO. https://doi.org/10.4060/cd6008en.

High Level Panel of Experts on Food Security and Nutrition (HLPE-FSN). 2019. *Agroecological and Other Innovative Approaches for Sustainable Agriculture and Food Systems that Enhance Food Security and Nutrition*. A Report by the HLPE-FSN. http://www.fao.org/3/ca5602en/ca5602en.pdf.

High Level Panel of Experts on Food Security and Nutrition (HLPE-FSN). 2020. *Food Security and Nutrition: Building a Global Narrative Towards 2030*. http://www.fao.org/3/ca9731en/ca9731en.pdf.

United Nations World Food Programme (WFP). n.d. *Sudan*. WFP. Accessed October 7, 2025. https://www.wfp.org/emergencies/sudan.

United Nations World Food Programme (WFP) and Food and Agriculture Organization (FAO). 2025. *Hunger Hotspots: FAO–WFP Early Warnings on Acute Food Insecurity: June to October 2025 Outlook*. WFP and FAO. https://doi.org/10.4060/cd5684en.

Webb, Patrick, Jennifer Coates, Edward A. Frongillo, Beatrice Lorge Rogers, Anne Swindale, and Paula Bilinsky. 2006. "Measuring Household Food Insecurity: Why It's So Important and Yet So Difficult to Do." *The Journal of Nutrition* 136: 1404S–08S https://doi.org/10.1093/jn/136.5.1404S.

Fossil Fuels

Beatriz Arnal-Calvo

Fossil fuels—primarily coal, oil, and gas—are hydrocarbon-rich, non-renewable energy sources formed in the geological strata from the decomposition of ancient organisms that died millions of years ago. They are also used to produce plastics, steel, agrochemicals, and pharmaceutical components.

Fossil fuel production, barely used before the Industrial Revolution, is the major driver of human-induced **climate change**. The **Intergovernmental Panel on Climate Change**'s Sixth Assessment Synthesis Report concludes that a century of fossil fuel use has "unequivocally caused global warming" (Intergovernmental Panel on Climate Change [IPCC] 2023, v). Burning fossil fuels releases large amounts of greenhouse gases into the air—predominantly CO_2, along with methane and nitrous oxide—that trap the sun's heat, driving rising temperatures and disrupting climate patterns.

Fossil fuel exploitation poses serious challenges to **peacebuilding** throughout its entire cycle (i.e., exploration, extraction, transport, distribution, industrial process of transformation, combustion, and consumption), with several studies identifying oil as a key factor increasing **conflict** risk (Fearon and Laitin 2003; Humphreys 2005; Lujala 2009; Colgan 2013). In recent decades, competition over fossil fuel resources has fueled socio-environmental conflicts, including ecological disasters and land expropriation, often affecting Indigenous and frontline communities worldwide (Kumar 2024). For instance, in Sudan, the military and allied militias displaced and killed civilians to clear the way for oil development (Patey 2012); in the Niger Delta and Ogoniland, oil revenues have perpetuated a transnational, oil-dependent system at the expense of its inhabitants (Obi 2010; Watts 2001); in Mozambique's Cabo Delgado province, communities near one of Africa's largest offshore gas sites were forcefully removed without fair consultation or compensation (Symons 2016; Namaganda et al. 2022), and in Colombia, human and **environmental defenders** opposing gas fracking, particularly women, face repeated threats (Corporación Compromiso 2022).

Fossil fuel production also threatens fundamental **human rights** (Office of the United Nations High Commissioner for Human Rights [OHCHR] 2014, 2019, 2024; OHCHR and UN Secretary-General 2024), particularly at extraction sites, for instance by polluting ecosystems, inducing forced **displacement**, and destroying traditional livelihoods. In Ecuador's Amazon, gas flaring harms Indigenous women's health (Hurtig and San Sebastián 2002; Sebastián and Hurtig 2004; Facchinelli et al. 2023) while Canada's pipeline projects disrupt Northern Indigenous women's **land usage** and weaken their territorial sovereignty (Lee et al. 2024). In response, human rights bodies urge States to limit fossil fuels and other extractive industries (Convention on the Elimination of All Forms of Discrimination Against Women 2018), advocating rapid phase-out through legal regulations and treaties like a fossil fuel non-proliferation treaty (Human Rights Council 2023). While global fossil fuel reliance impedes the implementation of Sustainable Development Goals (Daley and Lawrie 2022), global fossil fuel subsidies—which reached $7 trillion in 2022 (Black 2023)—along with increasing military expenditure, divert funds from **environmental**

peacebuilding (IPCC 2023). Lastly, fossil fuels perpetuate global systems of **power** and privilege, sustaining colonial and patriarchal fantasies of energy intensity and endless growth based on the extraction and exploitation of nature, labor, and care systems (Daggett 2018, 2021).

Environmental peacebuilding must prioritize addressing fossil fuels' role in the socio-ecological crisis, despite the pull of vast reserves across the planet promising enormous wealth and cheap energy. Transitioning away from fossil fuels offers substantial opportunities to foster **peace**, restore ecosystems, and promote global **justice**. However, a **just transition** to renewable alternatives must avoid placing undue burdens on vulnerable communities or worsening inequalities, especially in fragile, conflict-affected areas. More critically, environmental peacebuilding requires reimagining energy systems and power structures to achieve just, peaceful, and sustainable futures for all (Arnal-Calvo 2024).

References

Arnal-Calvo, Beatriz. 2024. *Hacia Una Política Climática Libre de Combustibles Fósiles*. Research Paper and Political Recommendations. WILPF España. https://wilpf.es/wp-content/uploads/2024/11/informe-fosiles_completo-i.pdf

Black, Simon. 2023. "IMF Fossil Fuel Subsidies Data: 2023 Update." *IMF Working Papers* (169): 1. https://doi.org/10.5089/9798400249006.001.

Colgan, Jeff D. 2013. "Fueling the Fire: Pathways from Oil to War." *International Security* 38 (2): 147–80. https://doi.org/10.1162/ISEC_a_00135.

Convention on the Elimination of All Forms of Discrimination Against Women (CEDAW). 2018. "General Recommendation No. 37 (2018) on the Gender-Related Dimensions of Disaster Risk Reduction in the Context of Climate Change." *International Legal Materials* 19 (1): 33–45. https://doi.org/10.1017/S0020782900043448.

Corporación Compromiso. 2022. *Amenazas e Incidentes de Seguridad y Violencia Basada en Género en el Departamento de Santander*. Boletín de Derechos Humanos Corporación Compromiso. Observatorio de Paz y Derechos Humanos, Corporación Compromiso. https://corporacioncompromiso.org/apc-aa files/6a67676b75363737383775363579367 9/boletin-de-derechos-humanos-enero-marzo-2022.pdf.

Daggett, Cara. 2018. "Petro-Masculinity: Fossil Fuels and Authoritarian Desire." *Millennium: Journal of International Studies* 47 (1): 25–44. https://doi.org/10.1177/0305829818775817.

Daggett, Cara. 2021. "Energy and Domination: Contesting the Fossil Myth of Fuel Expansion." *Environmental Politics* 30 (4): 644–62. https://doi.org/10.1080/09644016.2020.1807204.

Daley, Freddie, and Charlie Lawrie. 2022. *Fueling Failure: How Coal, Oil and Gas Sabotage all Seventeen Sustainable Development Goals*. University of Sussex. https://fossilfueltreaty.org/fuelling-failure.

Facchinelli, Francesco, Edoardo Crescini, Giuseppe Della Fera, and Massimo De Marchi. 2023. "The Apaguen Los Mecheros Campaign: Supporting Climate Justice in the Amazonian Cities of Ecuador by Estimating the Health Risks of Gas Flaring." *Landscape and Urban Planning* 240 (December): 104898. https://doi.org/10.1016/j.landurbplan.2023.104898.

Fearon, James D., and David D. Laitin. 2003. "Ethnicity, Insurgency, and Civil War." *American Political Science Review* 97 (1). https://www.jstor.org/stable/3118222.

Human Rights Council (HRC). 2023. *Impact of New Technologies Intended for Climate Protection on the Enjoyment of Human Rights Report of the Human Rights Council Advisory Committee*. A/HRC/54/47. https://documents.un.org/doc/undoc/gen/g23/141/86/pdf/g2314186.pdf.

Humphreys, Macartan. 2005. "Natural Resources, Conflict, and Conflict Resolution: Uncovering the Mechanisms." *The Journal of Conflict Resolution* 49 (4): 508–37. https://www.jstor.org/stable/30045129.

Hurtig, Anna-Karin, and Miguel San Sebastián. 2002. "Geographical Differences in Cancer Incidence in the Amazon Basin of Ecuador in Relation to Residence near Oil Fields." *International Journal of Epidemiology* 31 (5): 1021–27. https://doi.org/10.1093/ije/31.5.1021.

Intergovernmental Panel on Climate Change (IPCC). 2023. *Climate Change 2023: Synthesis Report. Contribution of Working Groups I, II and III to the Sixth Assessment Report of the Intergovernmental Panel on Climate Change.* Edited by Core Writing Team, Hoesung Lee, and José Romero. IPCC. https://doi.org/10.59327/IPCC/AR69789291691647.

Kumar, Mukul. 2024. "Violent Transitions: Towards a Political Ecology of Coal and Hydropower in India." *Climate and Development* 16 (9): 751–61. https://doi.org/10.1080/17565529.2023.2264259.

Lee, Hannah Grace, Tyler McCreary, and Joanne Connauton. 2024. "Where Are the Women? Exposing the Gap in Gender Analysis in Environmental Assessments of Fossil Fuel Pipeline Impacts on Indigenous Communities." *Health & Place* 89: 103338. https://doi.org/10.1016/j.healthplace.2024.103338.

Lujala, Päivi. 2009. "Deadly Combat over Natural Resources: Gems, Petroleum, Drugs, and the Severity of Armed Civil Conflict." *The Journal of Conflict Resolution* 53 (1): 50–71. https://www.jstor.org/stable/27638653.

Namaganda, Emilinah, Kei Otsuki, and Griet Steel. 2022. "Learning from Africana Critical Theory: A Historicized Contextualization of the Impacts of Mozambique's Natural Gas Project." *The Extractive Industries and Society* 10: 101075. https://doi.org/10.1016/j.exis.2022.101075.

Obi, Cyril I. 2010. "Oil Extraction, Dispossession, Resistance, and Conflict in Nigeria's Oil-Rich Niger Delta." *Canadian Journal of Development Studies / Revue Canadienne D'études du Développement* 30 (1–2): 219–36. https://doi.org/10.1080/02255189.2010.9669289.

Office of the United Nations High Commissioner for Human Rights (OHCHR). 2014. *Mapping Human Rights Obligations Relating to the Enjoyment of a Safe, Clean, Healthy and Sustainable Environment Focus Report on Human Rights and Climate Change.* https://doi.org/10.1163/9789004502758_044.

Office of the United Nations High Commissioner for Human Rights (OHCHR). 2019. "United Nations Climate Action Summit." https://www.ohchr.org/en/press-releasedreleases/2019/09/united-nations-climate-action-summit.

Office of the United Nations High Commissioner for Human Rights (OHCHR). 2024. *Promotion and Protection of All Human Rights, Civil, Political, Economic, Social and Cultural Rights, Including the Right to Development. Report of the Special Rapporteur on the Promotion and Protection of Human Rights in the Context of Climate Change, Elisa Morgera.* https://doi.org/10.1163/2210-7975_HRD-9970-2016149.

Office of the United Nations High Commissioner for Human Rights (OHCHR) and UN Secretary-General. 2024. *Analytical Study on the Impact of Loss and Damage from the Adverse Effects of Climate Change on the Full Enjoyment of Human Rights, Exploring Equity-Based Approaches and Solutions to Addressing the Same.* A /HRC/57/30. Human Rights Council. https://primarysources.brillonline.com/browse/human-rights-documents-online/promotion-and-protection-of-all-human-rights-civil-political-economic-social-and-cultural-rights-including-the-right-to-development;hrdhrd99702016149.

Patey, Luke A. 2012. "Lurking Beneath the Surface: Oil, Environmental Degradation, and Armed Conflict in Sudan." In *High-Value Natural Resources and Peacebuilding*, edited by Päivi Lujala and Siri Aas Rustad. Routledge.

San Sebastián, Miguel, and Anna Karin Hurtig. 2004. "Oil Exploitation in the Amazon Basin of Ecuador: A Public Health Emergency." *Revista Panamericana de Salud Pública* 15 (3). https://doi.org/10.1590/S1020-49892004000300014.

Symons, Kate. 2016. "Transnational Spaces, Hybrid Governance and Civil Society Contestation in Mozambique's Gas Boom." *The Extractive Industries and Society* 3 (1): 149–59. https://doi.org/10.1016/j.exis.2015.11.002.

Watts, Michael. 2001. "Petro-Violence: Community, Extraction, and Political Ecology of a Mythic Commodity." In *Violent Environments*, edited by Nancy Lee Peluso and Michael Watts. Cornell University Press.

Forests

Tom Cavanagh, Kelechi Eleanya, Tim Stork, Alice Taberner, Bernd Unger, and Pauliina Upla

Forests are complex and dynamic, primarily terrestrial ecosystems characterized by varying densities of tree/canopy cover that support diverse plant, animal, and human life and contribute to overall planetary health. They hold significant social, cultural, economic, ecological, and spiritual importance to local communities and societies at large, making them a strong foundation for peace but also a potential source of conflict.

The immense diversity of forest ecosystems makes it challenging—and often impractical—to establish a single, universally applicable definition of a "forest." This complexity arises from the global variation in geographical and ecological characteristics of forest habitats (such as boreal, temperate, and tropical), the wide range of forest uses (including primary forest **conservation**, community or subsistence practices, industrial management, and monoculture plantations), and the deeply contextual societal and cultural interpretations of what constitutes a "forest" and what it represents to different communities. This plurality of definitions positions forests as both potential sources of **conflict** and vital foundations for **peace**.

It is important to distinguish between "forests" (the ecosystems) and "forestry" (human management of these ecosystems). While not interchangeable, these two terms are closely intertwined, as forestry practices directly impact the health and **sustainability** of forests. Understanding this distinction is essential in **environmental peacebuilding**, where the human-forest interface can influence peace, livelihoods, and ecological stability in post-conflict settings.

From an institutional perspective, forests are defined based on their technical characteristics, such as land area covered, and tree density and height (Food and Agriculture Organization [FAO] 2018), or their ecological state (primary, secondary, plantation, old growth, and degraded), and the different management approaches applied to them (Convention on Biological Diversity 2006). Central to these definitions is the distinction between forests and other **land uses** (e.g., conversion to **agriculture**). Increasingly, forests are being recognized not only as ecological assets but also as critical components of global health (FAO 2020), climate stability, and **biodiversity** conservation (International Union for Conservation of Nature 2021). Governments have adopted variations of these institutional definitions in their land-use classifications, utilizing them to regulate access, conservation, and use.

Meanwhile, many Indigenous communities hold diverse and polysemic perspectives on forests. These often diverge significantly from institutional, Western, and colonial standpoints, which foreground the resource value of forests. These definitions reflect a broader understanding that forests serve as repositories of **Indigenous Knowledge** and cultural heritage, reinforcing the idea that their conservation is essential for maintaining cultural identity (Fa et al. 2020; Garnett et al. 2018). As a result, Indigenous perspectives on forests have underpinned resistance globally against highly extractive, colonial land uses (Le Billon and Lujala 2020).

The tension between resource-based and more holistic perspectives can make forests and forested land susceptible to conflicts. Their central role in providing for national economies and rural livelihoods, their importance as providers of global public goods supporting climate, biodiversity, and health, the shelter they provide to groups involved in the conflict, and the relative ease of extracting timber resources all place unique strains on forests in conflict-affected areas (Harwell 2010). Therefore, forests may reignite conflicts over land use, resource access, and management (Murillo-Sandoval et al. 2020). However, evidence from post-conflict settings, such as Liberia, suggests that effective management of forest resources can be a critical component of peacebuilding strategies, offering opportunities for **cooperation**, sustainable practices, and mutual benefit (Beevers 2015).

The challenge for environmental peacebuilders working in forests and forestry is to move beyond utilitarian notions of economic and resource-based value when considering the importance of forests. These limited perspectives can overlook the full spectrum of forest values, including spiritual and cultural dimensions (Meijaard et al. 2013). While progress has been made to incorporate evolving societal values and ecological considerations, as well as to acknowledge the non-utilitarian benefits of forests (Clark 2011), it is crucial to ensure that these advancements do not inadvertently contribute to new forms of commodification or resource extraction. Such approaches, including emerging market-based mechanisms for carbon and biodiversity conservation (e.g., on **REDD+** projects, see Alusiola et al. 2021), risk undermining and restricting access to and use of forests (Andoke Andoke et al. 2023), perpetuating injustices and colonial legacies.

References

Alusiola, Rowan A., Janpeter Schilling, and Paul Klär. 2021. "REDD+ Conflict: Understanding the Pathways between Forest Projects and Social Conflict." *Forests* 12: 748. https://doi.org/10.3390/f12060748.

Andoke Andoke, Levy, Eliran Arazi, Hernando Castro Suárez, Thomas F. Griffiths, and Esteban Gutiérrez Sánchez. 2023. "Amazonian Visions of Visión Amazonía: Indigenous Peoples' Perspectives on a Forest Conservation and Climate Programme in the Colombian Amazon." *Oryx* 57 (3): 335–49. https://doi.org/10.1017/S0030605322001636.

Beevers, Michael David. 2015. "Peace Resources? Governing Liberia's Forests in the Aftermath of Conflict." *International Peacekeeping* 22 (1): 26–42. https://doi.org/10.1080/13533312.2014.992568.

Clark, William. A. 2011. "Clarifying the Spiritual Values of Forests and Their Role in Sustainable Forest Management." *Journal for the Study of Religion, Nature and Culture* 5 (1): 18–38. https://doi.org/10.1558/jsrnc.v5i1.18.

Convention on Biological Diversity (CBD). 2006. *Forest Biodiversity – Definitions.* United Nations Environment Programme. https://www.cbd.int/forest/definitions.shtml.

Fa, Julia E., James E. M. Watson, Ian Leiper, et al. 2020. "Importance of Indigenous Peoples' Lands for the Conservation of Intact Forest Landscapes." *Frontiers in Ecology and the Environment* 18 (3): 135–40. https://doi.org/10.1002/fee.2148.

Food and Agriculture Organization of the United Nations (FAO). 2018. *Global Forest Resources Assessment 2020. Terms and Definitions FRA 2020.* FAO. https://openknowledge.fao.org/server/api/core/bitstreams/531a9e1b-596d-4b07-b9fd-3103fb4d0e72/content.

Food and Agriculture Organization of the United Nations (FAO). 2020. *Forests for Human Health and Well-Being – Strengthening the Forest–Health–Nutrition Nexus.* FAO. https://openknowledge.fao.org/items/250eccd3-d5e8-464a-81cd-a4dbde50a561

Garnett, Stephen T., Neil Burgess, Julia E. Fa, et al. 2018. "A Spatial Overview of the Global Importance of Indigenous Lands for Conservation." *Nature Sustainability* 1 (7): 369–74. https://doi.org/10.1038/s41893-018-0100-6.
Harwell, Emily. 2010. *Forests in Fragile and Conflict-Affected States*. Program on Forests (PROFOR). https://openknowledge.worldbank.org/server/api/core/bitstreams/79b95e9c-c3e6-56af-bc77-1fc10014c972/content.
International Union for the Conservation of Nature (IUCN). 2021. *Forests and Climate Change*. IUCN. https://iucn.org/sites/default/files/2022-04/forests_and_climate_change_issues_brief_2021.pdf.
Le Billon, Phillipe, and Päivi Lujala. 2020. "Environmental and Land Defenders: Global Patterns and Determinants of Repression." *Global Environmental Change* 65: 102163. https://doi.org/10.1016/j.gloenvcha.2020.102163.
Meijaard, Erik, Nicola K. Abram, Julie A. Wells, et al. 2013. "People's Perceptions about the Importance of Forests on Borneo." *PLoS One* 8 (9): e73008. https://doi.org/10.1371/journal.pone.0073008.
Murillo-Sandoval, Pedro J., Kelsey V. Dexter, Jamon Van Den Hoek, David Wrathall, and Robert E. Kennedy. 2020. "The End of Gunpoint Conservation: Forest Disturbance after the Colombian Peace Agreement." *Environmental Research Letters* 15 (3): 034033. https://doi.org/10.1088/1748-9326/ab6ae3.

Fragility

Milen Yishak, Ella Stack, and Jill Baggerman

Fragility is "a period of time during nationhood when sustainable socio-economic development requires greater emphasis on complementary peacebuilding and statebuilding activities such as building inclusive political settlements, security, justice, jobs, good management of resources, and accountable and fair service delivery" (g7+ 2014).

Literature on fragility in terms of state fragility has appeared in the political science realm since the 1990s, but the term is still widely debated in the international relations field and has been controversial throughout its use (g7+ 2014; Ficek 2021). Although there is a lack of consensus on what characterizes fragility (Ficek 2021; Guillaumont and Guillaumont Jeanneney 2011; Saeed 2020), fragile situations can largely be characterized by deep grievances and/or high levels of exclusion, lack of capacity, and limited provision of **basic services** to the population (World Bank 2020). According to the Organisation for Economic Co-operation and Development (OECD), fragility is "the combination of exposure to risk and insufficient coping capacity of the state, system and/or communities to manage, absorb, or mitigate those risks" (OECD 2016). The OECD framework recognizes that fragility is multidimensional, and its challenges are universal. The OECD's definition is referenced by other institutions such as the World Bank, International Monetary Fund, and Swiss Agency for Development and Cooperation, among others (Atashbar 2023; Diallo et al. 2022; World Bank 2020; Swiss Agency for Development and Cooperation n.d.). Alternative definitions tend to use different rhetoric but maintain conceptually within the boundaries of the OECD's definition: for example, the United States Government's Strategy to Prevent Conflict and Promote Stability describes fragility as "a country's or region's vulnerability to armed **conflict**, large-scale **violence**, or other instability, including an inability to manage transnational threats or other significant shocks" (U.S. Department of State 2020).

There are several internationally recognized fragility indices that produce mostly annual and biannual fragility lists and are used by different actors in the field of development, humanitarian assistance, and others. The g7+ is a group of over 20 member states affected by fragility and conflict (founding states were Afghanistan, Central African Republic, Cote d'Ivoire, Democratic Republic of the Congo, Haiti, Sierra Leone, and Timor-Leste) and which advocates for considering a "fragility spectrum" as a diagnostic tool for locally-led self-assessments about factors of fragility and **resilience** (g7+ 2014; g7+ n.d.). The fragility spectrum of the g7+ is meant to be a "counterweight to donor and expert-led efforts to classify and measure fragile states," so that states experiencing fragility are recognized as the experts "in terms of articulating what fragility has looked like, and continues to look like, in their experience and how to move to the next stage of resilience" (Government of the Republic of South Sudan 2012; g7+ 2017). The most popular fragility indices include the Fund for Peace's Fragile States Index (FSI), OECD's States of Fragility, the World Bank's Fragile and Conflict-affected Situations (FCS), and Country Policy and Institutional Assessment (CPIA). Relevant to **environmental peacebuilding**, the Economics and Peace's Ecological Threat Report focuses directly on risk factors like ecological threats, lack of resilience, and conflict dynamics (Institute for Economics & Peace 2024). The OECD's index and the CPIA also include environmental fragility as a measurement criterion or dimension. Fragility assessment data can support more tailored and aligned development, humanitarian aid, and **peace** programming to inform policy by providing evidence-based and contextualized information (International Organization for Migration 2023; States of Fragility 2025).

While the environmental peacebuilding field has long acknowledged the inherently complex linkages between **environment**, fragility, and peace, practitioners are now navigating the new challenge posed by the climate-fragility-peace nexus (CDA Collaborative Learning Projects n.d.). For instance, climate shocks can exacerbate fragility, increasing resource competition, transboundary water tensions, and livelihood insecurity, which challenge a state's resilience for addressing environmental and climate issues, creating a negative feedback loop (US Department of State and US Agency for International Development 2024; United States Institute of Peace 2021).

References

Atashbar, Tohid. 2023. "How Nations Become Fragile: An AI-Augmented Bird's-Eye View (with a Case Study of South Sudan)." IMF Working Papers No. 167. https://doi.org/10.5089/9798400252242.001.

CDA Collaborative Learning Projects. n.d. *Environment-Fragility-Peace Nexus*. CDA Collaborative. Accessed October 9, 2025. https://www.cdacollaborative.org/cdaproject/environment-fragility-peace/.

Diallo, Oumar, Gui-Diby, Steve Loris, and Imam, Patrick A. 2022. *Do Monetary Policy Outcomes Promote Stability in Fragile Settings?* International Monetary Fund, May 20. https://www.imf.org/en/Publications/WP/Issues/2022/05/20/Do-Monetary-Policy-Outcomes-Promote-Stability-in-Fragile-Settings-518203.

Ficek, Ryszard. 2021. "The Idea of a Fragile State: Emergence, Conceptualization, and Application in International Political Practice." *Stosunki Międzynarodowe–International Relations* 57: 93–110. https://doi.org/10.12688/stomiedintrelat.17468.1.

g7+. n.d. "Member States." Accessed October 9, 2025. https://www.g7plus.org/member-states/.

g7+. 2014. *Annual Report 2013 of the g7+*. g7+.

g7+. 2017. "Note on the Fragility Spectrum." https://www.g7plus.org/wpcontent/uploads/2021/09/fragility-spectrum-in-english.pdf.

Government of the Republic of South Sudan. 2012. *Fragility Assessment: Republic of South Sudan 2012 Summary Results*. Government of the Republic of South Sudan, December. https://www.g7plus.org/wp-content/uploads/2022/02/South-Sudan-Fragility-Assessment-Report-2012.pdf.

Guillaumont, Patrick, and Sylviane Guillaumont Jeanneney. 2011. *State Fragility and Economic Vulnerability: What Is Measured and Why?* Centre d'Études et de Recherches sur le Développement International, October 1. https://shs.hal.science/halshs-00554284v1.

Institute for Economics and Peace (IEP). 2024. "Ecological Threat Report 2024: Analysing Ecological Threats, Resilience & Peace." Accessed October 1, 2025. https://www.visionofhumanity.org/wp-content/uploads/2024/10/ETR-2024-web.pdf.

International Organization for Migration (IOM). 2023. "DTM South Sudan—Fragility Study Report: Understanding Multidimensional Fragility in South Sudan–2023." *Effective*, November 2. https://dtm.iom.int/reports/south-sudan-fragility-study-report-understanding-multidimensional-fragility-south-sudan.

Navone, Anthony. 2021. *The Intertwined Futures of Climate Action, Fragility and Peacebuilding*. United States Institute of Peace. Accessed November 9, 2024. https://web.archive.org/web/20241109065630/https://www.usip.org/publications/2021/04/intertwined-futures-climate-action-fragility-and-peacebuilding.

Organisation for Economic Co-operation and Development (OECD). 2016. *States of Fragility 2016: Understanding Violence*. http://dx.doi.org/10.1787/9789264267213-en.

Organisation for Economic Co-operation and Development (OECD). 2025. *States of Fragility 2025*. https://doi.org/10.1787/81982370-en.

Saeed, Rania. 2020. "The Ubiquity of State Fragility: Fault Lines in the Categorisation and Conceptualisation of Failed and Fragile States." *Social and Legal Studies* 29 (6): 767–89. https://doi.org/10.1177/0964663920906453.

Swiss Agency for Development and Cooperation (SDC). n.d. "Poverty, Instability and Violence in Fragile States." https://www.eda.admin.ch/deza/en/home/themes-sdc/fragile-contexts-and-prevention/fragile-states.html.

U.S. Department of State. 2020. "United States Strategy to Prevent Conflict and Promote Stability." https://www.state.gov/wp-content/uploads/2021/01/2020-US-Strategy-to-Prevent-Conflict-and-Promote-Stabilit-508c-508.pdf.

U.S. Department of State and U.S. Agency for International Development (USAID). 2024. "Climate Fragility Programming: Resource Guide." https://www.state.gov/wp-content/uploads/2024/10/Planning-Resource-Guide.pdf.

World Bank. 2020. "World Bank Group Strategy for Fragility, Conflict, and Violence 2020–2025." http://documents.worldbank.org/curated/en/844591582815510521/World-Bank-Group-Strategy-for-Fragility-Conflict-and-Violence-2020-2025.

G

Gender

Natalia Jiménez Galindo and Maryruth Belsey Priebe

Gender refers to the socially constructed roles, responsibilities, behaviors, expectations, and relations "along the continuum of masculinity, femininity, both, or neither" (Stanford Medicine 2025). As a social construct, gender varies across societies and can change over time (World Health Organization [WHO] n.d.; Kaufman et al. 2023).

Gender is distinct from biological sex, which relates to anatomical and physiological traits and distinguishes females and males (United Nations Women Training Centre 2020; Committee on Measuring Sex, Gender Identity, and Sexual Orientation et al. 2022). As a social construct, gender refers to social norms and roles based on what a society or culture deems appropriate according to a person's sex assigned at birth (Kaufman et al. 2023). Gender identity does not always correspond with a person's physiology or sex at birth (United Nations Women Training Centre 2020; WHO 2016). Gender influences how individuals experience rights, access to resources, decision-making power, and vulnerability to environmental, climate, and conflict-related risks (Anjum and Aziz 2025).

Environmental **change**, natural resource **governance**, and **conflict** often affect women, men, and gender-diverse people differently (United Nations Women 2025; Office of the High Commissioner for Human Rights n.d.). In the context of **environmental peacebuilding**, gender is a critical dimension because of these constructed roles, responsibilities, behaviors, expectations, and relationships. For example, women are frequently more vulnerable to resource scarcity and **displacement** due to existing inequalities, yet women play vital roles in environmental management and community **resilience**. Men may be more directly engaged in or vulnerable to resource-based armed conflict, while women may bear disproportionate burdens of increased caregiving or heightened risks of **gender-based violence** because of shifts in environmental health or resource availability (Camey et al. 2020). All genders, including men and boys, are vulnerable to gender- and sexual-based violence, whether in **peace**time or conflict-affected contexts, but how someone experiences that violence and access to healing services are often impacted by their gender (Office of the High Commissioner for Human Rights n.d.; Touquet 2021). Environmental degradation and resource scarcity create gender-specific vulnerabilities; for instance, in some contexts, women face increased time burdens and security risks during resource collection, while

DOI: 10.4324/9781003604143-9

men experience unemployment and economic marginalization, which may facilitate recruitment by extremist and armed groups (Sultana 2018).

Integrating gender analysis in environmental peacebuilding helps implementers understand the unequal impacts of **environmental degradation** and conflict, highlighting opportunities for inclusive and sustainable **peacebuilding.** It requires acknowledging diverse knowledge systems and ensuring meaningful **participation** and leadership of women and marginalized groups in environmental governance, climate adaptation, transparency, **violence** prevention, and conflict resolution processes.

Gender-inclusive environmental governance serves as a foundation for **sustainable peace** by addressing both environmental degradation and gender inequalities as interconnected drivers of insecurity (Yoshida et al. 2021). Countries with high levels of gender inequality are more likely to experience both environmental degradation and security instability, while women's participation in natural resource decision-making correlates with improved environmental outcomes and enhanced community resilience (Koengkan et al. 2021; Clement et al. 2019). Women's environmental networks function as platforms for cross-conflict dialogue and **cooperation,** providing mechanisms for peacebuilding that transcend traditional diplomatic channels (George 2014). Post-conflict environmental recovery programs that build on women's environmental expertise demonstrate enhanced effectiveness in both ecological restoration and **social cohesion** (Sawas et al. 2020). Institutional frameworks recognizing that women's environmental agency is essential to long-term security and peace represent critical innovations in environmental peacebuilding practice (Conca and Beevers 2018).

References

Anjum, Gulnaz, and Mudassar Aziz. 2025. "Climate Change and Gendered Vulnerability: A Systematic Review of Women's Health." *Women's Health* 12 (21): 17455057251323645. https://doi.org/10.1177/17455057251323645.

Camey, Itzá Castañeda, Laura Sabater, Cate Owren, and A. Emmett Boyer. 2020. *Gender-based Violence and Environment Linkages: The Violence of Inequality*. Edited by Jamie Wren. International Union for Conservation of Nature. https://portals.iucn.org/library/sites/library/files/documents/2020-002-En.pdf.

Clement, Floriane, Deepa Joshi, and Marlène Elias. 2019. "Feminist Political Ecologies of the Commons and Commoning." *International Journal of the Commons* 13 (1): 1–15. https://doi.org/10.18352/ijc.972.

Conca, Ken, and Michael D. Beevers. 2018. "Environmental Pathways to Peace." In *Routledge Handbook of Environmental Conflict and Peacebuilding*, edited by Ashok Swain and Joakim Öjendal. Routledge.

George, Nicole. 2014. "Promoting Women, Peace and Security in the Pacific Islands: Hot Conflict/Slow Violence." *Australian Journal of International Affairs* 68 (3): 314–32. https://doi.org/10.1080/10357718.2014.902032.

Kaufman, Michelle R., Evan L. Eschliman, and Tahilin Sanchez Karver. 2023. "Differentiating Sex and Gender in Health Research to Achieve Gender Equity." *Bulletin of the World Health Organization* 101 (10): 666–71. https://doi.lorg/10.2471/BLT.22.289310.

Koengkan, Matheus, José Alberto Fuinhas, Mônica Teixeira, et al. 2021. "Is Gender Inequality an Essential Driver in Explaining Environmental Degradation? Some Empirical Answers from the CO2 Emissions in European Union Countries." *Environmental Science and Pollution Research* 28 (37): 52327–43. https://doi.org/10.1007/s11356-021-14269-z.

National Academies of Sciences, Engineering, and Medicine. 2022. *Measuring Sex, Gender Identity, and Sexual Orientation*. Edited by Nancy Bates, Marshall Chin, and Tara Becker.

National Academies Press. https://nap.nationalacademies.org/catalog/26424/measuring-sex-gender-identity-and-sexual-orientation.
Office of the High Commissioner for Human Rights (OHCHR). n.d. *Women's Human Rights and Gender-related Concerns in Situations of Conflict and Instability*. OHCHR. Accessed October 6, 2025. https://www.ohchr.org/en/women/womens-human-rights-and-gender-related-concerns-situations-conflict-and-instability.
Sawas, Amiera, Christelle Soulard, Emilie Rees, and Judy El-Bushra. 2020. *Gender, Climate and Security: Sustaining Inclusive Peace on the Frontlines of Climate Change*. UN Environment Programme, UN Women, UNDP, UNDPPA/PBSO. https://wedocs.unep.org/bitstream/handle/20.500.11822/32638/GCS.pdf?sequence=1&isAllowed=y.
Stanford Medicine. 2025. "Gender." https://mededucation.stanford.edu/glossary/gender/.
Sultana, Farhana. 2018. "Gender and Water in a Changing Climate: Challenges and Opportunities." In *Water Security Across the Gender Divide*, edited by Christiane Fröhlich, Giovanna Gioli, Roger Cremades, and Henri Myrttinen. Springer International Publishing. https://doi.org/10.1007/978-3-319-64046-4_2.
Touquet, Helen et al. 2021. "From 'It Rarely Happens' to 'It's Worse for Men': Dispelling Misconceptions about Sexual Violence against Men and Boys in Conflict and Displacement." *Journal of Humanitarian Affairs* 2 (3). https://doi.org/10.7227/JHA.049.
United Nations Women. 2025. *How Gender Inequality and Climate Change are Interconnected*. United Nations Women. https://www.unwomen.org/en/articles/explainer/how-gender-inequality-and-climate-change-are-interconnected.
United Nations Women Training Centre. 2020. "Gender Equality Glossary." https://wrd.unwomen.org/practice/resources/gender-equality-glossary.
World Health Organization (WHO). n.d. "Gender and Health." https://www.who.int/health-topics/gender#tab=tab_1.
World Health Organization (WHO). 2016. "Frequently Asked Questions on Health and Sexual Diversity: An Introduction to Key Concepts." https://www.who.int/publications/i/item/WHO-FWC-GER-16.2.
Yoshida, Keina, Lina M. Céspedes-Báez, Laura J. Shepherd, and Elisa Randazzo. 2021. *Defending the Future: Gender, Conflict and Environmental Peace*. Centre for Women, Peace and Security at LSE. https://www.lse.ac.uk/women-peace-security/assets/documents/2021/Defending-the-Future.pdf.

Gender-Based Violence

Holly Dunn, Silja Halle, and Sidra Riaz

Gender-based violence is a form of violence directed at a person or group based on their affirmed or assumed gender. Harms can include, but are not limited to, physical, psychological, sexual, economic, emotional, and structural violence.

Gender-based **violence** (GBV) is a pervasive feature of armed **conflict** and other forms of political violence. Further, evidence shows that environmental pressures and GBV are directly linked and that **environmental degradation**, natural resource exploitation, and competition over land, water, and other resources can all exacerbate violence (International Union for Conservation of Nature 2020). **Climate change** further amplifies these risks, with highly gendered impacts (United Nations Environment Programme et al. 2020).

GBV in conflict manifests in distinct forms, often arising from systemic inequality, and has broader social, economic, and political consequences. GBV risks can be considered on a continuum, spanning from the period before armed conflict or political violence, through the period of transition, and extending into **peacebuilding** efforts

(Cockburn 2004). Armed forces and other groups in conflict use GBV as a weapon of **war** to assert control over resources or terrorize populations, taking advantage of a context of impunity. Those who have been victimized often lack avenues for reporting these violations or may hesitate to do so for a variety of reasons, with impacts often persisting through the transition to **peace**.

Analysis of the impacts of GBV in conflict is strengthened by considerations of intersectionality—that is, how multiple identity categories may overlap to disadvantage or privilege different groups within a given context (Kappler and Lemay-Hébert 2021). Factors such as race, class, age, and sexual orientation can compound vulnerability. Analyzes of GBV must also grapple with conceptual slippage between GBV, violence against women and girls, and sexual violence, which has at times led to imprecision in understanding the experiences or specific vulnerabilities of different populations during conflict and political violence. While women are more likely to be targeted for sexual violence, their experiences are multifaceted and diverse, including increased risks of domestic violence, sexual harassment, and trafficking (Kostovicova et al. 2020; Nordås and Cohen 2021). Similarly, men from certain demographic groups may be disproportionately targeted for lethal violence, but may also experience other forms of harm, including sexual violence, forced conscription, emotional abuse, and economic exploitation (Carpenter 2006; Greig n.d.). Further, research is limited on the experience of gender non-conforming populations. To the extent possible, assumptions about the manifestations and impacts of GBV should be avoided in favor of empirical analyzes.

Given the complexity and entrenched nature of GBV in many contexts, a focus on **positive peace** in **environmental peacebuilding** can help highlight the myriad of concerns, such as vulnerabilities based on gender-specific roles or coping strategies related to **natural resources**, as well as the gendered risks faced by **environmental defenders**. Positive peace requires not only stopping direct violence but also adopting a holistic approach to building safe, inclusive, and just environments that address the long-standing structural inequities that perpetuate violence. In this way, it reflects and furthers the objectives of **UN Security Council** Resolution 1325 on Women, Peace, and Security. In this context, positive peace emphasizes the necessity of not only protecting vulnerable populations but also addressing the broader, gender-specific impacts of environmental stress, climate change, and conflict.

There is a critical need for further research on the nexus of violence, intersectionality, and environmental peacebuilding. The continuum of violence demonstrates that GBV is not limited to periods of conflict; rather, it endures across pre-conflict, conflict, and post-conflict phases. Intersectional analysis illustrates how different identities, such as gender, ethnicity, and socioeconomic class, exacerbate vulnerability. These concepts are critical for comprehending the entire spectrum of GBV in conflict contexts, as well as devising gender-sensitive and intersectional environmental peacebuilding strategies.

References

Carpenter, R. Charli. 2006. "Recognizing Gender-Based Violence Against Civilian Men and Boys in Conflict Situations." *Security Dialogue* 37 (1): 83–103. https://doi.org/10.1177/0967010606064139.

Cockburn, Cynthia. 2004. "The Continuum of Violence." In *Sites of Violence: Gender and Conflict Zones*. Edited by Wenona Giles and Jennifer Hyndman. University of California Press.
Greig, Alan. n.d. *Men, Masculinities and Armed Conflict: Findings from a Four-Country Study by the Women's International League for Peace and Freedom*. Women's International League for Peace and Freedom. https://www.wilpf.org/wpcontent/uploads/2022/09/RL_19_WILPF-Men-Masculinities-Armed-Conflict-AW.pdf.
International Union for Conservation of Nature (IUCN). 2020. "Gender-Based Violence and Environment Linkages." https://www.iucn.org/resources/publication/gender-based-violence-and-environment-linkages.
Kappler, Stefanie, and Nicolas Lemay-Hébert. 2021. "From Power-Blind Binaries to the Intersectionality of Peace: Connecting Feminism and Critical Peace and Conflict Studies." In *Feminist Interventions in Critical Peace and Conflict Studies*. Edited by Laura McLeod and Maria O'Reilly. Routledge.
Kostovicova, Denisa, Vesna Bojicic-Dzelilovic, and Marsha Henry. 2020. "Drawing on the Continuum: A War and Post-War Political Economy of Gender-Based Violence in Bosnia and Herzegovina." *International Feminist Journal of Politics* 22 (2): 250–72.
Nordås, Ragnhild, and Dara Kay Cohen. 2021. "Conflict-Related Sexual Violence." *Annual Review of Political Science* 24 (1): 193–211. https://doi.org/10.1146/annurev-polisci-041719-102620.
United Nations Environment Programme (UNEP), United Nations Women, United Nations Department of Political and Peacebuilding Affairs (DPPA), and United Nations Development Programme (UNDP). 2020. "Gender, Climate and Security: Sustaining Peace on the Frontlines of Climate Change." https://www.undp.org/publications/gender-climate-security-sustaining-peace-frontlines-climate-change.

Gender Equality and Social Inclusion

Paula Gutitérrez Martínez and Martina Jaskolski

The Gender Equality and Social Inclusion, also known as GESI, approach focuses on redistribution of social, cultural, economic, and political power relations and aims at achieving the principle of leaving no one behind, which is at the center of the Sustainable Development Goals (SDGs).

GESI is based on two interconnected concepts: Gender Equality and Social Inclusion. Gender Equality is concerned with providing equal rights, opportunities, access to resources, protection, support, and **power** to all genders—women, men, and non-binary groups (Gender Equality and Social Inclusion Working Group 2017). Social Inclusion refers to the process of including traditionally excluded individuals and groups in the economy, society, and decision-making, regardless of their **gender**, ethnicity, religion, socioeconomic status, language, ability, location, age, or race (Global Green Growth Institute [GGGI] 2021).

The closing of economic and political gender gaps has been part of the feminist struggle for decades. In this context, academic and activist feminists have advocated for gender equity principles in a range of fields, including feminist environmentalism (Agarwal 1997), firmly positioning these struggles in the international development agenda (Coffey et al. 2022).

The concept of Gender Equality has had three different evolving paths: first, it has evolved from the realization of historical political, economic, and social inequalities, and exclusions affecting women, to the recognition of non-binary gender identities also traditionally excluded from society and decision-making (Dugarova 2020). Second, it has proposed a shift from equality to equity. While equality is defined as assigning the same rights or opportunities to all genders and social groups, the concept of equity recognizes that people have differing needs due to their different circumstances, which may require allocating rights and opportunities in ways that address each one's needs (Global Citizen 2021). For example, assigning women equal rights to men may not be enough to close the gender gap that has been established over centuries, and affirmative action might be required to close existing gaps. Third, it has increasingly integrated the concept of intersectionality (Kabeer 2010), which recognizes how multiple forms of discrimination, such as religion, ethnicity, race, location, and socioeconomic status, interplay with gender and have an impact on access to education, information, ownership of land and **natural resources**, and decision-making powers.

Over the last decades, GESI approaches have been increasingly integrated in the global development agenda, from the 2030 Agenda and its SDGs to guidelines of various development organizations. GESI in **sustainable development** initiatives and projects is organized along a range of four levels of ambition: from gender blind to gender transformative. Level 0 represents non-compliance where projects are GESI-blind and may unintentionally reinforce discrimination. Level 1 achieves minimum compliance by addressing basic needs and allowing marginalized groups to voice concerns. Level 2 focuses on **empowerment**, building capabilities, and enabling these groups to make choices and participate in decision-making. Level 3 represents transformation, where projects address power imbalances and give women and marginalized groups active control over resources and decisions, seeking broader institutional and societal change (UK Partnering for Accelerated Climate Transitions 2021).

In the context of **environmental peacebuilding**, GESI is of pivotal importance, as it helps understand "people's positions of relative power or vulnerability" (Myrttinen et al. 2016, 5) and recognizes that different genders and social groups are differently exposed to risks of **environmental change**, **climate change**, **conflict**, and **violence**. Gender-based violence (GBV), for example, is a form of violence specifically directed at people because of their gender. Environmental change and conflict around resources affect livelihood processes that in themselves are gendered and affected by structural social inequality, including **land tenure**, gender division of labor, and formal vs. informal employment. A better understanding of how inequality and social inclusion (or exclusion) shape, and are simultaneously shaped by, processes of social organization, environmental change, conflict, and **peace** is central to environmental peacebuilding.

Moreover, **peacebuilding** processes as well as environmental management and **governance** are gendered and connected to certain social groups. For example, women, elders, or members of indigenous communities play important roles in peacebuilding and peacekeeping, holding forms of knowledge, experiences, skills, or a social standing that enable them to foster and maintain social connections and environmental **sustainability**, as well as facilitating consultation and **mediation** processes (United Nations Development Programme [UNDP] 2021).

Mainstreaming GESI into environmental policy and program design and implementation can help ensure that they are as beneficial as possible to all social groups, fostering social sustainability, the achievement of the SDGs, and better peace outcomes. Several entry points for the inclusion of GESI have been identified to be integrated into environmental and peacebuilding initiatives: (1) building inclusive participatory decision-making processes to ensure that vulnerable and previously marginalized groups can have a voice and drive transformation processes, (2) gender-sensitive budgeting, (3) gender-inclusive financing, (4) the inclusion of sex-disaggregated **indicators** into **monitoring** and **evaluation** processes, (5) gender and social inclusion quotas, (6) strengthening environmental and social safeguards to avoid unintended negative impacts that would perpetuate structural inequalities, (7) conducting participatory GESI analysis to inform proposal development and implementation, and (8) fostering socially inclusive peacebuilding and peacekeeping processes (Gumucio et al. 2018; GGGI 2021; UNDP 2021).

References

Agarwal, Bina. 1997. "Environmental Action, Gender Equity and Women's Participation." *Development and Change* 28: 1–44. https://doi.org/10.1111/1467-7660.00033.

Coffey, Chloe, Gary Barker, Francisco Contreras-Urbina, et al. 2022. "Gender-Transformative Approaches in International Development: A Brief History and Five Uniting Principles." *Women's Studies International Forum* 158: 105975. https://doi.org/10.1016/j.worlddev.2022.105975.

Dugarova, Esuna. 2020. *Progress Towards Gender Equality and Women's Empowerment: The UNDP Journey.* United Nations Chronicle. https://www.un.org/en/un-chronicle/reflecting-progress-towards-gender-equality-and-women%E2%80%99s-empowerment-undp-journey?utm.

Gender Equality and Social Inclusion Working Group. 2017. *A Common Framework for Gender Equality & Social Inclusion.* United Nations. https://un.org.np/sites/default/files/doc_publication/2019-01/GESI%20Framework_English%2020.07.2017_opt.pdf.

Global Citizen. 2021. "Equity vs Equality: What's the Difference?" https://www.globalcitizen.org/en/content/equity-equality-whats-the-difference-global-goals/.

Global Green Growth Institute (GGGI). 2021. "Gender Equality and Social Inclusion Strategy 2021–2025." https://gggi.org/report/gggi-gender-equality-and-social-inclusion-strategy/.

Gumucio Tatiana, Sophia Juyer, James Hansen, Elisabeth Simelton, Samuel T. Partey, and Saroja Schwager. 2018. "Inclusion of Gender Equality in Monitoring and Evaluation of Climate Services." CCAFS Working Paper No. 249. CGIAR Research Program on Climate Change, Agriculture and Food Security (CCAFS).

Kabeer, Naila. 2010. "Can the MDGs Provide a Pathway to Social Justice? The Challenge of Intersecting Inequalities." IDS and MDG Achievement Fund. https://www.ids.ac.uk/files/dmfile/MDGreportwebsiteu2WC.pdf.

Myrttinen, Henri, Nicola Popovic, and Lana Khattab. 2016. *Measuring Gender in Peacebuilding: Evaluating Peacebuilding Efforts from Gender-Relational Perspective.* International Alert. https://www.international-alert.org/publications/measuring-gender-in-peacebuilding/.

United Nations Development Programme (UNDP). 2021. "Study on the Traditional and Changing Role of Gender and Women in Peacebuilding in South Sudan." https://www.undp.org/publications/study-traditional-and-changing-role-gender-and-women-peacebuilding-south-sudan.

UK Partnering for Accelerated Climate Transitions (UKPACT). 2021. "Guidance on Gender Equality and Social Inclusion (GESI)." https://f.hubspotusercontent10.net/hubfs/7376512/cp/general/UK%20PACT%20GESI%20Guidance.pdf?hsCtaTracking=2b59d319-333f-4053-a9ab-52492b2353a1%7C023f06ea-5c31-408a-99cd-c43abc2262b9.

Gender-Just Transition

Ambreen Tour Ben-Shmuel, Paula Gutiérrez Martínez, and Aseel Zahran

Gender-just transition is an approach ensuring that shifts toward sustainable practices are inclusive, particularly addressing gender-based disparities. In the context of environmental peacebuilding, this includes integrating gender considerations in efforts to address peace and sustainability simultaneously.

The concept of a gender-just transition is still evolving, and various sectors are refining their understanding and practices toward this approach. Drawing on insights from practitioners and researchers in the environmental field, while blending them with understandings of **gender** in inclusive **peace** strategies, helps to further a gender-just approach in **environmental peacebuilding.**

The term "**just transition**" emerged in the 1970s and 1980s, focusing on worker protections in polluting industries (Stevis and Felli 2014). Since then, it has broadened, looking to protect the rights and needs of those most impacted as societies move away from more extractive, polluting, and environmentally harmful economies (United Nations Women 2023). In principle, a gender-just transition entails employing a gender perspective in environmental transitions, from **agriculture** to energy and industry, among others. This approach recognizes that women and men often experience environmental challenges differently, including in **conflict**-affected contexts, due to existing social, economic, and cultural inequalities (Sultana 2013; Ben-Shmuel and Halle 2023). This is particularly evident in the case of women **environmental defenders**, who often face disproportionate risks when advocating for sustainable resource management and **environmental justice** in conflict-affected contexts (Kezie-Nwoha et al. 2021). A just transition emphasizes the need to address gender-specific impacts and embrace diverse perspectives in environmental management and **peacebuilding** (Smith et al. 2021). Additionally, an intersectional lens reveals how gender intersects with other aspects of identity and **power** relations, such as ethnicity, class, and geographic location, shaping how different groups experience and engage with environmental transitions (Yoshida and Céspedes-Báez 2021).

In the context of environmental peacebuilding, a gender-just transition can involve: (1) ensuring that initiatives address instead of exacerbate gender inequality, (2) recognizing and responding to how gender roles and power dynamics impact information and resource access and control in conflict and post-conflict settings, (3) promoting women's leadership in environmental **governance** and peacebuilding initiatives, (4) ensuring new economic opportunities created through environmental restoration or sustainable resource management are equally accessible to women and men, (5) addressing how **adaptation** and **mitigation** strategies for **climate change** in conflict-affected areas can promote gender equality and peace, and (6) strengthening systems for collecting, analyzing, and sharing gender-disaggregated data to inform evidence-based policymaking (Buckingham and Le Masson 2017; Mead and Jacobsson 2023; Ngum and Kim 2023).

While environmental peacebuilding has made strides in understanding gender dynamics, there is still work to be done, particularly regarding a gender-just transition. Research indicates that more robust institutional frameworks and gender mainstreaming efforts are needed to support gender-inclusive approaches, particularly in environmental initiatives where technical solutions often overshadow social considerations (Ben-Shmuel 2025). Future research and practice should better address how to integrate gender **justice** and gender-just transition considerations into environmental peacebuilding efforts.

References

Ben-Shmuel, Ambreen T. 2025. "When Peace Meets Environment: Gender Regimes, Ecotones, and the Future of Sustainability Work." *Academy of Management Proceedings* (1): 11961. https://journals.aom.org/doi/10.5465/AMPROC.2025.11961abstract.

Ben-Shmuel, Ambreen T., and Silja Halle. 2023. "Beyond Greenwashing: Prioritizing Environmental Justice in Conflict-Affected Settings." *Environment and Security* 1 (3–4): 209–18. https://doi.org/10.1177/27538796231186697.

Buckingham, Susan, and Virginie Le Masson. 2017. *Understanding Climate Change Through Gender Relations.* Routledge. https://doi.org/10.4324/9781315661605.

Kezie-Nwoha, Helen, Keina Yoshida, and Hannah Bond. 2021. "Defending the Future: Gender, Conflict and Environmental Peace." LSE Women, Peace and Security Policy Brief 03/2021. LSE Centre for Women, Peace and Security. https://www.lse.ac.uk/women-peace-security/publications/Defending-the-Future-Gender-Conflict-and-Environmental-Peace.

Mead, Sarah, and Marie Jacobsson. 2023. "Integrating Gender, Peace and Environment: The Gender Dimension of Environmental Peacebuilding." In *Research Handbook on International Law and Environmental Peacebuilding*, edited by Britta Sjöstedt and Michael Brzoska. Edward Elgar Publishing. https://doi.org/10.4337/9781789906929.00026.

Ngum, Sohna, and Luisa Kim. 2023. *Powering a Gender-Just Energy Transition.* Green Growth Knowledge Partnership Expert Group on Gender Working Paper 03/2023. Green Growth Knowledge Partnership. https://gggi.org/report/powering-a-gender-just-energy-transition/.

Smith, Jessica, Lauren Olosky, and Jennifer G. Fernández. 2021. *The Climate-Gender-Conflict Nexus: Amplifying Women's Contributions at the Grassroots.* Georgetown Institute for Women, Peace and Security. https://giwps.georgetown.edu/resource/the-climate-gender-conflict-nexus/.

Stevis, Dimitris, and Romain Felli. 2014. "Global Labour Unions and Just Transition to a Green Economy." *International Environmental Agreements: Politics, Law and Economics* 15(1): 29–43. https://doi.org/10.1007/s10784-014-9266-1.

Sultana, Farhana. 2013. "Gendering Climate Change: Geographical Insights." *The Professional Geographer* 66 (3): 372–81. https://doi.org/10.1080/00330124.2013.821730.

United Nations Women. 2023. *A Gender-Responsive Just Transition for People and Planet.* https://www.unwomen.org/en/digital-library/publications/2023/11/policy-brief-a-gender-responsive-just-transition-for-people-and-planet.

Yoshida, Keina, and Lina M. Céspedes-Báez. 2021. "The Nature of Women, Peace and Security: A Colombian Perspective." *International Affairs* 97 (1): 17–34. https://doi.org/10.1093/ia/iiaa173.

Governance

Gül Özerol and Liliana Pimentel

Governance is the processes, structures, and practices through which an organization or society operates, exercises power and authority, allocates and manages resources, and makes decisions.

Over the past decades, the notion of "governance replacing government" has gained increased attention, as the range of actors involved in decision-making has been expanding to involve both individual and organized public members as well as companies and other actors in the private sector. As a result, various definitions and meanings have been attributed to governance, reflecting its applications across diverse contexts and fields, such as environmental governance, **resource governance**, collaborative governance, wartime governance, and corporate governance. These applications are also typically linked to broader governance-related concepts. With regards to **environmental peacebuilding**, two concepts are largely relevant, namely good governance and multi-level governance.

Good governance encompasses a wide range of principles and practices that are universally applicable and endorsed by international organizations, governments, and scholars alike. It is participatory, consensus-oriented, accountable, transparent, responsive, effective, efficient, equitable, inclusive, and follows the rule of law. Keping (2018) emphasizes the need for governance systems to adapt to global changes, such as globalization and increasing uncertainty, arguing that good governance is crucial for economic competitiveness and sustainable social growth. Elinor Ostrom (1990) made groundbreaking contributions to governance, particularly in the management of common-pool resources. She challenged the conventional wisdom that common-pool resources could only be managed through state regulation or privatization and demonstrated that local communities could self-organize to manage these resources sustainably. In her seminal work, *Governing the Commons: The Evolution of Institutions for Collective Action*, Ostrom identified key principles for successful collective action, such as clearly defined boundaries, collective-choice arrangements, and **monitoring** and sanctioning systems. Her work has profoundly impacted the understanding of good governance, highlighting the importance of local knowledge, **participation**, and self-governance, influencing policies and practices in natural resource management, environmental protection, and **sustainable development**.

Multi-level governance is crucial for environmental peacebuilding as it emphasizes the integration of policies and the coordination of institutions across different governance levels, ranging from local to global. Multi-level governance facilitates the alignment of local priorities with national and international policies, fostering inclusive decision-making and **conflict** resolution (United Nations Development Programme 2016). By promoting collaboration and inclusiveness, multi-level governance can also help build trust among various **stakeholders**, manage environmental resources sustainably and equitably, and address grievances that arise because of conflict and its social and environmental impacts. In the long term, multi-level governance can also support the creation of strategic development alliances and enhance the **accountability** of public actors, ultimately contributing to **peace** and stability.

Overall, governance plays a vital role in **peacebuilding** by establishing and maintaining social order, **justice**, and the rule of law. Strong, multi-level governance structures and processes can reduce environmental vulnerability and enhance ecological **resilience**, fostering peaceful societies. In post-conflict settings, good governance is essential for rebuilding institutions, ensuring **security**, and promoting **social cohesion** (World Bank 1992; United Nations 2021). Effective governance in natural resource management and post-conflict settings can significantly contribute to long-term peace and stability by promoting sustainable practices and inclusive policies (Bruch et al. 2016).

References

Bruch, Carl, Carroll Muffett, and Sandra S. Nichols, eds. 2016. *Governance, Natural Resources and Post-conflict Peacebuilding*. Routledge.
Keping, Yu. 2018. *Governance and Good Governance*. Beijing University Press.
Ostrom, Elinor. 1990. *Governing the Commons: The Evolution of Institutions for Collective Action*. Cambridge University Press.
United Nations (UN). 2021. *Governance for Peace: Strengthening Inclusive, Just and Peaceful Societies Resilient to Future Crises*. United Nations.
United Nations Development Programme (UNDP). 2016. *Local Governance in Fragile and Conflict Affected Settings: Building a Resilient Foundation for Peace and Development.*
World Bank. 1992. *Governance and Development*. World Bank.

H

Human Rights

Stacey M. Mitchell and Eileen M. Angelini

Human rights are inalienable, universal, and non-reciprocal. Every person has these rights regardless of whether they are protected by the constitution and domestic laws of the state in which they reside. Human rights are broadly categorized into two sets: civil and political rights as well as economic, social, and cultural rights, both of which are widely seen as interlinked and interdependent and grounded in the concept of human dignity (Donnelly 2013; Sen 2000; Whelan 2010; United Nations General Assembly [UNGA] 1948).

In the field of **environmental peacebuilding**, human rights are a cornerstone of the process. The work of such scholars as Cederman and Vogt (2017), Ide (2020), Ide (2023), Ide et al. (2021), Bruch et al. (2022), Weinthal and Bruch (2023), Yoshida and Céspedes-Báez (2021), Lujala and Rustad (2012), and Vita (2018) emphasize that issues of **human security**, **sustainable development**, **conflict** resolution, resource management, and **gender** equity are embedded in the protection and promotion of both sets of human rights. Access to an adequate standard of living (Article 11), including food, shelter, and the efficient use of **natural resources**, in addition to the highest standard of physical and mental health, including a healthy **environment** (Article 12), is enshrined in the International Covenant on Economic, Social and Cultural Rights (1966). The International Covenant on Civil and Political Rights (1966) provides for the right to life (Article 6), **security** and liberty of persons (Article 9), and voting and participating in public affairs (Article 25). All these rights impact and are impacted by all types of **peace** missions.

The UN sees human rights and the environment as inseparable. From the 1992 Earth Summit in Rio de Janeiro, Brazil, to the present day, the UN has worked to create a more equitable global partnership toward sustainable development. The UN's 2000 Millennium Development Goals (UNGA 2000) and its 2015 replacement, the 17 Sustainable Development Goals (SDGs), meld peace, environmental **sustainability**, and human rights together in a comprehensive framework that all states can utilize for achieving a more livable world. SDG 16, which addresses the promotion of peaceful, inclusive societies for sustainable development, contains targets that directly pertain to environmental peacebuilding. For example, targets 16.6 and 16.7, which focus on the effectiveness, transparency, and representation of domestic

DOI: 10.4324/9781003604143-10

institutions of **governance**, have strong implications for a public's access to quality of life resources, including healthcare (SDG 3), clean water (SDG 6), food (SDG 2), freedom from chemical waste (SDG 12), and equitable access to affordable and clean sources of energy (SDG 7), all of which are essential components of **peacebuilding** processes. The SDGs were adopted as part of the UN's 2030 Agenda for Sustainable Development, which reinforces the principle that "[s]ustainable development cannot be realized without **peace** and security; and peace and security will be at risk without sustainable development" (UNGA 2015, 9).

Under the authority of the Human Rights Council, several thematic special procedures mechanisms were created on the environment, including Special Rapporteurs on Climate Change, Disposal and Management of Environmentally Toxic Substances and Wastes, Safe Drinking Water, and the Human Right to a Healthy Environment. The Committee on Economic, Social and Cultural Rights, which monitors implementation and enforcement of the ICESCR, recommends to member states methods by which they can improve upon the protection of these rights. These reporting procedures tend to run concurrently with environmental peacebuilding programs and represent an additional safeguard against the violation of these quality-of-life rights. Similarly, the Human Rights Committee, which oversees implementation and enforcement of the ICCPR, serves as another venue for **monitoring** human rights protections in member states.

References

Bruch, Carl, Marion Boulicault, Shuchi Talati, and David E. Jensen. 2022. "Peacebuilding and Natural Resources." *Environmental Forum* 29 (4): 48–55. https://www.researchgate.net/publication/292067972_Peacebuilding_and_natural_resources.

Cederman, Lars-Eric, and Manuel Vogt. 2017. "Dynamics and Logics of Civil War." *Journal of Conflict Resolution* 61 (9): 1992–2016. https://www.jstor.org/stable/26363974.

Donnelly, Jack. 2013. *Universal Human Rights in Theory and Practice*. Cornell University Press.

Ide, Tobias. 2020. "The Darkside of Environmental Peacebuilding." *World Development* 127 (1): 1–9. https://doi.org/10.1016/j.worlddev.2019.10477.

Ide, Tobias. 2023. "Rise or Recede? How Climate Disasters Affect Armed Conflict Intensity." *International Security* 47 (4): 50–78. https://doi.org/10.1162/isec_a_00459.

Ide, Tobias, Carl Bruch, Alexander Carius, et al. 2021. "The Past and Future(s) of Environmental Peacebuilding." *International Affairs* 97 (1): 1–16. https://doi.org/10.1093/ia/iiaa177.

Lujala, Päivi, and Siri Aas Rustad, eds. 2012. *High-Value Natural Resources and Post-Conflict Peacebuilding*. Routledge.

Roy, Vita. 2018. "Managing Resource-Related Conflict: A Framework of Lootable Resource Management and Postconflict Stabilization." *Journal of Conflict Resolution* 62 (5): 1044–71. https://doi.org/10.1177/0022002716669206.

Sen, Amyarta. 2000. *Development as Freedom*. Anchor Books.

United Nations General Assembly (UNGA). 1948. *Universal Declaration of Human Rights 217 A (III)*.

United Nations General Assembly (UNGA). 1966a. *International Covenant on Civil and Political Rights*. United Nations Treaty Series 999.

United Nations General Assembly (UNGA). 1966b. *International Covenant on Economic, Social and Cultural Rights*. United Nations Treaty Series, 993.

United Nations General Assembly (UNGA). 2000. *United Nations Millennium Declaration*, Resolution Adopted by the General Assembly. A/RES/55/2.

United Nations General Assembly (UNGA). 2015. *Transforming our World: The 2030 Agenda for Sustainable Development*. A/RES/70/1.

Weinthal, Erika, and Carl Bruch. 2023. "Protecting Nuclear Powerplants During War: Implications from Ukraine." *Environmental Law Reporter: News and Analysis* 43 (4): 10285–97. https://heinonline.org/HOL/LandingPage?handle=hein.journals/elrna53&div=39&id=&page=.
Whelan, Daniel J. 2010. *Indivisible Human Rights: A History*. University of Pennsylvania Press.
Yoshida, Keina, and Lina M. Céspedes-Báez. 2021. "The Nature of Women, Peace and Security: A Colombian Perspective." *International Affairs* 97 (1): 17–34. https://doi.org/10.1093/ia/iiaa173.

Human Right to a Clean Environment

Zerrin Savaşan, Atieh Khatibi, and Saeede Mokhtarzade

The human right to a clean environment is a universal and substantive right ensuring every person can live in a clean, healthy, and sustainable environment with access to its resources and services in sustainable and peaceful ways.

Environmental degradation, escalating pollution, and **climate change** profoundly affect all human beings, making the enjoyment of the right to a clean, healthy, and sustainable **environment** crucial for the realization of other fundamental **human rights** (Boyd and Knox 2018). Yet, states' obligation to fulfill other economic, social, and cultural rights must also be considered alongside ensuring no transboundary harm to the environment while fostering development (Jesse and Koppe 2013). Furthermore, businesses across all sectors—production, supply chain, and transportation—have a mutual influence on respecting the human right to a clean environment (Lagerqvist and Wåhlin 2022). This involves preventing environmental harm and recognizing the importance of **sustainable development**, particularly in developing countries. Therefore, recognizing this right in international human rights treaties can affirm states' obligation to comply with national policies and legislation, supporting judicial bodies in relevant litigation that reflects any aspects of this right (Office of the High Commissioner for Human Rights et al. 2023). This recognition is vital for holding states accountable and ensuring the enjoyment of other human rights, thus fostering more sustainable development.

Despite the growing urgency, the Universal Declaration of Human Rights does not contain any specific provision for environmental protection (United Nations General Assembly [UNGA] 1948). The proposals to add relevant protocols to the Declaration have been discussed over time, but no concrete results have been achieved to date (Savaşan 2021). While UN agencies have adopted an indirect approach in their regulations and resolutions (Commission on Human Rights 2005; UNGA 1990), documents like the World Nature Charter (prin. 23) and the Aarhus Convention provide procedural safeguards for environmental rights, primarily through the rights to information, **participation**, and access to **justice** (United Nations 1982; United Nations Economic Commission for Europe 1998). Similarly, regional instruments such as the African Charter and the Additional Protocol to the American Convention include relevant provisions but lack a direct reference or a separate listing of the right to the environment (African Union 1968; Organization of American States 1988). The European Convention on Human Rights (ECHR) text also does not contain a specific right to the environment (Council of Europe 1950). However, the European Court of Human Rights (ECtHR) has made indirect references in some cases through other rights, such as the right to life, the right to property, etc. More recently, significant strides toward adopting a substantive right to the environment have been made through a draft general comment on children's rights and the environment (CRC 2021), Resolution 48/13 by the Human Rights Council (2021), and a United Nations General Assembly declaration (2022). In recent years, many countries have also recognized some form of environmental protection in their constitutions, many specifically guaranteeing the right to a clean and healthy environment, like Portugal (1976), Spain (1978), and Brazil (1988) (United Nations Environment Programme 2019).

Environmental degradation, posing a non-traditional threat to **peace** and **human security**, is also a significant factor in increasing tensions and fueling **conflict**. The connection between environmental degradation and rising conflict disproportionately affects people, particularly the most vulnerable populations, whose livelihoods heavily depend on **natural resources** and a clean, healthy environment (Said 2021). The destruction of livelihoods caused by environmental pressures, overuse of natural assets, and weak **governance** can lead to competition over natural resources, potentially escalating into **violence** (Nitzschke and Studdard 2005). Therefore, respecting, protecting, and fulfilling the right to a clean environment not only contributes to safeguarding environmental human rights but also helps prevent conflicts based on environmental tensions (Geneva Academy 2024). Moreover, it can foster international **cooperation** and lead to multilateral and regional environmental treaties (Ide 2021). By enhancing public participation and access to justice, it also provides a valuable opportunity to enhance dialogue, build confidence, and encourage cooperation between groups as well as within or between states. This not only protects the environment but also advances **environmental peacebuilding** (United Nations Environment Programme 2009).

References

African Union (AU). 1968. *African Convention on the Conservation of Nature and Natural Resources.*

Boyd, David R., and John H. Knox. 2018. "Report of the Special Rapporteur on the Issue of Human Rights Obligations Relating to the Enjoyment of a Safe, Clean, Healthy and Sustainable Environment." A/73/188, 19 July 2018.

Commission on Human Rights (CHR). 2005. Human Rights Resolution 2005/60: Human Rights and the Environment as Part of Sustainable Development. E/CN.4/RES/2005/60.

Committee on the Rights of the Child (CRC). 2021. Draft General Comment No. 26 on "Children's Rights and the Environment with a Special Focus on Climate Change."

Council of Europe (CoE). 1950. *European Convention on Human Rights.*

Geneva Academy. 2024. *Environmental Human Rights As a Tool in Early Warning and Conflict Prevention: The Role of the Human Rights Council.* The Geneva Academy of International Humanitarian Law and Human Rights. https://geneva-academy.ch/joomlatools-files/docmanfiles/Environmental%20Human%20Rights%20as%20a%20Tool%20in%20Early%20Warning%20and%20Conflict%20Prevention.pdf.

Human Rights Council (HRC). 2021. "Promotion and Protection of all Human Rights, Civil, Political, Economic, Social and Cultural Rights, including the Right to Development." A/HRC/RES/48/13. October 18, United Nations General Assembly.

Ide, Tobias. 2021. *Environmental Peacebuilding.* Bogotá Instituto CAPAZ. https://www.instituto-capaz.org/wp-content/uploads/2021/03/DT-2-2021-Ingles-V3.pdf.

Jesse, Katinka D., and Erik V. Koppe. 2013. "Business Enterprises and the Environment." *Dovenschmidt Quarterly* 1 (4): 176–89. https://doi.org/10.5553/dq/221199812014001004004.

Lagerqvist, Yayoi, and Malena Wåhlin. 2022. "The Right to a Healthy Environment: What Now for Business?" *Swedwatch*. https://swedwatch.org/publication/policy-paper/the-right-to-a-healthy-environment-what-now-for-business/
Nitzschke, Heiko, and Kaysie Studdard. 2005. "The Legacies of War Economies: Challenges and Options for Peacemaking and Peacebuilding." *International Peacekeeping* 12 (2): 222–39. https://doi.org/10.1080/1353331050006651l.
Office of the High Commissioner for Human Rights (OHCHR), United Nations Environment Programme (UNEP), United Nations Development Programme (UNDP). 2023. "What Is the Right to a Healthy Environment?" Information Note. https://www.undp.org/sites/g/files/zskgke326/files/2023-01/UNDP-UNEP-UNHCHR-What-is-the-Right-to-a-Healthy-Environment.pdf
Organization of American States (OAS). 1988. Additional Protocol to the American Convention on Human Rights in the Area of Economic, Social and Cultural Rights (Protocol of San Salvador).
Said, Edward. 2021. "Environmental Degradation and Conflict Resolution." In *The Palgrave Handbook of Positive Peace*, edited by Kimberly Standish, Heather Devere, Analea Suazo, and Roger Rafferty. Palgrave Macmillan. https://doi.org/10.1007/978-981-16-0969-5.
Savaşan, Zerrin. 2021. "Çevre Hakları: İnsan Hakları Kuralları ile Korunmadan Asli Korumaya [Environmental Rights: From Protecting through Human Rights to Substantive Protection]." In *Çevre Hukuku ve Politikaları: Kavramlar, Teoriler ve Tartışmalar [Environmental Law and Policies: Concepts, Theories and Discussions*], edited by Zerrin Savaşan, Çağlar Söker, Fırat H. Yılmaz, 49–82. Seçkin Yayıncılık.
United Nations (UN). 1982. World Charter for Nature. A/RES/37/7.
United Nations Economic Commission for Europe (UNECE). 1998. *Convention on Access to Information, Public Participation in Decision-Making and Access to Justice in Environmental Matters*.
United Nations Environment Programme (UNEP). 2009. *From Conflict to Peacebuilding: The Role of Natural Resources and the Environment*. https://www.iisd.org/publications/conflict-peacebuilding-role-natural-resources-and-environment.
United Nations Environment Programme (UNEP). 2019. *Environmental Rule of Law: First Global Report*.
UN General Assembly (UNGA). 1948. *The Universal Declaration of Human Rights*.
UN General Assembly (UNGA). 1990. General Assembly Resolution 45/94. A/RES/45/94.
UN General Assembly (UNGA). 2022. "Promotion and Protection of Human Rights: Human Rights Questions, Including Alternative Approaches for Improving the Effective Enjoyment of Human Rights and Fundamental Freedoms." A/76/L.75, July 26.

Human Right to Water

Mara Tignino and Jessica Troell

The human right to water is defined as "sufficient, safe, acceptable, physically accessible and affordable water for personal and domestic uses" (UN Committee on Economic, Social and Cultural Rights [CESCR] 2003, para. 2). The right to safe, accessible water is crucial for human dignity and health (Boisson de Chazournes et al. 2023).

In 2010, the UN General Assembly (UNGA) officially recognized the human right to safe and clean drinking water and sanitation (UN General Assembly 2010), noting that the right is essential for the full enjoyment of life and all **human rights**. Since then, the UNGA and the UN Human Rights Council have regularly adopted resolutions clarifying both the normative content of the right and the responsibilities that States and other actors have for its realization. A UN Special Rapporteur on the Human Rights to Safe Drinking Water and Sanitation is also appointed by the UN Human Rights Council to further clarify the scope and application of the right to water and support its implementation. The legal basis of the human right to water is derived from the rights to an adequate standard of living and the right to the highest attainable standard of health and the right to food, enshrined in the International Covenant on Social, Economic and Cultural Rights (United Nations 1966).

The UN Committee on Economic, Social and Cultural Rights' General Comment 15 on the Right to Water also establishes that the implementation of the right to water should reflect the wider set of human rights that are dependent on water for their realization, including the rights to food, livelihoods, and cultural practices (CESCR 2003). Accordingly, States are encouraged to consider ways to ensure the provision of water for meeting basic human needs and the best mechanisms (e.g., legal instruments) for addressing other water-related human rights in ways that advance equity and non-discrimination in water resources allocation and management. General Comment 15 also clarifies that the human right to water must focus on **sustainability** and ensuring that implementation should meet the needs of present and future generations (CESCR 2003, para. 11).

The right to water has been affirmed at the international level by several human rights courts, and an increasing number of countries now either recognize a fundamental or human right to water in their constitutions or national legislation (e.g., Inter-American Court of Human Rights 2020; European Court of Human Rights 2020).

Access to water is a core humanitarian issue in conflict-affected environments and is being exacerbated by the impact of **climate change** (Troell and Keene 2022). The human right to water gives priority to basic human needs covering personal and domestic uses (de Chazournes et al. 2023). During armed conflict, the right to water and sanitation complements the prohibition of starvation and the prohibition on destroying **infrastructure** for water supply and the poisoning of water under **International Humanitarian Law** (United Nations 1977, Article 54). The human right to water also has a key role in the context of Indigenous Peoples, Afro-descendants, and other local communities that rely on access to sufficient water for food and livelihood security (Troell and Keene 2022). Some instruments, such as the UN Declaration on the Rights of Peasants and other People Working in Rural Areas (UN Human Rights Council 2018), affirm the human right to water, including in its scope of application the water necessary to ensure **food security** (Tignino et al. 2022), which can be essential for populations living in conflict-affected environments.

References

Boisson de Chazournes, Laurence, Mara Tignino, and Haoua Savadogo. 2023. "The Right to Water." In *The Roles of International Law in Development*, edited by Siobhan McInerney-Lankford and Robert McCorquodale. Oxford University Press. https://doi.org/10.1093/oso/9780192872906.003.0009.

European Court of Human Rights. 2020. *Hudorovič and Others v. Slovenia*. Application No. 24816/14. March 10, 2020.

Inter-American Court of Human Rights. 2020. *Indigenous Communities of the Lhaka Honhat (Our Land) Association v. Argentina*. Judgment. Series C No. 400. February 6, 2020.

International Covenant on Economic, Social and Cultural Rights, December 16, 1966, 993 U.N.T.S. 3.

Tignino, Mara, Tadesse Moges Kebebew, and Francesca Restifo. 2022. "The Rights to Water and Sanitation and the UNDROP: The Implementation Challenges of a Comprehensive Approach." In *The United Nations Declaration on Peasants Rights*, edited by Mariagrazia Alabrese, Adriana Bessa, Margherita Brunori, and Pier Filippo Giuggioli. Routledge.

Troell, Jessica, and Stephanie Keene. 2022. *Legal Recognition of Customary Water Tenure in Sub-Saharan Africa: Unpacking the Land-Water Nexus*. International Water Management Institute. https://doi.org/10.5337/2022.215.

United Nations. 1966. International Covenant on Economic, Social and Cultural Rights. December 16, 1966. 993 U.N.T.S. 3.

United Nations. 1977. Protocol Additional to the Geneva Conventions of 12 August 1949, and Relating to the Protection of Victims of Non-International Armed Conflicts (Protocol II). June 8, 1977. 1125 U.N.T.S. 609.

UN Committee on Economic, Social and Cultural Rights. 2003. "General Comment No. 15: The Right to Water (Arts. 11 and 12 of the Covenant)." E/C.12/2002/11.

UN General Assembly. 2010. "The Human Right to Water and Sanitation." Resolution 64/292. A/RES/64/292. July 28, 2010.

UN Human Rights Council. 2018. United Nations Declaration on the Rights of Peasants and Other People Working in Rural Areas. Resolution 39/12. A/HRC/RES/39/12. September 28.

Human Security

Stephen Lonergan and Shahab Khalid

Human security is safety from chronic threats such as hunger, disease, and oppression, and protection from the sudden and hurtful disruptions to the patterns of everyday life; the right of people to live in freedom and dignity, free from poverty and despair (United Nations General Assembly [UNGA] 2012), including the ability of individuals and communities to adapt to long-term environmental changes that threaten their economic, social, and ecological stability.

Initially, human security was interpreted as a threat to the physical **security** of the person. The Universal Declaration of Human Rights, adopted by the UN in 1948, states that "everyone has the right to life, liberty, and the security of a person." Today, the concept includes economic, health, and environmental concerns.

In 1994, the United Nations Development Programme (UNDP) attempted a more precise definition by noting that human security implied safety from chronic threats such as hunger, disease, and oppression, and protection from the sudden and hurtful disruptions in the patterns of everyday life. The UNDP expanded this definition in 2022 due to the new generation of interconnected threats. It recognized that solidarity—i.e., interdependence across all people and between people and the **environment**—must be considered a key component of human security (UNDP 2022; UNGA 2024). Human security is more than simply the absence of violent **conflict**; "it encompasses **human rights**, good **governance**, access to education and health care, and ensuring that each individual has opportunities and choices to fulfil his or her own potential" (Ogata and Sen 2003).

From its initial understanding as a protection against physical threats, human security has evolved to encompass the environmental, social, and economic dimensions of long-term threats (e.g., Adger 2000; Barnett and Adger 2007; Cannon 2008; Gemenne 2014). This perspective emphasizes how **climate change** and ecological disruptions pose significant risks to human security, particularly regarding **food security**, migration, and economic stability. It also includes the capacity of the socio-ecological system to adapt to long-term **environmental changes.**

Human security can only be achieved when individuals and communities: (1) have the options necessary to end, mitigate, or adapt to threats to their human, environmental, and social rights, (2) have the capacity and freedom to exercise these options, (3) actively participate in attaining these options, and (4) have the capacity to withstand and adapt to environmental change, such as climate shifts and resource shortages, which have broad impacts on livelihoods, food security, and social stability (Lonergan 1999; Gasper 2022).

Human security recognizes that **cooperation** in sharing environmental resources can mitigate conflict and instability arising from competition over resources and **environmental degradation.** By enhancing adaptive capacities and promoting collaborative approaches to **sustainable development**, **environmental peacebuilding** directly contributes to achieving broader human security objectives, including stability, **resilience**, and dignity for affected communities. In terms of environmental peacebuilding, a human security lens ensures that responses to environmental challenges, which may contribute to conflict and instability, prioritize the fundamental needs, rights, and capacities of communities, thereby reinforcing **sustainable peace.**

Acknowledgment

This work was carried out with support from the CGIAR Science Programs on Climate Action and Food Frontiers and Security. We would like to thank all funders who supported this research through their contributions to the CGIAR Trust Fund: https://www.cgiar.org/funders/.

References

Adger, W. Neil. 2000. "Social and Ecological Resilience: Are They Related?" *Progress in Human Geography* 24 (3): 347–64. https://doi.org/10.1191/030913200701540465.

Barnett, Jon, and W. Neil Adger. 2007. "Climate Change, Human Security and Violent Conflict." *Political Geography* 26 (6): 639–55. https://doi.org/10.1016/j.polgeo.2007.03.003.

Cannon, Terry. 2008. "Reducing People's Vulnerability to Natural Hazards Communities and Resilience." United Nations University World Institute for Development Economics Research No. 34. https://ciaotest.cc.columbia.edu/wps/unup/0001955/f_0001955_961.pdf.

Gasper, Des. 2022. "Rethinking Human Development And/as Human Security for the Anthropocene." *The International Journal of Social Quality* 12 (2): 1–24. https://doi.org/10.3167/ijsq.2022.120202.

Gemenne, François, et al. 2014. "Climate and Security: Evidence, Emerging Risks, and a New Agenda." *Climatic Change* 123 (1): 1–9. https://doi.org/10.1007/s10584-014-1074–7.

Lonergan, S. 1999. *Global Environmental Change and Human Security Science Plan.* International Human Dimensions Programme on Global Environmental Change.

Ogata, Sadako, and Amartya Sen. 2003. *Human Security Now.* Commission on Human Security.

United Nations Development Programme (UNDP). 2022. *New Threats to Human Security in the Anthropocene.* United Nations Development Programme. https://hdr.undp.org/system/files/documents/srhs2022overviewen.pdf.

United Nations General Assembly. 2012. Resolution 66/290, Follow-up to paragraph 143 on human security of the 2005 World Summit Outcome. A/Res/66/290. October 25.

United Nations General Assembly. 2024. *Human Security*. Report of the Secretary-General. A/78/665. January 2.

Human–Wildlife Conflict

Mirali Shukla and Rachel Weaver

Human–wildlife conflict (HWC) refers to the diverse perceived or actual negative interactions that arise from the proximity of wildlife and humans, often leading to disagreements between groups of people over how to manage or resolve the situation and negative impacts on people and/or wildlife (International Union for Conservation of Nature 2022; Gross et al. 2021).

There are continued debates over the term "human–wildlife **conflict**" as being problematic and detrimental to the long-term coexistence between humans and wildlife (Redpath et al. 2015). Many scholars and practitioners argue that this language depicts humans and wildlife as enemies, frames wildlife as sentient adversaries of humans (Peterson et al. 2010), and can lead to decreased cultural and traditional respect for some species (Fernando et al. 2005; Wilson et al. 2013). The word conflict itself attributes human motivations to non-human beings (Redpath et al. 2015).

Negative interactions between humans and wildlife occur globally and have been intensifying due to **climate change**, increasing human need for space, and **land use** changes driving habitat reduction, fragmentation, and degradation (Abrahms et al. 2023; IUCN 2022; Treves et al. 2006). The increasing geospatial overlap between humans and wildlife, and the resulting diminishing areas available for wildlife, denotes increased interactions and competition for space and resources like water and food (Gross et al. 2021; Ripple et al. 2014). Communities in these conditions experience negative impacts such as crop damage, livestock predation, or direct harm.

Cultural attitudes toward wildlife are complex and vary widely (Pimid et al. 2022; Redpath et al. 2013). Traditional practices, spiritual beliefs, and local knowledge play a crucial role in shaping human interactions with wildlife (Gross et al. 2021). While some communities hold certain animals in high regard and avoid harming them, others may view wildlife negatively due to historical conflicts. Contrasting values around resource use further complicate these dynamics. This is exemplified by contexts like photographic tourism, where external **conservation** and economic priorities may inform or hinder local needs and rights, as seen in communities surrounding protected areas (DeMotts and Hoon 2012). Understanding these sociocultural dimensions is essential for designing effective conservation strategies that resonate with local populations and reduce conflict.

HWC disproportionately affects local and Indigenous communities, especially those that are poor or marginalized, as they lack alternative incomes (Gross et al. 2021;Dickman 2010). This pattern is evident in many regions, including villages near protected areas in India, where communities frequently bear the brunt of conflict and may struggle to access timely or adequate support despite the presence of compensation schemes (Karanth et al. 2012). Conversely, wealthier communities may have greater access to **mitigation** tools, and this disparity often leads to varying rates of retaliatory wildlife killings, further exacerbating conservation challenges (Nyhus 2016).

The most direct negative impacts of HWC on people include loss of life, crops, livestock, and property (Graham et al. 2010). This can negatively impact subsistence farmers and fishers, threaten food and economic stability, and exacerbate poverty and inequality (Barua et al. 2013; Szteren 2002). HWC can also negatively affect the productivity of businesses, particularly companies relying on agricultural-based products (Othman et al. 2019). Societal responses to HWC are varied, and government responses and restrictions can create divisions among **stakeholders**, especially when there is an inequitable distribution of the costs and benefits of coexisting with wildlife (Madden 2004; Jordan et al. 2020). Unmanaged HWC can erode trust between local and Indigenous communities toward political actors, further decreasing tolerance for wildlife (Bond and Mkutu 2018; Jordan et al. 2020).

HWC poses a significant threat to species survival and can disrupt ecosystems, especially for species with a naturally low population density (Mekonen 2020). Preemptive or retaliatory killings in response to conflict or for self-defense can drive species closer to extinction, destabilizing ecosystems. The focus on charismatic species in HWC management can overshadow the damage caused by more common species and lead to misguided scapegoating or deallocation of conservation funds (Xu et al. 2024).

In an increasingly crowded world where HWC is accelerating (Abrahms et al. 2023), key stakeholders and practitioners must rethink how to address the issue and foster coexistence for the benefit of wildlife and people. **Environmental peacebuilding** offers a useful framework for addressing HWC by incorporating **conflict resolution**, negotiation, and resource management methods to facilitate more peaceful interactions between humans and wildlife (Gross et al. 2021). For example, compensation schemes, such as those implemented in India near protected areas, have been used to mitigate the negative impacts of HWC on local communities (Karanth et al. 2012). By fostering multistakeholder collaboration and dialogue, environmental peacebuilding can transform deep-rooted, complex HWC into opportunities for **peace**, **sustainable development**, and conservation (Dresse et al. 2019), acting as a neutral-third party.

Global trends in resource and land use, made even more complex by climate change, will bring increased challenges to coexistence (Abrahms et al. 2023; König et al. 2021). Managing HWC will require integrated, holistic, and collaborative approaches to ensure that both human and wildlife populations can thrive in shared landscapes (Pooley et al. 2017).

References

Abrahms, Briana, Neil H. Carter, T. J. Clark-Wolf, et al. 2023. "Climate Change as a Global Amplifier of Human–Wildlife Conflict." *Nature Climate Change* 13 (3): 224–34. https://doi.org/10.1038/s41558-023-01608-5.

Barua, Maan, Shonil A. Bhagwat, and Sushrut Jadhav. 2013. "The Hidden Dimensions of Human-Wildlife Conflict: Health Impacts, Opportunity and Transaction Costs." *Biological Conservation* 157: 309-316.https://doi.org/10.1016/j.biocon.2012.07.014.

Bond, Jennifer, and Kennedy Mkutu. 2017. "Exploring the Hidden Costs of Human-Wildlife Conflict in Northern Kenya." *African Studies Review* 61 (1): 33–54. https://doi.org/10.1017/asr.2017.134.

DeMotts, Rachel, and Parakh Hoon. 2012. "Whose Elephants? Conserving, Compensating, and Competing in Northern Botswana." *Society and Natural Resources* 25 (9): 837–51. https://doi.org/10.1080/08941920.2011.638362.

Dickman, Amy J. 2010. "Complexities of Conflict: The Importance of Considering Social Factors for Effectively Resolving Human-Wildlife Conflict." *Animal Conservation* 13 (5): 458–66. https://doi.org/10.1111/j.1469-1795.2010.00368.x.

Dresse, Anaïs, Itay Fischhendler, Jonas Ø. Nielsen, and Dimitrios Zikos. 2019. "Environmental Peacebuilding: Towards a Theoretical Framework." *Cooperation and Conflict* 54 (1): 99–119. https://doi.org/10.1177/0010836718808331.

Fernando, Prithiviraj, Eric Wikramanayake, Devaka Weerakoon, L. K. A. Jayasinghe, Manori Gunawardene, and H. K. Kanaja. 2005. "Perceptions and Patterns of Human–Elephant Conflict in Old and New Settlements in Sri Lanka: Insights for Mitigation and Management." *Biodiversity and Conservation* 14: 2465–81. https://doi.org/10.1007/s1 0531-004-0216-z.

Graham, Maximillian D., Benedikt Notter, William M. Adams, Phyllis C. Lee, and Tobias Nyumba Ochieng. 2010. "Patterns of Crop-Raiding by Elephants, *Loxodonta Africana*, in Laikipia, Kenya, and the Management of Human–Elephant Conflict." *Systematics and Biodiversity* 8 (4): 435–45. https://doi.org/10.1080/14772000.2010.533716.

Gross, Eva, Nilanga Jayasinghe, Ashley Brooks, Gert Polet, Rohan Wadhwa, and Femke Hilderink-Koopmans. 2021. "A Future for All: The Need for Human-Wildlife Coexist." World Wildlife Fund.

International Union for Conservation of Nature (IUCN). 2022. "Human-Wildlife Conflict." https://iucn.org/sites/default/files/2022-06/iucn-issues-brief-human-wildlife-conflict_final.pdf.

Jordan, Neil R., Bradley P. Smith, Robert G. Appleby, Lily M. van Eeden, and Hugh S. Webster. 2020. "Addressing Inequality and Intolerance in Human-wildlife Coexistence." *Conservation Biology* 34(4): 803-10. https://doi.org/10.1111/cobi.13471.

Karanth, Krithi K., Arjun M. Gopalaswamy, Ruth DeFries, and Natasha Ballal. 2012. "Assessing Patterns of Human-Wildlife Conflicts and Compensation around a Central Indian Protected Area." *PLoS One* 7 (12): e50433. https://doi.org/10.1371/journal.pone.0050433.

König, Hannes J., Silvia Ceauşu, Mark Reed, et al. 2021. "Integrated Framework for Stakeholder Participation: Methods and Tools for Identifying and Addressing Human–Wildlife Conflicts." *Conservation Science and Practice* 3 (3): e399. https://doi.org/10.1111/csp2.399.

Madden, Francine. 2004. "Creating Coexistence between Humans and Wildlife: Global Perspectives on Local Efforts to Address Human–Wildlife Conflict." *Human Dimensions of Wildlife* 9 (4): 247-57. https://doi.org/10.1080/10871200490505675.

Mekonen, Sefi. 2020. "Coexistence between Human and Wildlife: The Nature, Causes and Mitigations of Human-Wildlife Conflict around Bale Mountains National Park, Southeast Ethiopia." *BMC Ecology* 20 (51). https://doi.org/10.1186/s12898-020-00319-1.

Nyhus, Philip J. 2016. "Human–Wildlife Conflict and Coexistence." *Annual Review of Environment and Resources* 41 (1): 143–71. https://doi.org/10.1146/annurev-environ-110615-085634.

Othman, Nurzhafarina, Benoit Goossens., Cheryl Cheah, Sen Nathan, Renee Bumpus, and Marc Ancrenaz. 2019. "Shift of Paradigm needed towards improving human–elephant Coexistence in Monoculture Landscapes in Sabah." International Zoo Yearbook. 53 (1): 161-173. https://doi.org/10.1111/izy.12226.

Peterson, M. Nils. Jessie L. Birckhead, Kirsten Leong, Markus J. Peterson, and Tarla Rai Peterson. 2010. "Rearticulating the Myth of Human-Wildlife Conflict." *Conservation Letters* 3 (2): 74–82. https://doi.org/10.1111/j.1755-263X.2010.00099.x.

Pimid, Marcela, Mohammad Rusdi, Mohd Nasir, et al. 2022. "Understanding Social Dimensions in Wildlife Conservation; Multiple Stakeholder Views." *Animals* 12 (7): 811. https://doi.org/10.3390/ani12070811.

Pooley, Simon, Maan Barua, William Beinart, et al. 2017. "An Interdisciplinary Review of Current and Future Approaches to Improving Human–Predator Relations." *Conservation Biology* 31 (3): 513–23. https://doi.org/10.1111/cobi.12859.

Redpath, Steve M., Juliette Youn, Anna Evely, et al. 2013. "Understanding and Managing Conservation Conflicts." *Trends in Ecology and Evolution* 28 (2): 100–9. https://doi.org/10.1016/j.tree.2012.08.021.
Redpath, Stephen M., Saloni Bhatia, and Juliette Young. 2015. "Tilting at Wildlife: Reconsidering Human–Wildlife Conflict." *Oryx* 49 (2): 222–25. https://doi.org/10.1017/S0030605314000799.
Ripple, William J., James A. Estes, Robert L. Beschta, et al. 2014. "Status and Ecological Effects of the World's Largest Carnivores." *Science* 343 (6167): 1241484. https://doi.org/10.1126/science.1241484.
Szteren, Diana, and Enrique Páez. 2002. "Predation by Southern Sea Lions (*Ottaria flavescens*) on Artisanal Fishing Catches in Uruquay." *Marine and Freshwater Research* 53: 1161-7. https://doi.org/10.1071/MF02006.
Treves, Adrian, Robert B. Wallace, Lisa Naughton-Treves, and Andrea Morales. 2006. "Co-Managing Human–Wildlife Conflicts: A Review." *Human Dimensions of Wildlife* 11 (6): 383–96. https://doi.org/10.1080/10871200600984265.
Wilson, Scott, Tammy E. Davies, Nandita Hazarika, and Alexandra Zimmerman. 2015. "Understanding Spatial and Temporal Patterns of Human-Elephant Conflict in Assam, India." *Oryx* 49 (1): 140–49. https://doi.org/10.1017/S0030605313000513.
Xu, Wanyun, Lingxia Xu, Yuqi Cao, et al. 2024. "Community Perspectives of Flagship Species: Can Conservation Motivators Mitigate Human-Wildlife Conflict?" *Frontiers in Ecology and Evolution* 12. https://doi.org/10.3389/fevo.2024.1265694.

Hydro-Hegemony

Emmanuel Valax, Rebecca L. Farnum, and Alexandra Turgul

Hydro-hegemony is a conceptual framework providing an entry point for understanding how power asymmetries influence the outcomes of transboundary water interactions. Scholars and practitioners have used elements such as riparian position, material power, and discursive capabilities to explain "who gets how much of the water, how and why" (Zeitoun and Warner 2006).

> Initially defined as "hegemony at the river basin level, achieved through water resource control strategies such as resource capture, integration and containment" (Zeitoun and Warner 2006, 435), the original framework of hydro-hegemony proposed three pillars: riparian position, **power**, and exploitation potential. In 2010, Cascão and Zeitoun delineated material power (including exploitation potential), bargaining power, and ideational power as separate pillars, with geography (including riparian position) becoming a fourth pillar. While maintaining core assumptions, this revision allowed for a more nuanced analysis of power.
>
> In their "10 years later" retrospective, the London Water Research Group responded to three criticisms of the framework: (1) it is state-centric and prone to the territorial trap, (2) it assumes negative and immutable hegemonic power relations, and (3) it reifies hegemony and risks, potentially becoming a hegemonic concept itself (Warner et al. 2017). The authors conceded that their original framework did center the state, with the river basin as "the hegemonic, taken-for-granted scale of analysis." They also noted the negativity bias often associated with the framework, observing that empirical research had thus far yielded very limited positive cases. Though concurring that hydro-hegemony is not and "should not be considered inevitable or unchangeable," and grappling with whether a focus on hegemony "risks making it bigger, reifying it, and turning it into a monster," the authors cautioned against ignoring it entirely,

calling for more research on issues of scale and hegemony while avoiding absolutes that characterize any difference in power as "hegemonic" (Warner et al. 2017).

Several case studies have engaged with the framework in novel contexts. For example, Farnum (2013) and Greco et al. (2023) apply the framework of hydro-hegemony to issues of virtual water and international trade; others have considered the relationship between hydro-hegemony and international water law (Farnum et al. 2017; Woodhouse and Zeitoun 2008). Cascão (2008) and Cascão (2009) demonstrated techniques for contesting hydro-hegemony; Tawfik (2015) applied the concept of benefit sharing to evaluate the impact of counter-hegemony on hegemonic tactics.

Research inspired by the framework of hydro-hegemony has contributed greatly to our understanding of transboundary water interactions, and the concept has evolved away from its initial emphasis on the dominance of state actors to embrace the complexities of water **conflict** and **cooperation**. However, several gaps in its application present opportunities for further analysis. Studies using the framework still tend to focus on surface water, disregarding other kinds of freshwater and the diverse ecological services they provide (Hayat and Gupta 2016). The framework has yet to be used to conceptualize the role of power in relation to achieving the Sustainable Development Goals (SDGs), which provide benchmarks for assessing progress toward equitable and sustainable **resource governance**. The role of actors beyond the state, such as NGOs and transnational networks, in influencing state relations remains underexplored (Hayat et al. 2022), and the assumption within hydro-hegemony theory that state sovereignty is an uncontestable attribute prevents it from engaging fully with work that challenges concepts of state boundaries (Zeitoun and Warner 2006). A more robust framework would support more nuanced analysis of why water interactions play out the way they do, highlighting the need for further research (grounded in both theory and practice) to refine and expand the framework of hydro-hegemony.

As a versatile analytical framework that emphasizes power dynamics as well as geophysical factors, hydro-hegemony can support robust conflict analysis and inform **environmental peacebuilding** efforts.

References

Cascão, Ana Elisa. 2008. "Ethiopia–Challenges to Egyptian Hegemony in the Nile Basin." *Water Policy* 10 (S2): 13–28. https://doi.org/10.2166/wp.2008.206.

Cascão, Ana Elisa. 2009. "Changing Power Relations in the Nile River Basin: Unilateralism vs. Cooperation?" *Water Alternatives* 2 (2): 245–68. https://www.water-alternatives.org/index.php/alldoc/articles/vol2/v2issue2/52-a2-2-5/file.

Cascão, Ana Elisa, and Mark Zeitoun. 2010. "Power, Hegemony and Critical Hydropolitics." In *Transboundary Water Management: Principles and Practice*, edited by Anton Earle, Anders Jägerskog, and Joakim Öjendal. Taylor and Francis.

Farnum, Rebecca L. 2013. "Virtual Water, Equivocal Law, Actual Hegemony Expanding the Framework of Hydro-Hegemony to Inform Virtual Water Trade and International Law." MSc Thesis, University of East Anglia. https://kclpure.kcl.ac.uk/portal/en/studentTheses/virtual-water-equivocal-law-actual-hegemony.

Farnum, Rebecca L., Stephanie Hawkins, and Mia Tamarin. 2017. "Hydro-hegemons and International Water Law." In *Routledge Handbook of Water Law and Policy*, edited by Alistair Rieu-Clarke, Andrew Allan, and Sarah Hendry. Routledge.

Greco, Francesco, Martin Keulertz, and David Dent. 2023. "Virtual Water." In *Oxford Research Encyclopedia of Environmental Science*. Oxford University Press. https://doi.org/10.1093/acrefore/9780199389414.013.602.

Hayat, Shakeel, and Joyeeta Gupta. 2016. "Kinds of Freshwater and Their Relation to Ecosystem Services and Human Well-Being." *Water Policy* 18 (5): 1229–46. https://doi.org/10.2166/wp.2016.182.

Hayat, Shakeel, Joyeeta Gupta, Courtney Vegelin, and Hameed Jamali. 2022. "A Review of Hydro-Hegemony and Transboundary Water Governance." *Water Policy* 11: 1723–40. https://doi.org/10.2166/wp.2022.256.

Tawfik, Rawia. 2015. *Revisiting Hydro-Hegemony from a Benefit-Sharing Perspective: The Case of the Grand Ethiopian Renaissance Dam*. German Development Institute. https://hdl.handle.net/10419/199455.

Warner, Jeroen, Naho Mirumachi, Rebecca L. Farnum, Mattia Grandi, Filippo Menga, and Mark Zeitoun. 2017. "Transboundary 'Hydro-Hegemony': 10 Years Later." *WIREs Water* 4 (6): e1242. https://doi.org/10.1002/wat2.1242.

Woodhouse, Melvin, and Mark Zeitoun. 2008. "Hydro-hegemony and International Water Law: Grappling with the Gaps of Power and Law." *Water Policy* 10 (S2): 103–19. https://doi.org/10.2166/wp.2008.209.

Zeitoun, Mark, and Jeroen Warner. 2006. "Hydro-Hegemony – A Framework for Analysis of Trans-Boundary Water Conflicts." *Water Policy* 8 (5): 435–60. https://doi.org/10.2166/wp.2006.054.

I

Indicator

Amanda Woomer

An indicator is a variable, factor, characteristic, or attribute that can change in a context or situation or the performance of an intervention in terms of its activities, outputs, and outcomes. Indicators are often linked to an intervention's theory of change and the context in which it takes place.

While indicators may be qualitative or quantitative, both types should be simple, specific, clearly defined, able to be verified, and provide a reliable and realistic way of tracking change and performance over time (Organisation for Economic Co-operation and Development 2023). It is often recommended that indicators be specific, measurable, achievable, relevant, and time-bound. While indicators are often defined quantitatively and focused on activities, outputs, and outcomes along an intervention's **theory of change**, qualitative indicators have a specific role to play in **environmental peacebuilding** interventions as they can provide important information on why something is happening by capturing attitudes, judgments, feelings, etc. (Bruch and Woomer 2023). For instance, the Monitoring Dialogue in Conflict-Affected Contexts has recommendations for selecting and monitoring qualitative indicators, which are typically intangible but important in dialogue-based peacebuilding contexts, such as trust and empathy (Baggerman et al. 2025). To the extent possible, qualitative and quantitative indicators should be combined to provide a more holistic picture of change and context, including the linkages between the **environment** or climate on the one hand, and **peace** and **conflict** on the other. Given that there are often resource and **security** concerns with environmental peacebuilding interventions, it is also important to right-size indicators so that it is feasible, ethical, and safe to collect related information. Based on these considerations, it is often necessary to develop specific indicators for each environmental peacebuilding intervention rather than applying standard indicators across interventions. One way to do this is to ask **stakeholders** what to measure, for example, by asking how peace or security is experienced, such as in everyday peace indicators (Firchow 2018; Mac Ginty 2013). It is also important to develop contextual indicators to understand how the conflict context is changing and to identify and respond effectively to risks, as well as indicators of **conflict sensitivity** to understand whether the process of implementing an intervention is aware of and sensitive to conflict factors.

DOI: 10.4324/9781003604143-11

Examples of indicators for an environmental peacebuilding project may include indicators related to conflict (e.g., number of conflicts in a particular area or the percentage of intervention participants who have experienced conflict within a certain timeframe) or the environment (e.g., number of hectares demonstrating improved ecosystem condition). However, it is also essential to include indicators that connect conflict and environmental factors. This may include, for example, the percentage of community members who feel safer following the implementation of a natural resource management plan, or the percentage of people who participate in a water **conservation** intervention who can provide examples of improvements in intergroup relations. These examples demonstrate the importance of context-specific indicators that capture the mechanism by which environmental work is supposed to reduce conflict, or vice versa. In each case, it is important to consider whether certain indicators may create perverse incentives in intervention implementation or **monitoring.** For example, if an intervention tracks the number of conflicts in a particular area, there may be an incentive to underreport these to show the intervention is making progress. Additionally, to the extent that resources and time allow, it is important that interventions not limit their monitoring to pre-defined indicators, which will only capture information on what could be anticipated. If interventions only monitor anticipated changes, then it is likely they will miss some important unintended effects. Qualitative indicators about unintended outcomes can be incorporated to counter this challenge.

References

Baggerman, Jill, Esra Cuhadar, and Kristen Wall. 2025. *Monitoring Dialogue in Conflict-Affected Contexts: A Guide for Practitioners.* United States Institute of Peace. https://www.allianceforpeacebuilding.org/afp-publications/monitoring-dialogue-conflict-affected-contexts.

Bruch, Carl, and Amanda Woomer. 2023. *Toolkit on Monitoring and Evaluation of Environmental Peacebuilding.* Environmental Law Institute, Environmental Peacebuilding Association, and United States Institute of Peace. https://m-and-e.environmentalpeacebuilding.org/toolkit/.

Firchow, Pamina. 2018. *Reclaiming Everyday Peace: Local Voices in Measurement and Evaluation After War.* Cambridge University Press.

Mac Ginty, Roger. 2013. "Indicators+: A Proposal for Everyday Peace Indicators." *Evaluation and Program Planning* 36 (1): 56–63. https://doi.org/10.1016/j.evalprogplan.2012.07.001.

Organisation for Economic Co-operation and Development (OECD). 2023. *Glossary of Key Terms in Evaluation and Results-based Management (Second Edition).* OECD.

Indigenous Knowledge

Elena Kavanagh and Veronica Limeberry

Indigenous Knowledge (IK) is the collective term representing the myriad ways of knowing and practices of being in Indigenous cultural contexts across the globe (Jessen et al. 2022) that are "continually generated in relationship to place" (Simpson 2017).

Indigenous scholars point to knowledge creation and diffusion that challenge colonial and Western frames of science, inquiry, and practice (Carroll 2015; Clement 2019; Coulthard 2014; Estes 2019; Kimmerer 2013). Often, these knowledges are passed through oral history and cultural transmission over time, in what de la Cadena and

Blaser (2018) term "a concept of 'a world of many worlds.'" Or, as Escobar (2018) terms, "integrating pluriverses [that] speak to a diversity of ontologies."

IK is related to traditional ecological knowledge (TEK) which refers to the cumulative, culturally transmitted understanding, skills, and practices developed by Indigenous Peoples over generations, often in direct relationship with their local **environment** and ecologies (Berkes et al. 2000). TEK and IK typically include the sustainable management of **natural resources**, **agriculture**, **biodiversity**, and ecological stewardship, deeply embedded in cultural beliefs, values, and practices.

Within the context of **environmental peacebuilding**, IK is pivotal for fostering **sustainable peace** and environmental protection. These knowledge systems, often holistic and rooted in a deep connection to land and ecosystems, play a vital role in addressing environmental challenges, promoting biodiversity protection and **conservation**, and managing natural resources in ways that mitigate **conflict**. Their significance lies in the capacity to inform adaptive strategies to **climate change**, **land degradation**, and resource scarcity, making the systems highly relevant for contemporary environmental **governance** and conflict resolution.

IK has long been marginalized by colonial and Western scientific approaches. Despite this, it has survived and continues to co-evolve with practices and principles of Indigenous peoples worldwide. There are growing calls to integrate IK into **sustainable development** initiatives, biodiversity preservation, and climate change **adaptation/mitigation** (Argumedo 2013; Blackman et al. 2017; Cajete 2020; Campbell 2019; Zimmerer 2012). However, this presents new concerns of extracting the knowledge of Indigenous peoples under the continued hegemony of practices and goals of colonial/Western science (Escobar 1998; Fernández-Llamazares et al. 2024; Sousa Santos 2016; Zaitchik 2018). Indigenous scholars have offered ways of understanding the interactions and engagement between IK and Western sciences, such as "two-eyed seeing" or Etuaptmumk, championed by Mi'kmaq elder Albert Marshall (Institute for Integrative Science and Health [IISH] n.d.). This perspective

> refers to **learning** to see from one eye with the strengths of Indigenous knowledges and ways of knowing, and from the other eye with the strengths of Western knowledges and ways of knowing ... and learning to use both these eyes together, for the benefit of all.
>
> (IISH n.d.)

Research emphasizes the importance of integrating IK into environmental management and **peacebuilding** initiatives (Aikenheal and Ogawa 2007; Petzold et al. 2020). Of note, Indigenous communities must have the agency, support, and capacity to lead and make decisions on crucial environmental laws, policies, and practices. Inclusion and integration without leadership and decision-making authority can maintain political hierarchies of knowledge that diminish and reduce the efficacy of IK for environmental peacebuilding (Alfred 2005; Santos 2016; Vojno 2022).

The growing recognition of IK in global environmental agreements, such as the UN **Convention on Biological Diversity** and the recent UN Pact for the Future, further underscores its value for both environmental and peacebuilding agendas. As communities that reside on most of the world's most biodiverse lands and waters, who are marginalized and particularly vulnerable to conflict, Indigenous peoples and their knowledge are key to environmental peacebuilding (Bratspies 2020; Office of the High Commissioner for Human Rights [OHCHR] 2024; Vojno 2022; WWF et al. 2021). For example, as the Office of the United Nations High Commissioner for Human Rights report sums up: "Indigenous knowledge and ancestral wisdom are essential to achieving Colombia's aspirations for lasting and true Total Peace, as well as environmental protection" (OHCHR 2024). Thus, IK can foster more inclusive decision-making, promote ecological **sustainability**, and reduce tensions around land and resource use that lead to better **peace** outcomes.

References

Aikenhead, Glen S., and Masakata Ogawa. 2007. "Indigenous Knowledge and Science Revisited." *Cultural Studies of Science Education* 2: 539–620. https://doi.org/10.1007/s11422-007-9067-8.

Alfred, Gerald R. 2005. *Wasáse: Indigenous Pathways of Action and Freedom*. University of Toronoto Press.

Argumedo, Alejandro. 2013. *Collective Trademarks and Biocultural Heritage: Towards New Indications of Distinction for Indigenous Peoples in the Potato Park, Peru*. International Institute for Environment and Development, March. https://www.iied.org/16528iied

Berkes, Fikret, Johan Colding, and Carl Folke. 2000. "Rediscovery of Traditional Ecological Knowledge as Adaptive Management." *Ecological Applications* 10 (5): 1251–62. https://doi.org/10.1890/1051-0761(2000)010[1251:ROTEKA]2.0.CO;2.

Blackman, Allen, Leonardo Corral, Eirivelthon Santos Lima, and Gregory P. Asner. 2017. "Titling Indigenous Communities Protects Forests in the Peruvian Amazon." *Proceedings of the National Academy of Sciences* 114 (16): 4123–28. https://doi.org/10.1073/pnas.1603290114.

Bratspies, Rebecca. 2020. "'Territory Is Everything': Afro-Colombian Communities, Human Rights and Illegal Land Grabs." *Columbia Human Rights Law Review* 4 (2): 291–323. http://hrlr.law.columbia.edu/hrlr-online/territory-is-everything-afro-colombian-communities-human-rights-and-illegal-land-grabs/.

Cajete, Gregory A. 2020. "Indigenous Science, Climate Change, and Indigenous Community Building: A Framework of Foundational Perspectives for Indigenous Community Resilience and Revitalization." *Sustainability* 12 (22): 9569. https://doi.org/10.3390/su12229569.

Campbell, Jeffrey. 2019. "No Sustainable Development Without Indigenous Peoples." SDG Knowledge Hub, International Institute for Sustainable Development. Effective August 8. https://sdg.iisd.org:443/commentary/guest-articles/no-sustainable-development-without-indigenous-peoples/.

Carroll, Clint. 2015. *Roots of Our Renewal: Ethnobotany and Cherokee Environmental Governance*. University of Minnesota Press. http://ezproxy.uniandes.edu.co:8080/login?url=https://muse.jhu.edu/book/40477/.

Clement, Vincent. 2019. "Beyond the Sham of the Emancipatory Enlightenment: Rethinking the Relationship of Indigenous Epistemologies, Knowledges, and Geography through Decolonizing Paths." *Progress in Human Geography* 43 (2): 276–94. https://doi.org/10.1177/0309132517747315.

Coulthard, Glen. 2014. *Red Skin, White Masks: Rejecting the Colonial Politics of Recognition*. University of Minnesota Press.

de la Cadena, Marisol, and Mario Blaser. 2018. *A World of Many Worlds*. Duke University Press.

Escobar, Arturo. 1998. "Whose Knowledge, Whose Nature? Biodiversity, Conservation, and the Political Ecology of Social Movements." *Journal of Political Ecology* 5 (1). https://doi.org/10.2458/v5i1.21397.

Escobar, Arturo. 2018. *Designs for the Pluriverse: Radical Interdependence, Autonomy, and the Making of Worlds—New Ecologies for the Twenty-First Century*. Duke University Press.

Estes, Nick. 2019. *Our History Is the Future: Standing Rock Versus the Dakota Access Pipeline, and the Long Tradition of Indigenous Resistance*. Verso.

Fernández-Llamazares, Álvaro, Julia E. Fa, Dan Brockington, et al. 2024. "No Basis for Claim that 80% of Biodiversity is Found in Indigenous Territories." *Nature* 633 (8028): 32–5. https://doi.org/10.1038/d41586-024-02811-w.

Institute for Integrative Science and Health (IISH). n.d. *Guiding Principles (Two Eyed Seeing):Integrative Science*. IISH. Accessed April 23, 2025. http://www.integrativescience.ca/Principles/TwoEyedSeeing/.

Jessen, Tyler D., Natalie C. Ban, Nicholas XEMŦOLTW Claxton, and Chris T. Darimont. 2022. "Contributions of Indigenous Knowledge to Ecological and Evolutionary Understanding." *Frontiers in Ecology and the Environment* 20 (2): 93–101. https://doi.org/10.1002/fee.2435.

Kimmerer, Robin Wall. 2013. *Braiding Sweetgrass: Indigenous Wisdom, Scientific Knowledge and the Teachings of Plants*. Milkweed Editions.

Office of the High Commissioner for Human Rights (OHCHR). 2024. *UN Expert Urges Colombia to Prioritise Indigenous Peoples' Rights to Achieve 'Total Peace'*. Office of the United Nations High Commissioner for Human Rights, March 15. https://www.ohchr.org/en/media-advisories/2024/03/un-expert-urges-colombia-prioritise-indigenous-peoples-rights-achieve.

Petzold, Jan, Nadine Andrews, James D. Ford, Christopher Hedemann, and Julio C. Postigo. 2020. "Indigenous Knowledge on Climate Change Adaptation: A Global Evidence Map of Academic Literature." *Environmental Research Letters* 15 (11): 113007. https://doi.org/10.1088/1748-9326/abb330.

Santos, Boaventura de Sousa. 2016. *Epistemologies of the South: Justice Against Epistemicide*. Routledge.

Simpson, Leanne Betasamosake. 2017. *As We Have Always Done: Indigenous Freedom through Radical Resistance*. University of Minnesota Press.

Vojno, Natalija. 2022. "Braiding Knowledge Systems as Environmental Peacebuilding: A Four-Dimensional Analysis for Co-Applying Indigenous and Non-Indigenous Worldviews in Great Lakes Water Governance." Relating Systems Thinking and Design 2022 Symposium.

World Wildlife Fund (WWF), UN Environment Programme World Conservation Monitoring Centre (UNEP-WCMC) GEF Small Grants Programme ICCA-Global Support Initiative (SGP/ICCA-GSI), et al. 2021. *The State of Indigenous Peoples' and Local Communities' Lands and Territories: A Technical Review of the State of Indigenous Peoples' and Local Communities' Lands, Their Contributions to Global Biodiversity Conservation and Ecosystem Services, the Pressures They Face, and Recommendations for Actions*. WWF. https://wwflac.awsassets.panda.org/downloads/report_the_state_of_the_indigenous_peoples_and_local_communities_lands_and_territories_1.pdf.

Zaitchik, Alexander. 2018. "From Conservation to Colonization." *Foreign Policy*, July 16. https://foreignpolicy.com/2018/07/16/how-conservation-became-colonialism-environment-indigenous-people-ecuador-mining/.

Zimmerer, Karl S. 2012. "The Indigenous Andean Concept of Kawsay, the Politics of Knowledge and Development, and the Borderlands of Environmental Sustainability in Latin America." *Publications of the Modern Language Association of America* 127 (3): 600–6. https://doi.org/10.1632/pmla.2012.127.3.600.

Indigenous Rights

Alexandre Corriveau-Bourque, Veronica Limeberry, and Kimberly R. Marion Suiseeya

Indigenous Rights are Indigenous Peoples' inherent rights to self-determination, autonomy, and the preservation and flourishing of their cultural identities.

Indigenous identities emerge from the relationship between peoples and places through which distinct cultures—languages, religions, and **governance** institutions—have been created and continue to evolve over countless generations (Coulthard 2014). In many parts of the world, Indigenous Peoples' origin stories articulate their deep connections with place, as peoples emerging from the land itself, and refer to their relationships to these specific places and ecosystems "since time immemorial." Indigeneity from Indigenous perspectives is fundamentally rooted in sovereignty and self-determination—autonomous control over their lands, territories, and resources, and themselves as self-defined peoples, without interference from nation-states (Champagne 2008, 2013).

Many Indigenous Peoples have been forcibly displaced from the places central to their lifeways or have had their access severely restricted. This allows for colonial settlement on their territories and encroachment by extractive industries, including both illicit and industrial **agriculture**, timber, and mining, as well as **infrastructure** projects. Indigenous Peoples also experience **violence** and **displacement** from landscapes being "protected" by states and international organizations for **conservation** purposes (Tauli-Corpuz et al. 2018; Rights and Resources Initiative and McGill University 2021). Indigenous languages, cultural practices, and institutions continue to be deliberately suppressed through the actions of State and non-state actors in acts often internationally recognized as genocide (Minority Rights Group International 2000; National Inquiry into Missing and Murdered Indigenous Women and Girls 2019; Tagay et al. 2017; Wilkinson 2012). Indigenous communities and leaders continue to be violently targeted by these actors, especially when they resist and bring attention to the abuses being inflicted on their people and their **environment** (Global Witness 2024).

It is because of these systemic and ongoing violations that Indigenous communities and their allies have been working to enshrine Indigenous Rights into national and international laws, as well as within the norms and best practices of specific industries (e.g., Roundtable for Sustainable Palm Oil, Forest Stewardship Council, etc.) and international funding bodies, such as the International Finance Corporation. The United Nations Declaration on the Rights of Indigenous Peoples (United Nations 2007) and the Indigenous and Tribal Peoples Convention (International Labour Organization 1989) are foundational documents in international **human rights** law specifically addressing Indigenous Peoples. ILO 169 crucially affirms that Indigenous Peoples have inherent and distinct rights in international law, regardless of nation-state recognition.

Through these instruments and others, Indigenous Peoples have asserted individual and collective rights concerning minimum standards for survival, dignity, and well-being—such as housing, health, water, food, and education—as well as political rights, territorial and resource rights, cultural rights, economic rights, environmental rights, intellectual property rights, and more. However, many Indigenous scholars and activists argue that efforts by state structures and international institutions to define "Indigeneity" or Indigenous rights are inherently colonial. They contend that such definitions may serve to legitimize state sovereignty at the expense of Indigenous sovereignty (Coulthard 2014; Lightfoot 2016; Maldonado-Torres 2017; Simpson 2007).

Given the histories of dispossession and ongoing violations, Indigenous assertion of rights is inherently linked to struggles for **justice**—both through armed resistance as well as peaceful means. Rights, if only enshrined in law, can be meaningless unless they concretely contribute toward justice (Marion Suiseeya et al. 2022). As such, "there is a clear overlap between the international human rights-based protection of (Indigenous Peoples') rights to lands, resources and their environment and the discipline of **environmental peacebuilding** which can create a foundation for the pursuit of justice" (Rombouts 2023, 401).

"Indigenous Peoples" should be capitalized, denoting a distinct political and social identity—in the same way that "Europeans" or "Congolese" would not be written in lower-case. Where possible, authors referencing Indigenous Peoples within a specific area should use the names that those communities use for themselves, rather than using colonially imposed names or referring to them in the abstract.

References

Champagne, Duane. 2008. "The Indigenous Peoples' Movement: Theory, Policy, and Practice." 39th Annual Sorokin Lecture. University of Saskatchewan. https://artsandscience.usask.ca/sociology/documents/39th-annual-sorokin-lecture.pdf.

Champagne, Duane. 2013. "UNDRIP (United Nations Declaration on the Rights of Indigenous Peoples): Human, Civil, and Indigenous Rights." *Wicazo Sa Review* 28 (1): 9–22. link.gale.com/apps/doc/A329067271/LitRC?u=anon~ade02d52&sid=googleScholar&xid=d31a9014.

Coulthard, Glen Sean. 2014. *Red Skin, White Masks: Rejecting the Colonial Politics of Recognition*. University of Minnesota Press.

Global Witness. 2024. *Missing Voices: The Violent Erasure of Land and Environmental Defenders*.

International Labour Organization. 1989. "Indigenous and Tribal Peoples Convention, 1989 (No. 169)." https://normlex.ilo.org/dyn/nrmlx_en/f?p=NORMLEXPUB:12100:0::NO::P12100_ILO_CODE:C169.

Lightfoot, Sheryl. 2016. *Global Indigenous Politics: A Subtle Revolution*. Routledge.

Maldonado-Torres, Nelson. 2017. "On the Coloniality of Human Rights." *Revista Crítica de Ciências Sociais* 114: 117–36. https://doi.org/10.4000/rccs.6793.

Marion Suiseeya, Kimberly R., Laura Zanotti, and Kate Haapala. 2022. "Navigating the Spaces Between Human Rights and Justice: Cultivating Indigenous Representation in Global Environmental Governance." *The Journal of Peasant Studies* 49 (3): 604–28. https://doi.org/10.1080/03066150.2020.1835869.

Minority Rights Group International. 2000. *The Batwa Pygmies of the Great Lakes Region*.

National Inquiry into Missing and Murdered Indigenous Women and Girls. 2019. *Reclaiming Power and Place: The Final Report of the National Inquiry into Missing and Murdered Indigenous Women and Girls*. Government of Canada.

Rights and Resources Initiative and McGill University. 2021. *Status of Legal Recognition of Indigenous Peoples', Local Communities', and Afro-descendant Peoples' Rights to Carbon Stored in Tropical Lands and Forests*. Rights and Resources Initiative.

Rombouts, Bas. 2023. "Environmental Peacebuilding and Indigenous Peoples' Rights to Lands and Resources." In *Research Handbook on International Law and Environmental Peacebuilding*, edited by Daniella Dam-de Jong and Britta Sjöstedt. Edward Elgar Publishing Limited.

Simpson, Audra. 2007. "On Ethnographic Refusal: Indigeneity, Voice, and Colonial Citizenship." *Junctures: The Journal for Thematic Dialogue* 9: 67–80. https://junctures.org/index.php/junctures/article/view/66/60.

Tagay, Sefik, Dogan Ayhan, Claudia Catani, Ulrich Schnyder, and Martin Teufel. 2017. "The 2014 Yazidi Genocide and Its Effect on the Yazidi Diaspora." *The Lancet* 390 (10106): 1946. https://doi.org/10.1016/S0140-6736(17)32701-0.

Tauli-Corpuz, Victoria, Janis Alcorn, and Augusta Molnar. 2018. *Cornered by Protected Areas: Replacing 'Fortress' Conservation with Rights-based Approaches Helps Bring Justice for Indigenous Peoples and Local Communities, Reduces Conflict, and Enables Cost-effective Conservation and Climate Action*. Rights and Resources Initiative.

United Nations (UN). 2007. "United Nations Declaration on the Rights of Indigenous Peoples." https://www.un.org/development/desa/indigenouspeoples/wp-content/uploads/sites/19/2018/11/UNDRIP_E_web.pdf.

Wilkinson, Tracy. 2012. "Former Guatemalan Dictator Rios Montt May Face Genocide Trial." *Los Angeles Times*, January 24. https://www.latimes.com/archives/blogs/world-now/story/2012-01-24/former-guatemalan-dictator-rios-montt-may-face-genocide-trial.

Infrastructure

Silvia Danielak

Infrastructure is the physical structures that connect and allow for exchange between various elements. Initially used in the military context, infrastructure is now frequently referenced in the context of transportation, services, or communications and technology, but may extend more widely to encompass people and ideas to describe relations and hierarchies (Simone 2004; Carse 2016). Infrastructure is often considered "boring" or invisible but may become very visible when it breaks down (Star 1999; Petroski 2009).

Infrastructure is a crucial component of armed **conflict** and peacemaking. During conflict, both military and civilian infrastructure are frequently targeted, sometimes just as a new one is being built (Sowers and Weinthal 2021; Weinthal and Sowers 2019; Danielak 2022). The destruction, as well as conflict-driven neglect of infrastructure, has humanitarian and environmental consequences; from flooding-related damage due to destroyed dams to soil contamination or air and water pollution due to pervasive rubble. Infrastructure reconstruction, too, contributes to greenhouse gas emissions but is also an important element of post-conflict reconstruction and **environmental peacebuilding** (Pinera and Reed 2009).

Infrastructure is the backbone of processes of development, interaction, and—within the context of conflict and **peacebuilding**—one of the central elements of reconstruction and recovery. Material infrastructure for transportation, communication, energy, or water is complex and fragile, especially in the face of climate-related uncertainty (Harvey et al. 2017; Helmrich and Chester 2022). It is acutely exposed to environmental hazards and **climate change** and should possibly also be part of remediating them (Howe et al. 2016). For example, aspirations as technical fixes propel geoengineering just as they do climate **adaptation** and **mitigation** solutions, or the energy transition at large.

The reliance on infrastructure raises ethical questions and possibly conflicts, while also presenting opportunities for peacebuilding. Critical infrastructure studies have drawn attention to the unintended and/or harmful consequences of infrastructure: if infrastructure caters to one group or space, it might inadvertently disadvantage and marginalize another. Infrastructure is a tool to govern and control and can exert "**violence**"; for example, if a highway connects two sites but prevents access for in-between communities and possibly even raises their risk of traffic accidents or exposure to pollution (Scott 1998; Rodgers and O'Neill 2012). Even further, the dilemmas not only arise in the construction of infrastructure, but also in its operation, maintenance, and possibly even its closure or ruination, as communities might be differently impacted by infrastructure emissions, waste, or toxicity from the operation or leftovers of infrastructure (Gupta 2018). There is, thus, a (long) time component to infrastructure that creates lasting impacts on the built and natural **environment** and its inhabitants.

Designing infrastructure equitably and sustainably, thus, is both about the method and the output. Infrastructure planning can be done collectively and in a participatory way, thereby considering conflicting needs and environmental impact. For environmental peacebuilding, infrastructure can be a tool to achieve environmental **conservation** or decarbonization, just as it can contribute to improving community relationships, trust-building, and collaboration. In turn, environmental peacebuilding's mission can also be to preempt the possible harm from infrastructure and ensure a design that is both sensitive to conflict and ecology.

References

Carse, Ashley. 2016. "Keyword: Infrastructure: How a Humble French Engineering Term Shaped the Modern World." In *Infrastructures and Social Complexity: A Companion,* edited by Penelope Harvey, Casper Jensen, and Atsuro Morita. Routledge.

Danielak, Silvia. 2022. "The Infrastructure of Peace: Civil–Military Urban Planning in Mali." *International Peacekeeping* 29 (1): 115–38. https://doi.org/10.1080/13533312.2021.1996236.

Gupta, Akhil. 2018. "The Future in Ruins: Thoughts on the Temporality of Infrastructure." In *The Promise of Infrastructure*, edited by Nikhil Anand, Akhil Gupta, and Hannah Appel. Duke University Press.

Harvey, Penelope, Casper Jensen, and Atsuro Morita, 2017. "Introduction: Infrastructural Complications." In *Infrastructures and Social Complexity: A Companion.* Edited by Penny Harvey, Casper Jensen, and Atsuro Morita. Routledge.

Helmrich, Alysha M., and Mikhail V. Chester. 2022. "Reconciling Complexity and Deep Uncertainty in Infrastructure Design for Climate Adaptation." *Sustainable and Resilient Infrastructure* 7 (2): 83–99. https://doi.org/10.1080/23789689.2019.1708179.

Howe, Cymene, Jessica Lockrem, Hannah Appel, et al. 2016. "Paradoxical Infrastructures: Ruins, Retrofit, and Risk." *Science, Technology, and Human Values* 41 (3): 547–65. https://doi.org/10.1177/0162243915620017.

Petroski, Henry. 2009. "Engineering: Infrastructure." *American Scientist* 97 (5): 370–74. https://doi.org/10.1511/2009.80.370.

Pinera, Jean-François, and Robert A. Reed. 2009. "A Tale of Two Cities: Restoring Water Services in Kabul and Monrovia." *Disasters* 33 (4): 574–90. https://doi.org/10.1111/j.1467-7717.2008.01088.x.

Rodgers, Dennis, and Bruce O'Neill. 2012. "Infrastructural Violence: Introduction to the Special Issue." *Ethnography* 13 (4): 401–12. https://doi.org/10.1177/1466138111435738.

Scott, James C. 1998. *Seeing like a State: How Certain Schemes to Improve the Human Condition Have Failed.* Yale University Press.

Simone, AbdouMaliq. 2004. "People as Infrastructure: Intersecting Fragments in Johannesburg." *Public Culture* 16 (3): 407–29. https://doi.org/10.1215/08992363-16-3-407.

Sowers, Jeannie, and Erika Weinthal. 2021. "Humanitarian Challenges and the Targeting of Civilian Infrastructure in the Yemen War." *International Affairs* 97 (1): 157–77. https://doi.org/10.1093/ia/iiaa166.

Star, Susan Leigh. 1999. "The Ethnography of Infrastructure." *American Behavioral Scientist* 43 (3): 377–91. https://doi.org/10.1177/00027649921955326.

Weinthal, Erika, and Jeannie Sowers. 2019. "Targeting Infrastructure and Livelihoods in the West Bank and Gaza." *International Affairs* 95 (2): 319–40. https://doi.org/10.1093/ia/iiz015.

Integrated Water Resources Management

Zoe H. Rosenblum and Osama A. Abu-Libda

Integrated water resources management (IWRM) is a conceptual process for coordinating the management of water, land, and related resources and sectors to optimize sustainable economic, social, and ecological benefits.

Integrated water resources management (IWRM) aims to equitably maximize the economic and social benefits of water resource use without compromising vital ecosystems (Global Water Partnership 2024). It is an objective-driven strategy that aims to promote sustainable and equitable water use and utilization by fostering collaboration across sectors, involving **stakeholders**, and acknowledging ecological and hydrological patterns (Ibisch et al. 2016). A cross-sectoral strategy, IWRM aims to consider all factors affecting water resources. IWRM integrates hydrology, environmental science, economics, public policy, and sociology.

The origins of IWRM can be traced back to the 1977 United Nations Conference on Water in Mar del Plata, Argentina, where it was recommended for addressing the multiple competing uses of water resources (Biswas 2004). In the 1990s, there were growing concerns about global water scarcity and the unsustainable exploitation of water resources. The IWRM concept surfaced again at the 1992 International Conference on Water and the Environment in Dublin, Ireland, and remains informed by the Dublin Principles: (1) fresh water is a finite and vulnerable resource, essential to sustain life, development, and the **environment**; (2) water development and management should be based on a participatory approach, involving users, planners, and policymakers at all levels; (3) women play a central part in the provision, management, and safeguarding of water; and (4) water is a public good with a social and economic value in all its competing uses (World Meteorological Organization 1992).

IWRM was popularized in the 2000s to address global challenges such as **climate change** and **environmental degradation**. Currently, IWRM has evolved into a guiding principle for river basin organizations (Schmeier 2015). However, the practicality of implementing IWRM remains a question, due in part to the lack of specificity in its definition (Biswas 2004; Rahaman and Varis 2005; Swatuk and Qader 2023). The main critiques are that IWRM was conceptualized by the water sector alone, thus limiting its interdisciplinarity, and that it is a popular term to attract funding but is too abstract to meaningfully make a difference on the ground.

When done well, IWRM can contribute to **environmental peacebuilding** through collaborative water management to reduce potential **conflict**, particularly in areas facing water scarcity (United Nations Environment Programme [UNEP] 2020). In Sudan, IWRM projects linked water science, water planning, and **natural resources** development at the sub-catchment scale, empowered women, improved crop productivity, safeguarded water supplies, improved quality of life, and improved ecological conditions (UNEP 2020). In East, West, and Southern Africa, the likelihood of IWRM implementation and adoption is influenced by donor effect, water scarcity, transboundary water resources, and policy approach (Dirwai et al. 2021). With collaborative management structures and operational guidelines, IWRM can promote equitable and sustainable distribution of water resources and thus protect ecosystems and decrease the likelihood of water conflicts (UNEP 2009; Moriarty et al. 2004).

IWRM's holistic approach can enhance **water security** by enhancing resource distribution, encouraging effective water consumption, and protecting ecosystems that offer water services (UNEP 2009).

IWRM aims to overcome fragmented water management methods by highlighting connections between water resources and associated environmental, social, and economic systems (Stockholm International Water Institute 2020). Future trends in IWRM will increasingly integrate **sustainability** practices, collaborative **governance**, and community involvement to effectively address the interconnected challenges of water scarcity, pollution, and **biodiversity** loss. IWRM may play a role in enhancing participatory frameworks for conflict resolution and fostering environmental peacebuilding. This includes **adaptive management** practices, improved stakeholder engagement, and integration of climate **resilience** strategies, ultimately promoting **sustainable development** and **cooperation** among conflicting parties.

References

Biswas, Asit. 2004. "Integrated Water Resources Management: A Reassessment." *Water International* 29 (2): 248–56. https://doi.org/10.1080/02508060408691775.

Dirwai, Tinashe Lindel, Edwin Kimutai Kanda, Aidan Senzanje, and Toyin Isiaka Busari. 2021. "Water Resource Management: IWRM Strategies for Improved Water Management. A Systematic Review of Case Studies of East, West and Southern Africa." *PLoS One* 16 (5): e0236903. https://doi.org/10.1371/journal.pone.0236903.

Global Water Partnership (GWP). 2024. "IWRM Glossary." https://iwrmactionhub.org/iwrmglossary.

Ibisch, Ralf B., Janos J. Bogardi, and Dietrich Borchardt. 2016. "Integrated Water Resources Management: Concept, Research and Implementation." In *Integrated Water Resources Management: Concept, Research and Implementation*, edited by Dietrich Borchardt, Janos J. Bogardi, and Ralf B. Ibisch. Springer International Publishing.

Moriarty, Patrick, John Butterworth, and Charles Batchelor. 2004. *Integrated Water Resources Management and The Domestic Water and Sanitation Sub-Sector*. RC International Water and Sanitation Centre.

Rahaman, Mizanur Muhammad, and Olli Varis. 2005. "Integrated Water Resources Management: Evolution, Prospects and Future Challenges." *Sustainability: Science, Practice and Policy* 1 (1): 15–21. https://doi.org/10.1080/15487733.2005.11907961.

Schmeier, Susanne. 2015. "The Institutional Design of River Basin Organizations – Empirical Findings from around the World." *International Journal of River Basin Management* 13 (1): 51–72. https://doi.org/10.1080/15715124.2014.963862.

Stockholm International Water Institute (SIWI). 2020. *Principles and Practices of Integrated Water Resources Management: Workplace-Based Professional Training Manual 1*. https://siwi.org/wp-content/uploads/2020/06/IWRM_Manual1_final.pdf.

Swatuk, Larry A., and Adnan Qader. 2023. "IWRM: Ideology or Methodology?" In *The Oxford Research Encyclopedia of Water Management*, edited by Mike Muller. Oxford University Press. https://oxfordre.com/environmentalscience/view/10.1093/acrefore/978019389414.001.0001/acrefore-9780199389414-e-620.

United Nations Environment Programme (UNEP). 2009. "Water Security and Ecosystem Services–The Critical Connection." September 16. https://www.unepdhi.org/wp-content/uploads/sites/2/2020/08/The_critical_connection.pdf.

United Nations Environment Programme (UNEP). 2020. "Adapt for Environment and Climate Resilience in Sudan (ADAPT!)—Integrated Water Resources Management Good Practices in Sudan." February. https://wedocs.unep.org/handle/20.500.11822/31981.

World Meteorological Organization (WMO). 1992. "Dublin Statement on Water and Sustainable Development." January. https://unesdoc.unesco.org/ark:/48223/pf0000137373.

International Humanitarian Law

Nataliia Hendel, Róisín Burke, and Stavros Evdokimos Pantazopoulos

International humanitarian law (IHL) is a system of principles and norms of international law that distinguishes an independent branch of law on the protection of war victims and restriction of means and methods of warfare during international (IAC) and non-international armed conflicts (NIAC).

The four Geneva Conventions (1949) (GCI for the Amelioration of the Condition of the Wounded and Sick in Armed Forces in the Field; GCII on Wounded, Sick and Shipwrecked of Armed Forces at Sea; GCIII Relative to the Treatment of Prisoners of War; GCIV relative to the Protection of Civilian Persons in Time of War), Additional Protocol I relating to the Protection of Victims of International Armed Conflicts (1977) (API), Additional Protocol II relating to the Protection of Victims of Non-International Armed Conflict (1977) (APII), and the Hague Conventions (1899) concerning Respecting the Laws and Customs of War on Land, and Hague Convention (IV) Respecting the Laws and Customs of War on Land (1907), are the most significant IHL instruments. Many provisions are considered customary international law.

Armed **conflict** between states and situations of occupation are governed by the four Geneva Conventions (1949) and API (1977). NIACs are governed by Common Article 3 of the Geneva Conventions (1949) and APII (1977). The Martens Clause in the preamble to the Hague Conventions 1899 and 1907 offers some very basic humanitarian protection in conflicts, where other protections are not explicitly spelled out.

IHL contains both **environment**-specific and general rules that afford protection to the natural environment. Starting with the environment-specific rules, the use of methods or means of warfare that are intended, or may be expected, to cause widespread, long-term, and severe damage to the natural environment is prohibited (API, Article 35(3)). This rule sets a high threshold of application because of the cumulative articulation of the three qualifiers of prescribed environmental damage. The related obligation of taking care to protect the natural environment against widespread, long-term, and severe damage (API, Article 55(1)) is wider than just the means and methods of warfare but maintains the same high threshold.

The 1976 Convention on the Prohibition of Military or any other Hostile Use of Environmental Modification Techniques (United Nations 1976) obliges States Parties not to engage in military or any other hostile use of environmental modification techniques having widespread, long-lasting, or severe effects as the means of destruction, damage, or injury to any other State Party (Article I.1). Furthermore, customary IHL prohibits using destruction of the environment as a weapon (International Committee of the Red Cross Guidelines 2020, rule 3A).

IHL also provides general protection to parts of the natural environment because of its inherent civilian nature. This includes IHL principles and rules on distinction, indiscriminate attacks, proportionality, and precautions. Every attack must abide by the principle of distinction between military objectives and combatants, and civilian objects and non-combatants. Under API, it is a grave breach of IHL to attack civilians and civilian objects indiscriminately. Targeting decisions must also be guided by the IHL proportionality principle, which prohibits attacks that may cause incidental civilian harm or damage to civilian objects, including the natural environment and parts thereof, that would be excessive in relation to the concrete and direct military advantage anticipated. Beyond this, such action may constitute a **war** crime in IACs.

Parties to an armed conflict must take precautions to minimize harm to civilians, civilian objects, and the natural environment during hostilities.

In addition, attacking, destroying, removing, or rendering useless objects indispensable to the survival of the civilian population is prohibited, including when such objects are part of the natural environment (Gieseken and Murphy 2023). The natural environment is also protected by IHL rules on enemy property; therefore, "the destruction of any part of the natural environment is prohibited, unless required by imperative military necessity" (International Committee of the Red Cross Guidelines 2020, rule 13). This rule applies to situations of occupation as well. Pillage, including the use of **natural resources**, is prohibited, and this prohibition applies equally to situations of occupation.

References

Gieseken, Helen, and Vanessa Murphy. 2023. "The Protection of the Natural Environment under IHL: ICRC 2020 Guidelines." *International Review of the Red Cross* 105 (924): 1180–207. https://international-review.icrc.org/articles/protection-of-the-natural-environment-under-ihl-icrc-2020-guidelines-924.

International Committee of the Red Cross (ICRC). 2020. *Guidelines on the Protection of the Natural Environment in Armed Conflict*. ICRC. https://www.icrc.org/sites/default/files/document_new/file_list/guidelines_on_the_protection_of_the_natural_environment_in_armed_conflict_advance-copy.pdf.

United Nations. 1976. *Convention on the Prohibition of Military or Any Other Hostile Use of Environmental Modification Techniques*. https://treaties.un.org/doc/Treaties/1978/10/19781005%2000-39%20AM/Ch_XXVI_01p.pdf.

J

Justice

Theresa (Isa) Arriola, Megan Baker, and Sarah Njeri

Justice is a broad concept generally referring to the fundamental restoration of one's well-being and accountability for one's actions and is often associated with moral rightness and fairness.

Justice and its related principles, such as righteousness, dignity, and fairness, are values found in many religions, traditions, and societies. Not all justice systems are considered just. For instance, Western epistemologies and nations have dominated much of the discussion around concepts of justice as it is commonly referred to (Hunfeld 2022). These Western approaches have resulted in the reproduction of justice systems that uphold structural inequality, the destruction of lands, and disturbances to increasingly precarious ecosystems. Furthermore, because many Western justice systems rely on laws that were formed to systematically remove Indigenous Peoples from their lands (Bhandar 2018), a more genuine justice should recognize and center Indigenous sovereignty and forms of justice to attain restorative and **environmental justice**. Some Indigenous justice systems are also considered unjust and against customary and international law. For instance, "women for blood" are a type of forced marriage considered a traditional justice mechanism in some Indigenous societies (Aljabiri and Payton 2015; Inayat 2017).

There are many different types of justice systems. For instance, Indigenous justice and **restorative justice** are often guided by ancient knowledge systems which are intergenerational and center restoration and reconciliation over corporal punishment (Hand et al. 2012). Many of these ideologies center around knowledge taught by elders and respect for the plurality of differences in beliefs and the world around us, including the environment (Dhillon 2022). Elders often serve as mediators in times of **conflict**, as their life experiences and knowledge of the community help guide respectful and meaningful forms of justice to restore balance. For some Indigenous communities, healing with traditional medicine has remained a critical aspect of restoring harmony to the land and one another. Justice, regarding colonization, goes together with **decolonization**. Within colonial and imperial contexts, decolonization moves towards restoring Indigenous lifeways that had been fundamentally transformed by colonialism. The profound **violence** inflicted upon Indigenous peoples and their lands

DOI: 10.4324/9781003604143-12

often requires moving beyond mere acknowledgment of harm toward reconciliatory justice. Without significant structural changes that address the vast social inequality that Indigenous peoples have experienced historically and continue to experience, justice cannot be accomplished (Coulthard 2014).

Justice can be a contested notion, especially in conflict-affected contexts; people can be both the perpetrator and victim of harms, depending on one's perspective. Working toward justice is integral to **environmental peacebuilding.** Doing so requires attending to the impact of **war** on the most marginalized communities. Within the field of environmental peacebuilding, the approach to justice should be intersectional, considering systemic violence, histories of dispossession, and how communities experience injustice differently because of these histories (Fernández-Llamazares et al. 2020). Consequently, a more equitable theory of justice would be one that meaningfully incorporates the diverse definitions of Indigenous peoples across the world whose lands have borne the greatest impact.

Restoring justice and evenly distributing resources as part of environmental peacebuilding should also consider the period before, during, and after conflict. Environmental peacebuilding and **environmental justice** should rectify historic injustices through inclusive and ecologically restorative processes such as promoting and restoring land rights, bringing damaged ecosystems back to health, and creating management structures that focus on **cooperation.** Doing so prioritizes and promotes **peace** over control and extractive policies (Krampe et al. 2021).

References

Aljabiri, Janan, and Joanne Payton. 2015. "Woman-for-Blood Marriages Persist in Iraq." *Fair Observer*, December 29. https://www.fairobserver.com/region/middle_east_north_africa/woman-for-blood-marriages-persist-in-iraq-21101/.

Bhandar, Brenna. 2018. *Colonial Lives of Property: Law, Land, and Racial Regimes of Ownership*. Duke University Press. https://doi.org/10.1215/9780822371571.

Coulthard, Glen S. 2014. *Red Skin, White Masks: Rejecting the Colonial Politics of Recognition*. University of Minnesota Press. https://doi.org/10.5749/minnesota/9780816679645.001.0001.

Dhillon, Jaskiran, ed. 2022. *Indigenous Resurgence: Decolonialization and Movements for Environmental Justice*. Berghahn Books. http://www.jstor.org/stable/jj.15213782.

Fernández-Llamazares, Álvaro, Diana L. Ocampo-Peñuela, Eduardo S. Brondizio, Kate A. Brauman, Stephen T. Garnett, and Henrik Thorén. 2020. "A State-of-the-Art Review of Indigenous Peoples and Environmental Pollution." *Integrated Environmental Assessment and Management* 16 (3): 324–41. https://www.ncbi.nlm.nih.gov/pmc/articles/PMC7187223/.

Hand, Carol A., Judith Hankes, and Toni House. 2012. "Restorative Justice: The Indigenous Justice System." *Contemporary Justice Review* 15 (4): 449–67. https://doi.org/10.1080/10282580.2012.734576.

Hunfeld, Katharina. 2022. "The Coloniality of Time in the Global Justice Debate: De-Centring Western Linear Temporality." *Journal of Global Ethics* 18 (1): 100–17. https://doi.org/10.1080/17449626.2022.2052151.

Inayat, Naila. 2017. "Pakistan Cracks Down on Men Trading Daughters as Young as 1 in Custom Known as Vani." *USA Today*, May 16. https://www.usatoday.com/story/news/world/2017/05/16/pakistan-cracks-down-on-men-trading-daughters/101740256/

Krampe, Florian, Farah Hegazi, and Stacy D. VanDeveer. "Sustaining Peace Through Better Resource Governance. Three Potential Mechanisms for Environmental Peacebuilding." *World Development* 144: 105508. https://doi.org/10.1016/j.worlddev.2021.105508.

Just Transition

Larissa Jane Houston, Alexandre Corriveau-Bourque, and Gül Özerol

A just transition is, according to the United Nations Committee for Development Policy (2023, 3), "ensuring that no one is left behind or pushed behind in the transition to low-carbon and environmentally sustainable economies and societies, can enable more ambitious climate action and provide an impetus to attaining the Sustainable Development Goals." Ultimately, the just transition is the name given to the practice of formulating a framework or set of principles that ensures governments and industry conduct the shift away from "extractive" economic models in a fair and equitable manner, considering the needs and rights of workers, communities, and vulnerable populations whose lives and livelihoods are intertwined with these extractive models (Heffron 2021).

The just transition concept was first used within the context of the North American Labour market in the 1970s and 1980s by the Oil, Chemical, and Atomic Workers Union (OCAW), which sought to address the threat placed on workers in the mining and manufacturing industry with the development of environmental protection policies (Stevis and Felli 2015). Since then, the International Labour Organization (ILO) has played a crucial role in the development of guidelines for just transitions that address decent work, social dialogue, and rights at work (Olsen and Kemter 2012). However, this framework is increasingly being applied within various sectoral contexts, i.e., just energy transition, just transition for food systems and rural communities, and others (Just Rural Transition [JRT] 2023; Bundgård and Gallagher, 2023).

At the core of the just transition framework are the intertwined concepts of **justice** and inclusivity for peoples who have often been excluded and/or harmed by the previously dominant economic order, i.e., considerations of **gender**, race, ethnicity, and class (Abram et al. 2022; Wang and Lo 2021). The concept includes not only worker and producer rights but also promotes opportunities for poverty eradication and sustainable production and consumption (Wang and Lo 2021). It weaves together considerations of social justice, **environmental justice**, climate justice, food justice, land justice, and energy justice, to name a few (Müllerová et al. 2023; JRT 2021, 2023).

While the just transition has not explicitly been mentioned within the context of **environmental peacebuilding**, academics and practitioners have noted that efforts to achieve crosscutting elements of justice further promote the need to address environmental challenges while simultaneously seeking **peace** (Baker and Obradovic-Wochnik 2016; JRT 2021). Just transition frameworks have been formulated and implemented to ensure that social, economic, and environmental objectives align and that these are achieved.

A recent approach to just transition in **conflict**-affected and fragile contexts calls for applying a transformative, political economy lens (McCandless and Onbargi 2023). This approach aims to go beyond the focus on **adaptation** to address the root causes of the climate crisis and conflicts, support **accountability** and financial responsibility for the justice implications of the crises, and frame action around measures that build **resilience** across local, national, and global levels.

The just transition promotes a domino effect of environmental and socioeconomic development. Firstly, by addressing the socioeconomic impacts of changing economic systems on the livelihoods of workers and communities who have come to rely on extractive models, there is greater social acceptance of the other systemic changes needed to implement environmental policies (Minadakis and Vega-Araújo 2024). Secondly, considerations of justice within environmental, social, and economic objectives result in a holistic approach to development (Kaljonen et al. 2024). Lastly, just transition is becoming more relevant for international agreements and national policies, thus promoting international **cooperation** and global alignment of decarbonization goals (Houston and Ruppel 2022).

References

Abram, Simone, Ed Atkins, Alix Dietzel, et al. 2022. "Just Transition: A Whole-Systems Approach to Decarbonisation." *Climate Policy* 22 (8): 1033–49. https://doi.org/10.1080/14693062.2022.2108365.

Baker, Catherine, and Jelena Obradovic-Wochnik. 2016. "Mapping the Nexus of Transitional Justice and Peacebuilding." *Journal of Intervention and Statebuilding* 10 (3): 281–301. https://doi.org/10.1080/17502977.2016.1199483.

Bundgård, Annie, and Eileen Gallagher. 2023. "Advancing a Just Transition Across Sectors." *Business for Social Responsibility*, July 6. https://www.bsr.org/ja/blog/advancing-a-just-transition-across-sectors.

Heffron, Raphael J. 2021. "What Is the 'Just Transition'?" In *Achieving a Just Transition to a Low-Carbon Economy*, edited by Raphael J. Heffron. Springer International Publishing. https://doi.org/10.1007/978-3-030-89460-3_2.

Houston, Larissa Jane, and Oliver C. Ruppel. 2022. "Just Energy Transitions in Progress? The Partnership between South Africa and the EU." *Journal for European Environmental & Planning Law* 19 (1–2): 31–54. https://doi.org/10.1163/18760104-19010004.

Just Rural Transition (JRT). 2021. *Land Tenure in a Just Rural Transition: Restoring Our Relationships to Land and Natural Resources*. Meridian Institute.

Just Rural Transition (JRT). 2023. *Principles for Just Food System Transitions: Envisioning a More Equitable and Sustainable Future – and an Inclusive Path to Achieving It*. Meridian Institute.

Kaljonen, Minna, Ari Paloviita, Suvi Huttunen, and Teea Kortetmäki. 2024. "Policy Mixes for Just Transitions: A Holistic Evaluation Framework." *Environmental Innovation and Societal Transitions* 52: 100885. https://doi.org/10.1016/j.eist.2024.100885.

McCandless, Erin, and Alexia Faus Onbargi. 2023. "Just Transitions and Resilience in Contexts of Conflict and Fragility: The Need for a Transformative Approach." *Current Opinion in Environmental Sustainability* 65: 101360. https://doi.org/10.1016/j.cosust.2023.101360.

Minadakis, Paolo, and José Vega-Araújo. 2024. "Rethinking the Social License to Operate? A Theoretical Exploration of Its Synergies with Social Acceptance and Energy Justice for a Just Transition." *Energy Research and Social Science* 113: 103552. https://doi.org/10.1016/j.erss.2024.103552.

Müllerová, Hana, Eva Balounová, Oliver C. Ruppel, and Larissa Jane H. Houston. 2023. "Building the Concept of Just Transition in Law: Reflections on Its Conceptual Framing, Structure and Content." *Environmental Policy and Law* 53 (4): 275–88. https://doi.org/10.3233/EPL-230012.

Olsen, Lene, and Dorit Kemter. 2012. "The International Labour Organization and the Environment: The Way to a Socially Just Transition for Workers." In *Trade Unions in the Green Economy*, edited by Nora Räthzel and David Uzzell. Routledge.

Stevis, Dimitris, and Romain Felli. 2015. "Global Labour Unions and Just Transition to a Green Economy." *International Environmental Agreements: Politics, Law and Economics* 15 (1): 29–43. https://doi.org/10.1007/s10784-014-9266-1.

United Nations Committee for Development Policy (CDP). 2023. *Report on the Twenty-Fifth Session*. Official Records of the Economic and Social Council, Supplement No. 13 (E/2023/33). https://documents.un.org/doc/undoc/gen/n23/088/80/pdf/n2308880.pdf.

Wang, Xinxin, and Kevin Lo. 2021. "Just Transition: A Conceptual Review." *Energy Research and Social Science* 82: 102291. https://doi.org/10.1016/j.erss.2021.102291.

L

Land Degradation

Alan Channer and Mawa Karambiri

Land degradation is the reduction of the biological or economic productivity and the ecological complexity of land, caused by direct or indirect human-induced processes (Intergovernmental Science-Policy Platform on Biodiversity and Ecosystem Services 2018; Food and Agriculture Organization of the United Nations [FAO] 1999).

Wherever it occurs—on cropland, rangeland, woodland, or wetland—and however it occurs—because of wind erosion, flash flooding, over-intensive cropping, mismanaged grazing, salinization, or contamination by extractive industries—land degradation necessarily represents the loss of a key natural resource (Olsson et al. 2022). Depending on who is affected and how, and who is responsible and why, land degradation can cause **conflict** (United Nations Convention to Combat Desertification [UNCCD] 2024).

As much as 40% of the world's land is already degraded (UNCCD 2024). Each year, a further 12 million hectares are lost (FAO 2015). The impacts can be especially stark for households that depend on subsistence farming—that is, households that comprise about one quarter of the world's population.

Land degradation not only increases the marginalization of already vulnerable subsistence farmers but also disproportionately affects women within these households (Intergovernmental Panel on Climate Change 2022). Women in many regions, especially in the Global South, are the backbone of the rural workforce but often do not own land (Adebayo and Worth 2024), which limits their capacity to adapt to land degradation and their investment opportunities in land restoration. In Sahelian countries, women are often leased second-rate land to cultivate. Land degradation in these contexts increases intra-familial, intra-community, and inter-community tensions, notably herder-farmer conflict. It causes migration of rural populations to urban areas and is a push factor to **violence** and extremism, especially for young people.

Conversely, land restoration as a form of **environmental peacebuilding** can improve, create abundance for farmers and pastoralists, and reduce conflict. This has been observed, for example, with the **farmer-managed natural regeneration** of trees in Niger (Rinaudo et al. 2019)

DOI: 10.4324/9781003604143-13

Land degradation frequently requires inter-community dialogue processes to mitigate its impacts (Jahel and Lambin 2024). For example, upland deforestation and consequent land degradation can cause downslope landslides, lowland gully erosion, and river siltation, necessitating discussion between those more upstream and those more downstream of the causes and effects. **The United Nations Convention to Combat Desertification** (**UNCCD**) launched a Peace Forest Initiative to "foster cross-border **cooperation** on land and ecosystem restoration in conflict-affected and fragile regions by nurturing common approaches to **natural resources** management and **conservation** for trust-building and **conflict prevention** and resolution" (UNCCD 2024).

Land degradation caused by extractive industries—from oil rigs in the Niger Delta to open-pit gold mining in Brazil and strip mining for rare earths in Myanmar—sets up conflict dynamics between local communities on the one hand and governments and multinational companies on the other. **Environmental peacebuilding** approaches that contribute to restoring land and rehabilitating conflict-affected communities include post-diamond mining and post-**war** recovery in Sierra Leone (Engwicht 2018) and nascent examples fostered by the Catholic Church in Colombia, Peru, and the Philippines (Montevecchio and Powers 2022).

References

Adebayo, Johson A. and Stephen H. Worth. 2024. "Profile of Women in African Agriculture and Access to Extension Services." *Social Sciences and Humanities Open* 9: 100790. https://doi.org/10.1016/j.ssaho.2023.100790.

Engwicht, Nina. 2018. "Environmental Peacebuilding: What is it Good For?" *Climate Diplomacy*, May 29. https://climate-diplomacy.org/magazine/environment/environmental-peacebuilding-what-it-good.

Food and Agriculture Organization of the United Nations (FAO). 1999. "Poverty Alleviation and Food Security in Asia." In *Land Degradation*, Chapter 3. FAO. https://www.fao.org/4/x6625e/x6625e02b.htm#P358_65507.

Food and Agriculture Organization of the United Nations (FAO). 2015. *Agroecology to Reverse Soil Degradation and Achieve Food Security*. FAO. https://openknowledge.fao.org/server/api/core/bit/streams/bb2a86db-7f53-4e70-91ca-35ceb9d777db/content.

Intergovernmental Panel on Climate Change (IPCC). 2022. "Land Degradation." In *Climate Change and Land: IPCC Special Report on Climate Change, Desertification, Land Degradation, Sustainable Land Management, Food Security, and Greenhouse Gas Fluxes in Terrestrial Ecosystems*. Cambridge University Press. https://doi.org/10.1017/9781009157988.006.

Intergovernmental Science-Policy Platform on Biodiversity and Ecosystem Services (IPBES). 2018. *Summary for Policymakers of the Assessment Report on Land Degradation and Restoration of the Intergovernmental Science Policy Platform on Biodiversity and Ecosystem Services*. https://www.ipbes.dk/wp-content/uploads/2018/09/LandDegradation_SPM_2018.pdf.

Jahel, Camille and Eric F. Lambin. 2024. "Reversing Degradation of Social–Ecological Systems: Explaining the Outcomes of Interventions in Africa." *Sustainability Science* 20: 439-68. https://doi.org/10.1007/s11625-024-01568-5.

Montevecchio, Caesar A. and Gerard F. Powers, eds. 2022. *Catholic Peacebuilding and Mining: Integral Peace, Development, and Ecology*. Routledge.

Olsson, Lennart, Humberto Barbosa, Suruchi Bhadwal, et al. 2022. "Land Degradation." In *Climate Change and Lands,* edited by Intergovernmental Panel on Climate Change. Cambridge University Press.

Rinaudo, Tony, Alice Muller and Mary Morris. 2019. "Farmer Managed Natural Regeneration Manual." *World Vision Australia*. https://fmnrhub.com.au/wp-content/uploads/2024/05/FMNR-Field-Manual_DIGITAL_FA-2.pdf.

United Nations Convention to Combat Desertification (UNCCD). 2024. *Ground for Peace: Land Restoration for International Peace and Security.* https://www.unccd.int/sites/default/files/202412/Peace%20Forest%20Initiaitve_Ground%20for%20Peace_Summary_UNCCD.pdf.
United Nations Convention to Combat Desertification (UNCCD). 2024. *The Peace Forest Initiative Operational Guidelines.* https://www.unccd.int/sites/default/files/2024-12/Peace%20Forest%20Initiative%20Operational%20Guidelines_UNCCD.pdf.

Land Grabbing

Alexandre Corriveau-Bourque, Jacobo Grajales, and Ann J. Le

Land grabbing refers to the transfer of control over land without the informed consent of the pre-existing users and rights holders (International Land Coalition [ILC] 2011). This leads to prior users/rights holders being displaced/prevented from returning or significantly curtailed from access to places/resources that are essential to their livelihoods and cultural identities (ILC 2011). The term land grabbing has also been applied to natural resources, water, built infrastructure, or other properties.

The "grabbing" process can involve the use of tactics by "land grabbers" and their associates, such as force, coercion, and/or corruption (Borras and Franco 2024; Wolford et al. 2024). However, land grabs can also be the product of systemic failures in processes that are otherwise deemed to be "legal" (Anseeuw et al. 2012). Land grabs are also often associated with **land degradation** and other negative environmental and livelihood impacts that are disproportionately borne by pre-existing users and rights holders (ILC 2011).

Land grabs can have **reverberating effects** on four dimensions of concern to **environmental peacebuilding**: (1) when land grabbing is a consequence of armed **conflict** or a tactic of **war**, (2) when it is an opportunistic response to disasters, (3) as a way to describe a type of commercial transaction associated with elite accumulation, extractive industries, and increasingly, **renewable energy** developments, and (4) the removal of people from landscapes designated for "**conservation** and restoration" purposes. Multiple dimensions of land grabbing are common on the same landscapes.

Abundant research shows that land can become a prize of war in the hands of armed groups, as their leadership can use land grabs to accumulate wealth and/or to reward their supporters (Anwary 2023; Unruh 2016). Land grabs can also be connected to broader social transformations brought through conflict, such as ethnic cleansing, the forced **displacement** of populations, the weakening of **governance** institutions, and the transformation of social hierarchies (Cramer and Richards 2011). War opens opportunities for both elites to consolidate their domination and for excluded social groups to challenge their subordinate position. Similarly, land grabbing can follow the displacement and impoverishment of communities by disasters such as earthquakes, tsunamis, wildfires, droughts, and typhoons/hurricanes. Communities with insecure **land tenure** can be evicted, denied essential assistance for reconstruction, or compelled through legal and circumstantial pressure to transfer their rights (Brown 2020; Cohen 2011; Lakhani 2024). This, in turn, can drive migration, marginalization, and social unrest.

Land grabbing can also relate to commercial transactions, for instance, when local elites or companies associated with timber, agribusiness, or mineral extraction gain access to land that was previously under the control of Indigenous Peoples, nomadic communities, and/or smallholders (e.g., Brazil, Philippines, Indonesia, Guatemala, Colombia). A strong **indicator** that land grabbing may be taking place is when an actor "tak(es) possession and/or control (…) of land which is disproportionate in size in comparison with average land holdings in the region" (Offei 2014, 10). Commercial and/or elite-driven grabs are also highly likely when post-conflict and post-disaster economic policies encourage corporate investment and make land more easily available to investors, without addressing the legacies of forced displacement and dispossession (Grajales 2021).

Land grabbing can also be associated with the establishment of conservation areas, particularly in the form of the "fortress conservation" model and increasingly under market-based conservation schemes (Blomley et al. 2013). This approach to conservation can build on past instances of exclusion and dispossession, perpetuating the consequences of armed conflict and disaster-driven displacement.

By exacerbating social inequalities, dispossession, and **environmental degradation,** land grabbing undermines efforts to build **sustainable peace** and **resilience** in conflict-affected and fragile contexts. Addressing land grabs through inclusive governance, strengthening tenure **security**, and **conflict sensitivity** in environmental management is therefore critical to fostering long-term stability and ecological restoration in environmental peacebuilding initiatives.

References

Anseeuw, Ward, Liz Alden Wily, Lorenzo Cotula, and Michael Taylor. 2012. *Land Rights and the Rush for Land: Findings of the Global Commercial Pressures on Land Research Project.* International Land Coalition.

Anwary, Anowara. 2023. "Displacement of the Rohingyas of Myanmar, Land Grabbing, and Extractive Capital." *The Journal of Social Encounters* 7 (1): 26–48. https://doi.org/10.69755/2995-2212.1160.

Blomley, Tom, Dilys Roe, Fred Nelson, and Fiona Flintan. 2013. "'Land Grabbing': Is Conservation Part of the Problem or the Solution?" International Institute for Environment and Development. https://www.iied.org/sites/default/files/pdfs/migrate/17166IIED.pdf.

Borras, Saturnino M., Jr., and Jennifer C. Franco. 2024. "Land Rush." *The Journal of Peasant Studies* 1–16. https://doi.org/10.1080/03066150.2024.2317961.

Brown, Alleen. 2020. "Caribbean Islanders: Environmentalist Billionaire Building Resort on Protected Wetlands." *The Intercept*, December 9. https://theintercept.com/2020/12/09/barbuda-resorts-dejoria-de-niro/.

Cohen, Erik. 2011. "Tourism and Land Grab in the Aftermath of the Indian Ocean Tsunami." *Scandinavian Journal of Hospitality and Tourism* 11 (3): 224–36. https://doi.org/10.1080/15022250.2011.593359.

Cramer, Christopher, and Paul Richards. 2011. "Violence and War in Agrarian Perspective." *Journal of Agrarian Change* 11 (3): 277–97. https://doi.org/10.1111/j.1471-0366.2011.00312.x.

Grajales, Jacobo. 2021. *Agrarian Capitalism, War and Peace in Colombia: Beyond Dispossession.* Routledge.

International Land Coalition (ILC). 2011. Tirana Declaration: Securing Land Access for the Poor in Times of Intensified Natural Resources Competition. Official Conference Communication, ILC Global Assembly.

Lakhani, Nina. 2024. "First Came the Maui Wildfires. Now Come the Land Grabs: 'Who Owns the Land Is Key to Lahaina's Future.'" *The Guardian*, March 15. https://www.theguardian.com/us-news/2024/mar/15/maui-wildfires-community-land-trust.
Offei, Samuel, and Randy Stringer. 2014. "Land Grabbing and Sustainable Agriculture in Ghana." Paper presented at The Aotearoa New Zealand International Development Studies Network (DevNet) Biennial Conference, University of Otago.
Unruh, Jon D. 2016. "Mass Land Claims in Land and Property Following the Arab Spring: Lessons from Yemen." *Stability: International Journal of Security and Development* 5 (1): 6. http://doi.org/10.5334/sta.444.
Wolford, Wendy, Ben White, Ian Scoones, Ruth Hall, Marc Edelman, and Saturnino M. Borras Jr. 2024. "Global Land Deals: What Has Been Done, What Has Changed, and What's Next?" *The Journal of Peasant Studies*: 1–38. https://doi.org/10.1080/03066150.2024.2325685.

Land Tenure

Mohammed W. Bah

Land tenure refers to the legal or customary relationship between individuals or groups and land, including rules governing access to, and control over, land and natural resources. It defines who can use resources, under what conditions, and for how long (Food and Agriculture Organization of the United Nations 2002).

Land tenure is particularly significant in **environmental peacebuilding** because contested land rights often lie at the root of resource-based **conflict**. For instance, in Rwanda, the lack of clear tenure arrangements fueled tensions over land access after the genocide, complicating post-**war** recovery. Similarly, in Liberia, overlapping land claims and weak tenure systems have exacerbated tensions between returnees and host communities, thereby challenging **peacebuilding** efforts.

The tenures of land are crucial in land **governance**. At the heart of this is the distinction between secure and insecure land tenure, determining whether individuals and communities can confidently invest in, and benefit from, their land. Secure land tenure provides individuals and communities with the stability necessary for investment in land improvement and sustainable practices, thereby enhancing agricultural productivity, social equity, and environmental **sustainability**. In contrast, insecure land tenure results in a lack of legally recognized and enforceable rights to land, hindering individuals and communities from fully benefiting from land ownership (Deininger 2003).

There is a significant global trend toward recognizing rural communities as collective owners of their lands, reflecting a shift away from traditional individualistic views of property. This increasing statutory acknowledgment of community lands is essential for enhancing legal protection and stability for these communities (Wily 2018). Land tenure is a foundational element of land governance that significantly influences social and economic dynamics within communities. It is vital in protecting communities from evictions and fostering **social cohesion** and collective resource management.

Secure land tenure is vital for environmental peacebuilding because it underpins efforts to restore livelihoods, reduce resource-based grievances, and prevent renewed conflict in post-war settings. In conflict-affected countries where **agriculture** supports up to 80% of livelihoods, access to land is essential for **food security** and economic recovery, particularly for displaced populations and former combatants (Unruh and Williams 2013). Insecure or unclear land rights can intensify ethnic, religious, or political tensions and undermine reintegration, especially when pre-conflict grievances over land distribution persist. Conversely, effective land tenure systems can strengthen governance, promote social cohesion, and foster sustainable resource management, making land not only a source of identity and sustenance but also a cornerstone for building durable **peace** (Unruh and Williams 2013).

Acknowledgment

This definition has been adapted from the Food and Agriculture Organization of the United Nations (2002).

References

Deininger, Klaus. 2003. *Land Policies for Growth and Poverty Reduction.* World Bank and Oxford University Press. https://arellanolaw.edu/alpr/v5n1b.pdf.

Food and Agriculture Organization of the United Nations (FAO). 2002. *Land Tenure and Rural Development.* FAO Land Tenure Studies No. 3.

Unruh, Jon, and Rhodri C. Williams. 2013. "Land: A Foundation for Peacebuilding." In *Land and Post-Conflict Peacebuilding*, edited by Jon Unruh and Rhodri C. Williams. Routledge.

Wily, Liz Alden. 2018. *Collective Land Ownership in the 21st Century: Overview of Global Trends.* Van Vollenhoven Institute, Leiden Law School.

Land Use

Katharina Löhr, Anne Hennings, and Harifidy Rakoto Ratsimba

Land use refers to the relation between land and people. It is about how land is utilized, the impacts from using land for residential, agricultural, commercial, public, mining, environmental, or other purposes, and people's land management actions.

Worldwide, land functions are a key economic, social, and environmental variable (Fernandes and Reydon 2024). In recent years, land as a limited resource has come under increasing pressure with devastating effects on ecosystems and people's well-being. Rising populations, expanding economic activities, **climate change**, and **land degradation** contribute to land scarcity, degradation, **biodiversity** loss, and contested land use (Albertus 2025). Particularly, climate change poses a growing threat to food systems, distorting agricultural productivity, livelihoods, and **human security** (Birthal et al. 2021; International Food Policy Research Institute 2022; Ortiz-Bobea et al. 2021).

Land use strongly interrelates with **governance** and depends on context-dependent ecological, economic, and social factors that impact people's decision-making and practices (and vice versa). At the same time, land use decisions, especially when related to the production of commodities produced for global markets, such as timber, cocoa, and coffee, are also increasingly driven by factors in distant markets and international actors (Lambin et al. 2014). Questions of land reform, women's and customary tenure, **youth** access, as well as corporate interests and practices are major challenges that need to be considered, especially in countries that are affected by **conflict** (Hennings and Löhr 2026).

Land use governance tends to be complex, contested, and is deeply entrenched in sociopolitical interrelations and **power** structures from global to local scales (Swenson 2018a; Swenson 2018b). In legal pluralist settings, customary, local, religious, and/or state authorities may govern land use collaboratively, or spheres of authority can overlap (Unruh 2003). Inclusive and coordinated land governance plays a key role in addressing climate change-induced stress, **mitigation** and **adaptation** measures, providing sustainable livelihoods, and guaranteeing legal certainty over land for communities and investors. Land management, administration, and policy are increasingly important to ensure fairer and just access to resources, which is key to **peacebuilding** and **sustainable development** overall. That said, land governance is closely related to water governance, especially in the agricultural and mining sectors.

Land use is key in peacebuilding strategies and providing stability through inclusive territorial planning, considering socio-ecological balance between sectors and generations. **Environmental peacebuilding** efforts either lead to land use changes, i.e., through the establishment of transboundary protected areas, or **peace parks,** or aim at changing control over land use (Fias and Stoffelen 2024; Ramutsindela 2024). More recently, changes in land use in favor of climate adaptation and mitigation measures, such as for carbon deals in Liberia or wind parks in Kenya, led to the exclusion of local communities and contestation. Similar experiences related to biodiversity **conservation** priorities have also been observed in Madagascar concerning the limited consideration of legitimate tenure rights of local communities (Baba et al. 2023). Additionally, clarifying **land tenure** improves ownership, which is key to **sustainability** and social stability (De Schutter and Rolnik 2014).

References

Albertus, Michael. 2025. *Land Power: Who Has It, Who Doesn't, and How That Determines the Fate of Societies.* John Murray Press.

Baba, Check Abdel Kader, Washe Kazungu, Frederike Klümper, Jes Weigelt, and Harididy Rakoto Ratsmba. 2023. "The Risks of Decoupling Forest Conservation from Legitimate Tenure Rights in Madagascar." Global Soil Week Working Paper. TMG Research GmbH. https://assets.ctfassets.net/rrirl83ijfda/S29cEuby52OUCOINGNhnH/20e4859241333ea2d84f7fcc976d14e6/Global_Soil_Week_Madagascar_Report.pdf.

Birthal, Pratap S., Jaweriah Hazrana, and Digvijay S. Negi. 2021. "Effectiveness of Farmers' Risk Management Strategies in Smallholder Agriculture: Evidence from India." *Climatic Change* 169 (3): 30. https://doi.org/10.1007/s10584-021-03271-1.

De Schutter, Olivier, and Raquel Rolnik. 2014. "The Social Function of Land and Security of Tenure." In *Take Back the Land! The Social Function of Land and Housing, Resistance and Alternatives,* edited by Mathivet Charlotte. Ritmo.

Fernandes, Vitor Bukvar B., and Bastiaan Philip Reydon. 2024. "An Exploration of the Relationship Between Land Governance and Development Through a Systematic Review of Interventions." *Land Use Policy* 138: 107032. https://doi.org/10.1016/j.landusepol.2023.107032.

Fias, Mona, and Arie Stoffelen. 2024. "The Nature Conservation-Geopolitics Complex: Bridging Between Conservation Geopolitics and Peace Park Discourses." *Political Geography* 114: 103175. https://doi.org/10.1016/j.polgeo.2024.103175.

International Food Policy Research Institute (IFPRI). 2022. 2022 *Global Food Policy Report: Climate Change and Food Systems.* https://doi.org/10.2499/9780896294257.

Hennings, Anne, and Katharina Löhr. 2026. "Navigating Change: Effective and Legitimate Land Governance in Times of Climate Change, Conflict, and Economic Pressures." *Africa Spectrum.*

Lambin, Eric F., Patrick Meyfroidt, Ximena Rueda, et al. 2014. "Effectiveness and Synergies of Policy Instruments for Land Use Governance in Tropical Regions." *Global Environmental Change* 28: 129–40. https://doi.org/10.1016/j.gloenvcha.2014.06.007.
Ortiz-Bobea, Ariel, Toby R. Ault, Carlos M. Carrillo, Robert G. Chambers, and David B. Lobell. 2021. "Anthropogenic Climate Change has Slowed Global Agricultural Productivity Growth." *Nature Climate Change* 11 (4): 306–12. https://doi.org/10.1038/s41558-021-01000-1.
Ramutsindela, Maano. 2024. "Protecting Nature: Political Ecologies of Conservation through the Lens of Peace Parks." In *Doing Political Ecology*, edited by Gregory L. Simon and Kelly Kay. Routledge.
Swenson, Geoffrey. 2018a. "Corrigendum to 'Legal Pluralism in Theory and Practice.'" *International Studies Review* 20 (2): 342. https://doi.org/10.1093/isr/viy019.
Swenson, Geoffrey. 2018b. "Legal Pluralism in Theory and Practice." *International Studies* Review 20 (3): 438–62. https://doi.org/10.1093/isr/vix060.
Unruh, Jon. 2003. "Land Tenure and Legal Pluralism in the Peace Process." *Peace and Change* 28 (3): 352–77. https://doi.org/10.1111/1468-0130.00267.

Learning

Leonardo Medina and Michelle Bonatti

Learning is a dynamic and multifaceted process through which individuals and groups acquire, adapt, and reinforce knowledge, skills, attitudes, and values—leading to shifts in understanding, relationships, and behavior. Learning unfolds within and across individuals, collectives, and institutions in numerous forms and through a diversity of processes, such as direct experience, reflection and dialogue, embodied expression, and cultural immersion.

In the context of **environmental peacebuilding**, learning transcends the mere accumulation of knowledge. It becomes a vehicle for actors engaged in and affected by **conflict** to transform perspectives, strengthening collaborative capacities and cultivating new behaviors. This form of learning plays a critical role in bridging historical divides and fostering collective responses to environmental and sociopolitical challenges.

A key school of thought in this domain is social learning, which highlights how individuals and groups reshape mental models, perceptions, and relations in response to shared socio-environmental dilemmas. Rather than being limited to individual cognition, social learning involves shifts in collective understandings and norms (Reed et al. 2010). It emerges through processes such as inclusive dialogue, joint problem-solving, and cooperative resource management, enabling participants to reconsider others' intentions and identities, redefine socio-psychological boundaries, and expand their capacity for crosscutting collaboration (Medina et al. 2024).

Another related school, experiential learning, draws attention to the importance of direct, context-rich engagement. Activities like joint restoration efforts or resource management simulations enable **stakeholders** to engage in iterative cycles of action and reflection—gaining insights through direct involvement, reflecting on impacts, forming new strategies, and testing approaches (Kolb 2015). Through place-based interactions, participants move from abstract knowledge to embodied understanding, fostering empathy, shared meaning, and **resilience**.

Closely related is collaborative learning, which emphasizes co-construction of knowledge and shared responsibility among diverse actors, often navigating asymmetrical **power** relations (Akkerman et al. 2007). In environmental peacebuilding, collaborative learning facilitates trust-building and the development of inclusive strategies that surface and address latent tensions. It emphasizes the interdependence of stakeholders, encouraging a shift from adversarial positions to mutual recognition and shared purpose—key ingredients for sustained **cooperation** and environmental resilience.

At a deeper level, transformative learning offers a lens for understanding how individuals and groups critically re-evaluate entrenched worldviews, including beliefs about identity, historical grievances, and resource entitlements (Mezirow 1997). This reflective process can generate powerful shifts in how people perceive the underlying structures that shape how they relate to each other and to **natural resources**, opening space for reconciliation and **sustainability**.

Importantly, learning in environmental peacebuilding also extends to the institutional level, shaping policies and organizational norms. Drawing on the single-, double-, and triple-loop learning model (Argyris and Schön 1997), institutions may first adapt actions in response to feedback (single-loop), then reconsider underlying assumptions (double-loop), and ultimately reframe the processes through which they can change (triple-loop). This progression allows institutions to move beyond technical adjustments toward systemic transformation—essential in addressing complex, deeply rooted environmental-conflict dynamics.

Practitioners of environmental peacebuilding must also learn how to operationalize these diverse forms of learning. Yet the field faces an evidence gap on how learning processes unfold and their potential contribution to **peacebuilding. Monitoring** and **evaluation** (M&E) can serve not only as **accountability** tools but also as catalysts for learning. When designed with a learning orientation, M&E creates structured opportunities for reflection, dialogue, and **adaptation** across scales (Morales-Muñoz et al. 2021). By capturing stakeholder experiences, documenting emergent outcomes, and enabling critical inquiry, M&E helps practitioners recognize patterns, revisit assumptions, and adjust strategies. Participatory M&E approaches can further include localized collaboration to foster experiential learning and enhance social learning by valuing local knowledge, fostering mutual accountability, and strengthening ownership. Iterative feedback cycles can also trigger institutional learning loops, allowing organizations to evolve in response to complexity and change.

Acknowledgment

This work was carried out with support from the CGIAR Initiative on Climate Resilience, ClimBeR, and the CGIAR Initiative on Fragility, Conflict, and Migration. We would like to thank all funders who supported this research through their contributions to the CGIAR Trust Fund: https://www.cgiar.org/funders/.

References

Akkerman, Sanne, Piet Van Den Bossche, Wilfried Admiraal, et al. 2007. "Reconsidering Group Cognition: From Conceptual Confusion to a Boundary Area Between Cognitive and

Socio-Cultural Perspectives?" *Educational Research Review* 2 (1): 39–63. https://doi.org/10.1016/j.edurev.2007.02.001.
Argyris, Ch., and D. A. Schön. 1997. "Organizational Learning: A Theory of Action Perspective." *Reis* (77/78): 345. https://doi.org/10.2307/40183951.
Kolb, David A. 2015. *Experiential Learning: Experience as the Source of Learning and Development*. Pearson Education, Inc.
Medina, Leonardo, Marisa O. Ensor, Frans Schapendonk, et al. 2024. "Community Voices on Climate, Peace and Security: A Social Learning Approach to Programming Environmental Peacebuilding." *Environment and Security* 2 (1): 75–104. https://doi.org/10.1177/27538796231207030.
Mezirow, Jack. 1997. "Transformative Learning: Theory to Practice." *New Directions for Adult and Continuing Education* 1997 (74): 5–12. https://doi.org/10.1002/ace.7401.
Morales-Muñoz, Héctor, Katharina Löhr, Michelle Bonatti, Luca Eufemia, and Stefan Sieber. 2021. "Assessing Impacts of Environmental Peacebuilding in Caquetá, Colombia: A Multistakeholder Perspective." *International Affairs* 97 (1): 179–99. https://doi.org/10.1093/ia/iiaa175.
Reed, Mark S., Anna C. Evely, Georgina Cundill, et al. 2010. "What Is Social Learning?" *Ecology and Society* 15 (4): resp1. https://doi.org/10.5751/ES-03564-1504r01.

Legal Pluralism

Leonardo Medina and Roísín Burke

Legal pluralism is the coexistence of two or more legal systems in the same social and/or geographical space. Legal pluralism is present in many states, especially because of colonial and imperial rule, which often introduced formal legal systems that coexisted, and frequently clashed, with Indigenous or customary justice systems, and sometimes religious orders (Le Sage 2005). Perceptions of justice systems will necessarily deviate from Western conceptions of justice systems in many parts of the world, including in conflict-affected states.

Legal pluralism plays a significant role in **peacebuilding**, particularly in **conflict**-affected or fragile regions. It recognizes that formal state law is not the only legal framework governing people's lives; customary laws, religious laws, and local traditions often play an equally critical, if not more critical, role, especially in areas where state presence is weak or contested. In such contexts, a range of non-state actors may contribute to the administration of **justice** or the resolution of disputes. These actors can include vigilante groups, community elders, religious leaders, and informal arbitration bodies, each performing distinct roles. For instance, insurgency groups may enforce local norms or provide **security** in the absence of the state (Podder 2013), while community-based arbitration mechanisms may resolve disputes without recourse to formal courts (Kifle and Mekonnen 2024).

Informal justice systems, not least customary systems, often have a much longer history than formal legal systems, the latter of which were often a product of colonial regimes and legal transplants. During conflict, these informal and customary systems have frequently proven more resilient and accessible to local populations than formal legal systems, as they are often considered better understood, more legitimate, and malleable (Chirayath et al. 2005).

Nevertheless, informal justice systems may be unregulated or under-regulated, be structurally biased, or undermine the rights of marginalized groups, particularly where

power differentials exist and are subject to elite capture (Burke 2022). Moreover, in plural legal contexts, there may be jurisdictional overlap and contradictions embedded in the system, making it problematic to navigate. In some cases, the relationship between legal actors is partially codified, for instance in South Africa, and in others not (Organisation for Economic Co-operation and Development 2025). Medina et al. (2025) suggest that, in the context of post-conflict Mindanao, local leaders strategically maneuver across systems of authority based on customary, rebel, and formal **governance** structures in pursuit of political aims, which can be both a source of conflict and cohesion depending on local perceptions of legitimacy.

In contexts of resource-related conflicts, legal pluralism can be particularly relevant because it can shape the legitimacy, accessibility, and contestation of claims over land and **natural resources**. Customary laws often govern land and resource use, especially in Indigenous or rural communities. Natural resource management in fragile, post-conflict, or post-colonial settings is often characterized by institutional multiplicity, whereby customary or traditional laws regulate access to and use of resources, while the state enforces formal laws that may contradict or ignore these local practices (Meinzen-Dick and Pradhan 2002). This can lead to disputes over resource ownership, usage rights, and environmental management. Furthermore, overlapping, hierarchical, and competing sources of political power are often strategically manipulated by elites to consolidate or amplify their influence (Hilhorst 2013). Understanding and engaging with this institutional diversity is essential for fostering **sustainable peace**, especially in areas affected by **environmental degradation**, resource scarcity, and **climate change**.

Legal pluralism can serve as a **governance** strategy to mitigate the conflict-related risks of institutional multiplicity, which acknowledges that different groups may rely on various legal frameworks, depending on their social, economic, and cultural background. In **environmental peacebuilding**, incorporating these legal systems into natural **resource governance** mechanisms ensures that resource conflicts are managed in a way that respects the rights, knowledge, and traditions of all stakeholders, thereby preventing **violence** and promoting **cooperation** (Le Sage 2005). Integrating these legal frameworks into peacebuilding processes can prevent or address conflicts over natural resources, which are often a key driver of violence.

Environmental peacebuilding also often involves restoring justice and equity in contexts where environmental degradation has fueled conflicts. Here, legal pluralism plays a role in ensuring **environmental justice** by creating space for marginalized or vulnerable groups who might not have access to formal legal systems to defend their environmental rights. Customary or Indigenous legal systems often have stronger connections to environmental stewardship and can provide alternative forms of justice, such as through community reconciliation principles, that align with local cultural values.

Acknowledgment

This work was carried out with support from the CGIAR Initiative on Climate Resilience, ClimBeR, and the CGIAR Initiative on Fragility, Conflict, and Migration. We would like to thank all funders who supported this research through their contributions to the CGIAR Trust Fund: https://www.cgiar.org/funders/.

References

Burke, Roisin. 2022. "Somalia and Legal Pluralism: Advancing Gender Justice Through Rule of Law Programming in Times of Transition." *International Law Review* 16 (2): 177. https://lawecommons.luc.edu/lucilr/vol16/iss2/3.

Chirayath, Leila, Caroline Mary Sage, and Michael Woolcock. 2005. *Customary Law and Policy Reform: Engaging with the Plurality of Justice Systems*. Working Paper 33655. World Bank. https://documents.worldbank.org/en/publication/documents-reports/documentdetail /675681468178176738/customary-law-and-policy-reform-engaging-with-the-plurality-of-justice -systems.

Hilhorst, Thea, ed. 2013. *Disaster, Conflict and Society in Crises: Everyday Politics of Crisis Response*. Routledge.

Kifle, Alagaw Ababu, and Kindeye Fenta Mekonnen. 2024. "The Relevance of Traditional Mechanisms of Conflict Resolution to End Localized Communal Conflicts in Africa." *Journal of Asian and African Studies*. https://doi.org/10.1177/00219096241275418.

Le Sage, Andre. 2005. *Stateless Justice in Somalia: Formal and Informal Rule of Law Initiatives*. Centre for Humanitarian Dialogue. https://www.files.ethz.ch/isn/20303/Somalia_stateless _justice.pdf.

Medina, Leonardo, Grazia Pacillo, Peter Läderach, Stefan Sieber, and Michelle Bonatti. 2025. "Adapting to Climate Change under Threats of Violence: A Comparative Institutional Analysis of Incentives for Conflict and Collaboration." *Current Research in Environmental Sustainability* 9: 100276. https://doi.org/10.1016/j.crsust.2024.100276.

Meinzen-Dick, Ruth, and Rajendra Pradhan. 2002. *Legal Pluralism and Dynamic Property Rights*. CAPRi Working Papers. International Food Policy Research Institute (IFPRI).

Organisation for Economic Co-operation and Development (OECD). 2025. *The Codification of Customary Laws in South Africa*. https://www.oecd.org/en/publications/gender-equality _cfc9cac7-en/the-codification-of-customary-laws-in-south-africa_3022b119-en.html.

Podder, Sukanya. 2013. "Non-State Armed Groups and Stability: Reconsidering Legitimacy and Inclusion." *Contemporary Security Policy* 34 (1): 16–39. https://doi.org/10.1080/13523260 .2013.771029.

M

Maladaptation

Christina Greene and E. Lisa F. Schipper

Maladaptation occurs when climate adaptation inadvertently triggers vulnerability to climate change. This can be through reinforcing existing vulnerabilities, redistributing vulnerabilities, or introducing new vulnerabilities.

Maladaptation to **climate change** emerged as a theoretical idea based on a concept in evolutionary biology. There, maladaptation describes biological processes that respond to **environmental change** but create new problems. For instance, human bodies develop responses to adapt to new situations, including the brain development of "phantom pain" in lost limbs. However, this development can create physical and psychological trauma rather than help with recovery. The idea that a measure to prevent damage can encourage behavior that increases risk is encapsulated in the term "moral hazard," which stems from the insurance literature (Rowell and Connelly 2012). These studies found that insurance made homeowners or others change their risk attitudes because they felt invincible with financial guards.

Starting in the 1990s, climate change **adaptation** scholars began adopting the concept to refer to adaptation responses that could create counter-incentives that would lead to behavior that would make people worse off (Smit 1993). For example, crop irrigation is an adaptation strategy to changing rainfall, especially for the poorest farmers dependent on rain-fed **agriculture**. Yet, drawing surface or groundwater often decreases water availability elsewhere. Further, because agriculture in many places is unlikely to remain a viable income and food source in a changing climate, investments in irrigation postpone inevitable livelihood shifts and close off other adaptation options. Investment in irrigation can introduce new problems and lock people into a nonviable livelihood (Albizua et al. 2019).

Much of the initial discussion on maladaptation was based on experiences of development interventions, not necessarily on climate change adaptation. Some of these ideas were, therefore, largely theoretical. However, the growth in empirical evidence from adaptation projects since the late 2000s has provided insights into maladaptation. The current literature on maladaptation identifies three main types of maladaptation: (1) rebounding vulnerability (or reinforcement of existing vulnerabilities), (2) shifting vulnerability to new groups or regions, and (3) introduction of new

DOI: 10.4324/9781003604143-14

negative environmental, social, and economic inequalities and risks (Juhola et al. 2016; Schipper 2020; Eriksen et al. 2021).

Adaptation projects can reinforce existing vulnerabilities by entrenching existing societal inequalities, such as interventions that only benefit landowners and not farm labor (Mikulewicz 2020). Adaptation interventions can also increase vulnerability by focusing on current or recent climate variability instead of adapting to future climate change (Dilling et al. 2015). The redistribution of vulnerability risk often accumulates among groups that are already marginalized, increasing global inequities (Atteridge and Remling 2018).

Maladaptation outcomes are increasingly used to understand how adaptation and **environmental peacebuilding** efforts can exacerbate **conflict** dynamics and hinder **peacebuilding** (Ide 2020; Dabelko et al. 2022). Conflict considerations are largely left out of adaptation planning, and adaptation interventions that redistribute resources and exacerbate inequalities can heighten existing conflict dynamics (Remling and Meijer 2025). For example, a resettlement program in Ethiopia intended to safeguard households from floods and droughts increased sociopolitical tensions with conflicts over **land tenure** (Milman and Arsano 2014). Similarly, a dam in Afghanistan built to improve **water security** contributed to growing grievances and insecurity around water access (Krampe et al. 2021). There is a growing number of considerations and frameworks for designing and implementing climate-resilient peacebuilding interventions and conflict-sensitive climate change adaptation interventions to avoid the impacts of maladaptation (Magnan et al. 2016; Crawford et al. 2015; Tänzler and Scherer 2019). Efforts to integrate and mainstream climate interventions into peacebuilding efforts can draw from the lessons learned on the maladaptation outcomes from development interventions. This includes considering inequalities and **power** relations that drive vulnerability and ensuring collaborations with communities to define meaningful and equitable vulnerability reduction (Eriksen et al. 2021).

Although maladaptation and the task of identifying maladaptation before it happens are increasingly popular topics, there is also growing frustration and fear in policy circles that an overemphasis on maladaptation can suggest that adaptation is too risky and consequently convince donors to stop providing funding for adaptation (Schipper and Mukherji 2024).

References

Albizua, Amaia, Esteve Corbera, and Unai Pascual. 2019. "Farmers' Vulnerability to Global Change in Navarre, Spain: Large-Scale Irrigation as Maladaptation." *Regional Environmental Change* 19: 1147–58. https://doi.org/10.1007/s10113-019-01462-2.

Atteridge, Aaron, and Elise Remling. 2018. "Is Adaptation Reducing Vulnerability or Redistributing It?" *WIREs Climate Change* 9: e500. https://doi.org/10.1002/wcc.500.

Crawford, Alec, Angie Dazé, Anne Hamill, Jo-Ellen Parry, and Alicia Natalia Zamudio. 2015. *Promoting Climate-Resilient Peacebuilding in Fragile States*. International Institute for Sustainable Development. https://www.iisd.org/system/files/publications/promoting-climate-resilient-peacebuilding-fragile-states.pdf.

Dabelko, Geoffrey D., Anniek Barnhoorn, Noah Bell, et al. 2022. *Navigating a Just and Peaceful Transition: Environment of Peace (Part 3)*. Stockholm International Peace Research Institute. https://doi.org/10.55163/BHYR7656.

Dilling, Lisa, Meaghan E. Daly, William R. Travis, Olga V. Wilhelmi, and Roberta A. Klein. 2015. "The Dynamics of Vulnerability: Why Adapting to Climate Variability Will Not Always Prepare us for Climate Change." *WIREs Climate Change* 6: 413–25. https://doi.org/10.1002/wcc.341.

Eriksen, Siri, E. Lisa F. Schipper, Morgan Scoville-Simonds, et al. 2021. "Adaptation Interventions and Their Effect on Vulnerability in Developing Countries: Help, Hindrance or Irrelevance?" *World Development* 141: 105383. https://doi.org/10.1016/j.worlddev.2020.105383.

Ide, Tobias. 2020. "The Dark Side of Environmental Peacebuilding." *World Development* 127: 104777. https://doi.org/10.1016/j.worlddev.2019.104777.

Juhola, Sirkku, Erik Glaas, Björn-Ola Linnér, and Tina-Simone Neset. 2016. "Redefining Maladaptation." *Environmental Science and Policy* 55: 135–40. http://dx.doi.org/10.1016/j.envsci.2015.09.014.

Krampe, Florian, Elizabeth S. Smith, and Mohammad Daud Hamidi. 2021. "Security Implications of Climate Development in Conflict-affected States: Implications of Local-level Effects of Rural Hydropower Development on Farmers in Herat." *Political Geography* 90: 102454. https://doi.org/10.1016/j.polgeo.2021.102454.

Magnan, A. K., E. Lisa F. Schipper, M. Burkett, et al. 2016. "Addressing the Risk of Maladaptation to Climate Change." *WIREs Climate Change* 7: 646–65. https://doi.org/10.1002/wcc.409.

Mikulewicz, Michael. 2020. "Disintegrating Labor Relations and Depoliticized Adaptation to Climate Change in Rural São Tomé and Príncipe." *Area* 53 (3): 422–30. https://doi.org/10.1111/area.12630.

Milman, Anita, and Yacob Arsano. 2014. "Climate Adaptation and Development: Contradictions for Human Security in Gambella, Ethiopia." *Global Environmental Change* 29: 349–59. http://dx.doi.org/10.1016/j.gloenvcha.2013.11.017.

Remling, Elise, and Karen Meijer. 2025. "Conflict Considerations in the United Nations Framework Convention on Climate Change's National Adaptation Plans." *Climate and Development* 17 (1): 37–52. https://doi.org/10.1080/17565529.2024.2321156.

Rowell, David, and Luke B. Connelly. 2012. "A History of the Term 'Moral Hazard.'" *Journal of Risk and Insurance*: 1051–75. https://doi.org/10.1111/j.1539-6975.2011.01448.x.

Schipper, E. Lisa F. 2020. "Maladaptation: When Adaptation to Climate Change Goes Very Wrong." *One Earth* 3: 409–14. https://doi.org/10.1016/j.oneear.2020.09.014.

Schipper, E. Lisa F., and A. Mukherji. 2024. "Misguided Negative Adaptation Narratives are Hurting the Poor." *Science* 386 (6722): 624–26. https://doi.org/10.1126/science.adq7821.

Smit, Barry, ed. 1993. "Adaptation to Climatic Variability and Change: Report of the Task Force on Climatic Adaptation." Occasional Paper, Department of Geography, University of Guelph: Canadian Climate Program.

Tänzler, Dennis, and Nikolas Scherer. 2019. *Guidelines for Conflict-sensitive Adaptation to Climate Change*. adelphi.

Mediation

Joseph A. Siegel

Mediation is a process that uses confidentiality and an impartial third party to enable disputants, through communication and negotiation, to voluntarily make their own free and informed choices regarding their dispute (American Arbitration Association et al. 2005).

The mediation process can vary considerably across cultures and from one field or context to another. While mediation has existed in a wide variety of forms since biblical times (Moore 2003), the definition above has elements of a more contemporary Western view of mediation. In general, there are several stages of mediation during which the mediator helps the parties communicate, exchange information, identify issues and interests, consider options for settlement, and reach closure (Moore 2003).

Mediation can be used for disputes between private individuals as well as matters in which a local, national, or international governmental body is a party. At the international level, the 1945 Charter of the United Nations incorporates mediation as a strategy by which parties to the Charter can resolve disputes (United Nations [UN] 1945). In conformity with the Charter, mediations are expected to incorporate eight fundamentals, including the following: (1) preparedness of the parties, (2) consent, (3) impartiality, (4) inclusivity, (5) national ownership of the agreement and its implementation, (6) consistency with international laws and norms, (7) coherence, coordination, and complementarity of mediation efforts among the various entities involved in mediation, and (8) quality agreements with clear modalities for implementation (UN 2012).

The first contemporary use of mediation in the environmental field can be traced to a dispute in 1973 over a proposal to build a flood control dam on a river in the State of Washington, in the United States (Singer 1994). In the decades that followed, mediation has become an effective approach for clarifying issues and reducing the time and cost of litigation for parties engaged in environmental and other disputes (Siegel 2007). There is a wide range of applications and contexts for environmental mediation. For example, it has been used to address a **conflict** over the sale of timber and potential resulting impacts on endangered species habitat (Udall Foundation n.d.). Mediation has also been used to address **land use** and waste-related conflicts between Indigenous peoples, other underserved communities, and governmental authorities (Moore 2003). It has been used in the development of environmental regulations (Madrid and Martin 2002) and to address **climate change**-related conflicts between nomadic and sedentary communities in a tri-border region of Africa (Centre for Humanitarian Dialogue 2021).

Environmental mediation can be viewed as part of a broader spectrum of environmental collaboration and conflict resolution processes, such as facilitation, conflict assessment, collaboration, and negotiated rulemaking, among others (Emerson 2003). The processes along the spectrum share several characteristics with mediation, including voluntary **participation**, the ability to withdraw from the process, presence of an impartial third party with no decision-making authority, and formulation of solutions by the parties (Emerson 2003).

In the environmental field, disputes can emerge due to the complexity of environmental public policy issues, the frequent involvement of multiple **stakeholders** and organizations, scientific uncertainty and technical complexity, and the challenge of navigating unequal **power** and resources often within a highly public and/or political context (Emerson 2003). Given the broad spectrum of environmental collaboration and conflict resolution, and the complexity of environmental public policy decisions, it is not uncommon to see a significant overlap between the role and tools used by environmental mediators and the approaches of facilitators engaged with stakeholders who are not in a formal dispute (Moore 2013). That said, environmental mediation, like mediation in other fields, can be simply stated as facilitated negotiation between parties where a dispute already exists (Oretskin 2002; Daniels and Walker 2001).

References

American Arbitration Association, American Bar Association, and Association for Conflict Resolution. 2005. *Model Standards of Conduct for Mediators*. https://cdn.ymaws.com/acrnet.org/resource/resmgr/docs/MODEL_STANDARDS_OF_CONDUCT.pdf.
Centre for Humanitarian Dialogue (CHD). 2021. *Agro-Pastoral Mediation in the Sahel*. https://www.hdcentre.org/wp-content/uploads/2021/09/Agro-pastoral-mediation-in-the-Sahel.pdf.
Daniels, Steven E., and Gregg B. Walker. 2001. *Working Through Environmental Conflict: The Collaborative Learning Approach*. Praeger Publishers.
Emerson, Kirk, Tina Nabatchi, Rosemary O'Leary, and John Stephens. 2003. "The Challenges of Environmental Conflict Resolution." In *The Promise and Performance of Environmental Conflict Resolution*, edited by Rosemary O'Leary and Lisa B. Bingham. RFF Press.
Madrid, Jay J., and Jay G. Martin. 2002. "Advantages of Using Mediation and Arbitration to Settle Environmental Disputes." In *Environmental Dispute Resolution: An Anthology of Practical Solutions*, edited by Ann L. MacNaughton and Jay G. Martin. American Bar Association.
Moore, Christopher. 2003. *The Mediation Process: Practical Strategies for Resolving Conflict*, 3rd ed. Jossey-Bass.
Moore, Lucy. 2013. *Common Ground on Hostile Turf: Stories from an Environmental Mediator*. Island Press.
Oretskin, Nancy A., and Ann L. MacNaughton. 2002. "Is Mediation a Better Alternative for the Resolution of International Environmental Disputes?" In *Environmental Dispute Resolution: An Anthology of Practical Solutions*, edited by Ann L. MacNaughton and Jay G. Martin. American Bar Association.
Siegel, Joseph A. 2007. "Alternative Dispute Resolution in Environmental Enforcement Cases: A Call for Enhanced Assessment and Greater Use." *Pace Environmental Law Review* 24 (187): 196–99. https://doi.org/10.58948/0738-6206.1058.
Singer, Linda R. 1994. *Settling Disputes: Conflict Resolution in Business, Families, and the Legal System*. Westview Press.
Udall Foundation. n.d. *Udall Foundation*. Accessed July 1, 2025. https://www.udall.gov/OurPrograms/Institute/ProjectCaseSummary.aspx?Project=912.
United Nations (UN). 1945. Charter of the United Nations. art. 33, par. 1.
United Nations (UN). 2012. *Guidance for Effective Mediation*. https://unrcca.unmissions.org/sites/default/files/dpa_msu_guidance_english_web.pdf.

Militarism

Nico Edwards and Beatriz Arnal-Calvo

Militarism refers to the processes by which war and armed conflict become possible and military practices/rationalities bleed into civilian life (Mann 1987; Mabee and Vucetic 2018; Stavrianakis and Selby 2012). In short, militarism signifies "the preparation for war, its normalization and legitimation" (Stavrianakis and Stern 2018, 4).

As a practice, militarism depends upon and reinforces interconnected systems of oppression based on **gender**, class, race, ethnicity, and sexual orientation, among others (Rossdale 2019; Sjoberg and Via 2010). As a concept, it elucidates cases where military actors and interests enjoy an undue influence over society and politics, undermining the possibilities of resolving **conflict** or building **sustainable peace** (Cohn and Duncanson 2020). This influence often translates into disproportionate investments in the buildup for armed conflict at the expense of activities that address socio-environmental needs and help *prevent* conflict (Ide et al. 2021; MacKenzie and Wegner 2021; Rogaly 2024). Key examples are the interrelated militarization of the police (Weitzberg 2022; Lemieux 2023), border control (Campaign Against

Arms Trade 2021; Transnational Institute 2024), and climate policy (Akkerman et al. 2024; Buxton and Hayes 2015; Edwards 2023; Vogler 2023). Notably, militarism is not a natural or inevitable response but rather a deliberate choice with fundamental political, ethical, and ecological implications.

Attention to militarism has bearing for environmental peacebuilders, in theory and practice. **Environmental peacebuilding** recognizes the role and importance of environmental protection and natural resource management in triggering violent conflict or building the conditions for sustainable peace (Environmental Peacebuilding Association n.d.; Environmental Law Institute n.d.). Without healthy ecosystems and ensuring people's equal access to the means of life (**human security**), the prospects for maintaining healthy communities and intra- or interstate relations are slim. A common misunderstanding, however, is the idea that environmental conditions (change, degradation, "scarcity") cause conflict rather than the underlying social, economic, and political systems productive of those environmental conditions (Selby 2023; Selby et al. 2022). Militarism, in its socio-material facilitation of armed conflict and military **power**, is a central root cause behind **climate change**, **environmental degradation**, and resource competition (Crawford 2022; Edwards 2023).

Indeed, armed conflicts are often driven by the desire to control resource access, extraction, and use, or are triggered by socio-environmental inequities and injustices (Dunlap 2019, 2022; Dunlap and Brock 2022; Martinez-Alier 2002). The making and maintenance of weapons and military power require huge amounts of **natural resources**—from energy to minerals (Belcher et al. 2020; Selwyn 2020, 2022)—and exact heavy socio-environmental costs (Griffiths 2022; Griffiths and Redwood 2024): from military emissions (Bigger and Neimark 2017; Neimark et al. 2023; Parkinson and Cottrell 2022) to the environmental conflicts facing mining communities to the toxic legacies of weapons use and infrastructural destruction during conflict, which travel far beyond active battlefields and remain for generations after conflicts end (Griffiths and Rubaii 2024; Rubaii 2020).

Considering militarism in environmental peacebuilding requires recognizing the role of militaries and armed conflict in exacerbating climate disasters and ecosystem harm, and looking for the social, political, and economic root causes behind social inequity, resource conflicts, and environmental destruction. Militarism is one such root cause, along with other patriarchal, colonial, capitalist, racist, and classist norms upon which militarism depends. Attention to militarism requires asking tough questions about how "**security**" often means different things to state elites, military actors, and marginalized communities (Masco 1999; Nixon 2011), recognizing that defense policies, military spending, and arms production can work to exacerbate rather than mitigate socio-environmental inequities and vulnerabilities (Bigger et al. 2023). Finally, it means taking calls for arms control, disarmament, and demilitarization seriously as avenues for building sustainable peace and **environmental justice** at once (Edwards 2023).

References

Akkerman, Mark, Nick Buxton, and Todd Miller. 2024. "Global Climate Wall: How the World's Wealthiest Nations Prioritise Borders over Climate Action." https://www.tni.org/en/publication/global-climate-wall.

Belcher, Oliver, Patrick Bigger, Ben Neimark, and Cara Kennelly. 2020. "Hidden Carbon Costs of the 'Everywhere War': Logistics, Geopolitical Ecology, and the Carbon Boot-print of the US Military." *Transactions of the Institute of British Geographers* 45 (1): 65–80. https://doi.org/10.1111/tran.12319.

Bigger, Patrick, and Benjamin D. Neimark. 2017. "Weaponizing Nature: The Geopolitical Ecology of the US Navy's Biofuel Program." *Political Geography* 60: 13–22. https://doi.org/10.1016/j.polgeo.2017.03.007.

Bigger, Patrick, Khem Rogaly, Nick Pearce, and Ketaki Zodgekar. 2023. *Less War, Less Warming: A Reparative Approach to US and UK Military Ecological Damages | Report | Common Wealth*. Climate + Community Project and Common Wealth. https://www.common-wealth.org/publications/less-war-less-warming-a-reparative-approach-to-us-and-uk-military-ecological-damages.

Buxton, Nick, and Ben Hayes, eds. 2015. *The Secure and the Dispossessed: How the Military and Corporations Are Shaping a Climate-Changed World*. Illustrated edition. Pluto Press.

Campaign Against Arms Trade (CAAT). 2021. "Militarising Borders." January 21. https://caat.org.uk/challenges/borders/.

Cohn, Carol, and Claire Duncanson. 2020. "Women, Peace and Security in a Changing Climate." *International Feminist Journal of Politics* 22 (5): 742–62. https://doi.org/10.1080/14616742.2020.1843364.

Crawford, Neeta. 2022. *The Pentagon, Climate Change, and War: Charting the Rise and Fall of U.S. Military Emissions*. The MIT Press. https://mitpress.mit.edu/9780262047487/the-pentagon-climate-change-and-war/.

Dunlap, Alexander. 2019. "'Agro Sí, Mina NO!' The Tía Maria Copper Mine, State Terrorism and Social War by Every Means in the Tambo Valley, Peru." *Political Geography* 71: 10–25. https://doi.org/10.1016/j.polgeo.2019.02.001.

Dunlap, Alexander. 2022. "Weaponizing People in Environmental Conflicts: Capturing 'Hearts,' 'Minds,' and Manufacturing 'Volunteers' for Extractive Development." *Current Sociology* 71 (2): 275–98. https://doi.org/10.1177/00113921221086828.

Dunlap, Alexander, and Andrea Brock, eds. 2022. *Enforcing Ecocide: Power, Policing and Planetary Militarization*. Springer International Publishing. https://doi.org/10.1007/978-3-030-99646-8.

Edwards, Nico. 2023. *Resisting Green Militarism: Building Movements for Peace and Eco-Social Justice*. World Peace Foundation. https://worldpeacefoundation.org/wp-content/uploads/2024/04/Resisting-green-militarism-final.pdf.

Environmental Peacebuilding Association (EnPAx). n.d. "Books–Environmental Peacebuilding." Accessed 16 November 2024. https://www.environmentalpeacebuilding.org/library/books/.

Environmental Law Institute (ELI). n.d. "Environmental Peacebuilding." Accessed 16 November 2024. https://www.eli.org/environmental-peacebuilding.

Griffiths, Mark. 2022. "The Geontological Time-Spaces of Late Modern War." *Progress in Human Geography* 46 (2): 282–98. https://doi.org/10.1177/03091325211064266.

Griffiths, Mark, and Henry Redwood. 2024. "Late Modern War and the Geos." *International Political Sociology* 18 (2). https://doi.org/10.1093/ips/olae011.

Griffiths, Mark, and Kali Rubaii. 2024. "Late Modern War and the Geos: The Ecological 'Beforemaths' of Advanced Military Technologies." *Security Dialogue* 56 (1): 38–57. https://doi.org/10.1177/09670106241265636.

Ide, Tobias, Carl Bruch, Alexander Carius, et al. 2021. "The Past and Future(s) of Environmental Peacebuilding." *International Affairs* 97 (1): 1–16. https://doi.org/10.1093/ia/iiaa177.

Lemieux, Frederic. 2023. "Militarization of Law Enforcement in America." In *Oxford Research Encyclopedia of Criminology and Criminal Justice*. https://doi.org/10.1093/acrefore/9780190264079.013.62.

Mabee, Bryan, and Srdjan Vucetic. 2018. "Varieties of Militarism: Towards a Typology." *Security Dialogue* 1–2 (49): 96–108. https://www.jstor.org/stable/26395259.

MacKenzie, Megan, and Nicole Wegner, eds. 2021. *Feminist Solutions for Ending War*. Pluto Press. https://directory.doabooks.org/handle/20.500.12854/74939.

Mann, Michael. 1987. "The Roots and Contradictions of Modern Militarism." *New Left Review* I/162: 35–50. https:///doi.org/10.64590/avy.

Martinez-Alier, Joan. 2002. *The Environmentalism of the Poor: A Study of Ecological Conflicts and Valuation*. Edward Elgar Publishing.

Masco, Joseph. 1999. "States of Insecurity: Plutonium and Post-Cold War Anxiety." In *Cultures of Insecurity: States, Communities, and the Production of Danger*, edited by Jutta Weldes. University of Minnesota Press.

Neimark, Benjamin, Oliver Belcher, Kirsti Ashworth, and Reuben Larbi. 2023. "Concrete Impacts: Blast Walls, Wartime Emissions, and the US Occupation of Iraq." *Antipode* 56 (3): 983–1005. https://doi.org/10.1111/anti.13006.

Nixon, Rob. 2011. *Slow Violence and the Environmentalism of the Poor*. Harvard University Press.

Parkinson, Stuart, and Linsey Cottrell. 2022. *Estimating the Military's Global Greenhouse Gas Emissions*. Scientists for Global Responsibility and Conflict and Environment Observatory. https://ceobs.org/wp-content/uploads/2022/11/SGR-CEOBS_Estimating_Global_MIlitary_GHG_Emissions.pdf.

Rogaly, Khem. 2024. *A Lucas Plan for the Twenty First Century From Asset Manager Arsenal to Green Industrial Strategy*. Common Wealth. https://www.common-wealth.org/publications/a-lucas-plan-for-the-twenty-first-century-from-asset-manager-arsenal-to-green-industrial-strategy.

Rossdale, Chris. 2019. "Conceptualising Militarism." In *Resisting Militarism*, edited by Chris Rossdale. Edinburgh University Press. https://doi.org/10.3366/edinburgh/9781474443036.003.0003.

Rubaii, Kali. 2020. *Birth Defects and the Toxic Legacy of War in Iraq*. Middle East Report 296. MERIP. https://merip.org/2020/10/birth-defects-and-the-toxic-legacy-of-war-in-iraq-296/.

Selby, Jan. 2023. "Climate Change and Conflict." *Nature Reviews Earth and Environment*, January. https://www.academia.edu/106384064/Climate_change_and_conflict.

Selby, J., G. Daoust, and C. Hoffmann. 2022. *Divided Environments: An International Political Ecology of Climate Change, Water and Security*. Cambridge University Press. https://books.google.co.uk/books?id=_Up6EAAAQBAJ.

Selwyn, Daniel. 2020. *Martial Mining: The Links between Global Extractivism, the Arms Trade, and Warfare*. London Mining Network. https://londonminingnetwork.org/project/martial-mining-2020/.

Selwyn, Daniel. 2022. "Global Britain and London's Mega-Mining Corporations: Colonial Ecocide, Extractive Zones, and Frontiers of Martial Mining." In *Enforcing Ecocide: Power, Policing and Planetary Militarization*, edited by Alexander Dunlap and Andrea Brock. Springer International Publishing. https://doi.org/10.1007/978-3-030-99646-8_5.

Sjoberg, Laura, and Sandra Via. 2010. *Gender, War, and Militarism: Feminist Perspectives*. Praeger. https://pure.royalholloway.ac.uk/en/publications/gender-war-and-militarism-feminist-perspectives.

Stavrianakis, Anna, and Jan Selby. 2012. *Militarism and International Relations: Political Economy, Security, Theory*. Taylor & Francis Group. http://ebookcentral.proquest.com/lib/suss/detail.action?docID=1101439.

Stavrianakis, Anna, and Maria Stern. 2018. "Militarism and Security: Dialogue, Possibilities and Limits." *Security Dialogue* 49 (1–2): 3–18. https://doi.org/10.1177/0967010617748528.

Transnational Institute (TNI). 2024. "Border Wars." https://www.tni.org/en/topic/border-wars.

Vogler, Anselm. 2023. "Barking up the Tree Wrongly? How National Security Strategies Frame Climate and Other Environmental Change as Security Issues." *Political Geography* 105: 102893. https://doi.org/10.1016/j.polgeo.2023.102893.

Weitzberg, Keren. 2022. *A Very British Problem: The Evolution of Britain's Militarised Policing Industrial Complex*. NetPol and CAAT. https://netpol.org/2022/08/08/new-report-exposes-the-militarised-nature-of-british-policing/.

Mitigation

Stephanie Martinez

Mitigation in a climate change context refers to human interventions that limit emissions of greenhouse gases or reduce the levels of greenhouse gases in the atmosphere (International Panel on Climate Change 2023).

Mitigation takes many forms and is focused on addressing the drivers of **climate change** (IPCC 2022). Examples of interventions might include improving energy efficiency in buildings (reducing new emissions) or increasing the capacity of carbon sinks through forest restoration (reducing existing emissions) (Grubb et al. 2022; United Nations Development Programme [UNDP] 2024). In the context of environmental **governance** agreements, actors set and work toward emissions-reduction targets. The landmark Paris Agreement is a global example, while city-wide decisions such as the city of San Diego's Climate Action Plan in the United States are ways that climate mitigation efforts are operationalized locally (City of San Diego 2022; United Nations 2015). In the example of San Diego, the city committed to net zero greenhouse gas emissions by 2035 through actions such as phasing out natural gas usage in half of its municipal facilities, reducing waste generated by city operations, and capturing landfill methane (City of San Diego 2022).

Climate change and **conflict** share deep roots in terms of both drivers and solutions (Morales-Muñoz et al. 2023; Rüttinger et al. 2015). **Environmental peacebuilding** engages with mitigation by exploring how activities that lessen the risk of environmental conflict may also do so for climate change. A prominent example of this is Ousseyni Kalilou's investigation of gum arabic cultivation in Niger and the opportunities for community cohesion and **peacebuilding** that could be expanded around this nitrogen-fixing tree (Kalilou 2021). The purview of environmental peacebuilding extends beyond the usual bounds of climate change mitigation, encompassing complex socio-environmental dynamics that are not always directly attributed to the global crisis. Environmental peacebuilding activities historically excluded climate change despite compelling data that point to overlap in cases of countries coming out of **war** and countries facing elevated climate change vulnerabilities (Matthew 2014). The research base at this intersection is growing, particularly around indirect causal pathways between climate variability and conflict risk (von Uexkull and Buhaug 2021). While there is not yet consensus on whether climate disasters increase conflict risk, scholars like Ide et al. (2020) have produced research that identifies increased risk of armed conflicts immediately after climate-related disasters under highly contextual factors such as large populations, ethnic segregation, and underdevelopment (e.g., life expectancy, schooling, and standard of living as measured by the Human Development Index) (Ide et al. 2020; UNDP 2025).

Mitigation and environmental peacebuilding share common ground, especially around the centrality of public **participation**. Within climate mitigation scholarship, there is a call to mobilize demand-side approaches that consider the ways the public engages with emissions-producing services such as transportation, communication, and housing (Creutzig et al. 2022). Approaching mitigation via demand requires broad and equitable participation from all parts of society, which in turn facilitates decisions that can both address the climate crisis and human well-being (Creutzig et al. 2022). However, scholars also encourage an engagement with complexity when operationalizing what public engagement can look like in these contexts. For example, Sändig et al. (2024) argue that while local knowledge, narratives, networks, and collective identities are critical components of public participation, romanticizing the "local community" can homogenize actors and render invisible diverse and often marginalized segments of the population, such as women and migrant workers. In considering mitigation in environmental peacebuilding contexts, there is also a need for the field to pay attention to the ways that interventions might threaten local communities. Scholars highlight examples of hydropower and agricultural expansion projects in Myanmar, which have rationalized dispossession while increasing tension in conflict-affected regions (Borras et al. 2020; Krampe et al. 2024).

References

Borras, Saturnino M., Jennifer C. Franco, and Zau Nam. 2020. "Climate Change and Land: Insights from Myanmar." *World Development* 129: 104864. https://doi.org/10.1016/j.worlddev.2019.104864.

City of San Diego. 2022. *City of San Diego Climate Action Plan.*

Creutzig, Feliz, Joyashree Roy, Patrick Devine-Wright, et al. 2022. *Demand, Services, and Social Aspects of Mitigation.* Cambridge University Press. https://doi.org/10.1017/9781009157926.007.

Grubb, Michael, Chukwumerije Okereke, Jun Arima, et al. 2022. *Introduction and Framing.* Cambridge University Press. https://doi.org/10.1017/9781009157926.003.

Ide, Tobias, Michael Brzoska, Jonathan F. Donges, and Carl-Friedrich Schleussner. 2020. "Multi-Method Evidence for When and How Climate-Related Disasters Contribute to Armed Conflict Risk." *Global Environmental Change* 62: 102063. https://doi.org/10.1016/j.gloenvcha.2020.102063.

International Panel on Climate Change (IPCC). 2022. "Summary for Policymakers." In *Climate Change 2022–Mitigation of Climate Change: Working Group III Contribution to the Sixth Assessment Report of the Intergovernmental Panel on Climate Change*, edited by Priyadarshi R. Shukla, Jim Skea, Andy Reisinger, et al. Cambridge University Press. https://doi.org/10.1017/9781009157926.001.

International Panel on Climate Change (IPCC). 2023. *2023 Annex I: Glossary.* Edited by Andy Reisinger, Diego Cammarano, Andreas Fischlin, et al. https://doi.org/10.59327/IPCC.

Kalilou, Ousseyni. 2021. "Climate Change and Conflict in the Sahel: The Acacia Gum Tree as a Tool for Environmental Peacebuilding." *International Affairs* 97 (1): 201–18. https://doi.org/10.1093/ia/iiaa178.

Krampe, Florian, Dylan O'Driscoll, McKenzie Johnson, Dahlia Simangan, Farah Hegazi, and Cedric de Coning. 2024. "Climate Change and Peacebuilding: Sub-Themes of an Emerging Research Agenda." *International Affairs* 100 (3): 1111–30. https://doi.org/10.1093/ia/iiae057.

Matthew, Richard. 2014. "Integrating Climate Change into Peacebuilding." *Climatic Change* 123 (1): 83–93. https://doi.org/10.1007/s10584-013-0894-1.

Morales-Muñoz, Héctor, Leigh Martens, Katharina Löhr, et al. 2023. "Integrating Climate Mitigation and Environmental Peacebuilding Objectives through Sustainable Land Use Systems: Theory of Change and Indicators." *PLOS Climate* 2 (5): e0000075. https://doi.org/10.1371/journal.pclm.0000075.

Rüttinger, Lukas, Dan Smith, Gerald Stang, Dennis Tänzler, and Janani Vivekananda. 2015. *A New Climate for Peace: Taking Action on Climate and Fragility Risks: An Independent Report Commissioned by the G7 Members.* adelphi, International Alert, Wilson Center, and European Union Institute for Security Studies.

Sändig, Jan, Natalia Dalmer, Tobias Ide, and Anselm Vogler. 2024. "From Climate Conflicts to Environmental Peacebuilding: Exploring Local Dimensions." *Environment and Security* 2 (1): 3–20. https://doi.org/10.1177/27538796241231090.

United Nations Development Programme (UNDP). 2024. *What Is Climate Change Mitigation and Why Is It Urgent? | UNDP Climate Promise*. https://climatepromise.undp.org/news-and-stories/what-climate-change-mitigation-and-why-it-urgent.
United Nations Development Programme (UNDP). 2025. "Human Development Index (HDI)." In *Human Development Reports*. https://hdr.undp.org/data-center/human-development-index.
United Nations (UN). 2015. *Paris Agreement to the United Nations Framework Convention on Climate Change*. https://unfccc.int/sites/default/files/english_paris_agreement.pdf.
von Uexkull, Nina, and Halvard Buhaug. 2021. "Security Implications of Climate Change: A Decade of Scientific Progress." *Journal of Peace Research* 58 (1): 3–17. https://doi.org/10.1177/0022343320984210.

Monitoring

Amanda Woomer

Monitoring is an ongoing and systematic process of collecting, analyzing, and using information about an intervention's activities, effects, and context (Bruch and Woomer 2023).

Monitoring is generally composed of tracking changes in quantitative and qualitative **indicators** to understand progress (or lack of progress) toward their associated targets and baselines (Lai 2012). Data collection methods could include reviews, reflection on information, and open-ended, qualitative information about the results of an intervention or the broader context in which the intervention takes place. While there is often some overlap with **evaluation**, monitoring is generally descriptive and relies on tracking indicators at various points during an intervention rather than broader evaluative questions. Monitoring information is used continuously in the day-to-day management of an intervention to track progress against plans, functioning as evidence for strategic decision-making, **learning**, and results achievement (United Nations Development Group [UNDG] 2017). Monitoring can reveal whether the **theory of change** is still valid or needs to be adapted during implementation or as the context changes (UNDAF 2017; O'Flynn and Moberly 2017). Monitoring is a crucial tool to assess both the intended and unintended effects of an intervention, and whether the intervention needs to be reconsidered (Lemon and Pinet 2018).

Environmental peacebuilding interventions should rely on the full range of monitoring tools available, capturing information on the results of environmental or climate-related efforts as well as **peace** and **conflict** activities. For instance, tracking intangible but essential aspects such as changes in community perceptions or trust could incorporate indicators from social psychology research, which the practitioner guide "Monitoring Dialogue in Conflict-Affected Contexts" is based on (Baggerman et al. 2025). This guide also proposes a new framework for more strategic monitoring of dialogue-based peacebuilding processes. Given the insecure contexts in which environmental peacebuilding interventions take place, monitoring is particularly important as an early warning system, providing indications of potential problems before they escalate (UNDAF 2017). When collecting and sharing monitoring information, practitioners can utilize participatory ways of assessing an intervention's information **environment** and address concerns around transparency and data sensitivity (US Global Development Lab 2019). If done well, the monitoring process itself can support the objectives of an environmental peacebuilding intervention, such as by building relationships and trust between **stakeholders**. To do so, it is essential that monitoring is undertaken with **conflict sensitivity**, and that the safety and **security** of **stakeholders** is balanced with **participation** and transparency.

Acknowledgment

This definition has been adapted from Bruch and Woomer (2023).

References

Baggerman, Jill, Esra Cuhadar, and Kristen Wall. 2025. *Monitoring Dialogue in Conflict-Affected Contexts: A Guide for Practitioners*. United States Institute of Peace. https://www.alliancefor peacebuilding.org/afp-publications/monitoring-dialogue-conflict-affected-contexts.

Bruch, Carl, and Amanda Woomer. 2023. *Toolkit on Monitoring and Evaluation of Environmental Peacebuilding*. Environmental Law Institute, Environmental Peacebuilding Association, and United States Institute of Peace. https://m-and-e.environmentalpeacebuilding.org/toolkit/.

Lai, K. C. 2012. *Stocktaking of M&E and Management Information Systems: Selected Agricultural and Rural Development Projects in South Asia*. Food and Agricultural Organization Investment Centre. https://www.fao.org/3/i2883e/i2883e.pdf.

Lemon, Adrienne, and Mélanie Pinet. 2018. "Measuring Unintended Effects in Peacebuilding: What the Field of International Cooperation Can Learn from Innovative Approaches Shaped by Complex Contexts." *Evaluation and Program Planning* 68: 253–61. https://doi.org/10.1016/j.evalprogplan.2017.09.007.

O'Flynn, Maureen, and Clare Moberly. 2017. *Theory of Change*. International NGO Training and Research Centre. https://www.intrac.org/app/uploads/2017/01/Theory-of-Change.pdf.

U.S. Global Development Lab. 2019. *Considerations for Using Data Responsibly at USAID*. USAID. https://www.usaid.gov/sites/default/files/documents/15396/USAID-UsingDataResponsibly.pdf.

United Nations Development Group. 2017. *UNDAF Companion Guidance: Monitoring and Evaluation*. United Nations.

N

Natural Resources

Tom Cavanagh, Guido Trivellini, and Bernd Unger

Natural resources are the elements of the environment that are used by or are potentially useful to humans. These elements include land, water, forests, and biodiversity, as well as high-value resources like metals, minerals, gemstones, fuels, oil, diamonds, iron ore, rubber, and timber, alongside the ecosystem services that sustain them, such as nutrient and hydrological cycles, and the components of the Earth system in which they are embedded, such as the climate system and biosphere.

The category of "resource" is not a static or neutral descriptor. Instead, it is socially constructed, emerging through political, economic, and technological processes that render aspects of nature legible, extractable, and valuable for human use. In this sense, resources are inseparable from human labor, **governance** systems, and market relations—they exist in nature in various forms but only become "resources" through their integration into human systems of use and exchange. Thus, "resource" can be used in two registers: (1) an ecological sense that recognizes stocks, flows, reproduction, and interdependence within ecosystems, and (2) a sociopolitical sense in which things become "resources" through human valuation, institutions, and **power** relations. While policy debates often emphasize human use-value, we underline that ecological functions and dynamics are not reducible to human benefit; rather, "resources" are co-produced at the interface of ecological processes and social relations.

The value ascribed to natural resources is historically contingent and subject to social debate. One debate exists between the traditional resource use patterns and extractive practices, often framed in terms of "economic modernization" (Le Billon 2014). Alternative valuation and accounting models have emerged to resolve this tension, ranging from economically quantifying the avoided exploitation of resources to assessing payments for **ecosystem services** (PES).

The diverse characteristics of natural resources influence their role in **conflict** and **peace**. Natural resources can be renewable (e.g., **forests** and fisheries) or non-renewable (e.g., oil and minerals); non-renewable resources with high economic value are often linked to armed conflict (Johnson et al. 2021) while renewable are more often the subject of disputes and grievances which are rarely, if ever, the sole cause

DOI: 10.4324/9781003604143-15

of violent conflict (UN Interagency Framework Team for Preventive Action 2012). Also, the geographical location (whether a resource is proximate or distant) and the concentration (whether a resource has a point source or is diffuse) shape the types of conflicts that emerge around them (Le Billon 2001). These factors shape the political economy of extracting and transforming these resources into useful commodities. Furthermore, conflicts may be linked to ancillary aspects of the **environment** and natural resources, such as production, labor, trade routes, and other factors that make natural resources "valuable" (Fairhead 2001).

Production within this broader political economy is a central component of resources. While they exist in various forms within nature, it is through their human use that they become resources and, therefore, inseparable from human processes (Le Billon 2014). This resource transformation is embedded in transnational political economies, where extraction sites are linked to global commodity chains, financial flows, and regulatory regimes (Bebbington et al. 2018). These connections mean that resource-related conflicts are rarely confined to the local scale but are shaped by, and in turn shape, global markets and governance systems.

Natural resources and the politics surrounding them sit at the center of **environmental peacebuilding**. Moving beyond the simplistic notions of a **resource curse** or resource scarcity, natural resources are seen as an incentive for **cooperation** and peace (Dresse et al. 2021). Environmental peacebuilding recognizes the importance of economic development while ensuring resource **sustainability**, promoting post-**war** recovery through sustainable resource use and enhanced cooperation (Chavez-Miguel et al. 2022). The sustainable management of natural resources, both before and after conflict, is recognized as contributing to peace by promoting economic development, establishing institutions for collective action, and fostering trust and understanding between conflicting parties (Ide 2023).

At the same time, natural resources continue to bring risks to **peacebuilding** processes. Despite intentions, efforts to improve natural **resource governance** in post-conflict settings can reflect historical tensions and create new sources of conflict, particularly where communities have a high reliance on natural resources for their livelihoods (Beevers 2015). Furthermore, in its efforts to achieve win-win solutions around natural resource management, environmental peacebuilding risks depoliticizing underlying issues and overlooking complex power dynamics and socioeconomic conditions (Ide 2020).

Understanding natural resources as socially constructed and politically embedded is essential for analyzing their role in both conflict and peacebuilding. This perspective moves beyond resource determinism to consider how a broader political economy shapes the ways resources contribute to either renewed contestation or long-term stability. In post-conflict contexts, environmental peacebuilding provides a framework for harnessing natural resources to support recovery; however, its success hinges on addressing structural inequalities and ensuring that local communities maintain meaningful control over their resource futures.

References

Bebbington, Anthony. 2018. *Governing Extractive Industries: Politics, Histories, Ideas*. Oxford University Press.
Beevers, Michael D., 2015. "Governing Natural Resources for Peace: Lessons from Liberia and Sierra Leone." *Global Governance* 21 (2):227–46. https://doi.org/10.5555/1075-2846-21.2.227.
Chavez-Miguel, Giovanna, Michelle Bonatti, Álvaro Ácevedo-Osorio, Stefan Sieber, and Katharina Löhr. 2022. "Agroecology as a Grassroots Approach for Environmental Peacebuilding: Strengthening Social Cohesion and Resilience in Post-Conflict Settings with Community-Based Natural Resource Management." *GAIA–Ecological Perspectives for Science and Society* 31 (1): 36–45. https://doi.org/10.14512/gaia.31.1.9.
Dresse, Anaïs, Jonas Østergaard Nielsen, and Itay Fischhendler. 2021. "From Corporate Social Responsibility to Environmental Peacebuilding: The Case of Bauxite Mining in Guinea." *Resources Policy* 74: 102290. https://doi.org/10.1016/j.resourpol.2021.102290.
Fairhead, James. 2001. "International Dimensions of Conflict over Natural and Environmental Resources." In *Violent Environments*, 1st ed. Cornell University Press. https://books.google.com/books?hl=en&lr=&id=efcw5qe7mygC&oi=fnd&pg=PA213&dq=international+dimensions+of+conflict+over+natural+and+environmental+resources&ots=dOOnymfWtE&sig=-L9vPqXejVsZ9pNQhSSb5wDlqL0.
Ide, Tobias. 2020. "The Dark Side of Environmental Peacebuilding." *World Development* 127: 104777. https://doi.org/10.1016/j.worlddev.2019.104777.
Ide, Tobias. 2023. "Sampling Bias in Environmental Peacebuilding Research." *Environment and Security* 1 (1–2): 10–23. https://doi.org/10.1177/27538796221143850.
Johnson, McKenzie F., Luz A. Rodríguez, and Manuela Quijano Hoyos. 2021. "Intrastate Environmental Peacebuilding: A Review of the Literature." *World Development* 137: 105150. https://doi.org/10.1016/j.worlddev.2020.105150.
Le Billon, Philippe. 2001. "The Political Ecology of War: Natural Resources and Armed Conflicts." *Political Geography* 20 (5): 561–84. https://doi.org/10.1016/S0962-6298(01)00015-4.
Le Billon, Philippe. 2014. *Wars of Plunder: Conflicts, Profits and the Politics of Resources*. Oxford University Press. https://doi.org/10.1093/acprof:oso/9780199333462.001.0001.
UN Interagency Framework Team for Preventive Action. 2012. *Toolkit and Guidance for Preventing and Managing Land and Natural Resources Conflict. Renewable Resources and Conflict*. https://www.un.org/en/land-natural-resourcesconflict/pdfs/GN_Renew.pdf.

Natural Resource Conflict

Erik Post, Liliana Pimentel, and Andres Suarez

A natural resource conflict is a dispute or incompatibility related to natural resources that is caused by the actual or perceived opposition of needs, values, and interests that can involve the use of force and multiple forms of violence (United Nations Environment Programme 2009).

Natural resources are inextricably linked to many **conflict** situations. Between 1946 and 2008, at least 40% of intrastate conflicts were triggered, funded, or sustained by natural resources. Since 1990, revenues generated from the exploitation of natural resources in at least 35 countries have financed armed conflicts (Ahmadnia et al. 2022). In addition, many natural resource conflicts involve individuals and collectives known as **environmental defenders**, who protest unjust and unsustainable natural resource uses because of social and environmental reasons. The Environmental Justice Atlas documents over 4,000 cases of conflicts between civil society actors and state, corporate, and illicit actors. Because Indigenous communities are overrepresented in natural resource conflicts as land and environmental defenders, they often comprise those killed during conflicts, such as in Colombia and the Philippines (Scheidel et al. 2023).

Dominant approaches to natural resource conflicts often draw a direct causal link between conflicts and land, minerals, **forests**, water, fisheries, or other natural resources. These perspectives typically focus on resource scarcity or abundance as the primary driver of conflict and tend to refer to the so-called "**resource curse**," which posits that nations rich in natural resources are inherently prone to conflict. However, this oversimplifies the complex relationship between socio-**environmental change** and conflicts. It overlooks the social construction of resources and ignores how conflicts and resource management are embedded in historical and structural contexts. Moreover, these views obscure the connections between local conflicts in the "Global South" and global processes.

Starting in the 1990s, scholars from diverse backgrounds, such as in **political ecology**, critical **security** studies, and **peace** and conflict studies, but also practitioners and policymakers, started to challenge this direct causal conceptualization. From their critiques, the approaches of **Human Security** and **Ecological Distribution Conflicts** emerged, which are more attentive to the dynamic relationships between conflicts and natural resources and the interplay and **mediation** of local conflicts and global dynamics (Le Billon and Duffy 2018).

The framing of Ecological Distribution Conflicts highlights how conflicts are struggles over the burdens of pollution or sacrifices made to extract resources that are embedded in social and political economic inequalities and part of more expansive **gender**, class, caste, and ethnic struggles. Importantly, conflicts impact populations with differential access to natural resources in class-based, gendered, and racialized ways that can differ greatly between places. Beyond direct forms of **violence**, these conflicts can exhibit structural, cultural, ecological, and even epistemological and ontological dimensions of violence. These can be expressed through clashing valuation languages used to value natural resources and populations, such as incommensurabilities between economic values and territorial or **human rights**, sacredness, as well as ecological, aesthetic, and cultural values (Martinez-Aliers et al. 2010). Conflicts over natural resources are increasing due to a variety of factors that are driving global demand for natural resources.

Crucially, natural resource conflicts are not necessarily inherently negative but also represent potential avenues for attaining peace and **justice**. The **United Nations Environment Programme** (2016) has conceptualized the various roles that natural resources can play across the conflict cycle. They cause tensions and grievances, spark early warning/preventive diplomacy, ease tensions, fund armed conflict, function as an entry point for dialogue, represent economic incentives for peace or for undermining peace, enable confidence-building through **cooperation** on resource management, create jobs and support livelihoods, as well as generate revenues for **peacebuilding**. **The United Nations Security Council** has also recognized that natural resources have an important role in both conflict and reducing tensions and contributing to peace (Kapoor and Lewis 2023).

References

Ahmadnia, Shaade, Agathe Marie Christien, Phoebe Spencer, Tracy Hart, and Caio Cesar de Araujo Barbosa. 2022. *Defueling Conflict: Environment and Natural Resource Management as a Pathway to Peace.* World Bank.

Kapoor, Radhika, and Dustin A. Lewis. 2023. "The Practice of the UN Security Council Pertaining to the Environment and Armed Conflict, 1945–2021." *International Review of the Red Cross* 105 (924): 1646–67. https://doi.org/10.1017/S1816383123000425.

Le Billon, Philippe, and Rosaleen Duffy. 2018. "Conflict Ecologies: Connecting Political Ecology and Peace and Conflict Studies." *Journal of Political Ecology* 25 (1): 239–60. https://doi.org/10.2458/v25i1.22704.

Martinez-Alier, Joan, Giorgos Kallis, Sandra Veuthey, Mariana Walter, and Leah Temper. 2010. "Social Metabolism, Ecological Distribution Conflicts, and Valuation Languages." *Ecological Economics* 70 (2): 153–158. https://doi.org/10.1016/j.ecolecon.2010.09.024.

Scheidel, Arnim, Álvaro Fernández-Llamazares, Anju Helen Bara, et al. 2023. "Global Impacts of Extractive and Industrial Development Projects on Indigenous Peoples' Lifeways, Lands, and Rights." *Science Advances* 9 (23): 1–9. https://doi.org/10.1126/sciadv.ade9557.

United Nations Environment Programme (UNEP). 2009. *From Conflict to Peacebuilding.*

United Nations Environment Programme (UNEP). 2016. *Environmental Cooperation for Peacebuilding Programme: Final Report.*

Nature-Based Solutions

Liliana Pimentel, Marie Schellens, and Alan Channer

Nature-based solutions (NbS) are actions to protect, conserve, restore, sustainably use, and manage natural or modified terrestrial, freshwater, coastal, and marine ecosystems that effectively and adaptively address social, economic, and environmental challenges while simultaneously providing human well-being, ecosystem services, resilience, and biodiversity benefits (United Nations Environment Assembly [UNEA] 2022).

Nature-based solutions emerged in the early 2000s, gaining more structured attention around 2008 when the World Bank mentioned it in a major report (Cooney and Smith 2020; World Bank 2008). The **International Union for Conservation of Nature (IUCN)** further solidified this concept in 2016 by defining NbS and outlining its principles. Drawing from the IUCN's NbS definition, the United Nations Environment Assembly (UNEA-5), in its fifth session, endorsed a global NbS definition (UNEA 2022). This milestone resulted from years of negotiations between nations and civil society, providing an official reference for the **Convention on Biological Diversity** (CBD) and other international agreements (Cooney and Smith 2020). While ancestral knowledge has long embraced techniques akin to NbS, the formal recognition and structured framework provided by the IUCN helped mainstream the concept by offering clear guidelines and benchmarks, making it easier for governments, businesses, and NGOs to adopt and implement these solutions effectively. At the 16th Conference of the Parties (COP 16) of the CBD, delegates and civil society representatives aimed to advance inclusive and equitable NbS to enhance climate **resilience**, **gender** equality, and **biodiversity** net gains, working to implement the Kunming-Montreal Global Biodiversity Framework (Convention on Biological Diversity [CBD] 2022), and integrating **Indigenous knowledge** and rights into **conservation** and restoration to meet global **sustainability** goals and foster a **peacebuilding** culture (CBD 2024; International Institute for Sustainable Development 2024).

NbS implementation varies significantly across regions. In the Global North, NbS is usually integrated into urban planning, policy for **sustainable development**, and climate **adaptation**. Meanwhile, in the Global South, NbS is generally implemented to tackle immediate environmental challenges like flooding and **food security**, with an emphasis on local knowledge and community-based projects (Raymond et al. 2017; McPhearson et al. 2022; Kabisch and Frantzeskaki 2023; Kauark-Fontes et al. 2023).

Growing awareness of the environmental dimensions of **conflict** has prompted peacebuilding efforts to address conflict-related ecological degradation. NbS provided a welcome framework for responding to interwoven environmental, climatic, socioeconomic, and political challenges in fragile, conflict-affected contexts, aiming for **sustainable peacebuilding** efforts. NbS can play a role throughout the conflict cycle. Preventive measures include improved legal protection of the **environment** during armed conflicts (International Law Commission 2022), armed forces training on military guidelines for environmental protection (International Committee of the Red Cross 2020), and improved environmental analysis and response in humanitarian settings. During early conflict phases, improved environmental management and land rights can reduce competition and prevent tensions over **natural resources**, fostering **social cohesion**. During the escalation of conflict, NbS can guarantee vulnerable communities' access to essential resources. Environmental issues can be nonpolitical, more technical topics that facilitate dialogue among conflicting parties and help build trust. In post-conflict settings, NbS supports ecosystem recovery, pollution cleanup, sustainable resource use, and livelihoods while creating economic opportunities and reducing poverty and economic disparities. NbS also helps preserve cultural and social values tied to natural landscapes, usually a cross-boundary feature, fostering a sense of common identity and community cohesion. These strategies improve environmental stability, promote social cohesion, and enhance community resilience to environmental stresses, such as **climate change** and natural disasters—fundamental for long-lasting **peace**.

Critics highlight several controversies and shortcomings of NbS, including a lack of clarity in their definition, leading to misunderstandings, **stakeholder** misalignment, and potentially promoting false or dangerous solutions (Sowińska-Świerkosz and García 2022). Some oppose NbS for its instrumental view of nature, contrasting it with ecocentric or pluralistic perspectives that value diverse knowledge systems (Woroniecki et al. 2020; Indigenous Peoples Major Group 2023). Additionally, there are knowledge gaps regarding their effectiveness, divergent stakeholder values, and insufficient political support. Critics warn against NbS being used for greenwashing or dispossessing vulnerable communities if not implemented with justice and equity (Anguelovski and Corbera 2023; Nelson et al. 2020). Other risks include reliance on carbon offsets over emission reductions, lack of meaningful community consent and **participation**, normalization of political injustices, **human rights** violations, militarized conservation, and potential environmental harm (Wolters and Schellens 2024).

To ensure effective, conflict-sensitive implementation of NbS and peaceful local impacts in vulnerable communities worldwide, it is crucial to advance NbS for peace based on lessons learned from successful experiences and failures (Ecosystem for Peace 2024), as well as based on open discussions between marginalized voices where NbS for peace are implemented, influential international organizations, and all actors in between.

Acknowledgment

The definition adopted in this work is based on the global definition endorsed by the United Nations Environmental Assembly (UNEA-5) and used as the official reference for the CBD and other multilateral environmental agreements.

References

Anguelovski, Isabelle, and Esteve Corbera. 2023. "Integrating Justice in Nature-Based Solutions to Avoid Nature-Enabled Dispossession." *Ambio* 52 (1): 45–53. https://doi.org/10.1007/s13280-022-01771-7.

Convention on Biological Diversity (CBD). 2022. "Kunming-Montreal Global Biodiversity Framework." https://www.cbd.int/doc/decisions/cop-15/cop-15-dec-04-en.pdf.

Convention on Biological Diversity (CBD). 2024. "COP 16 Opens in Cali, Colombia." https://Www.Cbd.Int/Article/Cop-16-Opens-in-Cali-Colombia-2024.

Cooney, Rosie, and Mark Stafford Smith. 2020. "Nature-Based Solutions—A Short Overview for GEF STAP." https://cdn.unenvironment.org/stapgef/public/documents/NBS%20review%202020-05-15final.pdf.

Ecosystem for Peace. 2024. *Nature-Based Solutions for Peace and Security*. Edited by Pax and UNEP. https://solutions.ecosystemforpeace.org/.

Indigenous Peoples Major Group (IPMG). 2023. *An Approach to Nature-Based Solutions*. United Nations Environment Programme. https://wedocs.unep.org/bitstream/handle/20.500.11822/42532/Indigenous_Peoples_NbS_Submission.pdf?sequence=1&isAllowed=y.

International Committee of the Red Cross. 2020. *Guidelines on the Protection of the Natural Environment in Armed Conflict*. https://www.icrc.org/sites/default/files/document_new/file_list/guidelines_on_the_protection_of_the_natural_environment_in_armed_conflict_advance-copy.pdf.

International Institute for Sustainable Development (IISD). 2024. "Conference COP 16 | Can We Have It All? Approaches to Equitable and Inclusive Nature-Based Solutions Aiming for Climate Resilience, Gender Equality and Biodiversity Net Gain." https://www.iisd.org/events/cop16-equitable-inclusive-nature-based-solutions.

International Law Commission. 2022. *Draft Principles on Protection of the Environment in Relation to Armed Conflicts*. https://legal.un.org/ilc/texts/instruments/english/draft_articles/8_7_2022.pdf.

Kauark-Fontes, Beatriz, Livia Marchetti, and Fabio Salbitano. 2023. "Integration of Nature-Based Solutions (NBS) in Local Policy and Planning toward Transformative Change. Evidence from Barcelona, Lisbon, and Turin." *Ecology and Society* 28 (2). https://doi.org/10.5751/ES-14182-280225.

McPhearson, Timon, Nadja Kabisch, and Niki Frantzeskaki. 2023. *Nature-Based Solutions for Cities*. Edward Elgar Publishing. https://doi.org/10.4337/9781800376762.

Nelson, Donald R., Brian P. Bledsoe, Susana Ferreira, and Nathan P. Nibbelink. 2020. "Challenges to Realizing the Potential of Nature-Based Solutions." *Current Opinion in Environmental Sustainability* 45: 49–55. https://doi.org/10.1016/J.COSUST.2020.09.001.

Raymond, Christopher M., Niki Frantzeskaki, Nadja Kabisch, et al. 2017. "A Framework for Assessing and Implementing the Co-Benefits of Nature-Based Solutions in Urban Areas." *Environmental Science and Policy* 77: 15–24. https://doi.org/10.1016/J.ENVSCI.2017.07.008.

Sowińska-Świerkosz, Barbara, and Joan García. 2022. "What Are Nature-Based Solutions (NBS)? Setting Core Ideas for Concept Clarification." *Nature-Based Solutions* 2: 100009. https://doi.org/10.1016/J.NBSJ.2022.100009.

United Nations Environment Assembly. 2022. "Nature-Based Solutions for Supporting Sustainable Development." https://wedocs.unep.org/bitstream/handle/20.500.11822/39864

/NATUREBASED%20SOLUTIONS%20FOR%20SUPPORTING%20SUSTAINABLE%20DEVELOPMENT.%20English.pdf?sequence=1&isAllowed=y.
Wolff, Erich, Hanna A. Rauf, Loan Diep, Boonanan Natakun, Kris Kelly, and Perrine Hamel. 2022. "Implementing Participatory Nature-Based Solutions in the Global South." *Frontiers Sustainable Cities* 4. https://doi.org/10.3389/frsc.2022.956534.
Wolters, Elise, and Marie Schellens. 2024. "Nature in Action for Peace: Challenges and Opportunities to Address Environmental Dimensions of Conflict through Nature-Based Solutions." *Perspectives* by United Nations Environment Programme Civil Society Group 48. https://wedocs.unep.org/bitstream/handle/20.500.11822/44841/Issue48_Nature_in_Action_for_Peace_2024-02-06_V4.pdf?sequence=1&isAllowed=y.
World Bank. 2008. "Biodiversity, Climate Change, and Adaptation Nature-Based Solutions from the World Bank Portfolio." https://openknowledge.worldbank.org/server/api/core/bitstreams/5a3ca700-5c7e-5670-bb6c-dd9247a60d7c/content.
Woroniecki, Stephen, Hausner Wendo, Ebba Brink, et al. 2020. "Nature Unsettled: How Knowledge and Power Shape 'Nature-Based' Approaches to Societal Challenges." *Global Environmental Change* 65: 102132. https://doi.org/10.1016/J.GLOENVCHA.2020.102132.

Negative Peace

Jeannie Shroads and Nancy Boyer

Negative peace is the absence of war and direct, armed violence.

Coined by Johan Galtung (1964), negative peace describes the absence of a state of **war**, direct **violence**, or overt oppression. Examples of negative peace include a ceasefire agreement during a war or the presence of an armed force to dissuade unrest (Jakubowski 2021). Under Galtung's definition, direct violence may include psychological or physical harm, where the perpetrator(s) and victim(s) are known, and the harm is intentional. However, the **environmental peacebuilding** literature generally views negative peace as the absence of armed **conflict** in which "at least one social group [is] using physical violence in an organized way against another social group" (Ide 2021, 11). In both the **peace** field broadly and in environmental peacebuilding literature, negative peace is considered one end of a continuum that ranges from the cessation of hostilities on one side (Ide 2021; Conca and Dabelko 2002), to the presence of **positive peace** on the other side. Negative peace should be understood in conjunction with Galtung's complementary notion of positive peace. Galtung evolved his definition of positive peace from "the integration of human society" to the absence of structural violence (Grewal 2003). Negative peace is "the absence of tension while positive peace is the presence of **justice**" (King 1963). Historically, the orientation of indexes, study, and research focused on the achievement and management of negative peace; however, the tide has been turning more toward an orientation of positive peace (Diehl 2019).

Negative peace typically follows the cessation of armed conflict and is sometimes only short-term, reflecting a tendency to view a given conflict as resolved once the direct violence or fighting stops (Shields 2017). However, this viewpoint may undermine efforts for a broader peace by freezing the status quo without resolving underlying issues and tensions. For instance, the situation in Afghanistan under Taliban rule represents negative peace with historically low levels of armed conflict, yet **human security** and **human rights** are also at record lows (Ahmadi et al. 2024). The absence of direct violence and war does not ensure the absence of cultural violence, structural violence, injustice, repression, or rights restrictions, nor does it indicate how long-term peace may be strengthened or maintained (Council of Europe n.d.).

Krampe (2021) noted a few examples where **indicators** of positive peace were facilitated by environmental peacebuilding in the presence of negative peace: increased trust between conflicting ethnic groups in South Sudan after receiving facilitated veterinary services and resource management assistance (Food and Agriculture Organization 2017) and a judicial tribunal from Colombia's 2016 peace agreement recognizing the environmental and cultural harm caused by FARC against an Indigenous community (Yoshida 2021). While these examples may not demonstrate environmental peacebuilding as a determinant or lasting factor in a transition from a state of negative peace toward positive peace, they do provide some basis for environmental peacebuilding serving a supporting role as part of that transition.

References

Ahmadi, Belquis, Joyana Richer, Jill Baggerman, William Byrd, and Scott Worden. 2024. "Where is Afghanistan Three Years into Taliban Rule?" *United States Institute of Peace,* February 24, 2025. https://web.archive.org/web/20250224212925/https://www.usip.org/publications/2024/09/where-afghanistan-three-years-taliban-rule.

Conca, Ken, and Geoffrey Dabelko, eds. 2002. *Environmental Peacebuilding.* Johns Hopkins University Press.

Council of Europe. n.d. "Peace and Violence." Accessed January 27, 2025. https://www.coe.int/en/web/compass/peace-and-violence.

Diehl, Paul F. 2019. "Peace: A Conceptual Survey." *Oxford Research Encyclopedia of International Studies,* April 26. https://oxfordre.com/internationalstudies/view/10.1093/acrefore/9780190846626.001.0001/acrefore-9780190846626-e-515.

Food and Agriculture Organization (FAO). 2017. "Linking Community-Based Animal Health Services with Natural Resource Conflict Mitigation in the Abyei Administrative Area |KORE-Knowledge Sharing Platform on Resilience." Accessed August 2, 2025. https://www.fao.org/in-action/kore/good-practices/good-practices-details/en/c/1026219/.

Galtung, Johan. 1964. "An Editorial." *Journal of Peace Research* 1 (1): 1–4. https://doi.org/10.1177/002234336400100101.

Grewal, Baljit Singh. 2003. "Johan Galtung: Positive and Negative Peace." *Building Peace*, August 30. https://www.buildingpeaceforum.com/no/fred/Positive_Negative_peace.pdf

Ide, Tobias. 2021. *Environmental Peacebuilding.* Bogotá Instituto CAPAZ. https://www.instituto-capaz.org/wp-content/uploads/2021/03/DT-2-2021-Ingles-V3.pdf.

Jakubowski, Benjamin. 2021. *Positive and Negative Peace.* Growing Democracy. https://growingdemocracyoh.org/wp-content/uploads/2021/06/Positive-and-Negative-Peace.docx.pdf.

King, Jr., Martin Luther. 1963. "Martin Luther King, Jr: Open Letter." In *King, Martin Luther Jr.*, edited by Ali B. Ali-Dinar, April 16. https://www.africa.upenn.edu/Articles_Gen/Letter_Birmingham.html.

Krampe, Florian, Farah Hegazi, and Stacy D. VanDeveer. 2021. "Sustaining Peace through Better Resource Governance: Three Potential Mechanisms for Environmental Peacebuilding." *World Development* 144: 105508. https://doi.org/10.1016/j.worlddev.2021.105508.

Shields, Patricia M. 2017. "Limits of Negative Peace, Faces of Positive Peace." *The US Army War College Quarterly: Parameters* 47 (3). https://doi.org/10.55540/0031-1723.2868.

Yoshida, Keina, and Lina M. Céspedes-Báez. 2021. "The Nature of Women, Peace and Security: A Colombian Perspective." *International Affairs* 97 (1): 17–34. https://doi.org/10.1093/ia/iiaa173.

P

Participation

Annika Erickson-Pearson, Rebecca L. Farnum, and Asiem Sanyal

Participation refers to the intentional involvement of individuals, communities, and organizations in a particular process or approach. Within the peacebuilding field, participation most often refers to the active engagement of the individuals and institutions most affected by a particular project or intervention; in practice, active participation can act as a vehicle for the redistribution of power. Engagement can occur in the conceptualization, design, implementation, monitoring, and evaluation phases (Bruch and Woomer 2023).

Participation is a blanket term that is often applied widely. Sherry Arnstein's 1969 Ladder of Citizen Participation serves as one widely cited model for understanding participation in practice. The ladder itself has eight rungs, or levels, that transition from "nonparticipation" to "degrees of citizen power." Under "nonparticipation," the first two rungs are "Manipulation" and "Therapy." These rungs are forms of engagement directed back on the individual with no **power** conferred or a transformative outcome. The next three rungs, "informing," "consultation," and "placation," fall into a category Arnstein labels "degrees of tokenism," conferring what she calls "counterfeit power." These modes of engagement once again do not result in transformative outcomes, creating tokenistic one-way communication and little assurance that the results of consultations will be incorporated.

In Arnstein's Ladder, the three levels under "degrees of citizen power" are the following: "partnership," "delegated power," and "citizen control." These three rungs of the ladder engage actors in decision-making and planning processes. Power can be shared through partnership, delegated, or yielded entirely.

The ladder reminds us that not all forms of participation meaningfully engage individuals, communities, or other **stakeholders**. Examples of weak citizen participation in environmental policymaking and natural resource management appear in the literature, citing examples in which citizens were either not consulted or were consulted in a manner where their perspectives were not influential in decisions or outcomes (Aarts 2010; Simmons 2008). Genuine participation is both a tool for and an outcome of sustainable, **positive peace**; however, superficial or performative approaches are unlikely to be productive and may even exacerbate a problem or contribute to **conflict**. A truly transformative redistribution of power moves beyond predetermined

DOI: 10.4324/9781003604143-16

participation, as participation implies that an actor (participant) acts within (participates) a space or process developed by a power-holder.

The 2018 Agreement on Access to Information, Public Participation and Justice in Environmental Matters in Latin America and the Caribbean (Escazú Agreement) and the 1998 UNECE Convention on Access to Information, Public Participation in Decision-Making and Access to Justice in Environmental Matters (Aarhus Convention) offer examples of positive progress toward the protection of public participation in the environmental sphere. The Escazú Agreement, which specifically highlights the needs of vulnerable communities, is the first legally binding treaty requiring states to actively respond to attacks on **environmental defenders** through investigation and sanction. The Escazú Agreement also includes many provisions like the Aarhus Convention, which obliges parties in Europe and Central Asia to guarantee public participation in decision-making for projects, programs, and plans affecting the **environment**.

Within the **peacebuilding** sphere, participatory methods are often used to foster more inclusion. Pauls (2023) highlights ways to push participatory approaches beyond performative action and into genuine partnership through intentional project management by being conscious of power dynamics and hierarchies; budgeting for participatory methodologies from the outset (including potentially paying for participants' access to participate, such as for travel or childcare during activities), building in extra time and participant remuneration; and adjusting donor and project team expectations, along with the program outcomes, as needed to protect the integrity of the **participatory process**.

Future work on participation in the **environmental peacebuilding** arena might explore (1) how to move from 'participation' to power-transformative collaboration and (2) how to meaningfully engage with strategically undervalued communities, future generations, and non-human actors such as nature itself.

References

Aarts, Noelle, and Cees Leeuwis. 2010. "Participation and Power: Reflections on the Role of Government in Land Use Planning and Rural Development." *The Journal of Agricultural Education and Extension* 16 (2): 131–45. https://doi.org/10.1080/13892241003651381.

Arnstein, Sherry R. 1969. "A Ladder of Citizen Participation." *Journal of the American Institute of Planners* 35 (4): 216–24. https://doi.org/10.1080/01944366908977225.

Bruch, Carl, and Amanda Woomer. 2023. *Toolkit on Monitoring and Evaluation of Environmental Peacebuilding*. Environmental Law Institute, Environmental Peacebuilding Association, and United States Institute of Peace. https://m-and-e.environmentalpeacebuilding.org/toolkit/.

Pauls, Evelyn. 2023. "Participatory Methods in Peacebuilding Work." *Berghof Foundation*, March 20. https://berghof-foundation.org/library/participatory-methods-in-peacebuilding-work.

Simmons, W. Michele. 2008. *Participation and Power: Civic Discourse in Environmental Policy Decisions*. SUNY Press.

Participatory Process

Larissa Jane Houston and Jeannie Shroads

Participatory process is the practice of consulting and involving relevant stakeholders in the agenda-setting, decision-making, and policy-forming activities of organizations or institutions responsible for policy development (Hassenforder et al. 2015). It also emphasizes the inclusion of diverse perspectives, especially from vulnerable and marginalized communities and voices (Wheeler et al. 2020).

Participatory processes have been included and considered across various fields and have developed more recently in terms of environmental policy design. However, the practice of participatory engagement became popular from the 1930s to the 1970s with the introduction of **participation** in decision-making regarding **cities** and housing developments (Guldi 2017). The introduction of these participatory processes sought to bring about greater **empowerment** of local communities and voices by challenging traditional structures that forced top-down decision-making.

Participatory processes can be useful in providing greater fairness in decision-making among **stakeholders** and better acceptance of final outcomes among those who do not agree with final decisions (Werner and Marien 2022).

Within the context of **environmental peacebuilding**, participatory processes are used to foster dialogue, build trust, and create shared visions for sustainable resource management (Das 2023). Participatory processes can take various forms, such as community-based natural resource management (Chavez-Miguel et al. 2022), participatory environmental impact assessments (Jong and Sjöstedt 2023), and collaborative watershed management (Tawfik 2021). Through seeking to incorporate local and **Indigenous knowledge** with scientific expertise, participatory processes can lead to more contextually appropriate and sustainable solutions (Vojno 2022). Participatory processes are thus a crucial element in **peacebuilding**, especially considering the potential to transform and build relationships between conflicting parties (Jarraud and Lordos 2012). By engaging many **stakeholders** in structured dialogues, participatory processes can assist in building understanding, identifying common interests, and developing mutually beneficial solutions to environmental challenges.

The significance of participatory processes in environmental peacebuilding cannot be understated and provides a myriad of benefits (Hachmann et al. 2023). First, they enhance the legitimacy and local ownership of environmental initiatives, resulting in a greater likelihood of success in implementation and long-term **sustainability**. Second, they provide a platform for capacity building, contributing to the development of skills in negotiation, problem-solving, and environmental management within the community. Moreover, they contribute to broader goals of peacebuilding by promoting social inclusion, equity, and democratic **governance**.

Implementing effective participatory processes, especially within the context of environmental peacebuilding, can, however, be a challenge (Hassenforder et al. 2015); for example, issues such as **power** imbalances, cultural differences, and the technical complexity of environmental issues may hinder meaningful participation. Participatory processes also take substantial time and other resources, including those of the participants themselves. Addressing these challenges requires careful design of participatory mechanisms, skilled facilitation, and a commitment to ongoing **learning** and **adaptation**.

References

Chavez-Miguel, Giovanna, Michelle Bonatti, Álvaro Ácevedo-Osorio, Stefan Sieber, and Katharina Löhr. 2022. "Agroecology as a Grassroots Approach for Environmental Peacebuilding: Strengthening Social Cohesion and Resilience in Post-Conflict Settings with Community-Based Natural Resource Management." *GAIA—Ecological Perspectives for Science and Society* 31 (1): 36–45. https://doi.org/10.14512/gaia.31.1.9.

Das, Onita. 2023. "Sustainable Development and Environmental Peacebuilding." In *Research Handbook on International Law and Environmental Peacebuilding*. Edward Elgar Publishing. https://www.elgaronline.com/edcollchap-oa/book/9781789906929/book-part-9781789906929-9.xml.

Guldi, Jo. 2017. "A History of the Participatory Map." *Public Culture*, January 1. https://doi.org/10.1215/08992363-3644409.

Hachmann, Samyra, Katharina Löhr, Héctor Morales-Muñoz, Luca Eufemia, Stefan Sieber, and Michelle Bonatti. 2023. "Conceptualizing Community-Based Environmental Peacebuilding in Cesar, Colombia." *Human Ecology* 51 (2): 221–35. https://doi.org/10.1007/s10745-023-00399-9.

Hassenforder, Emeline, Alex Smajgl, and John Ward. 2015. "Towards Understanding Participatory Processes: Framework, Application and Results." *Journal of Environmental Management* 157: 84–95. https://doi.org/10.1016/j.jenvman.2015.04.012.

Jarraud, Nicolas Stephane, and Alexandros Lordos. 2012. "Participatory Approaches to Environmental Conflict Resolution in Cyprus." *Conflict Resolution Quarterly* 29 (3): 261–81. https://doi.org/10.1002/crq.21040.

Jong, Daniëlla Dam-de, and Britta Sjöstedt. 2023. "The International Legal Dimensions of Environmental Peacebuilding." In *Research Handbook on International Law and Environmental Peacebuilding*. Edward Elgar Publishing. https://www.elgaronline.com/edcollchap-oa/book/9781789906929/book-part-9781789906929-7.xml.

Tawfik, Rawia. 2021. "The Limits of Environmental Peacebuilding: Challenges to Cooperation in the Eastern Nile." In *The Politics of Peacebuilding in Africa,* edited by Thomas Kwasi Tieku, Amanda Coffie, Mary Boatemaa Setrana, Akin Taiwo. Routledge.

Vojno, Natalija. 2022. "Braiding Knowledge Systems as Environmental Peacebuilding: A Four-Dimensional Analysis for Co-Applying Indigenous and Non-Indigenous Worldviews." *Systemic Design Association*, October. https://rsdsymposium.org/environmental-peacebuilding/.

Werner, Hannah, and Sofie Marien. 2022. "Process vs. Outcome? How to Evaluate the Effects of Participatory Processes on Legitimacy Perceptions." *British Journal of Political Science* 52 (1): 429–36. https://doi.org/10.1017/S0007123420000459.

Wheeler, Joanna, Jackie Shaw, and Jo Howard. 2020. "Politics and Practices of Inclusion: Intersectional Participatory Action Research." *Community Development Journal* 55 (1): 45–63. https://doi.org/10.1093/cdj/bsz036.

Pastoralism/Transhumance

Toni-Giovanni Pegurri, Emmanuelle Veuillet, and Jeremy Allouche

Pastoralism is a livelihood system centered on the extensive breeding and seasonal movement of herds (transhumance), enabling the use of natural pastures following seasonal variations and resource availability. Transhumance may occur within or across national borders.

Pastoralism is frequently described as being "in crisis" across various contexts, due to recent economic, political, and environmental transformations. Transhumance, as a particular form of pastoralism and nomadism, finds itself generally at odds with mainstream modes of capitalist agricultural production, regimes of private property (Scoones 2021), and nation-state policies around borders, taxation, and migration that are adopted in many countries (Allouche et al. 2024). Within the African continent, whereas pastoralism and transhumance are still widespread and commonly practiced, they are increasingly associated with **conflict** and environmental harm. The presence of herders and their transhumance activities is likely to become a source of tensions and violent conflicts whenever people's livelihoods, production systems, or resource distribution are affected (Brottem 2020). Moreover, pastoralists are also considered potential threats to states and governments, regarding their mobility and supposed implication in international crime networks. The concept of "New Fringe Pastoralism" emerged to emphasize the active involvement of certain pastoralist groups in criminal activities and, in the case of West Africa, jihadist movements. The term appeared in 2017 in the working networks of the United Nations Economic Commission for Africa (United Nations Economic Commission for Africa 2017) and has since gained traction within the expertise sphere and policy forums.

Previous debates around farmers and herders' conflicts were very much influenced by a substantial body of analytical work which has linked the occurrence of conflicts to resource scarcity and **climate change** (Schwarz and Randall 2003; Homer-Dixon 2010; Adano et al. 2012; Ralston 2015). These neo-Malthusian and environmental approaches first emerged in the 1990s (Bonfiglioli 1991; Bernus 1995) and gained significant ground within the development aid community and policymaking circles. However, the prevailing consensus in the scientific literature suggests that this interpretation should be qualified (Benjaminsen and Ba 2019). Current debates over farmer-herder conflicts underline the necessity to adopt a more multidimensional and complex view of the actors involved and the genesis of these conflicts. This allows for going beyond using monolithic, sometimes normative, static categories such as "herders," "farmers," or "non/native" (Krätli and Toulmin 2020) and to better acknowledge the political economy in which they are embedded and inequalities that exist among and within these groups (Turner 2004; Veuillet 2020). Moreover, analyzing local resource access and distribution, as well as social and political endowments of these actors, is key to understanding potential marginalization and stereotyping of nomadic groups (Eke 2020). Pastoralists are not merely opportunistic actors but are also a vulnerable group (Ejiofor 2022). Similarly, considering broader sociopolitical and historical dynamics, the literature shows how relations between pastoralist communities and non-state armed groups tend to be fluid and evolutive (Bobbo 2022; Assanvo 2023).

Following these considerations, two predominant interpretations have emerged. The first, advocated by Tor Benjaminsen and Boubacar Ba, is rooted in the **political ecology** of conflicts. It stresses the role of land management policies, government interventions, and the unequal distribution of resources in the emergence of conflicts (Benjaminsen and Ba 2019; Benjaminsen and Ba 2021). The second perspective, primarily supported by Lief Brottem, highlights that climate change, combined with agricultural expansion, does not lead to conflicts in a direct and linear way. However, it unsettles local agreements and norms and weakens the ability of authorities to regulate the disputes, thereby increasing the risk of **violence** (Brottem 2016; Brottem and Turner 2024).

This new body of literature helps to follow a non-essentialist and historically situated approach to study pastoralism. It also underlines how climate change and environmental disasters could strategically be employed by various actors to justify and legitimize unequal access to **natural resources**, exclusion, or even large-scale dispossession. Indeed, (un)willingly, climate change could become a political instrument, used at the expense of both farmers and herders. Thus, more in-depth and systematic research on pastoral mobility, its continuous transformation, and its relation to climate change is crucial for **environmental peacebuilding**.

References

Adano, Wario, Ton Dietz, Karen Witsenburg, and Fred Zaal. 2012. "Climate Change, Violent Conflict and Local Institutions in Kenya's Drylands." *Journal of Peace Research* 49 (1): 65–80. https://doi.org/10.1177/0022343311427344.

Allouche, Jeremy, Cyprien Yao Yao, and Kando Soumahoro Amédée. 2024. "Rethinking 'Farmer–Herder' Conflicts in the Ivorian Internal Frontier." *African Affairs* 123 (493): 449–67. https://doi.org/10.1093/afraf/adae027.

Assanvo, William. 2023. *Links Between Violent Extremism and Illicit Activities in Côte d'Ivoire*. ISS Africa report. https://issafrica.org/research/west-africa-report/links-between-violent-extremism-and-illicit-activities-in-cote-divoire.

Benjaminsen, Tor, and Boubacar Ba. 2021. "Fulani-Dogon Killings in Mali: Farmer-Herder Conflicts as Insurgency and Counterinsurgency." *African Security* 14 (1): 1–23. https://doi.org/10.1080/19392206.2021.1925035.

Benjaminsen, Tor, and Boubacar Ba. 2019. "Why do Pastoralists in Mali Join Jihadist Groups? A Political Ecological Explanation." *The Journal of Peasant Studies* 46 (1): 1–20. https://doi.org/10.1080/03066150.2018.1474457.

Bernus, Edmond. 1995. "Pasteurs Face à la Sécheresse: Rebondir ou Disparaître ? / Shepherds Faced with Drought: Survival or Disappearance?" *Géocarrefour* 70 (3–4): 255–59. https://doi.org/10.3406/geoca.1995.4220.

Bobbo, Moussa. 2022. *Boko Haram dans la Région de l'Extrême-Nord du Cameroun: L'arbre Qui Cache la Forêt*. Notes de l'Ifri. https://www.ifri.org/fr/notes/boko-haram-dans-la-region-de-lextreme-nord-du-cameroun-larbre-qui-cache-la-foret#how-to-use.

Bonfiglioli, Angelo. 1991. "Mobilité et Survie. Les Pasteurs Sahéliens Face aux Changements de Leur Environnement." In *Savoirs Paysans et Développement*, edited by Georges Dupré. Karthala.

Brottem, Leif. 2020. "Pastoral Resource Conflict in the Context of Sudano–Sahelian Security Crises: A Critical Review of Research." *African Security* 13 (4): 380–402. https://doi.org/10.1080/19392206.2020.1871291.

Brottem, Leif. 2016. "Environmental Change and Farmer-Herder Conflict in Agro-pastoral West Africa." *Human Ecology* 44 (5): 547–63. https://doi.org/10.1007/s10745-016-9846-5.

Brottem, Lief and Matthew D. Turner. 2024. "Violent Farmer-Herder Conflicts in West Africa: Landscape-Livelihood Interactions and the Political Ecology of Customary Dispute Resolution." *The Journal of Peasant Studies* 52 (3): 574–96. https://doi.org/10.1080/03066150.2024.2399139.

Ejiofor, Promise Frank. 2022. "'We Don't Have Anything': Understanding the Interaction Between Pastoralism and Terrorism in Nigeria." *Conflict, Security and Development* 22 (4): 345–85. https://doi.org/10.1080/14678802.2022.2122695.

Eke, Surulola. 2020. "'Nomad Savage' and Herder–Farmer Conflicts in Nigeria: The (Un)making of an Ancient Myth." *Third World Quarterly* 41 (5): 745–63. https://doi.org/10.1080/01436597.2019.1702459.

Hagmann, Tobias. 2005. "Confronting the Concept of Environmentally Induced Conflict." *Peace, Conflict and Development* 6: 1–22. https://forskning.ruc.dk/en/publications/confronting-the-concept-of-environmentally-induced-conflict.

Homer-Dixon, Thomas. 2010. *Environment, Scarcity, and Violence*. Princeton University Press.

Krätli, Saverio, and Camilla Toulmin. 2020. *Farmer-Herder Conflict in Africa: Re-thinking the Phenomenon?* International Institute for Environment and Development. https://www.iied.org/17753iied.

Krätli, Saverio, and Camilla Toulmin. 2020. *Farmer-Herder Conflict in Africa: Re-thinking the Phenomenon?* International Institute for Environment and Development. https://www.iied.org/17753iied.
Ralston, Laura. 2015. *Conflict and Climate: A Micro-level Analysis*. World Bank.
Schwarz, Peter, and Doug Randall. 2003. *An Abrupt Climate Change Scenario and its Implications for United States National Security*. California Institute of Technology.
Scoones, Ian. 2021. "Pastoralists and Peasants: Perspectives on Agrarian Change." *The Journal of Peasant Studies* 48 (1): 1–47. https://doi.org/10.1080/03066150.2020.1802249
Turner, Matthew. 2004. "Political Ecology and the Moral Dimensions of 'Resource Conflicts': The Case of Farmer–Herder Conflicts in the Sahel." *Political Geography* 23 (7): 863–89. https://doi.org/10.1016/j.polgeo.2004.05.009.
United Nations Economic Commission for Africa (UNECA). 2017. *New Fringe Pastoralism: Conflict and Insecurity and Development in the Horn of Africa and the Sahel*. https://repository.uneca.org/handle/10855/23727.
Veuillet, Emmanuelle. 2020. "Le Grain Contre le Bétail: La Contestation d'un Ordre Socio-Politique Imposé. Relecture des Conflits Agropastoraux de la Région de Mundri, au Soudan du Sud." *Revue Internationale des Études du Développement* 243 (3): 89–113. https://doi.org/10.3917/ried.243.0089.

Peace

Laura E. R. Peters, Emily Sample, and Ty Terlaak

Peace is a dynamic, context-specific, and ongoing process that supports the absence of violence in all its forms and fosters improved attitudes, behaviors, relationships, and structures toward coexistence and mutual understanding on multiple levels.

Peace studies emerged as an academic discipline in the 1950s, building on long-term interest in peace and **war** in the field of international relations and an increased policy and practice focus on peace following the end of World War II. Despite this history, the concept of peace has received less scholarly attention than topics related to war and **violence** (Gleditsch et al. 2014), and conceptualizations and measurements of peace are still debated.

Johan Galtung, a pioneer of peace research, conceptualized peace into the two broad categories of **negative peace** and **positive peace**, denoting the absence of direct violence and the absence of direct and structural violence, respectively (Galtung 1969). Other **peacebuilding** scholars have emphasized that peace is an ongoing process rather than a state, event, or outcome (Lederach 1998). Recent advancements have been made to explore the qualities or varieties of peace, including the ideational, relational, and situational dimensions of peace (Jarstad et al. 2019). For instance, on the local level, "everyday peace **indicators**" are derived from the daily lived experiences of affected peoples (Pamina 2018; Mac Ginty 2013). Change spectrums are a new way to track progress, stagnation, or regression of peace outcomes during dialogue processes (Baggerman et al. 2025). Globally, peace has also been measured with indices such as the Institute for Economics & Peace's annual Global Peace Index or Ecological Threat Report (Institute for Economics and Peace 2025; Institute for Economics and Peace 2024).

As the field of peace studies has evolved, the assumptions underlying peace have been increasingly challenged, leading to an expanded diversity of understandings of peace.

For example, peace theories rooted in liberal democracies and free-market capitalism—including liberal, democratic, and capitalist peace theories—are critiqued for implicitly valuing system stability, including through oppression and other forms of violence to maintain stability, over the more emancipatory potentials of peace practice (Kühn 2012; Leonardsson 2024). Moreover, practitioner-academics have pointed out that top-down imposed peace often has little representation in the local and everyday experience of people affected by ongoing, recent, and/or historical situations of violence and war (Autesserre 2014; Mac Ginty 2014; Pamina 2018).

Feminist peace scholars advocate for a reimagining of peace, seeking a complex and comprehensive approach to peacebuilding that also works to address and dismantle interconnected systems of oppression (Sapiano 2024). This research examines peace and violence as interconnected, ongoing processes shaped by systems of **power** operating at multiple, interrelated scales. While **gender** is a central lens, this framework also interrogates how intersecting structures of power and positionality—such as race, class, sexuality, and disability—interact to reinforce or challenge dynamics of violence and peacebuilding (Väyrynen et al. 2021). These interlocking dynamics are critical to **environmental peacebuilding**, which is deeply connected to how people experience and interact with their local, national, and global **environment**. Additionally, there has been a recent recognition of Indigenous peace and **Indigenous knowledge**, which approach the relationship between peace and violence differently than Western tradition (Nepram 2017), including related to environmental peacebuilding (Rodríguez and Inturias 2018).

Numerous environmental peacebuilding scholars have advocated for the field to adopt positive peace framings (e.g., Fisher et al. 2021; Simangan et al. 2023). Barron (2024) suggests the environmental peacebuilding field must commit to deeper structural analyzes of societies experiencing **conflict**, so they can address how structural environmental injustices are being used to maintain existing oppression. In a meta-analysis of empirical environmental peacebuilding literature, Ide (2019) represented the continuum from negative to positive peace progressing from the absence of violent conflict to symbolic rapprochement to substantial integration. A literature review tracking environmental peacebuilding in intrastate conflicts identified four dimensions of peace—absence of violence, shared identity, capability, and substantial integration—and found that bottom-up peacebuilding initiatives addressing economic livelihood development, political inclusion, trust and **cooperation**, and resource **sustainability** were most effective at fostering sustainable, positive peace (Johnson et al. 2021).

References

Autesserre, Séverine. 2014. *Peaceland: Conflict Resolution and the Everyday Politics of International Intervention.* Cambridge University Press. https://doi.org/10.1017/CBO9781107280366.

Baggerman, Jill, Esra Cuhadar, and Kristen Wall. 2025. *Monitoring Dialogue in Conflict-Affected Contexts: A Guide for Practitioners.* United States Institute of Peace. https://www.allianceforpeacebuilding.org/afp-publications/monitoring-dialogue-conflict-affected-contexts.

Barron, Elsa. 2024. "On Dialogue and Beyond: Positive Environmental Peacebuilding in Palestine." *The Journal of Social Encounters* 8 (1): 21–28. https://doi.org/10.69755/2995-2212.1239.

Firchow, Pamina. 2018. *Reclaiming Everyday Peace: Local Voices in Measurement and Evaluation After War.* Cambridge University Press.

Fisher, Joshua, Poonam Arora, Siqi Chen, Sophia Rhee, Tempest Blaine, and Dahlia Simangan. 2021. "Four Propositions on Integrated Sustainability: Toward a Theoretical Framework to Understand the Environment, Peace, and Sustainability Nexus." *Sustainability Science* 16: 1125–45. https://doi.org/10.1007/s11625-021-00925-y.

Galtung, Johan. 1969. "Violence, Peace, and Peace Research." *Journal of Peace Research* 6 (3): 167–91. https://doi.org/10.1177/002234336900600301.

Gleditsch, Nils Petter, Jonas Nordkvelle, and Håvard Strand. 2014. "Peace Research – Just the Study of War?" *Journal of Peace Research* 51 (2): 145–58. http://dx.doi.org/10.1177/0022343313514074.

Ide, Tobias. 2019. "The Impact of Environmental Cooperation on Peacemaking: Definitions, Mechanisms, and Empirical Evidence." *International Studies Review* 21 (3): 327–46. https://doi.org/10.1093/isr/viy014.

Institute for Economics and Peace (IEP). 2024. *Ecological Threat Report 2024: Analysing Ecological Threats, Resilience & Peace*. IEP. Accessed October 1, 2025. https://www.visionofhumanity.org/wp-content/uploads/2024/10/ETR-2024-web.pdf.

Institute for Economics and Peace (IEP). 2025. *Global Peace Index 2025: Identifying and Measuring the Factors that Drive Peace*. IEP. Accessed June 15, 2025. https://www.economicsandpeace.org/wp-content/uploads/2025/06/GPI-2025-web.pdf.

Jarstad, Anna, Niklas Eklund, Patrik Johansson, et al. 2019. *Three Approaches to Peace: A Framework for Describing and Exploring Varieties of Peace*. Umeå University. https://urn.kb.se/resolve?urn=urn:nbn:se:umu:diva-165979.

Johnson, McKenzie F., Luz A. Rodríguez, and Manuela Quijano Hoyos. 2021. "Intrastate Environmental Peacebuilding: A Review of the Literature." *World Development* 137: 105150. https://doi.org/10.1016/j.worlddev.2020.105150.

Kühn, Florian P. 2012. "The Peace Prefix: Ambiguities of the Word 'Peace.'" *International Peacekeeping* 19 (4): 396–409. https://doi.org/10.1080/13533312.2012.709785.

Lederach, John Paul. 1998. *Building Peace: Sustainable Reconciliation in Divided Societies*. United States Institute of Peace.

Leonardsson, Hanna. 2024. "Peace Through Violence and Violence Through Peace: Peacebuilding Practices and a Conflictual Peace in Lebanon." *Geopolitics* 29 (5): 1538–58. https://doi.org/10.1080/14650045.2023.2270936.

Mac Ginty, Roger. 2013. "Indicators+: A Proposal for Everyday Peace Indicators." *Evaluation and Program Planning* 36 (1): 56–63. https://doi.org/10.1016/j.evalprogplan.2012.07.001.

Mac Ginty, Roger. 2014. "Everyday Peace: Bottom-Up and Local Agency in Conflict-Affected Societies." *Security Dialogue* 45 (6): 548–64. https://doi.org/10.1177/0967010614550899.

Nepram, Binalakshmi. 2017. "Indigenous Women of Northeast India at the Forefront of a Strong Non-Violent Peace Movement." In *Indigenous Peoples' Rights and Unreported Struggles: Conflict and Peace*, edited by Elsa Stamatopoulou. Institute for the Study of Human Rights, Columbia University.

Rodríguez, Iokiñe, and Mirna Liz Inturias. 2018. "Conflict Transformation in Indigenous Peoples' Territories: Doing Environmental Justice with a 'Decolonial Turn'." *Development Studies Research* 5 (1): 90–105. https://doi.org/10.1080/21665095.2018.1486220.

Sapiano, Jenna. 2024. "What is Feminist Peace? Half the Answer is Asking the Right Question." *International Feminist Journal of Politics* 27 (1): 255–61. https://doi.org/10.1080/14616742.2024.2407595.

Simangan, Dahlia, Srinjoy Bose, John Lee Candelaria, Florian Krampe, and Shinji Kaneko. 2023. "Positive Peace and Environmental Sustainability: Local Evidence from Afghanistan and Nepal." *Environment and Security* 1 (3–4): 142–62. https://doi.org/10.1177/27538796231185677.

Väyrynen, Tarja, Swati Parashar, and Elise Féron, eds. 2021. *Routledge Handbook of Feminist Peace Research*. Routledge.

Peace Agreements

Róisín Burke, Thomas De Lathauwer, and Laura Higgins

Peace agreements are critical mechanisms in the process of the settlement of violent conflicts. In effect, they set out the terms under which warring parties agree to end conflict, or at least a phase thereof.

Peace agreements are often negotiated in several phases, from ceasefires to pre-negotiations informal agreements (which may include **confidence-building measures**), to framework agreements, to implementation agreements (Bell and O'Rourke 2010). They can be partial or more comprehensive in scope (Forster et al. 2019). Peace agreements are often perceived by the parties thereto as binding documents, if not legally, then at least strongly politically, and the language used therein is usually legalistic in nature. These agreements, when signed by states in interstate disputes, take on the character of a legal instrument. The legal status of peace agreements is varied and subject to disagreement in the literature and depends on factors such as actor legitimacy. At an intrastate level, these agreements are more political in nature and may involve a plurality of non-state actors, and sometimes customary and/or religious leaders, in their negotiation and form (Forster et al. 2019; Ozcelik 2020). Moreover, some of the obligations contained in the negotiated agreement may be incorporated into domestic law (Wallensteen et al. 2009). The length and complexity of peace agreements vary greatly (Bell and O'Rourke 2010). These processes tend to be dominated by male elite actors and warring parties, often failing to apply a critical **gender** lens to their provisions (United Nations Women 2015).

The content of peace agreements is varied, but they can contain clauses relating to **security**, **power**, transitional **justice**, the economy, borders, development, socioeconomic issues, implementation, and natural resource management (Palmiano-Federer and Gasser 2016). Peace agreements often touch on environmental concerns in mitigating possible **conflict** drivers, such as the management of **natural resources**, including equitable division and **infrastructure** (e.g., dams, irrigation systems) (Dam-de Jong 2020; Troell and Weinthal 2014; Epple 2025). For example, clauses on access to water were included in the Agreement on Social and Economic Aspects and Agrarian Situation in Guatemala (1996) and the Darfur Peace Agreement (2006) (Tignino and Kebebew 2023).

During times of conflict, it is common for natural resources to be politicized or even be associated with the outbreak of conflict (Aggestam and Sundell 2016). For example, water was a conflict multiplier in the **war** between Israel and its Arab neighbors from 1964 to 1967. This can be due to issues with supply, dominion, or the lack thereof, or lack of critical resources. The importance of natural resources remains relevant during post-conflict, wherein the management of natural resources should be relevant to peacebuilding efforts, whether this be at an interstate or intrastate level. Where peace agreements include natural resource management, it may foster better relations between parties and address the underlying and potential causes of tension (Dam-de Jong 2020; Cooley 1984).

With increasing issues related to renewable and **extractive resources,** strong environmental **governance** will continue to play a larger part in ensuring local or regional security (Dam-de Jong 2020). When correctly engaged and governed, clauses within peace agreements regarding the **environment** or natural resources have the potential to mitigate or remove environmental conflict drivers.

References

Aggestam, Karin, and Anna Sundell. 2016. "Depoliticizing Water Conflict: Functional Peacebuilding in the Red Sea–Dead Sea Water Conveyance Project." *Hydrological Sciences Journal* 61 (7): 1302–12. https://doi.org/10.1080/02626667.2014.999778.

Bell, Christine, and Catherine O'Rourke. 2010. "Peace Agreements or Pieces of Paper? The Impact of UNSC Resolution 1325 on Peace Processes and Their Agreements." *International and Comparative Law Quarterly* 59: 941–80. https://www.jstor.org/stable/40983611.

Cooley, John K. 1984. "The War over Water." *Foreign Policy* (54): 3–26. https://doi.org/10.2307/1148352.

Dam-de Jong, Daniëlla. 2020. "Building a Sustainable Peace: How Peace Processes Shape and Are Shaped by the International Legal Framework for the Governance of Natural Resources." *Review of European, Comparative and International Environmental Law* 29 (1): 21–32. https://doi.org/10.1111/reel.12307.

Epple, Tim. 2025. "Peace Agreements in a Changing Climate: Three Ways in Which Climate Change and Peace Processes Interact." *Environment and Security* 3 (3). https://doi.org/10.1177/27538796241310298.

Forster, Robert A. 2019. "Peace Agreements." In *The Palgrave Encyclopedia of Global Security Studies*, edited by Scott Romaniuk and Péter Marton. Palgrave Macmillan. https://doi.org/10.1007/978-3-319-74336-3_94-1.

Ozcelik, Asli. 2020. "Entrenching Peace in Law: Do Peace Agreements Possess International Legal Status?" *Melbourne Journal of International Law* 21 (1). https://law.unimelb.edu.au/data/assets/pdf_file/0006/3577434/06Ozcelik-unpaginated.pdf.

Palmiano-Federer, Julia, and Rachel Gasser. 2016. *International Peace Mediation and Gender: Bridging the Divide.* BRICS Policy Center, Global South Centre for Mediation, November. https://bricspolicycenter.org/download/7735/.

Tignino, Mara, and Tadesse Kebebew. 2023. "Access to Water and Peacebuilding." In *Research Handbook on International Law and Environmental Peacebuilding*, edited by Daniëlla Dam-de Jong and Britta Sjöstedt. Edward Elgar Publishing. https://doi.org/10.4337/9781789906929.00024.

Troell, Jessica, and Erika Weinthal. 2014. "Harnessing Water Management for More Effective Peacebuilding." In *Water and Post-Conflict Peacebuilding*, edited by Erika Weinthal, Jessica Troell, and Mikiyasu Nakayama. Routledge.

United Nations Women. 2015. *Women's Participation and a Better Understanding of the Political.* https://wps.unwomen.org/participation/.

Wallensteen, Peter, and Mikael Eriksson. 2009. *Negotiating Peace: Lessons from Three Comprehensive Peace Agreements.* Department of Peace and Conflict Research, Uppsala University; Mediation Support Unit, Department of Political Affairs, United Nations.

Peacebuilding

Richard Matthew

Peacebuilding refers to a broad range of actions taken to promote enduring peace, especially by strengthening civil and state capacity to manage, modify, or replace the structures, institutions, practices, and beliefs that trigger, cause, or amplify violent conflict.

Johan Galtung is generally credited with coining the term peacebuilding around 1975. Galtung's objective was to describe a process beyond peacemaking (agreements to cease hostilities) and peacekeeping (ensuring the terms of **peace agreements** were implemented) for mobilizing resources with the objective of creating the conditions for enduring **peace** by addressing the "root causes of **conflict**"—its social and structural factors (Galtung 1976; United Nations 1992, para. 15).

The concept was operationalized in the 1990s as part of a transformative and interlocking set of events and initiatives, including the end of the Cold War in 1991, the first Earth Summit in 1992, and the genocide in Rwanda in 1994. Within this context, UN Secretary-General Boutros Boutros-Ghali determined that the end of the Cold War had created an unprecedented opportunity for the UN to address an alarming increase in armed **violence**, especially in Africa and the Middle East, that was reversing the global trend toward peace experienced in the years after World War II (Cravo 2018). Building on the work of Galtung, Boutros-Ghali released *An Agenda for Peace*, which marked a new direction for the United Nations (UN General Assembly 1992).

In the decade following the launch of the report, UN peacebuilding was typically organized into four general categories of action: professionalizing **security** forces; developing an economic plan attractive to investors; strengthening **governance** institutions; and implementing a national reconciliation process (Cravo 2018). Critics of this approach underscored its top-down, cookie-cutter tendencies, which were implicitly guided by the idea that countries like Canada and Sweden were good models to emulate, and worried that peacebuilding was essentially recolonizing spaces coming out of **war** (Chandler 2017; Cravo 2018; Tanabe 2017). Their calls for other ways of imagining peacebuilding that would embody greater sensitivity to local context, more realistic time frames for change, the broad inclusion of civil society, and reconstruction and visioning processes led by local **stakeholders**, together with on-the-ground experience that made it clear that peace could be supported but not imposed from the outside, have catalyzed the past two decades of reimagining and evolution witnessed within the UN system.

In 2010, peacebuilding was described by the UN as:

> a range of measures targeted to reduce the risk of lapsing or relapsing into conflict by strengthening national capacities at all levels for conflict management, and to lay the foundations for **sustainable peace** and development. Peacebuilding strategies must be coherent and tailored to the specific needs of the country concerned, based on national ownership, and should comprise a carefully prioritized, sequenced, and therefore relatively narrow set of activities aimed at achieving the above objectives.
>
> (United Nations Peacebuilding Support Office 2010, 5)

While the term peacebuilding—and many of its related terms like national or local ownership—continues to be contested, it remains a core function of the UN. In this regard, its operational forms are evolving as the UN seeks to reimagine itself for the challenges of the 21st century.

As the concept of peacebuilding was maturing through research and practice, a parallel effort was unfolding to expand the purview of the UN that focused on coordinating a global response to shared environmental challenges. This effort, which had roots in the 1960s and 1970s when the UN established the **United Nations Environment Programme** in response to mounting evidence of global **environmental change**, was powerfully amplified when the first Earth Summit (1992) highlighted the gravity of trends in **climate change, biodiversity** loss, and pollution, and called for broad adoption of a paradigm of **sustainable development.** The release of *An Agenda for Peace* in 1992 provided a novel pathway for mobilizing support around the importance of the **environment** by linking it to the UN's security agenda.

Much initial focus was given to the ways in which global environmental change might cause, trigger, or amplify violent conflict (Homer-Dixon 1994). This line of analysis quickly generated widespread policy interest and support, but it simultaneously received significant criticism and counter-arguments (Deudney and Matthew 1999). For example, Mark Halle, who directed the **International Institute for Sustainable Development**'s European office, commissioned a state-of-the-art review assessing research linking **environment**, security, and development (1999). When Achim Steiner became the Director General of the **International Union for the Conservation of Nature (IUCN)**, he became interested in this work and commissioned another study with Halle, *Conserving the Peace*, which focused on exploring the potential role of the **conservation** community in peacebuilding activities (Matthew et al. 2002). Steiner later became the Executive Director of the United Nations Environment Programme and initiated the formal integration of environment and peacebuilding, beginning with the policy justification document *From Conflict to Peacebuilding: The Role of Natural Resources and the Environment* (Matthew et al. 2009).

As these efforts were unfolding in the UN, Conca and Dabelko were systematically making the case for an environmentally sensitive approach to the entire spectrum of peace-related activities, presented in their 2002 study *Environmental Peacemaking.*

References

Chandler, David. 2017. *Peacebuilding: The Twenty Years' Crisis, 1997–2017.* Springer.

Conca, Ken, and Geoffrey D. Dabelko, eds. 2002. *Environmental Peacemaking.* Woodrow Wilson Center Press.

Cravo, Teresa Almeida. 2018. "Peacebuilding: Assumptions, Practices and Critiques." *ASPJ: Africa and Francophonie* 8 (1): 4–20. https://www.airuniversity.af.edu/Portals/10/ASPJ_French/journals_E/Volume-09_Issue-1/cravo_e.pdf.

Daniel H. Deudney and Richard A. Matthew (eds.). 1999. *Contested Grounds: Security and Conflict in the New Environment Politics.* SUNY Press.

Galtung, Johan. 1976. "Three Approaches to Peace: Peacekeeping, Peacemaking, and Peacebuilding." In *Peace, War and Defense: Essays in Peace Research*, Vol. II, edited by Johan Galtung. Christian Ejlers.

Halle, Mark, Geoffrey D. Dabelko, Stephen Lonergan, and Richard A. Matthew.. 1999. *State-of-the-Art Review on Environment, Security and Development Co-operation.* International Institute for Sustainable Development.

Homer-Dixon, Thomas F. 1994. "Environmental Scarcities and Violent Conflict: Evidence from Cases." *International Security* 19 (1): 5–40. https://muse.jhu.edu/article/447316.

Matthew, Richard A., Oli Brown, and David Jensen. 2009. *From Conflict to Peacebuilding: The Role of Natural Resources and the Environment.* UNEP.

Matthew, Richard A., Mark Halle, and Jason Switzer, eds. 2002. *Conserving the Peace: Resources, Livelihoods and Security.* International Institute for Sustainable Development.

Tanabe, Juichiro. 2017. "Beyond Liberal Peace: A Critique of Liberal Peacebuilding and Exploring a Postmodern Post-Liberal Hybrid Model of Peacebuilding." *International Relations and Diplomacy* 5 (8): 447–59. https://doi.org/10.17265/2328-2134/2017.08.001.
United Nations General Assembly. 1992. *An Agenda for Peace: Preventive Diplomacy, Peacemaking and Peace-Keeping*. Report of the Secretary-General Pursuant to the Statement Adopted by the Summit Meeting of the Security Council on 31 January 1992, A/47/277, June 17.
United Nations Peacebuilding Support Office (UNPSO). 2010. *UN Peacebuilding: An Orientation*, September. https://www.un.org/peacebuilding/sites/www.un.org.peacebuilding/files/documents/peacebuilding_orientation.pdf.

Peace Dividend

Gabriel Gomes Couto and Shreya Singh

Peace dividends are commonly defined as the economic and social benefits that arise when conflict ceases, allowing resources previously allocated to military expenditures to be redirected toward development and public welfare to foster long-term stability (Knight et al. 1995).

The "peace dividend" holds two primary interpretations: first, as a common good that enhances the quality of life by reducing international tensions, threats, and destruction, and second, as an economic concept—here, a **"positive peace" indicator**—where military spending is redirected toward civilian goods and services (Hartley 1997). This shift goes beyond budgetary maneuvers; it involves redistributing physical and human resources, such as factories, workers, and land, toward purposes like reconstruction, healthcare, and environmental **conservation**. While the effects are most visible in sectors like **infrastructure** (e.g., more schools and hospitals), its potential in **environmental peacebuilding** remains underexplored (UNDP and Climate Finance Mechanism 2021).

Post-**conflict** societies, healing from decades of environmental damage, tend to suffer severe ecological consequences long after the political end of the conflict. It leaves legacies of inter-community resource competition, deforestation, water contamination, and soil degradation, exacerbated by displaced populations and destroyed infrastructure (United Nations Framework Convention on Climate Change 2022). In the aftermath of violent conflict, **natural resources** such as land, timber, minerals, oil, and gas are often the primary assets that governments can access. How they use these assets can fundamentally alter the course of **peacebuilding** (Brown and Nicolucci-Altman 2022). These environmental costs hinder post-conflict recovery, slow **sustainable development**, and pose an obstacle to climate action by leading to the—sometimes deliberate—destruction of crucial assets needed for **adaptation**, energy access, and resilient, low-carbon development (UNDP and Climate Finance Mechanism 2021). Conflict drives environmental damage, while **climate change** often exacerbates conflict, mostly through indirect pathways such as low **governance** capacity, making the redirection of resources toward these interlinkages in peace dividends essential.

The "peace dividend" offers a powerful lens through which we can view environmental peacebuilding. By investing in reforestation, sustainable water management, **renewable energy** infrastructure, and sustainable livelihood projects, it is possible not only to restore vital ecosystems but also to create economic opportunities that can improve the access and governance of scarce natural resources, thereby lowering the likelihood of future conflict. For instance, clean energy projects can stabilize post-conflict regions, create jobs, and lessen dependence on contested resources, while adaptation strategies like sustainable **agriculture** and disaster preparedness enhance **resilience** and reduce the potential for resource-based tensions (Wong 2022).

However, despite its promise, the peace dividend remains underutilized in climate adaptation compared to **mitigation**, underscoring the need for integrated financing within peacebuilding frameworks (United Nations Environment Programme 2023). Moreover, for environmental peacebuilding to reach its full potential, it must continue to actively incorporate **Indigenous Knowledge** and entrepreneurship—long-standing pillars of ecological balance and **social cohesion**. Indigenous communities, often underrepresented in global policy discussions, bring invaluable expertise in sustainable stewardship, making their inclusion essential for fostering both lasting **peace** and environmental resilience (Center for International Relations and International Security 2024). Encouragingly, environmental peacebuilding is gaining momentum and further recognizing the relevance of the peace dividend, with legal frameworks such as the UN Human Rights Council's 2021 resolution on the right to a clean, healthy, and sustainable **environment** and the **International Law Commission**'s draft principles on environmental protection in armed conflict reinforcing the peace dividend as a concrete policy tool (International Law Commission 2022).

If leveraged effectively, the peace dividend is a strategic imperative to reallocate post-**conflict resources** toward long-term environmental and livelihood resilience, reducing the risk of renewed tensions driven by resource scarcity.

References

Brown, Oli and Giuliana Nicolucci-Altman. 2022. *The Future of Environmental Peacebuilding: Nurturing an Ecosystem for Peace – A White Paper*. Geneva Peacebuilding Platform, International Union for Conservation of Nature, PeaceNexus Foundation, Environmental Law Institute, and Environmental Peacebuilding Association. https://dam.gcsp.ch/files/doc/the-white-paper-on-the-future-of-environmental-peacebuilding.

Center for International Relations and International Security (CIRIS). 2024. "Peace Dividend." CIRIS. https://www.ciris.info/learningcenter/peace-dividend/.

Hartley, Keith. 1997. "The Economics of the Peace Dividend." *International Journal of Social Economics* 24 (1/2/3): 28–45. https://doi.org/10.1108/03068299710161124.

International Law Commission. 2022. Draft Principles on Protection of the Environment in Relation to Armed Conflicts. https://legal.un.org/ilc/texts/instruments/english/draft_articles/8_7_2022.pdf.

Knight, Malcom, Norman Loayza, and Delano Villanueva. 1995. "The Peace Dividend: Military Spending Cuts and Economic Growth." *IMF WorkingPapers* 1995 (053): 1-37. International Monetary Fund. . https://doi.org/10.5089/9781451847338.001.

Malcolm Knight, Norman Loayza and Delano Villanueva, 1996. "The Peace Dividend: Military Spending Cuts and Economic Growth." *Staff Papers* 43(1): 1–37. Palgrave Publishers.

UNDP and Climate Security Mechanism, 2021.Climate Finance for Sustaining Peace. Making climate finance work for conflict-affected and fragile contexts. UNDP.

United Nations Framework Convention on Climate Change (UNFCCC). 2022. *Conflict and Climate*. https://unfccc.int/news/conflict-and-climate.

United Nations Environment Programme (UNEP). 2023. *Adaptation Gap Report 2023: Underfinanced. Linderprepared. Inadequate Investment and Planning on Climate Adaptation Leaves World Exposed*. https://www.unep.org/resources/adaptation-gap-report-2023.

Wong, Catherine. 2022. *Climate Finance and the Peace Dividend: A Co-benefits Argument*. Green Finance Platform. https://www.greenfinanceplatform.org/blog/climate-finance-and-peace-dividend-co-benefits-argument.

Peace Parks

Elaine (Lan Yin) Hsiao and Saleem H. Ali

Peace parks have been conceived of in many ways, ranging from inspirationally named urban spaces, commemorative landmarks, and green spaces to cross-border conservation areas dedicated simultaneously to the lofty goals of strengthening biodiversity and peaceful relations between countries.

Within **environmental peacebuilding** literature, the term initially gained currency to promote environmental **conservation** collaborations that can intervene to address existing **conflicts** between jurisdictions–either national or international (Quinn et al. 2012; Brock 1991). Cases where the formation of a peace park can help to resolve territorial conflicts, such as the demilitarized zone between the Koreas, are regularly championed. In post-apartheid South Africa, President Mandela stated that "In a world beset by conflicts and division, **peace** is one of the cornerstones of the future. Peace parks are a building block in this process" (Ramutsindela 2007, 6). The most concrete example of such a case is the establishment of a transboundary conservation area, La Cordillera del Condor, between Ecuador and Peru in the 1990s to help resolve a longstanding territorial dispute that dated back to Spanish colonization (Alcalde et al. 2005; Ali 2018).

Prior to the Cordillera del Condor, the world's first officially named international peace park with a clear mandate for conservation and **cooperation** was Waterton-Glacier International Peace Park, which was established in 1932 through Canadian and American Rotary Clubs lobbying for its legislation in both countries (Waterton-Glacier International Peace Park Association n.d.). The peace park allows transboundary access for visitors in a unique way, including mobile app registration, rather than border checkpoints. Urban peace parks, such as the Global Peace Parks Project started by the International Institute for Peace through Tourism, offer civic zones of peace, while places like Hiroshima Peace Memorial Park highlight the importance of memory in healing after an atrocity (International Institute for Peace through Tourism n.d.). Larger-scale transboundary peace parks emphasize the **peacebuilding** potential of developing "habits of cooperation" through harmonizing cross-border conservation efforts, while Indigenous peace parks may emphasize Indigenous sovereignty, healing of the land and people, and interspecies kinship and duties of care (Conca and Dabelko 2002; Ali 2007; Cowan 2023; Hsiao 2018).

Although real-world examples of peace parks include local interpretations, such as Montreal's skateboard Peace Park, environmental peacebuilding and transboundary conservation literature focus on peace parks as a type of protected area. While at the UN University for Peace, Professor Gerardo Budowski referred to peace parks as "protected areas where there is a significant conflictive past" and did not require them to be transboundary, so they would not exclude places such as conflict-afflicted islands (Sandwith and Besançon 2010, 3; Budowski 2004, 30). The International Union for the Conservation of Nature (IUCN) World Commission on Protected Areas (WCPA) has referred to peace parks as a kind of "special designation" that can apply to all three categories of transboundary conservation areas: (1) protected areas, (2) conservation landscapes or seascapes, and (3) migration conservation areas (Vasilijević et al. 2015, xi). The special designation can be accorded if the area is "dedicated to the promotion, celebration, and/or commemoration of peace and cooperation" (IUCN n.d.). Building on the IUCN's definition of peace parks, the Peace and Biodiversity Dialogue Initiative under the Convention on Biological Diversity (CBD) Secretariat advises that peace parks should be designated where they serve as "both a protected area and a no conflict zone," introducing the concept that they might also be protected from conflict activities (CBD Secretariat 2017). Cumulatively, these different definitions speak to the main goals of peace parks toward commemorating friendly relations between countries that may have had a history of conflict in the past, collaborative conservation as a form of environmental peacebuilding, and the potential for peace parks to contribute to the cessation of violent or armed conflict.

While the aspirations of peace parks are laudable, their implementation comes with the numerous challenges most protected or conserved areas face (e.g., **climate change**, lack of resources or political will, and impacts of extractive industries, including tourism), as well as the complexities of cross-border collaboration, regional harmonization, and the insecurity of armed conflict or post-conflict recovery. Researchers have observed that peace park initiatives or management are often stalled or break down when hostilities between countries escalate (Waisová 2015; Hsiao 2020), suffer from hegemonic geopolitical influences that shape **power** dynamics between cooperating countries (Büscher 2013; Duffy 2007; Zbicz 2003), including donor interests (Sjöstedt 2017; Hsiao and Wilson 2013), and do not always adhere to non-violent mechanisms for conservation or conflict resolution (e.g., application of extrajudicial killing as anti-poaching policy) (Massé 2020; Massé and Lunstrum 2015).

Critiques of peace parks are also predicated on the view that the term can be co-opted to justify state-sanctioned encroachment on community land and dispossession of customary rights. Scholars have noted this concern with reference to the work of the Peace Parks Foundation in South Africa (Büscher and Ramutsindela 2015; King 2010). The Foundation has responded by offering local employment to local ecotourism operations run on these conservation areas and allowing for free use of the lands by local pastoralists who adhere to specific management plans.

Security personnel have also voiced concerns that demilitarization of border areas for peace park designation can heighten security concerns and be exploited by traffickers of illegal goods or armed groups. However, through close collaboration with regional intelligence and security sectors and clearly differentiated roles and responsibilities, there can be security protocols added to peace parks without turning rangers into an extension of the armed forces (Braack et al. 2006). If peace parks are to be considered conflict "no-go zones," as proposed under the concept of **protected zones** recognized in the **International Law Commission**'s Principles on the Protection of the Environment in Relationship to Armed Conflict, they cannot be utilized for military operations, including military occupation (Hsiao 2020). As potential protected zones, the neutrality of conservation actors must be recognized and operationalized to sustain environmental protection in armed conflict zones (Hsiao et al. 2023).

This requires peace parks to embrace conflict-sensitive practices to contribute toward effective conservation and non-violent conflict resolution or peace.

References

Alcalde, Martin, Carlos F. Ponce, and Yanitza Curonisy. 2005. "Peace Parks in the Cordillera Del Cóndor Mountain Range and Biodiversity Conservation Corridor." *Environmental Change and Security Program Report* 11: 63–4. https://www.wilsoncenter.org/sites/default/files/media/documents/event/ponce.pdf.

Ali, Saleem H., ed. 2007. *Peace Parks: Conservation and Conflict Resolution*. MIT Press.

Ali, Saleem H. 2018. "A Casualty of Peace? Lessons on De-Militarizing Conservation in the Cordillera Del Condor Corridor." In *Collateral Values of Natural Capital*. Springer. https://ssrn.com/abstract=2951307.

Braack, Leo, Trevor Sandwith, David Peddle, and Thomas Petermann. 2006. *Security Considerations in the Planning and Management of Transboundary Conservation Areas*. International Union for Conservation of Nature (IUCN).

Brock, Lothar. 1991. "Peace through Parks: The Environment on the Peace Research Agenda." *Journal of Peace Research* 28 (4): 407–23. https://doi.org/10.1177/0022343391028004006

Budowski, Gerardo. 2004. "PEACE through Parks." *Our Planet* 14 (4): 30–31.

Büscher, Bram. 2013. *Transforming the Frontier: Peace Parks and the Politics of Neoliberal Conservation in Southern Africa*. Duke University Press.

Büscher, Bram, and Maano Ramutsindela. 2015. "Green Violence: Rhino Poaching and the War to Save Southern Africa's Peace Parks." *African Affairs* 115 (458): 1–22. https://doi.org/10.1093/afraf/adv058.

Conca, Ken, and Geoffrey D. Dabelko, eds. 2002. *Environmental Peacemaking*. Woodrow Wilson Center Press.

Cowan, Carolyn. 2023. "Award-Winning, Indigenous Peace Park Dragged into Fierce Conflict in Myanmar." Mongabay, May 15. https://news.mongabay.com/2023/05/award-winning-indigenous-peace-park-dragged-into-fierce-conflict-in-myanmar/.

Duffy, Rosaleen. 2007. "Peace Parks and Global Politics: The Paradoxes and Challenges of Global Governance." In *Peace Parks: Conservation and Conflict Resolution*, edited by Saleem H. Ali. MIT Press.

Hsiao, Elaine C. 2018. "Protecting Places for Nature, People, and Peace: A Critical Socio-Legal Review of Transboundary Conservation Areas." PhD Dissertation, University of British Columbia. https://hdl.handle.net/2429/67561.

Hsiao, Elaine C. 2020. "Protecting Protected Areas in Bello: Learning From Institutional Design and Conflict Resilience in the Greater Virunga and Kidepo Landscapes." *Goettingen Journal of International Law* 10 (1): 67–110. https://doi.org/10.3249/1868-1581-10-1-hsiao.

Hsiao, Elaine C., and Cory Wilson. 2013. "Transcending Boundaries: Perspectives from the Central Albertine Rift Transfrontier Protected Area Network." IPPExpeditions. Video, 24min., 27 sec. https://vimeo.com/62257328.

Hsiao, Elaine (Lan Yin), Adrian Garside, Doug Weir, and Andrew J. Plumptre. 2023. "Protected Zones in Context: Exploring the Complexity of Armed Conflicts and Their Impacts on the Protection of Biodiversity." *International Review of the Red Cross* 105 (924): 1412–40. https://doi.org/10.1017/S1816383123000450.

International Institute for Peace Through Tourism. n,d. "Global Peace Parks Project." NGO. *IIPT Global Peace Parks Project*. Accessed August 14, 2025. https://peacetourism.org/peace-parks-project/.

International Union for Conservation of Nature (IUCN). n.d. "IUCN WCPA Transboundary Conservation Specialist Group." Accessed August 14, 2025. https://iucn.org/our-union/commissions/group/iucn-wcpa-transboundary-conservation-specialist-group.

King, Brian. 2010. "Conservation Geographies in Sub-Saharan Africa: The Politics of National Parks, Community Conservation and Peace Parks." *Geography Compass* 4 (1): 1–14. https://doi.org/10.1111/j.1749-8198.2009.00288.x.

Massé, Francis. 2020. "Conservation Law Enforcement: Policing Protected Areas." *Annals of the American Association of Geographers* 110 (3): 758–73. https://doi.org/10.1080/24694452.2019.1630249.
Massé, Francis, and Elizabeth Lunstrum. 2015. "Accumulation by Securitization: Commercial Poaching, Neoliberal Conservation, and the Creation of New Wildlife Frontiers." *Geoforum* 69: 227–37.
Peace and Biodiversity Dialogue. 2017. "Transboundary Conservation" Convention on Biological Diversity. Effective March 10. https://www.cbd.int/peace/about/peace-parks.
Quinn, Michael S., Wayne A. Freimund, and Len Broberg, eds. 2012. *Parks, Peace, and Partnership: Global Initiatives in Transboundary Conservation*. Energy, Ecology, and the Environment Series No. 4. University of Calgary Press. http://www.deslibris.ca/ID/444836.
Ramutsindela, Maano. 2007. *Transfrontier Conservation in Africa: At the Confluence of Capital, Politics and Nature*. CABI.
Sandwith, Trevor, and Charles Besançon. 2010. "Making Peace: Protected Areas Contributing to Conflict Resolution." In *Arguments for Protected Areas: Multiple Benefits for Conservation and Use*, edited by Sue Stolton and Nigel Dudley. Earthscan.
Sjöstedt, Britta. 2017. "Environmental Governance and Peacebuilding as a Joint Enterprise." In *Protecting Nature in Conflicts and Building Peace: Success Stories in Conflicts and Their Aftermath*, 12. Unpublished.
Vasilijević, Maja, Kevan Zunckel, Matthew McKinney, Boris Erg, Michael Schoon, and Tatjana Rosen Michel. 2015. *Transboundary Conservation: A Systematic and Integrated Approach*, edited by Adrian Phillips. International Union for the Conservation of Nature. https://portals.iucn.org/library/efiles/documents/PAG-023.pdfhttps://portals.iucn.org/library/efiles/documents/PAG-023.pdf.
Waisová, Šárka. 2015. "Environmental Cooperation as Instrument of Conflict Transformation in Conflict-Prone Areas: Where Does It Start, How Deep It Can Be and What Effects It Can Have." *Politické Vedy* 2: 105–26. http://www.politickevedy.fpvmv.umb.sk/userfiles/file/2_2015/WAISOVA.pdf.
Waterton-Glacier International Peace Park (WGIPP) Association. n.d. "Waterton-Glacier International Peace Park Association History." NGO. *WGIPPA History*. Accessed August 14, 2025. https://www.watertonglacierpeacepark.org/history.
Zbicz, Dorothy C. 2003. "Imposing Transboundary Conservation: Cooperation Between Internationally Adjoining Protected Areas." *Journal of Sustainable Forestry* 17 (1–2): 21–37. https://doi.org/10.1300/J091v17n01_03.

Political Ecology

Philippe Le Billon, Francis Massé, and Nancy Peluso

Political ecology is an interdisciplinary field of study examining uneven power relations over natural and agrarian resources and the environment, seeking to advance environmental justice.

Political ecology critically examines **power** relations in human-**environment** interactions. Since its emergence as a field in the mid-1980s, political ecologists have been particularly concerned with inequities in access to and control of land and **natural resources**, responsibility for and exposure to environmental harms, inclusion in decision-making about environmental management and governing, and struggles seeking just and sustainable futures (Watts 1983; Blaikie 1985; Hecht 1989; Peluso 1992; Le Billon 2001; Batterbury 2018). Political ecology is often understood as both a critique and a **justice**-oriented practice, advocacy, and activism (Sultana 2023).

Political ecology's intellectual and interdisciplinary roots can be traced through critical agrarian studies (Borras 2023; Guthman 2000), political economy, human geography, and environmental history (Robbins 2019). Political ecologists research with feminist (Elmhirst 2011; Rocheleau et al. 2013; Lamb et al. 2019; Harcourt et al. 2023), critical race (Loftus 2020; Mollett and Faria 2018, Van Sant et al. 2021), post- and de-colonial (Carroll 2015; Ekowati et al. 2023; Ferdinand 2021), science and technology studies (Forsyth 2004; Latour 2004), more-than-human (Sundberg 2014; Choi 2016; Whatmore 2013), and relational approaches (Abdurrahim et al. 2023).

Political ecologists study the multi-scalar connections between **environmental change** and **conflicts** over resources and socionatures ranging from extreme to slow **violence** around "deforestation," land and resource grabs, **conservation**, pollution, mining, ongoing settler colonial dynamics, and other resource or nature-based conflicts (Blaikie 1985; Le Billon 2001; Carroll 2015; Enns and Bersaglio 2024; Hecht 1989, 2013; Peluso and Watts 2001; Nixon 2011; Lunstrum 2014). Multi-sited and multi-scalar conflicts require complex analyzes examining the interplay of social forces, relations, and structures, as well as the diverse imaginaries, memories, and materialities differentially shaping these relations, socionatures, and environmental access, conflicts, and claims (Carroll 2015; Escobar 2006; Swyngedouw 2015; Sultana 2021). Political ecologists oppose environmental determinism and apolitical understandings of environmental change and conflict. Rather, they focus on complex social and cultural processes, including conflict dynamics, conceptualized through historical, geographic, and intersectional subjectivities, including race, class, **gender**, generation, and region.

Empirically, political ecology follows the site specific, trans-scalar, and historically grounded circumstances surrounding conflicts, often tracing "webs of connection" (Zhu 2022), sites of accumulation (Smith 1983; McCarthy and Prudham 2004; Brockington et al. 2012), social movements (Borras and Franco 2013; Wolford 2010), and modes of difference (Katz 2004; Van Sant et al. 2021). Political ecologists have been at the forefront of solidarities with communities and marginalized groups, often collaborating to address environmental injustices and pursue just socio-ecological futures for humans and non-humans alike (Batterbury 2018; Ramutsindela et al. 2022). The scope of empirical issues examined is constantly shifting and expanding; grounded as it is, in the resource and environmental conflicts of their times. Common topics include conflicts and contestations related to soil erosion, **forests**/deforestation, exclusionary and violent forms of conservation, water, **land grabbing**, mining, **agriculture** and **pastoralism**, human and environmental health, **food security**/sovereignty, **climate change,** and broad and ongoing regimes of dispossession and accumulation shaped by neoliberal capitalist, patriarchal, colonial, and racialized regimes and forms of power (Robbins 2019; Benjaminsen and Svarstad 2021).

Importantly, conflicts over land, livelihoods, the environment, and resources are understood as political struggles potentially enabling vulnerable actors to challenge and transform dominant power structures and subjectivities. This requires paying attention to forms of resistance and repression, and a wide range of socio-environmental relations, including consent, collusion, and cooptation, brutal force, and the discursive violence that often informs environmental policy, interventions, and management (Massé 2022). This position of understanding conflicts as emancipatory struggles for **environmental justice** means that political ecology can be at odds with

environmental peacebuilding when it attempts to prevent or end conflicts, as this can be interpreted as "pacifying" efforts that risk undermining the actual and diverse interests of aggrieved communities and factions (Ide et al. 2023). However, political ecology can be useful to ensure greater environmental justice within environmental peacebuilding by exposing unequal power relations, re-politicizing environmental conflicts and injustices, and supporting the environmental rights of marginalized communities (Davis et al. 2023; Le Billon and Duffy 2018).

References

Abdurrahim, Ali Y., Arya H. Dharmawan, Soeryo Adiwibowo, Herry Yogaswara, and Meine van Noordwijk. 2023. "Relational and Instrumental Values of Tropical Peat Landscapes: Morality and Political Ecology in Indonesia." *Current Opinion in Environmental Sustainability* 64: 101318. https://doi.org/10.1016/j.cosust.2023.101318.

Batterbury, Simon. 2018. "Political Ecology." *In Companion to Environmental Studies*, edited by Noel Castree, Mike Hulme, and James D. Proctor. Routledge.

Benjaminsen, Tor A., and Hanne Svarstad. 2021. *Political Ecology: A Critical Engagement with Global Environmental Issues*. Springer Nature.

Blaikie, Piers. 1985. *The Political Economy of Soil Erosion in Developing Countries*. Routledge. https://doi.org/10.4324/9781315637556.

Borras, Saturnino M. 2023. "Politically Engaged, Pluralist and Internationalist: Critical Agrarian Studies Today." *The Journal of Peasant Studies* 50 (2): 449–89. https://doi.org/10.1080/03066150.2022.2163164.

Borras, Saturnino M., Jr., and Jennifer C. Franco. "Global Land Grabbing and Political Reactions 'From Below.'" *Third World Quarterly* 34 (9): 1723–47. https://doi.org/10.1080/01436597.2013.843845.

Brockington, Dan, Rosaleen Duffy, and Jim Igoe. 2012. *Nature Unbound: Conservation, Capitalism and the Future of Protected Areas*. Routledge.

Carroll, Clint. 2015. *Roots of our Renewal: Ethnobotany and Cherokee Environmental Governance*. University of Minnesota Press.

Choi, Myung. 2016. "More-Than-Human Geographies of Nature: Toward a Careful Political Ecology." *Journal of the Korean Geographical Society* 51 (5): 613–632. https://koreascience.kr/article/JAKO201607959405067.page.

Davis, Katy, Laura E. Peters, Jamon Van Den Hoek, and Ken Conca. 2023. "Power in Environmental Peacebuilding." *World Development Sustainability* 3: 100110. https://doi.org/10.1016/j.wds.2023.100110.

Ekowati, Dian, Marlene Gómez, Iliana Monterroso, and Ankita Shrestha. 2023. "Perspectives on Decoloniality for FPE." In *Contours of Feminist Political Ecology*, edited by Wendy Harcourt, Ana Agostino, Rebecca Elmhirst, Marlene Gómez, and Panagiota Kotsila. Springer International Publishing. https://doi.org/10.1007/978-3-031-20928-4.

Elmhirst, Rebeca. 2011. "Introducing New Feminist Political Ecologies." *Geoforum* 42 (2): 129–32. https://doi.org/10.1016/j.geoforum.2011.01.006.

Enns, Charis, and Brock Bersaglio. 2024. *Settler Ecologies: The Enduring Nature of Settler Colonialism in Kenya*. University of Toronto Press.

Escobar, Arturo. 2006. "Difference and Conflict in the Struggle Over Natural Resources: A Political Ecology Framework." *Development* 49 (3): 6–13. https://doi.org/10.1057/palgrave.development.1100267.

Ferdinand, Malcom. 2021. *Decolonial Ecology: Thinking from the Caribbean World*. John Wiley and Sons.

Forsyth, Timothy. 2004. *Critical Political Ecology: The Politics of Environmental Science*. Routledge.

Guthman, Julie H. 2000. *Agrarian Dreams? The Paradox of Organic Farming in California*. University of California Press.

Harcourt, Wendy, Ana Agostino, Rebecca Elmhirst, Marlene Gómez, and Panagiota Kotsila, eds. 2023. *Contours of Feminist Political Ecology*. Springer Nature.

Hecht, Susanna, and Alexander Cockburn. 1989. *The Fate of the Forest: Developers, Destroyers, and Defenders of the Amazon*. University of Chicago Press.

Hecht, Susanna. 2013. *The Scramble for the Amazon and the "Lost Paradise" of Euclides da Cunha*. University of Chicago Press.

Ide, Tobias, McKenzie F. Johnson, Jon Barnett, et al. 2023. "The Future of Environmental Peace and Conflict Research." *Environmental Politics* 32 (6): 1077–103. https://doi.org/10.1080/09644016.2022.2156174.

Katz, Cindy. 2004. *Growing Up Global: Economic Restructuring and Children's Everyday Lives*. University of Minnesota Press.

Lamb, Vanessa, Laura Schoenberger, Carl Middleton, and Borin Un. 2019. "Gendered Eviction, Protest, and Recovery: A Feminist Political Ecology Engagement with Land Grabbing in Rural Cambodia." In *Gender and Generation in Southeast Asian Agrarian Transformations*, edited by Clara Mi Young Park and Ben White. Routledge.

Latour, Bruno. 2004. *Politics of Nature*. Harvard University Press.

Le Billon, Philippe. 2001. "The Political Ecology of War: Natural Resources and Armed Conflicts." *Political Geography* 20 (5): 561584. https://doi.org/10.1016/S0962-6298(01)00015-4.

Le Billon, Philippe, and Rosaleen V. Duffy. 2018. "Conflict Ecologies: Connecting Political Ecology and Peace and Conflict Studies." *Journal of Political Ecology* 25 (1): 239–60. https://doi.org/10.2458/v25i1.22704.

Loftus, Alex. 2020. "Political Ecology III: Who Are 'the People'?" *Progress in Human Geography* 44 (5): 981–90. https://doi.org/10.1177/0309132519884632.

Lunstrum, Elizabeth. 2014. "Green Militarization: Anti-Poaching Efforts and the Spatial Contours of Kruger National Park." *Annals of the Association of American Geographers* 104 (4): 816–32. https://doi.org/10.1080/00045608.2014.912545.

Massé, Francis. 2022. "Police Power in Green: Furthering Political Ecologies of the State." *Political Geography* 97: 102627. https://doi.org/10.1016/j.polgeo.2022.102627.

McCarthy, James, and Scott Prudham. 2004. "Neoliberal Nature and the Nature of Neoliberalism." *Geoforum* 35 (3): 275–83. https://doi.org/10.1016/j.geoforum.2003.07.003.

Mollett, Sharlene, and Caroline Faria. 2013. "Messing with Gender in Feminist Political Ecology." *Geoforum* 45: 116–25. https://doi.org/10.1016/j.geoforum.2012.10.009.

Nixon, Rob. 2011. *Slow Violence and the Environmentalism of the Poor*. Harvard University Press.

Peluso, Nancy Lee. 1992. *Rich Forests, Poor People: Resource Control and Resistance in Java*. University of California Press.

Peluso, Nancy Lee, and Michael Watts. 2001. *Violent Environments*. Cornell University Press.

Ramutsindela, Maano, Frank Matose, and Tafadzwa Mushonga, eds. 2022. *The Violence of Conservation in Africa: State, Militarization and Alternatives*. Edward Elgar Publishing.

Robbins, Paul. 2019. *Political Ecology: A Critical Introduction*. John Wiley and Sons.

Rocheleau, Diane, Barbara Thomas-Slayter, and Esther Wangari, eds. 2013. *Feminist Political Ecology: Global Issues and Local Experience*. Routledge.

Smith, Neil. 1983. *Uneven Development: The Production of Nature Under Capitalism*. Johns Hopkins University.

Sultana, Farhana. 2021. "Political Ecology 1: From Margins to Center." *Progress in Human Geography* 45 (1): 156–65. https://doi.org/10.1177/0309132520936751.

Sultana, Farhana. 2023. "Political Ecology III: Praxis-Doing, Undoing, and Being in Radical Political Ecology Research." *Progress in Human Geography* 47 (5): 728–37. https://doi.org/10.1177/03091325231157360.

Sundberg, Juanita. 2014. "Decolonizing Posthumanist Geographies." *Cultural Geographies* 21 (1): 33–47. https://doi.org/10.1177/1474474013486067.

Swyngedouw, Erik. 2015. *Liquid Power: Contested Hydro-modernities in Twentieth-century Spain*. MIT Press.

Van Sant, Levi, Richard Milligan, and Sharlene Mollett. 2021. "Political Ecologies of Race: Settler Colonialism and Environmental Racism in the United States and Canada." *Antipode* 53 (3): 629–42. https://doi.org/10.1111/anti.12697.

Watts, Michael J. 1983. *Silent Violence: Food, Famine, and Peasantry in Northern Nigeria*. University of California Press.

Whatmore, Sarah J. 2013. "Political Ecology in a More-than-Human World: Rethinking 'Natural' Hazards." In *Anthropology and Nature*, edited by Kirsten Hastrup. Routledge.
Wolford, Wendy. 2010. *This Land is Ours Now: Social Mobilization and the Meanings of Land in Brazil*. Duke University Press.
Zhu, Annah L. 2022. *Rosewood: Endangered Species Conservation and the Rise of Global China*. Harvard University Press.

Polycrisis

Michael Lawrence and Thomas Homer-Dixon

A polycrisis occurs when crises in multiple systems become causally entangled in ways that cause major human harms. The interactions between constituent crises are significant enough to produce emergent harms that are different from, and usually greater than, the sum of the harms those crises would produce separately.

The term "polycrisis" has become increasingly popular amidst the proliferation of global crises in recent years. It reflects a growing concern that the world's crises interrelate in ways that amplify their conjoined harms and make crisis resolution especially difficult. The Cascade Institute has refined the term from a loose buzzword into the conceptual foundation for productive research into crisis interactions by using the above definition (Lawrence et al. 2022, 2024a, 2024b):

This formulation is often quoted but not universally accepted. Scholars (e.g., Swilling 2020; Davies and Hobson 2023; Albert 2024; Helleiner 2024), practitioners (e.g., United Nations Development Programme Regional Bureau for Asia and the Pacific 2022; World Economic Forum 2023; Juncker 2018), and commentators (e.g., Tooze 2022; Mackenzie and Sahay 2022; Summers and Ahmed 2022) use the term "polycrisis" with growing frequency but for varying purposes (for a list of polycrisis definitions, see Matlovic and Matlovičová 2024). Nonetheless, they generally agree on several core features of the polycrisis phenomenon (Lawrence et al. 2024c):

I. Emergent harms: interconnected crises condition, exacerbate, and reshape one another so that they have different impacts when they occur together than they would separately
II. Multiple causes: a polycrisis is not reducible to a single root cause. It involves multiple causes acting at different spatial and temporal scales across several systems
III. Deep uncertainty: complex crisis interconnections strain comprehension and often exceed our ability to anticipate future developments
IV. Systemic context: from its origins in complexity science (Morin and Kern 1999), polycrisis thinking recognizes that crises arise within complex systems and, therefore, require systems thinking
V. New knowledge and action: established frameworks, institutions, and practices are ill-equipped to address crisis interactions; polycrises require new modes of research and practice that overcome disciplinary and policymaking silos

A polycrisis can occur at any scale, from local to global, that hosts multiple systems susceptible to interrelated crises. The world has experienced many historical instances of polycrisis at various scales, but the present global polycrisis is unprecedented in the density of global interconnectivity involved and because human activities are pushing planetary boundaries like never before (Homer-Dixon 2023).

Although the polycrisis concept originated outside the field of **environmental peacebuilding**, the two resonate in significant ways. Both constitute efforts to break down disciplinary silos that impede understanding and action on complex, multifaceted phenomena. Environmental peacebuilding does so by integrating concerns of **peace**, **security**, **environment**, and development, and by recognizing the indispensable interdependence between human and ecological systems (Bruch et al. 2022). Likewise, polycrisis thinking posits that systems of economy, food production, transportation, health, and **governance** depend on the well-being of the planet's ecological systems, and that those ecological systems are profoundly affected by social systems and their conflicts.

Environmental peacebuilding and polycrisis thinking also generate similar practical strategies. The nexus between human and ecological systems offers new opportunities for more effective **peacebuilding**, whether by addressing the environmental causes of **conflict** or promoting stewardship as a cooperative, superordinate endeavor for erstwhile enemies. Participatory approaches, adaptive governance, and social **justice** concerns advanced in peacebuilding can foster more resilient (that is, crisis-resistant) socio-ecological systems, a key goal of polycrisis practice.

And finally, polycrisis thinking and environmental peacebuilding can further enrich one another going forward. Polycrisis analysis (Lawrence et al. 2024a) offers a sophisticated systemic lens on the intersecting stresses and crises that underpin conflicts and must be addressed by peacebuilding efforts. Peacebuilding knowledge can advance polycrisis analysis by highlighting the role of agency and the operation of **power** within crises (e.g., Kumar and Cousens 2001; Ide et al. 2021), both of which are often underemphasized in systems thinking.

References

Albert, Michael J. 2024. *Navigating the Polycrisis: Mapping the Futures of Capitalism and the Earth*. The MIT Press.

Bruch, Carl, Erika Weinthal, McKenzie Johnson, and Tobias Ide. 2022. "Towards a Definition of Environmental Peacebuilding." Ecosystem for Peace. https://www.ecosystemforpeace.org/compendium/towardsadefinitionofenvironmentalpeacebuilding.

Davies, Mathew, and Christopher Hobson. 2023. "An Embarrassment of Changes: International Relations and the COVID-19 Pandemic." *Australian Journal of International Affairs* 77 (2): 150–168. https://doi.org/10.1080/10357718.2022.2095614.

Helleiner, Eric. 2024. "Economic Globalization's Polycrisis." *International Studies Quarterly* 68 (2): 1–9. https://doi.org/10.1093/isq/sqae024.

Homer-Dixon, Thomas. 2023. "Why so Much Is Going Wrong at the Same Time." *Vox*. October 18. https://www.vox.com/future-perfect/23920997/polycrisis-climate-pandemic-population-connectivity.

Ide, Tobias, Carl Bruch, Alexander Carius, et al. 2021. "The Past and Future(s) of Environmental Peacebuilding." *International Affairs* 97 (1): 1–16. https://doi.org/10.1093/ia/iiaa177.

Juncker, Jean-Claude. 2018. "Speech by President Jean-Claude Juncker at the Opening Plenary Session of the Ideas Lab 2018 'Europe–Back on Track' of the Centre for European Policy Studies." February 22. https://ec.europa.eu/commission/presscorner/detail/en/SPEECH_18_1121.

Kumar, Chetan, and Elizabeth M. Cousens, eds. 2001. *Peacebuilding as Politics: Cultivating Peace in Fragile Societies*. Lynne Rienner Publishers.

Lawrence, Michael, Thomas Homer-Dixon, Scott Janzwood, Johan Rockström, Ortwin Renn, and Jonathan F. Donges. 2024a. "Global Polycrisis: The Causal Mechanisms of Crisis Entanglement." *Global Sustainability* 7 (6). https://doi.org/10.1017/sus.2024.1.

Lawrence, Michael, Scott Janzwood, and Thomas Homer-Dixon. 2022. "What Is a Global Polycrisis? And How Is It Different from a Systemic Risk?" Discussion Paper 2022–24. Version 2.0. Cascade Institute. https://cascadeinstitute.org/technical-paper/what-is-a-global-polycrisis/.

Lawrence, Michael, Megan Shipman, and Thomas Homer-Dixon. 2024b. "Introduction to Polycrisis Analysis." Cascade Institute. https://cascadeinstitute.org/technical-paper/introduction-to-polycrisis-analysis-guide/.

Lawrence, Michael, Megan Shipman, Scott Janzwood, et al. 2024c. "Polycrisis Research and Action Roadmap: Gaps, Opportunities, and Priorities for Polycrisis Research and Action." Technical paper 2024–4. Cascade Institute, Potsdam Institute for Climate Impact Research, Research Institute for Sustainability–Helmholtz Centre Potsdam, and Centre for the Study of Existential Risk. https://cascadeinstitute.org/technical-paper/polycrisisroadmap/.

Mackenzie, Kate, and Tim Sahay. 2022. "An Introduction." The Polycrisis. October 20. https://www.phenomenalworld.org/analysis/an-introduction/.

Matlovic, Rene, and Kvetoslava Matlovičová. 2024. "Polycrisis in the Anthropocene as a Key Research Agenda for Geography: Ontological Delineation and the Shift to a Postdisciplinary Approach." *Folia Geographica* 66 (1): 5–33. https://www.unipo.sk/public/media/47641/700Polycrisis%20in%20Anthropocene%20as%20Key%20Research%20Agenda%20for%20Geography%20Ontological%20Delineation%20and%20Shift%20to%20Postdisciplinary%20Approach.pdf.

Morin, Edgar, and Anne Brigitte Kern. 1999. *Homeland Earth: A Manifesto for the New Millenium.* Hampton Press.

Summers, Lawrence H., and Masood Ahmed. 2022. "World Bank-IMF Meetings Are the Last Stop Before a Coming Economic Storm." *The Washington Post*, October 5. https://www.washingtonpost.com/opinions/2022/10/05/imf-world-bank-meetings-prepare-economic-downtown/.

Swilling, Mark. 2020. *The Age of Sustainability: Just Transitions in a Complex World.* Routledge.

Tooze, Adam. 2022. "Welcome to the World of the Polycrisis." *Financial Times*, October 28. https://www.ft.com/content/498398e7-11b1-494b-9cd3-6d669dc3de33.

United Nations Development Programme (UNDP) Regional Bureau for Asia and the Pacific. 2022. "Polycrisis and Long-Term Thinking." Reimagining Development in Asia and the Pacific Foresight Brief. UNDP. https://www.undp.org/asia-pacific/publications/polycrisis-and-long-term-thinking-reimagining-development-asia-and-pacific-foresight-brief.

World Economic Forum (WEF). 2023. *The Global Risks Report 2023.* https://www3.weforum.org/docs/WEF_Global_Risks_Report_2023.pdf.

Positive Peace

Randall Amster, Elsa Barron, and Liliana Pimentel

Positive peace is the presence of values, norms, and actions that promote justice and well-being within a society. It goes beyond **negative peace**, which is defined as the absence of violence, by emphasizing the cultivation of conditions that support fairness, sustainability, and resilience to create "an optimal environment for human potential to flourish" (Institute for Economics and Peace [IEP] 2024).

> The field of peace studies recognizes both negative and positive peace as foundational concepts. Positive peace is understood as an active concept requiring more than merely contesting violent, unjust, or unsustainable conditions. Environmental peacebuilding calls for efforts to foster peace and justice at all levels and within societal structures.

The positive peace concept was initially developed by Jane Addams in the early 1900s, later refined in the speeches of Martin Luther King, Jr. in the mid-1900s, and widely discussed in the publications of Johan Galtung that spanned from the 1960s to the mid-1990s (Pureza 2000). In a published sermon, Dr King stated that, "Peace is not merely the absence of some negative force—war, tensions, confusion but it is the presence of some positive force—justice, goodwill, the **power** of the kingdom of God" (King 1956). This differentiation was elaborated substantively in the academic literature by Galtung when he defined positive peace as the absence of structural **violence**, or put more simply, the presence of social justice (Galtung 1969). Today, these concepts and categories are applied across the field of peace studies, including **environmental peacebuilding.**

The Institute of Economics and Peace developed a Global Peace Index (GPI) that has been published since 2007 and has included a positive peace assessment since 2012. Later, they released a Positive Peace Index (PPI) that identifies the eight pillars of positive peace as: Well-functioning Government; Sound Business Environment; Equitable Distribution of Resources; Acceptance of the Rights of Others; Good Relations with Neighbors; Free Flow of Information; High Levels of Human Capital; and Low Levels of Corruption (IEP 2022). The GPI and PPI provide unique insights into the state of global peace, highlighting the foundations necessary for long-term peace in ways that comport with many of the practices and policies of environmental peacebuilding, connecting **governance** and **conflict** resolution with resources and sustainability.

Environmental peacebuilding includes a reactive element that seeks to ameliorate the effects of conflict through direct intervention, as well as proactive initiatives to mitigate further conflict by promoting structural conditions for a society to become more just and sustainable. These latter components often include promoting gender equity, access to resources, increased **participation** in governance, economic opportunities, and invitations to dialogue. Through these mechanisms, environmental peacebuilding encompasses several values and practices that align well with the systemic dynamics of positive peace.

Positive peace also provides a framework for a critique of environmental peacebuilding initiatives that fail to address underlying structural violence or injustice. For example, in the case of Palestine, many environmental peace initiatives rooted in cross-border and cross-community dialogue are implemented but can fall short of achieving positive peace without addressing the underlying structural inequities relating to resource access and **environmental violence** (Barron 2024). Programs focused on transboundary water management may improve relations between neighboring communities and even nation-states in some cases, yet can fail to address inequality in water access and patterns of environmental injustice as they intersect with day-to-day lives.

Building positive peace is systemic and complex; it is a continuous process that relies both on state action and promotes societal-level, attitudinal changes, helping to prevent or better manage conflicts. By seeking to implement or strengthen the pillars of positive peace, several other benefits are pursued by extension that relate to environmental peacebuilding, including a freer flow of information that can be a

confidence-building measure between different actors. Similarly, equitable distribution of resources can reduce injustice in each society, creating a better environment for participation in decision-making processes, potentially helping reduce the chance of intractable conflicts and enhancing governance and transparency. Higher levels of resource access and economic opportunities can help ease some conflict drivers by fostering innovative solutions to environmental crises and improving **conservation** and resiliency.

References

Barron, Elsa. 2024. "On Dialogue and Beyond: Positive Environmental Peacebuilding in Palestine." *The Journal of Social Encounters* 8 (1): 21-8. https://doi.org/10.69755/2995-2212.1239.

Galtung, Johan. 1969. "Violence, Peace, and Peace Research." *Journal of Peace Research* 6 (3): 167–91. https://doi.org/10.1177/002234336900600301.

Institute for Economics and Peace (IEP). 2022. *Positive Peace Report*. January. https://www.visionofhumanity.org/wp-content/uploads/2022/01/PPR-2022-web.pdf.

Institute for Economics and Peace (IEP). 2024. *Positive Peace Report*. March. https://www.economicsandpeace.org/wp-content/uploads/2024/04/PPR-2024-web.pdf.

King, Jr., Martin Luther. 1956. "When Peace Becomes Obnoxious." Stanford University King Institute, March 18. https://kinginstitute.stanford.edu/king-papers/documents/when-peace-becomes-obnoxious

Pureza, João Manuel. 2000. "Estudos Sobre a Paz e Cultura da Paz." Speech given at the colloquium on Conflict Prevention and Culture of Peace, National Defense Institute. Lisbon, Portugal, July. https://core.ac.uk/download/pdf/62685955.pdf.

Power

Radhika Singh, Alexandra Turgul, and Gina Gilson

Power is the ability of one actor to influence or control another, either using overt force or through subtle persuasion and framing. By itself, power is neither "bad" nor "good"; rather, it is a tool that can be used (either intentionally or unintentionally) to achieve certain ends. Some types of power are formalized and clearly designated, but other types of power are informal and embedded in everyday practices, culture, and social norms (Hobbes 1998; Weber 1947; Foucault 1991; Waltz 1986).

Classical theorists conceptualize power as having two faces (Bachrach and Baratz 1962). The first is the view of Dahl (1957, 203), who states that power is the ability of "A to get B to do something that B would not otherwise do." This form of "hard power" may involve overt uses of force, including **violence** or coercion, through the targeted deployment of resources (Zeitoun and Warner 2006; Zeitoun and Mirumachi 2008). The second is "soft" power, which aims at "getting others to want what you want" (Nye 1990, 167) through more hidden or subtle means. This type of power may (re)frame problems and attempt to influence other actors' perceptions of the problem and of each other (Zeitoun et al. 2011). Building on these foundations, scholars have proposed new ways of understanding power. For example, Lukes (2005) proposed a "third face" that is ideological, where issues could be suppressed (or promoted) through social influence (Baldwin 2021), for example, by influencing thoughts and desires through targeted exposure to certain types of information or knowledge (Menga 2016). As another example, Partzsch (2017) draws attention to how power is exercised in environmental politics, including power with (**learning** and **cooperation**), power to (resistance and **empowerment**), and power over (coercion and manipulation).

In the field of **environmental peacebuilding**, power is often considered in the context of actors' ability to determine how **natural resources** will be used and distributed. Environmental peacebuilding interventions, and related research, examine the effects of power asymmetries, marginalization of certain groups in political processes, and hegemony of certain actors (Ide et al. 2023). The role of **peacebuilding** processes in neglecting, reinforcing, shifting, or transforming power dynamics is also scrutinized (Dresse et al. 2019; Davis et al. 2023; Teixeira and Nicoson 2024). A conceptualization of both hard and soft power, and an understanding of their complex manifestations in a particular context, is essential to understanding the reasons for and barriers to equitable and sustainable uses of natural resources (Richmond et al. 2021). In transboundary river basins, asymmetrical distributions of power stemming from a combination of factors such as riparian position and access to resources can drive unequal water sharing arrangements. For example, Turkey's upstream position in the Euphrates-Tigris basin and its superior technical capacity have enabled it to construct an extensive dam network on the two rivers at the expense of downstream states (e.g., Warner 2008; Turgul 2025), while in the Nile basin, Egypt wields political and economic power to impose a favorable status quo (e.g., Cascão and Zeitoun 2010).

However, scholars observe a lack of engagement with, and sometimes deliberate obfuscation of, power in environmental peacebuilding research (Ide 2020; Medina et al. 2024). Often, minimal efforts are made to understand local dynamics, initiatives, and historical processes and how they may give rise to peacebuilding processes that are not necessarily aligned with dominant, liberal, Western agendas and institutions (Richmond 2006; Pogodda 2020; Ide et al. 2023; Davis et al. 2023). **Political ecology** literature suggests that by neglecting local histories and power dynamics driving root causes of **conflict**, certain actors and alternative ideas of **peace** may be marginalized, and dynamics contributing to inequity and environmental destruction may be reinforced (Peluso and Watts 2001; Mac Ginty and Richmond 2013; Ide 2020; Sultana 2021). Being explicit about the ways that research and practice are engaging with different conceptualizations, forms, and types of power is important for the effective analysis and **evaluation** of environmental peacebuilding processes (Davis et al. 2023).

Acknowledgment

This work was carried out with support from the Udall Foundation, the CGIAR Initiative on Climate Resilience, ClimBeR, and the CGIAR Initiative on Fragility, Conflict, and Migration. We would like to thank all funders who supported this research through their contributions to the CGIAR Trust Fund: https://www.cgiar.org/funders/.

References

Bachrach, Peter, and Morton S. Baratz. 1962. "Two Faces of Power." *American Political Science Review* 56 (12): 947–52. https://doi.org/10.2307/1952796.

Baldwin, David A. 2021. "The Faces of Power Revisited." *Journal of Political Power* 14 (1): 85–96. https://doi.org/10.1080/2158379X.2021.1878407.

Cascão, Ana Elisa, and Mark Zeitoun. 2010. "Power, Hegemony and Critical Hydropolitics." In *Transboundary Water Management*, 27–42. Routledge. https://www.taylorfrancis.com/chapters/edit/10.4324/9781849776585-4/power-hegemony-critical-hydropolitics-ana-elisa-casc%C3%A3o-mark-zeitoun.

Dahl, Robert A. 1957. "The Concept of Power." *Behavioral Science* 2 (3): 201–15.

Davis, Katy, Laura E. R. Peters, Jamon Van Den Hoek, and Ken Conca. 2023. "Power in Environmental Peacebuilding." *World Development Sustainability* 3: 100110. https://doi.org/10.1016/j.wds.2023.100110.

Dresse, Anaïs, Itay Fischhendler, Jonas Østergaard Nielsen, and Dimitrios Zikos. 2019. "Environmental Peacebuilding: Towards a Theoretical Framework." *Cooperation and Conflict* 54 (1): 99-119. https://www.jstor.org/stable/48596439.

Foucault, Michel. 1991. *Discipline and Punish: The Birth of a Prison*. Penguin.

Hobbes, Thomas. 1998. *Hobbes: On the Citizen*. Cambridge: Cambridge University Press.

Ide, Tobias. 2020. "The Dark Side of Environmental Peacebuilding." *World Development* 127: 104777. https://doi.org/10.1016/j.worlddev.2019.104777.

Ide, Tobias, McKenzie F. Johnson, Jon Barnett, et al. 2023 "The Future of Environmental Peace and Conflict Research." *Environmental Politics* 32 (6): 1077-103. https://doi.org/10.1080/09644016.2022.2156174.

Lukes, Steven. 2005. *Power: A Radical View*. 2nd ed. Palgrave Macmillan.

Mac Ginty, Roger, and Oliver P. Richmond. 2013. "The Local Turn in Peace Building: A Critical Agenda for Peace." *Third World Quarterly* 34 (5): 763-83. https://doi.org/10.1080/01436597.2013.800750.

Medina, Leonardo, Marisa O. Ensor, Frans Schapendonk, et al. 2024. "Community Voices on Climate, Peace and Security: A Social Learning Approach to Programming Environmental Peacebuilding." *Environment and Security* 2 (1): 75–104. https://doi.org/10.1177/27538796231207030.

Menga, Filippo. 2016. "Reconceptualizing Hegemony: The Circle of Hydro-Hegemony." *Water Policy* 18 (2): 401418. https://doi.org/10.2166/wp.2015.063.

Nye, Joseph S. 1990. "Soft Power." *Foreign Policy* 80: 153–71. https://doi.org/10.2307/1148580.

Partzsch, Lena. 2017. "'Power With' and 'Power To' in Environmental Politics and the Transition to Sustainability." *Environmental Politics* 26 (2): 193-211. https://doi.org/10.1080/09644016.2016.1256961.

Peluso, Nancy Lee, and Michael Watts, eds. 2001. *Violent Environments*. Cornell University Press.

Pogodda, Sandra. 2020. "Revolutions and the Liberal Peace: Peacebuilding as Counterrevolutionary Practice?" *Cooperation and Conflict* 55 (3): 347-64. https://doi.org/10.1177/0010836720921881.

Richmond, Oliver P. 2006. "The Problem of Peace: Understanding the 'Liberal Peace'." *Conflict, Security and Development* 6 (3): 291–314. https://doi.org/10.1080/14678800600933480.

Richmond, Oliver P., Roger Mac Ginty, Sandra Pogodda, and Gëzim Visoka. 2021. "Power or Peace? Restoration or Emancipation Through Peace Processes." *Peacebuilding* 9 (3): 243–57. https://doi.org/10.1080/21647259.2021.1911916.

Sultana, Farhana. 2021. "Political Ecology 1: From Margins to Center." *Progress in Human Geography* 45 (1): 156–65. https://doi.org/10.1177/0309132520936751.

Teixeira, Barbara Magalhães, and Christie J. Nicoson. 2024. "Transforming Environmental Peacebuilding: Addressing Extractivism in Building Climate Resilient Peace." *Ecology and Society* 29 (3): 2. https://doi.org/10.5751/ES-14860-290302.

Turgul, Alexandra C. 2025. "Power, Water, and Peace in the Middle East: A Multi-Scalar Analysis of Power and Hydropolitics in the Jordan and Euphrates-Tigris River Basins." PhD diss., Oregon State University.

Waltz, Kenneth. 1986. "Reductionist and Systemic Theories." In *Neorealism and Its Critics*, edited by Robert O. Keohane. Columbia University Press.

Warner, Jeroen. 2008. "Contested Hydrohegemony: Hydraulic Control and Security in Turkey." *Water Alternatives* 1 (2): 271–88. https://www.water-alternatives.org/index.php/volume1/v1issue2/32-a1-2-6/file.

Weber, Max. 1947. *The Theory of Economic and Social Organization*. Translated by A. M. Henderson and Talcott Parsons. Oxford University Press.

Zeitoun, Mark, and Jeroen Warner. 2006. "Hydro-Hegemony–A Framework for Analysis of Trans-Boundary Water Conflicts." *Water Policy* 8 (5): 435–60. https://doi.org/10.2166/wp.2006.054.

Zeitoun, Mark, and Naho Mirumachi. 2008. "Transboundary Water Interaction I: Reconsidering Conflict and Cooperation." *International Environmental Agreements: Politics, Law and Economics* 8 (4): 297–316. https://doi.org/10.1007/s10784-008-9083-5.

Zeitoun, Mark, Naho Mirumachi, and Jeroen Warner. 2011. "Transboundary Water Interaction II: The Influence of 'Soft' Power." *International Environmental Agreements: Politics, Law and Economics* 11: 159–78. https://doi.org/10.22459/GW.05.2014.10.

Protected Zones

Britta Sjöstedt, Elaine (Lan Yin) Hsiao, and Doug Weir

Protected zones are areas of major ecological importance that are recognized to have a certain protection in relation to armed conflict. Such areas of ecological importance may include national parks, wildlife reserves, natural World Heritage Sites, and transboundary conservation areas, which are often negatively affected in times of armed conflict and during post-conflict recovery (Hsiao 2020; Sjöstedt 2021).

The term "protected zones" has been used by the UN **International Law Commission** in Principles 4 and 18 on the Principles on **Protection of the Environment in Relation to Armed Conflicts (PERAC)**. Protected zones may also include areas of cultural importance, such as Indigenous Peoples' Territories of Life (Sajeva et al. 2019). The cultural connection is affirmed in PERAC Principle 5, the Convention Concerning the Protection of the World Cultural and Natural Heritage (the World Heritage Convention), the **Convention on Biological Diversity (CBD)**, and the UN Declaration on the Rights of Indigenous Peoples. The PERAC Principles on protected zones build upon existing international law (Pantazopoulos 2020). Article 60 of Additional Protocol I to the Geneva Conventions provides for establishing demilitarized zones through agreements between warring parties wherein they abstain from conducting military operations in these zones. These zones can be established during peacetime, and they may be created for the purpose of safeguarding natural **environments**. However, if used for certain military purposes, their protected status is revoked. The International Committee of the Red Cross also recommended in its Guidelines on the Protection of the Natural Environment in Armed Conflict that areas of environmental significance or **fragility** could be designated as demilitarized zones (International Committee of the Red Cross 2020).

The idea of protecting ecologically important areas during armed **conflict** is not new. It was discussed during the drafting of the Additional Protocols to the Geneva Conventions but was not included in the final text. Similarly, the San Remo Manual urged warring parties to avoid hostile actions in marine areas with rare ecosystems or endangered species. Article 32 in the International Union for the Conservation of Nature's Draft International Covenant on Environment and Development and Article 9 in the Draft Code for Transboundary Protected Areas also call for protecting natural and cultural sites and avoiding military activities in such areas, including potential demilitarization.

Several multilateral environmental agreements provide for area-based environmental protection that could be considered for protected zone designation, for example, the CBD, the World Heritage Convention, and the Convention on Wetlands of International Importance (the Ramsar Convention). Sites affected by or at risk from armed conflict have been listed under both the World Heritage Convention and the Ramsar Convention.

Protected zones could play a crucial role in protecting environmental and cultural areas that often face secondary impacts during armed conflicts, such as the collapse of institutions, **displacement** of people, and increased exploitation of **natural resources**. Given the overlap between conflict displacement (e.g., refugee and internally displaced populations) and Key Biodiversity Areas, specific recommendations include mapping and **monitoring** these zones, establishing clear protocols for permissible activities during conflict, protecting local **conservation** actors as neutral forces, and fostering international coordination among policymakers, donors, and organizations. These efforts aim to ensure protected zones serve as active tools in **environmental peacebuilding**, promoting **cooperation**, stability, and long-term **peace** (Hsiao et al. 2023).

References

Hsiao, Elaine, Adrian Garside, Doug Weir, and Andrew J. Plumptre. 2023. "Protected Zones in Context: Exploring the Complexity of Armed Conflicts and their Impacts on the Protection of the Environment." *International Review of the Red Cross* 105 (924): 1412–40. https://doi.org/10.1017/S1816383123000450.

Hsiao, Elaine. 2020. "Protecting Protected Areas in Bello: Learning From Institutional Design and Conflict Resilience in the Greater Virunga and Kidepo Landscapes." *Goettingen Journal of International Law* 10 (1): 67-110. https://doi.org/10.3249/1868-1581-10-1-hsiao.

International Committee of the Red Cross (ICRC). 2020. *Guidelines on the Protection of the Natural Environment in Armed Conflict.* https://www.icrc.org/sites/default/files/document_new/file_list/guidelines_on_the_protection_of_the_natural_environment_in_armed_conflict_advance-copy.pdf.

Pantazopoulos, Stavros. 2020. "An Overview of Area-Based Environmental Protection in Relation to Armed Conflicts." Conflict and Environment Observatory. https://ceobs.org/conflicts-and-conservation-the-promise-and-perils-of-protected-zones/.

Sajeva, Giulia, Grazia Borrini-Feyerabend, and Thomas Niederberger. 2019. "Meanings and More..." Policy Brief of the ICCA Consortium No. 7. ICCA Consortium in collaboration with Cenesta. https://www.iccaconsortium.org/2019/11/25/icca-consortium-policy-brief-no-7-meanings-and-more/ .

Sjöstedt, Britta. 2021. *The Role of Multilateral Environmental Agreements: A Reconciliatory Approach to Environmental Protection in Armed Conflict.* Hart Publishing.

Protection of the Environment in Relation to Armed Conflicts

Stavros Evdokimos Pantazopoulos, Kendra Dupuy, and Doug Weir

Protection of the Environment in Relation to Armed Conflicts, also known as PERAC Principles, is a process to strengthen the legal framework protecting the environment from war by gathering knowledge of how the environment is affected by war, and by strengthening or developing the law to make it better at protecting it. The process seeks to change how the environment is treated throughout the cycle of conflicts, to the benefit of both people and ecosystems.

> Legal regulation can help enhance the protection of the **environment** in relation to armed **conflict**. For example, in 1976, the international community adopted the Convention on the Prohibition of Military or Any Other Hostile Use of Environmental Modification Techniques (ENMOD). The convention was a reaction to the growing access to technologies that would allow belligerents to devastate components of the environment and was driven in part by the US use of chemical defoliants during the Vietnam War. Such sensitivities were also visible in the environmental component of Additional Protocol I to the Geneva Conventions, which was agreed upon the following year.
>
> Since then, other legal measures have sought to enhance environmental protection in relation to armed conflict. Following Iraq's invasion of Kuwait in 1990, the **UN Security Council (UNSC)** established the United Nations Compensation Commission (UNCC) to address **reparation** claims, including those for environmental harms. During the 2000s, sources such as international courts, fact-finding missions, and the **United Nations Environment Programme (UNEP)** post-conflict assessments increasingly highlighted the importance of environmental protection throughout conflict cycles. This period reflects growing global environmental awareness and the significant role of judicial and quasi-judicial bodies in advancing environmental protection in the context of armed conflicts (Pantazopoulos 2023).
>
> The period between 2009 and 2022 witnessed the most important developments in the PERAC normative landscape since the 1970s, certainly as far as international instruments are concerned. It began with UNEP's publication of an influential report detailing the law relevant to environmental protection during conflicts (UNEP 2009).
>
> Recommendations in UNEP's 2009 report led to the UN **International Law Commission** beginning its work on PERAC in 2013, which culminated in the adoption of 27 principles on the "Protection of the Environment in Relation to Armed Conflicts" during its 2022 session (International Law Commission 2022). The PERAC principles outline how the environment should be protected before, during, and after armed conflicts, and in situations of occupation (Conflict and Environment Observatory [CEOBS] 2022). They vary in strength from reflecting existing international law to proposals for its progressive development. Examples of key provisions of the principles include: the obligation of States to provide full reparation for environmental harm caused by an illegal act in relation to armed conflict, the obligation of occupying **powers** to respect and protect the environment of the occupied territory, calling on parties to conflicts to actively protect the environment and **natural resources**, and to include environmental provisions in **peace agreements** and post-conflict interventions.
>
> In parallel with the ILC's work to further develop the PERAC principles, the International Committee of the Red Cross (ICRC) updated its environmental guidelines on the protection of the natural environment in armed conflict (ICRC 2020). The ICRC's process complemented that of the ILC, with its guidelines broadly mirroring the PERAC principles applicable during conflicts. Other PERAC projects

during the same period examined the rules protecting water **infrastructure** (Geneva List 2019) and the standards for assisting victims affected by **toxic remnants of war** (HLS International Human Rights Clinic and CEOBS 2020). Another project that emerged during this time and that will become increasingly relevant is the effort to define the crime of **ecocide** in international law (Stop Ecocide Foundation 2021).

Besides the relevant international law-making processes, the environmental dimensions of armed conflicts have assumed higher importance in the international policy agenda (CEOBS 2025). This is demonstrated by the passage of relevant resolutions in the UN Environment Assembly in 2016, 2017, and 2024, and developing debates on climate and **water security**, and environmental damage in the UNSC. In 2016, the Office of the Prosecutor of the International Criminal Court issued a Policy Paper on Case Selection and Prioritisation that evinces the OTP's heightened interest in "prosecuting Rome Statute crimes that are committed by means of, or that result in, inter alia, the destruction of the environment, the illegal exploitation of natural resources or the illegal dispossession of land" (Office of the Prosecutor, International Criminal Court 2016, para 41). In addition to the crime of ecocide, international crimes under the ICC Statute can potentially address environmental harm (Gillett 2022). Notably, in February 2024, the OTP launched a public consultation on a new policy to strengthen **accountability** for **environmental crimes** under the Rome Statute.

References

Conflict and Environment Observatory (CEOBS). 2025. "How Does War Damage the Environment?" May 5. https://ceobs.org/how-does-war-damage-the-environment.

Conflict and Environment Observatory (CEOBS). 2022. "The PERAC Legal Framework, Frequently Asked Questions." https://ceobs.org/perac-principles-frequently-asked-questions/#1663073205542-17413d76-0fcd.

Geneva Water Hub. 2019. *Geneva List of Principles on the Protection of Water Infrastructure.* https://www.genevawaterhub.org/sites/default/files/atoms/files/gva_list_of_principles_protection_water_infra_www.pdf.

Gillett, Matthew. 2022. *Prosecuting Environmental Harm before the International Criminal Court.* Cambridge University Press.

Harvard Law School International Human Rights Clinic and Conflict and Environment Observatory (HLS and CEOBS). 2020. "Confronting Conflict Pollution: Principles for Assisting Victims of Toxic Remnants of War." http://hrp.law.harvard.edu/wp-content/uploads/2020/09/Confronting-Conflict-Pollution.pdf.

International Committee of the Red Cross (ICRC). 2020. *Guidelines on the Protection of the Natural Environment in Armed Conflict.* https://www.icrc.org/sites/default/files/document_new/file_list/guidelines_on_the_protection_of_the_natural_environment_in_armed_conflict_advance-copy.pdf.

International Law Commission (ILC). 2022. Draft Principles on Protection of the Environment in Relation to Armed Conflicts: Texts and Titles of the Draft Preamble and the Draft Principles Adopted by the Drafting Committee on Second Reading. UN Doc. A/CN.4/L.968, May 20, 2022.

Office of the Prosecutor, International Criminal Court. 2016. "Policy Paper on Case Selection and Prioritization." https://www.icc-cpi.int/sites/default/files/itemsDocuments/20160915_OTP-Policy_Case-Selection_Eng.pdf.

Pantazopoulos, Stavros-Evdokimos. 2023. "Greening the Law of Environmental Protection in Armed Conflicts." In *Netherlands Yearbook of International Law 2021*, edited by Daniëlla Dam-de Jong and Fabian Amtenbrink. Springer. https://doi.org/10.1007/978-94-6265-587-4_4.

Stop Ecocide Foundation. 2021. "Independent Expert Panel for the Legal Definition of Ecocide: Commentary and Core Text." https://static1.squarespace.com/static/5ca2608ab914493c64ef1f6d/t/60d1e6e604fae2201d03407f/1624368879048/SE+Foundation+Commentary+and+core+text+rev+6.pdf.
United Nations Environment Programme (UNEP). 2009. *Protecting the Environment during Armed Conflict: An Inventory and Analysis of International Law.* https://www.unep.org/resources/report/protecting-environment-during-armed-conflict-inventory-and-analysis-international.

R

REDD+

Shaadee Ahmadnia

REDD+ is a climate mitigation framework developed to incentivize reduced emissions from deforestation and forest degradation in tropical developing countries.

The acronym REDD+ stands for "Reducing Emissions from Deforestation and Degradation," with the "+" reflecting the role of sustainable forest management, **conservation**, and the enhancement of forest carbon stocks (United Nations Framework Convention on Climate Change n.d.). Results-based payments are provided for implementing these activities and achieving verified carbon emissions reductions, while adhering to standards and safeguards that continue to be refined to ensure positive environmental and social outcomes.

Early REDD+ efforts focused on site-specific projects, which delivered mixed results for communities and the **environment** (Duchelle et al. 2018). As the mechanism has evolved, greater emphasis has been placed on equity, rights, and the inclusion of Indigenous peoples and local communities. Safeguards have been strengthened to address risks such as elite capture and land grabs by "carbon cowboys." The impacts of project-based REDD+ have been scrutinized due to the difficulty of accurately measuring emissions reductions without accounting for the broader landscape. Issues such as the displacement of deforestation to other areas (leakage), baseline manipulation, and limited scalability have led to growing interest in "jurisdictional REDD+" (JREDD+), which implements REDD+ across an entire jurisdiction, such as a country, state, or Indigenous Territory. Buyer countries want to transact JREDD+ carbon credits under Article 6 of the Paris Agreement, while companies seek to engage through the Voluntary Carbon Market. According to Morgan Stanley, the Voluntary Carbon Market is expected to grow from US$2 billion to US$250 billion by 2050 (Morgan Stanley 2023).

Over half of all REDD+ countries are impacted by **conflict** and **fragility** (Blundell et al. 2016). Considering the need for climate finance in fragile contexts, building on the existing REDD+ **infrastructure** offers a key opportunity for **environmental peacebuilding** and pathways to scale climate **mitigation** efforts from local actions to national and global impacts and finance. The development of REDD+ technical, social, and institutional infrastructure, especially at the jurisdictional scale, builds **resilience** to climate and conflict risks. Effectively designed and implemented,

DOI: 10.4324/9781003604143-17

JREDD+ can support strengthened **governance**, **social cohesion**, **empowerment** of local and Indigenous communities, and natural resource management (UN-REDD Programme 2022). At the same time, it is vital to integrate **conflict sensitivity** into jurisdictional REDD+ as the carbon market shifts in favor of these types of credits.

References

Blundell, Harwell, Emily Harwell, and Kerstin Canby. 2016. *REDD+ & Post Conflict*. Oslo REDD-Exchange. Forest Trends & Natural Capital Advisors. https://www.forest-trends.org/wp-content/uploads/imported/oslo-one-pager-v2-pdf.pdf.

Duchelle, Amy E., Gabriela Simonet, William D. Sunderlin, and Sven Wunder. 2018. "What is REDD+ Achieving on the Ground?" *Current Opinion in Environmental Sustainability* 32: 134–40. https://doi.org/10.1016/j.cosust.2018.07.001.

Morgan Stanley. 2023. "Where the Carbon Offset Market is Poised to Surge." https://www.morganstanley.com/ideas/carbon-offset-market-growth#:~:text=With%203%2C800%20more%20projects%20listed,around%20%24250%20billion%20by%202.

United Nations Framework Convention on Climate change (UNFCCC). n.d. "What Is REDD+?" Accessed October 7, 2025. https://unfccc.int/topics/land-use/workstreams/redd/what-is-redd.

UN-REDD Programme. 2022. *Recognizing and Empowering Indigenous Peoples and Local Communities as Critical Partners in Forest Solutions to the Climate Emergency*. UN-REDD Programme, Food and Agriculture Organization of the United Nations, United Nations Development Programme, and UN Environment Programme. https://www.un-redd.org/sites/default/files/2022-03/Final%20IP%20Brief.pdf.

Remote Sensing

Ariel Ozdemir, Marie Schellens, and Emnet Negash

Remote sensing is the science and technology of obtaining information about objects or areas from a distance, typically using satellites, aircraft, or drones.

Remote sensing involves capturing data through sensors that detect and record electromagnetic radiation, such as visible light, infrared, or microwave signals, reflected or emitted from the Earth's surface (Read and Torrado 2009). This data is then processed and analyzed to understand various physical characteristics of the land, water, or atmosphere, including vegetation health, **land use**, climate, and disaster impacts. Remote sensing is widely used in **agriculture**, environmental **monitoring**, **climate change** monitoring, urban planning, and disaster management (McCabe et al. 2017). Although often used interchangeably with the term "**earth observation**," remote sensing is the technique of collecting information about an object, area, or phenomenon, while earth observation is more focused on the study of the Earth itself (Lillesand et al. 2015; Committee on Earth Observation Satellites 2023).

Remote sensing technologies have emerged as valuable tools for **environmental peacebuilding** and **conflict** monitoring (Witmer 2015; Sticher et al. 2023). **Peacebuilding** initiatives involving the environmental dimensions of conflicts and the ensuing ecological and anthropological impacts should consider the environmental data from such conflict-affected areas (Weir et al. 2019). However, given that the sociopolitical insecurities of a conflict and post-conflict contexts frequently impede access to traditional data collection, the use of remote sensing technologies for environmental assessments is often an effective mechanism to sidestep such difficulties (Avtar et al.

2021). These technologies are particularly useful in assessing **environmental degradation** in conflict-affected areas, including deforestation, **land degradation**, and resource exploitation (Garzón and Valánszki 2020).

In regions affected by armed conflict, ground-truthing becomes difficult or impossible. In practice, environmental peacebuilding practitioners often triangulate remote sensing data with other methodologies, including open-source intelligence (OSINT) and civilian (citizen) science (Suman et al. 2024; PAX 2024). The triangulation of remotely sensed data allows for the interpretation of satellite imagery by supporting reliable image analysis, calibration, and accuracy assessment. Some technical limitations to remote sensing capabilities include issues of accessibility due to paywalls; trade-offs between spectral, temporal, and spatial resolution and coverage; the need for clear skies (free of clouds, fog, or smoke) for optical imagery; as well as difficulties penetrating dense surfaces like **forests** or water bodies, limiting effectiveness in capturing subsurface details (Jensen 2016). Lastly, there remains a lack of ethical considerations around the potentially harmful impacts of earth observation data generation, analysis, and publication on already vulnerable communities and ecosystems. The complexities and sensitivities of conflict and environmental contexts may unintentionally expose sensitive information that can be exploited by military, criminal, commercial, armed, or governmental actors, resulting in severe ecological damage and intensified human suffering (Bennett et al. 2024).

References

Avtar, Ram, Asma Kouser, Ashwani Kumar, et al. 2021. "Remote Sensing for International Peace and Security: Its Role." *Remote Sensing* 13: 439. https://doi.org/10.3390/rs13030439.

Bennett, Mia M., Colin J. Gleason, Beth Tellman, Luis F. Alvarez Leon, Hannah K. Friedrich, Ufuoma Ovienmhada, and Adam J. Mathews. 2024. "Bringing Satellites down to Earth: Six Steps to More Ethical Remote Sensing." *Global Environmental Change Advances* 2: 100003. https://doi.org/10.1016/j.gecadv.2023.100003.

Committee on Earth Observation Satellites (CEOS). 2023. *Earth Observation Handbook: Space Data for the Global Stocktake*. https://eohandbook.com/gst/.

Garzón, Fernando Arturo Mendez, and István Valánszki. 2020. "Environmental Armed Conflict Assessment Using Satellite Imagery." *Journal of Environmental Geography* 13 (3–4): 1–14. https://doi.org/10.2478/jengeo-2020-0007.

Jensen, John R. 2016. *Introductory Digital Image Processing: A Remote Sensing Perspective*. Pearson Series in Geographic Information Science.

Lillesand, T., R. W. Kiefer, and J. Chipman. 2015. *Remote Sensing and Image Interpretation*. John Wiley and Sons.

McCabe, Matthew F., Matthew Rodell, Douglas E. Alsdorf, et al. 2017. "The Future of Earth Observation in Hydrology." *Hydrology and Earth System Sciences* 21 (7): 3879–914. https://doi.org/10.5194/hess-21-3879-2017.

Read, Jane M., and Marla J. Torrado. 2009. "Remote Sensing." *International Encyclopedia of Human Geography*: 335–46. https://doi.org/10.1016/B978-008044910-4.00508-3.

PAX. 2024. *Strengthening Collective Environment Action in Iraq*. PAX for Peace. https://paxforpeace.nl/what-we-do/programmes/strengthening-collective-environment-action in-iraq/#content.

Sticher, Valerie, Jan D.Wegner, and Birke Pfeifle. 2023. "Toward the Remote Monitoring of Armed Conflicts." *PNAS Nexus* 2 (6). https://doi.org/10.1093/pnasnexus/pgad181.

Suman, Anna Berti, Gabriel Rojas, and Francesca Panico. 2024. "Mapping Gaps and Opportunities: Community Engagement Through Civic Environmental Monitoring and Environmental Citizen Science in Ukraine." *GROMADA*. https://gromada-erasmus.eu/publications/mapping-gaps-and-opportunities-wp2a1/.

Weir, Doug, Dan McQuillan, and Robert A. Francis. 2019. "Civilian Science: The Potential of Participatory Environmental Monitoring in Areas Affected by Armed Conflicts." *Environmental Monitoring and Assessment* 191 (618): 1–17. https://doi.org/10.1007/s10661-019-7773-9.
Witmer, Frank. 2015. "Remote Sensing of Violent Conflict: Eyes from Above. *International Journal of Remote Sensing* 36 (9): 2326–52. https://doi.org/10.1080/01431161.2015.1035412.

Renewable Energy

Ann J. Le

Renewable energy sources (RERs), also known as clean energy, are derived from natural sources that are replenished at a faster rate than they are depleted. RERs emit significantly less greenhouse gas emissions and air pollutants than fossil fuels and are considered a healthier, more affordable, and environmentally sustainable alternative to fulfilling global energy demands (United Nations Economic Commission for Europe 2022; de Chalendar and Beson 2019; Hussain et al. 2017).

The transition from fossil fuels to RERs—across **power**, heat, and transport sectors—is central to keeping the rise in average global temperatures below the critical climate tipping point of 1.5°C (Intergovernmental Panel on Climate Change 2023; McKay et al. 2022). Developments in both mainstream (e.g., wind, hydropower, solar, biomass, and geothermal) and emerging renewable energy sources (e.g., marine, artificial photosynthesis, cellulosic ethanol, concentrated solar photovoltaics) point toward a portfolio of solutions to curb carbon emissions and reduce reliance on fossil fuels (Olabi and Abdelkareem 2022). However, while the International Renewable Energy Agency (IRENA) estimates that electricity generation can be decarbonized entirely through RERs by 2050, different barriers to their adoption and diffusion remain. These include, but are not limited to, the financial, administrative, social, and technical requirements to make energy transitions to RERs possible (International Renewable Energy Agency 2025;Olabi and Abdelkareem 2022; Hussain et al. 2017).

These challenges have spurred calls for a just energy transition, wherein the needs and rights of people are placed at the forefront of RER development to confer benefits to all. For example, renewable energy projects are a strategic resource to promote **peace** and stability in many fragile and **conflict**-afflicted states with limited energy access. In contexts where local militias or conflict actors control fossil fuel supply chains, decentralized renewable energy (DRE) systems, such as solar micro-grids and small wind turbines, offer a way for local communities to diversify their energy sources, create new economic opportunities, power health and education facilities, and provide entry points for dialogue among conflict parties (Priestley 2020; IRENA 2019). In the long term, increased DRE developments may confer a positive multiplier effect on **peacebuilding** efforts, as improved socioeconomic realities driven by renewable energy investments, in addition to a reduction in the viability of illicit fossil fuel markets, may disincentivize individuals to join armed groups (Lentschig et al. 2024).

References

Armstrong McKay, D. I., Jesse Abrams, Ricarda Winkelmann, et al. 2022. "Exceeding 1.5 °C Global Warming Could Trigger Multiple Climate Tipping Points." *Science* 377 (6611): eabn7950.

de Chalendar, Jacques A., and Sally M. Benson. 2019. "Why 100% Renewable Energy Is Not Enough." *Joule* 3 (6): 1389–93.
Hussain, Akhtar, Syed Muhammad Arif, and Muhammad Aslam. 2017. "Emerging Renewable and Sustainable Energy Technologies: State of the Art." *Renewable and Sustainable Energy Reviews* 71: 12–28. https://doi.org/10.1016/j.rser.2016.12.033
International Renewable Energy Agency. 2019. *A New World: The Geopolitics of the Energy Transformation.* https://www.irena.org/media/files/irena/agency/publication/2019/jan/global_commission_geopolitics_new_world_2019.pdf/.
International Renewable Energy Agency. 2025. *Renewable Power Generation Costs in 2024.* https://www.irena.org/Publications/2025/Jun/Renewable-Power-Generation-Costs-in-2024.
Intergovernmental Panel on Climate Change (IPCC) 2023. *Climate Change 2023: Synthesis Report. Contribution of Working Groups I, II and III to the Sixth Assessment Report of the Intergovernmental Panel on Climate Change.* Edited by H. Lee and J. Romero. Core Writing Team.
Lentschig, Hannah, Louise van Schaik, David Mozersky, Giovanni D'Amico, and Rei Tang. 2024. *Why Renewable Energy Matters in the Context of Peace and Stability.* Clingendael Institute, Energy Peace Partners and Stanley Center for Peace and Security, November.
Olabi, A. G., and Mohammad Ali Abdelkareem. 2022. "Renewable Energy and Climate Change." *Renewable and Sustainable Energy Reviews* 158: Article 112111. https://doi.org/10.1016/j.rser.2022.11211.
Priestley, Cara. 2020. *The Peacebuilding Implications of Energy Transitions to a Carbon-Neutral Future.* Working Paper. Quaker United Nations Office. https://quno.org/sites/default/files/resources/QUNO%20The%20Peacebuilding%20Implications%20%20of%20Energy%20Transitions%20to%20a%20%20Carbon-Neutral%20Future_12OCT2020.pdf.
United Nations Economic Commission for Europe (UNECE). 2022. *Carbon Neutrality in the UNECE Region: Integrated Life-Cycle Assessment of Electricity Sources.* ECE Energy Series Report No. ECE/ENERGY/141. UNECE.

Reparations

Stavros Evdokimos Pantazopoulos and Elsa Barron

Reparations in the context of wartime environmental damage refer to the measures taken by a state, entity, or individual responsible for environmental harm during armed conflict to restore the affected environment or compensate for the damage done. These reparations aim to address harm to ecosystems, natural resources, and livelihoods that depend on a healthy environment. As global understanding of environmental harm deepens, the scope of reparations is increasingly seen as essential for post-conflict recovery and the rights of communities reliant on natural resources.

In the context of **climate change**, reparations refer to climate **justice** demands that fall within a framework of positive **peacebuilding**, even in non-**conflict** scenarios. International climate negotiations within the **United Nations Framework Convention on Climate Change** (**UNFCCC**) have responded to activist demands for reparations in part by creating a fund for loss and damage (United Nations Framework Convention on Climate Change 2022). However, as a voluntary fund lacking an **accountability** mechanism, many advocates argue that a more robust framework is required to make reparations a true **peace** and justice solution. Demands for debt justice are also increasingly cited in connection with demands for climate reparations (Perry 2021).

Reparations may include restitution, compensation, satisfaction, rehabilitation, and guarantees of non-repetition. Restitution seeks to return the **environment** to its pre-damage state, though this is often challenging in practice. Compensation involves financial payments to support environmental recovery or offset losses experienced by affected populations. Satisfaction entails symbolic acts, such as acknowledgments of wrongdoing, apologies, or memorials that contribute to public awareness and healing. Rehabilitation focuses on ecological restoration to enable damaged ecosystems to recover. Finally, guarantees of non-repetition involve measures to prevent future harm, such as legislative reforms, **monitoring** mechanisms, and training on compliance with **international humanitarian law** (IHL) for military forces.

Drawing on general international law, the **International Law Commission**, in its recently adopted Principles on the **Protection of the Environment in Relation to Armed Conflicts (PERAC)**, dedicated Principle 9 to state responsibility (International Law Commission 2022), whose first paragraph stipulates that:

[a]n internationally wrongful act of a State, in relation to an armed conflict, that causes damage to the environment entails the international responsibility of that State, which is under an obligation to make full reparation for such damage, including damage to the environment in and of itself.

The first paragraph of PERAC Principle 9, which reflects customary international law, adjusts the general international law principle of full reparation to the specific case of violations of international law that result in environmental harm in relation to armed conflict. Rule 28 of the 2020 International Committee of the Red Cross Guidelines details further obligations to repress **war** crimes that concern the natural environment (International Committee of the Red Cross 2020). Nonetheless, these provisions neither clarify whether individuals have a right to reparations nor define the suitable forms that reparations should take (Pantazopoulos 2023).

The principle of full reparation notably includes recognition of pure environmental damage—ecosystem harm without direct human or property impact—as a basis for reparations, marking an important step forward. While this concept is relatively new, it aligns with the United Nations Compensation Commission (UNCC)'s earlier stance that environmental harm alone warrants compensation, a precedent established following the Gulf War (Payne 2017).

One of the fundamental goals of comprehensive restitution for environmental damage caused by conflict is to restore the environment's overall ecological functioning (Kong and Zhao 2023, 1461).

The case-law of the **International Court of Justice** reinforces this perspective in cases such as *Costa Rica v. Nicaragua* (International Court of Justice [ICJ] 2015), albeit in a peacetime context, by ruling that environmental damage can indeed require compensation. The ICJ's decision encompassed both compensation and satisfaction as reparation forms, stating that harm impairing an environment's capacity to provide resources or services justifies compensation. For this, the ICJ used a comprehensive "overall assessment" method, valuing the damage from an ecosystem-wide viewpoint (ICJ 2018).

This legal approach is now extending into armed conflict cases, illustrated by the Democratic Republic of the Congo's claim against Uganda for environmental damage during occupation. These developments signal a strengthening consensus in international law that pure environmental damage—especially from wartime activity—deserves meaningful reparations, thereby advancing the framework for environmental accountability in conflict settings alongside international conversations about climate reparations in both conflict and non-conflict settings. Integrating environmental reparations into post-conflict recovery strategies and climate reparations into international diplomacy can strengthen **environmental peacebuilding** by promoting ecological restoration, enhancing community trust, responding to demands for justice, and contributing to inclusive and **sustainable development** outcomes, ultimately building **positive peace.**

References

International Court of Justice (ICJ). 2015. *Certain Activities Carried out by Nicaragua in the Border Area (Costa Rica v. Nicaragua) and Construction of a Road in Costa Rica Along the San Juan River (Nicaragua v. Costa Rica), Merits.* Judgment, December 16.

International Court of Justice (ICJ). 2018. *Certain Activities Carried out by Nicaragua in the Border Area (Costa Rica v. Nicaragua), Compensation Owed by the Republic of Nicaragua to the Republic of Costa Rica.* Judgment, February 2.

International Committee of the Red Cross (ICRC). 2020. *Guidelines on the Protection of the Natural Environment in Armed Conflict.* https://www.icrc.org/sites/default/files/document_new/file_list/guidelines_on_the_protection_of_the_natural_environment_in_armed_conflict_advance-copy.pdf.

International Law Commission (ILC). 2022. "Protection of the Environment in Relation to Armed Conflicts." In *Report of the International Law Commission on the Work of Its 73rd Session (18 April–3 June and 4 July–5 August 2022).* UN Doc. A/77/10, ch. V. A/CN.4/L.968.

Kong, Lingjie, and Yuqing Zhao. 2023. "Remedying the Environmental Impacts of War: Challenges and Perspectives for Full Reparation." *International Review of the Red Cross* 105 (924): 1441–62. https://doi.org/10.1017/S1816383123000280.

Pantazopoulos, Stavros Evdokimos. 2023. "Greening the Law of Environmental Protection in Armed Conflicts." In *Netherlands Yearbook of International Law 2021*, edited by Daniëlla Dam-de Jong and Fabian Amtenbrink. Springer. https://doi.org/10.1007/978-94-6265-587-4_4.

Payne, Cymie R. 2017. "Developments in the Law of Environmental Reparations: A Case Study of the UN Compensation Commission." In *Environmental Protection and Transitions from Conflict to Peace*, edited by Carsten Stahn, Jennifer S. Easterday, and Jens Iverson. Oxford University Press.

Perry, Keston K. 2021. "The New 'Bond-Age', Climate Crisis and the Case for Climate Reparations: Unpicking Old/New Colonialities of Finance for Development within the SDGs." *Geoforum* 126: 361–71. https://doi.org/10.1016/j.geoforum.2021.09.003.

United Nations Framework Convention on Climate Change (UNFCCC). 2022. *COP27 Reaches Breakthrough Agreement on New 'Loss and Damage' Fund for Vulnerable Countries.* UNFCCC. https://unfccc.int/news/cop27-reaches-breakthrough-agreement-on-new-loss-and-damage-fund-for-vulnerable-countries.

Resilience

Adam Savelli, Elise Doumergue, and Adriana Melgar

Resilience is the capacity of individuals, households, communities, or societies to "resist, absorb, accommodate, adapt to, transform, and recover from the effects of a hazard in a timely and efficient manner, including through the preservation and restoration of its essential basic structures and functions through risk management" (United Nations Office for Disaster Risk Reduction [UNDRR] 2015). This multi-scalar phenomenon operates across different timeframes—before, during, and after emergencies—encompassing both long-term adaptive capacity and short-term coping mechanisms (Juncos and Joseph 2020).

Historically, resilience has referred to the ability of a system to preserve normal functioning by absorbing and mitigating the impact of shocks, or for households, communities, or societies to maintain normal daily life amid severe social, political, economic, or environmental hazards. The concept originated in ecological science, exploring how systems maintain stability when faced with shocks, and later expanded to social-ecological systems, examining complex human-**environment** interactions. The inclusion of resilience in the 2015 Sendai Framework for Disaster Risk Reduction (UNDRR 2015) elevated the concept in environmental **governance** discussions, linking it to **climate change**, natural hazards, economic shocks, **conflict**, and **violence**.

In **environmental peacebuilding**, resilience addresses the interplay between environmental challenges and conflict dynamics (Juncos and Joseph 2020). Its significance lies in its multifaceted nature—ecological, social, economic, and institutional—bridging environmental **sustainability** and **human security**. Resilience is fostered through sustainable livelihoods, equitable **resource governance**, climate **adaptation**, basic service provision, and **social cohesion** efforts. The development of social capital plays an important role in building trust, resolving disputes, and facilitating collective action (Aldrich and Meyer 2015). Scholarship around environmental peacebuilding has called for increased scrutiny of the causal pathways linking natural resource management to **peace** and the ways resilience influences this relationship (Johnson et al. 2023). Understanding the conditions under which capacity building efforts result in peace is necessary to unlock the peace-amplifying potential of resilience and to address the structural causes of vulnerability and inequality.

Contemporary resilience thinking focuses on the transformative potential of shocks and opportunities to support systemic transformation (Few et al. 2017; Hellin et al. 2022). The notion of "building back better" by improving rather than simply preserving systems has grown in popularity alongside complementary developmental concepts like transformative adaptation. As resilience is influenced by the quality of governance, the strength of civil society, and the capacity for cross-sector collaboration (Chandler and Reid 2016), this approach is particularly relevant in conflict-affected areas where traditional systems may be unsustainable. Environmental peacebuilders argue that crises can productively influence resource management by stimulating **learning**, adaptation, and renewal (Berkes et al. 2003).

Key resilience actors include households and community-level civil society organizations, which share financial resources, leverage social networks, and provide local support during crises (Juncos and Joseph 2020; Adger 2000; Chandler 2015). Public institutions at subnational and national levels contribute through basic service provision, the effective governance of resources, and emergency preparedness and response mechanisms that enhance adaptive capacity. In severe cases where the impact of a hazard exceeds local capacity, international humanitarian organizations and global funding bodies may provide additional operational and financial support.

Acknowledgment

This work was carried out with support from the CGIAR Science Programs on Climate Action and Food Frontiers and Security. We would like to thank all funders who supported this research through their contributions to the CGIAR Trust Fund: https://www.cgiar.org/funders/.

References

Adger, W. Neil. 2000. "Social and Ecological Resilience: Are They Related?" *Progress in Human Geography* 24 (3): 347–64. https://doi.org/10.1191/030913200701540465.

Aldrich, Daniel P., and Michelle A. Meyer. 2015. "Social Capital and Community Resilience." *American Behavioral Scientist* 59 (2): 254–69. https://doi.org/10.1177/0002764214550299.

Berkes, Fikret, Johan Colding, and Carl Folke. 2003. "Introduction." In *Navigating Social-Ecological Systems: Building Resilience for Complexity and Change*, edited by Fikret Berkes, Johan Colding, and Carl Folke. Cambridge University Press.

Chandler, David. 2015. "Rethinking the Conflict-Poverty Nexus: From Securitising Intervention to Resilience." *Stability: International Journal of Security and Development* 4 (1): 1–14. https://stabilityjournal.org/articles/10.5334/sta.fb.

Chandler, David, and Julian Reid. 2016. *The Neoliberal Subject: Resilience, Adaptation and Vulnerability*. Rowman and Littlefield.

Few, Roger, Daniel Morchain, Dian Spear, Adelina Mensah, and Ramkumar Bendapudi. 2017. "Transformation, Adaptation and Development: Relating Concepts to Practice." *Palgrave Communications* 3 (1): 1–9. https://doi.org/10.1057/palcomms.2017.92.

Hellin, Jon, Giriraj Amarnath, Andrew Challinor, et al. 2022. "Transformative Adaptation and Implications for Transdisciplinary Climate Change Research." *Environmental Research*: Climate 1 (2): 1–9. https://doi.org/10.1088/2752-5295/ac8b9d.

Johnson, McKenzie F., Tobias Ide, and Jesann Gonzalez Cruz. 2023. "Conceptualizing Resilience within Environmental Peacebuilding." *Current Opinion in Environmental Sustainability* 65: 1–7. https://doi.org/10.1016/j.cosust.2023.101362.

Juncos, Ana E., and Jonathan Joseph. 2020. "Resilient Peace: Exploring the Theory and Practice of Resilience in Peacebuilding Interventions." *Journal of Intervention and Statebuilding* 14 (3): 289–302. https://doi.org/10.1080/17502977.2020.1745445.

United Nations Office for Disaster Risk Reduction (UNDRR). 2015. "Sendai Framework for Disaster Risk Reduction 2015–2030." https://www.undrr.org/publication/sendai-framework-disaster-risk-reduction-2015-2030.

Resource Curse

Philippe Le Billon, Erika Weinthal, and Cyril Obi

Resource curse is the argument that natural resource wealth can negatively affect a country's economy and institutions.

The mainstream perspective on the "resource curse" describes a paradoxical phenomenon whereby countries rich in **natural resources** tend to experience poor economic performance, political instability, social inequality, and weakly institutionalized states. This counterintuitive relationship between abundant natural resources and negative socioeconomic outcomes has been a subject of extensive academic research and policy debates (Alssadek and Benhin 2023; Henri 2019; Ross 2015). The concept has its origins in the fields of development economics and political science, with early studies drawing attention to the negative correlation between resource wealth and economic development (Auty 1993; Beblawi 1990; Sachs and Warner 1995; Karl 1997; Savoia and Sen 2021). Explanations have included the "Dutch disease" (i.e., where a surge in natural resource exports leads to a decline in other sectors of the economy, such as manufacturing) and the "rentier effect" (i.e., where governments become reliant on export resource revenues and lose incentives to invest in strong fiscal institutions). The concept has drawn criticisms for promoting resource-determinism while glossing over certain structural and systemic factors that go beyond the 'natural' resource endowment of a country and connect to the political economy of its control, access, and distribution (Obeng-Odoom 2020; Mlambo 2022; Obi 2023). Many studies have emphasized the role of institutions in shaping the impact of resource wealth, suggesting that countries with strong democratic institutions are better able to manage the resource curse (Mehlum et al. 2006; Ross 2001); other studies have emphasized the link between ownership structure and negative political and economic outcomes (Jones Luong and Weinthal 2010; Ross 2013). The "curse" is thus not inevitable but rather a result of specific conditions, including historical and colonial legacies, weak institutions, types of resource endowment involved, and ownership structures (Badeeb et al. 2017; Savoia and Sen 2021; Van der Ploeg 2011).

The significance of the resource curse for **peace** is multifaceted. Natural resources can be a significant factor in **conflicts**, with competition for control over resources exacerbating existing tensions and resource revenues financing belligerents, fueling corruption, and undermining **governance** structures (Le Billon 2014; Lujala 2009). Additionally, **environmental degradation** associated with resource extraction often negatively affects ecosystems and communities, creating a vicious cycle of conflict and environmental harm. Yet, investments in resource sectors, revenue redistribution, as well as foreign incentives to maintain production, can also contribute to peace (Basedau and Lay 2009; Wennmann 2011). Whether a country is at peace or not with respect to resource curse factors also reflects broader geopolitical dimensions and inequities in global transnational relations, underscoring the logics of natural resource extractivism (Alberdi and Barroso 2021; Cottle and Collins 2016; Paarlberg-Kvam 2021).

The resource curse has implications for **environmental peacebuilding** in several ways (Lujala and Rustad 2012). First, it highlights the need for sustainable and equitable management of natural resources to prevent conflicts and promote peace. Second, it underscores the importance of good governance and institutional capacity in mitigating the negative effects of resource wealth. Third, the scholarship on natural resources and post-conflict **peacebuilding** has emphasized the need to invest in economic diversification, natural resource funds, social **justice** and equity, and transparency and **accountability** measures such as the Extractive Industries Transparency Initiative (EITI) so as to avoid the pitfalls of the resource curse and recreating, through post-conflict reforms, "the very systems that led to exploitation, exclusion, and corruption" (Beevers 2021, 169; Lujala and Rustad 2012). Lastly, it emphasizes the interconnectedness of environmental, economic, and social issues, and the need for integrated approaches to resource management, development, and environmental peacebuilding (Weinthal and Johnson 2018).

The resource curse is a complex phenomenon with significant implications for environmental peacebuilding. While the debate over its causes and consequences continues, environmental peacebuilding, engaging with the sustainable management of natural resources and sound management of resource revenues, can help promote good governance, prevent conflicts, foster economic growth, and build a just and lasting peace.

References

Alberdi, Jokin, and Manuel Barroso. 2021. "Broadening the Analysis of Peace in Mozambique: Exploring Emerging Violence in Times of Transnational Extractivism in Cabo Delgado." *Global Society* 35 (2): 229–46. https://doi.org/10.1080/13600826.2020.1772730.

Alssadek, Marwan, and James Benhin. 2023. "Natural Resource Curse: A Literature Survey and Comparative Assessment of Regional Groupings of Oil-Rich Countries." *Resources Policy* 84: 103741. https://doi.org/10.1016/j.resourpol.2023.103741.

Auty, Richard M. 1993. *Sustaining Economic Growth in Mineral Economies: The Resource Curse Thesis*. Routledge.

Badeeb, Ramez A., Hooi Hooi Lean, and Jeremy Clark. 2017. "The Evolution of the Natural Resource Curse Thesis: A Critical Literature Survey." *Resources Policy* 51: 123–34. https://doi.org/10.1016/j.resourpol.2016.10.015.

Basedau, Matthias, and Jann Lay. 2009. "Resource Curse or Rentier Peace? The Ambiguous Effects of Oil Wealth and Oil Dependence on Violent Conflict." *Journal of Peace Research* 46 (6): 757–76. https://doi.org/10.1177/0022343309340500.

Beblawi, Hazem, and Giacomo Luciani. 1990. *The Rentier State*. Routledge.

Beevers, Michael D. 2021. "Natural Resource Governance Reform and Environmental Peacebuilding." In *Routledge Handbook of Environmental Security*, edited by Richard A. Matthew, Evgenia Nizkorodov, and Crystal Murphy. Routledge.

Cottle, Drew, and Joe Collins. 2016. "From Client State to Rentier State?: New Compradors, Transnational Capital, and the Institutionalization of Globalizing Dynamics in Australia, 1990–2013." In *Globalization and Transnational Capitalism in Asia and Oceania*, edited by Jeb Sprague. Routledge.

Henri, Atangana Ondoa. 2019. "Natural Resources Curse: A Reality in Africa." *Resources Policy* 63: 101406. https://doi.org/10.1016/j.resourpol.2019.101406.

Jones Luong, Pauline, and Erika Weinthal. 2010. *Oil Is Not a Curse: Ownership Structure and Institutions in Soviet Successor States*. Cambridge University Press.

Karl, Terry Lynn. 1997. *The Paradox of Plenty: Oil Booms and Petro-States*. University of California Press.

Le Billon, Philippe. 2014. *Wars of Plunder: Conflicts, Profits and the Politics of Resources*. Oxford University Press.

Lujala, Päivi. 2009. "Deadly Combat over Natural Resources: Gems, Petroleum, Drugs, and the Severity of Armed Civil Conflict." *Journal of Conflict Resolution* 53 (1): 50–71. https://doi.org/10.1177/0022002708327644.

Lujala, Päivi, and Siri Aas Rustad. 2012. *High-Value Natural Resources and Post-Conflict Peacebuilding*. Routledge.

Mehlum, Halvor, Karl Moene, and Ragnar Torvik. 2006. "Institutions and the Resource Curse." *The Economic Journal* 116 (508): 1–20. https://doi.org/10.1111/j.1468-0297.2006.01045.x.

Mlambo, Courage. 2022. "Politics and the Natural Resource Curse: Evidence from the Selected African Countries." *Cogent Social Sciences* 8 (1). https://doi.org/10.1080/23311886.2022.2035911.

Obeng-Odoom, Franklin. 2020. *Property, Institutions, and Social Stratification in Africa*. Cambridge University Press.

Obi, Cyril. 2023. "Resource Curse." In *Handbook on Alternative Global Development*, edited by Franklin Obeng-Odoom. Edward Elgar Publishing.

Paarlberg-Kvam, Kate. 2021. "Open-Pit Peace: The Power of Extractive Industries in Post-Conflict Transitions." *Peacebuilding* 9 (3): 289–310. https://doi.org/10.1080/21647259.2021.1897218.

Ross, Michael L. 2001. "Does Oil Hinder Democracy?" *World Politics* 53 (3): 325–61. https://doi.org10.1353/wp.2001.0011.
Ross, Michael L. 2013. *The Oil Curse: How Petroleum Wealth Shapes the Development of Nations*. Princeton University Press.
Ross, Michael L. 2015. "What Have We Learned About the Resource Curse?" *Annual Review of Political Science* 18 (1): 239–59. https://doi.org/10.1146/annurev-polisci-052213040359.
Sachs, Jeffrey D., and Andrew M. Warner. 1995. *Natural Resource Abundance and Economic Growth*. Harvard Institute for International Development. Development Discussion Paper No. 517a.
Savoia, Antonio, and Kunal Sen. 2021. "The Political Economy of the Resource Curse: A Development Perspective." *Annual Review of Resource Economics* 13 (1): 203–23. https://doi.org/10.1146/annurev-resource-100820-092612.
Van der Ploeg, Frederick. 2011. "Natural Resources: Curse or Blessing?" *Journal of Economic Literature* 49 (2): 366–420. https://doi.org/10.1257/jel.49.2.366.
Weinthal, Erika, and Marcela Johnson. 2018. "Post-War Environmental Peacebuilding: Navigating Renewable and Non-Renewable Resources." In *Routledge Handbook of Environmental Conflict and Peacebuilding*, edited by Ashok Swain and Joakim Öjendal. Routledge.
Wennmann, Achim. 2011. "Breaking the Conflict Trap? Addressing the Resource Curse in Peace Processes." *Global Governance* 17: 265–79.

Restorative Justice

Madeleine Loll and Larry A. Swatuk

Restorative justice addresses harm or conflict by emphasizing the repairing of relationships between people, communities, and their environments (Zehr 2002). At its core, restorative justice is relational, participatory, inclusive, and flexible, functioning either as a complement to or an alternative to the formal legal system.

By emphasizing inclusive dialogue, **accountability**, and repair, restorative **justice** provides mechanisms, such as community forums and truth-telling processes, that enable harm to be acknowledged and addressed while co-creating solutions for social restoration (Llewellyn and Philpott 2014; Ide 2019). The concept draws heavily from traditions of justice and peacemaking within some Indigenous communities, many of which view harm as a disruption not only among people but also between communities and the natural world (Cunneen 2002; McCaslin 2005). These cultural foundations are exemplified in the Māori practice of *whakawhanaungatanga*, the process of building relationships and trust within a community, and the use of *hui*, or community assemblies, to resolve disputes (Cunneen 2002). These practices emphasize collective responsibility, a restoration of trust within community networks, and an acknowledgment of the interdependence of all life (McCaslin 2005). Contemporary restorative justice frameworks, while often adapted for institutional contexts, continue to reflect these Indigenous principles by prioritizing dialogue, cooperative decision-making, and relational repair.

Scholars, however, caution against oversimplifying Indigenous cultures and practices. While restorative approaches are evident in communities such as the Māori or Diné (sometimes externally known as Navajo), Indigenous methods of maintaining social order and resolving disputes are diverse and context-specific. Early restorative justice scholarship often sought to validate the field by tracing its roots to traditions said to span all human history. Scholars like Braithwaite (1999) argued that practices resembling restorative justice could be found in ancient civilizations worldwide, suggesting it reflected an enduring human orientation toward relational forms of justice. Yet while modern restorative justice frameworks build on these precedents, they are not identical to them, and their contemporary application reflects specific institutional, cultural, and political contexts.

A renewed interest in Indigenous forms of justice in the 1970s also coincided with Indigenous rights movements in settler-colonial states, such as Australia and New Zealand/Aotearoa. These movements highlighted Indigenous social structures and often employed restorative justice language in advocacy, intertwining the frameworks with indigeneity (Cunneen 2002). Critical scholars such as Cunneen (2008) further warn against the selective appropriation of Indigenous practices by state institutions, noting that this can strip them of cultural integrity and obscure the colonial histories and ongoing struggles of Indigenous peoples. In this sense, while restorative justice carries important insights from Indigenous epistemologies, it must also be understood within broader debates about power, cultural integrity, and **decolonization**.

Beyond questions of appropriation, critics highlight broader risks. Poorly designed restorative processes may reinforce existing inequalities or marginalize vulnerable voices (Roche 2003), while institutionalization can reduce the approach to a procedural tool rather than a transformative paradigm (Gavrielides 2007). These critiques underscore the importance of situating restorative justice in context, resisting universalizing claims, and remaining attentive to power dynamics.

In **environmental peacebuilding**, restorative justice views environmental harm, such as **land degradation** or pollution, as a rupture in social and ecological relationships (Zehr 2002; Forsyth 2022). This harm, therefore, requires processes of acknowledgment, accountability, and repair, which may include participatory spaces for communities to engage in dialogue, share grievances, and collaborate on remedies. These remedies may include ecological restoration, land restitution, recognition of Indigenous rights to land or resources, or the implementation of legal and political safeguards to prevent recurring harm (Llewellyn and Philpott 2014; Ide 2019). Reparative outcomes that intersect culture and the **environment**, such as developing programs that restore customary ecological management practices, can also be seen as a form of restorative justice in action.

An interesting example of restorative justice principles applied in environmental **peacebuilding** is the Maya Biosphere Reserve (MBR), created in 1990 in Guatemala following decades of armed conflict, which left Indigenous and rural communities marginalized from decisions about forest use. Following the 1996 peace accords brokered by international negotiators working with rebel groups, Indigenous communities, and the government of Guatemala, a community forestry concession system was introduced within the MBR, granting communities legal rights to manage and benefit from the forest (Radachowsky et al. 2012). By regaining the right to various forest-based livelihood activities, this model repaired the broken trust between state and communities while helping to restore the ecological health of the forest—embodying restorative justice through its empowerment of affected communities to co-create reparative action and by integrating ecological repair with social reconciliation. By transforming a conflict-ridden frontier into a space where local people co-manage forests, the MBR helped reduce tensions over land and resources, thereby constituting a preventive peacebuilding strategy. However, given ongoing conflicts, it should be noted that restorative justice is not an endpoint, but the beginning of improved social relations (Cifuentes et al. 2025)

When critically applied with careful attention to **power**, equity, and context, restorative justice provides an important relational and dialogical dimension to environmental peacebuilding. It recognizes the interdependence of communities and ecosystems on which **sustainable peace** depends. By addressing harms through fostering cooperative relationships around essential ecosystems, restorative justice provides a path for long-term **peace**, **human security**, and **resilience**.

References

Braithwaite, John. 1999. "Restorative Justice: Assessing Optimistic and Pessimistic Accounts." In *Crime and Justice: A Review of Research*, edited by Michael Tonry. University of Chicago Press.

Cifuentes, Silvia Ximena Ureta, Fabian Marcelo Zamora Mejia, Pedro Arnulfo Pineda Cotzojay, and Marie Ange Ngo Bieng. 2025. "Thirty Years of Community Forest Management in the Maya Biosphere Reserve: A Review of its Successes and Challenges for Conservation and Governance of Tropical Forest Socio-Ecological Systems." *Trees, Forests and People* 21: 100949. https://doi.org/10.1016/j.tfp.2025.100949.

Cunneen, Chris. 2002. "Restorative Justice and the Politics of Decolonization." In *Restorative Justice: Theoretical Foundations*, edited by E. Weitekamp and H. J. Kerner. Willan Publishing.

Cunneen, Chris. 2008. "Reviving Restorative Justice Traditions?" In *The Handbook of Restorative Justice*, edited by Gerry Johnstone and Daniel Van Ness. Willan Publishing.

Forsyth, Melanie, Brunilda Pali, and Felicity Tepper. 2022. "Environmental Restorative Justice: An Introduction and an Invitation." In *The Palgrave Handbook of Environmental Restorative Justice*, edited by Miranda Forsyth, Brunilda Pali, and Felicity Tepper. Palgrave Macmillan. https://doi.org/10.1007/978-3-031-04223-2.

Gavrielides, Theo. 2007. *Restorative Justice Theory and Practice: Addressing the Discrepancy*. European Institute for Crime Prevention and Control.

Ide, Tobias. 2019. "The Impact of Environmental Cooperation on Peacemaking: Definitions, Mechanisms, and Empirical Evidence." *International Studies Review* 21 (3): 327–46. https://doi.org/10.1093/isr/viy014.

International Center for Transitional Justice. 2024. *Transforming Social Relations: Restorative Responses to Massive Human Rights Violations*. https://www.ictj.org/sites/default/files/2024-04/ictj_report_restorative-justice_2024_en_1.pdf.

Llewellyn, Jennifer J., and Daniel Philpott, eds. 2014. *Restorative Justice, Reconciliation, and Peacebuilding*. Studies in Strategic Peacebuilding. Oxford University Press. https://doi.org/10.1093/acprof:oso/9780199364862.001.0001.

McCaslin, Wanda D., ed. 2005. *Justice as Healing: Indigenous Ways*. Living Justice Press.

Radachowsky, Jeremy, Victor H. Ramos, Roan McNab, Erick H. Baur, and Nikolay Kazakov. 2012. "Forest Concessions in the Maya Biosphere Reserve, Guatemala: A Decade Later." *Forest Ecology and Management* 268: 18–28. https://doi.org/10.1016/j.foreco.2011.08.043.

Roche, Declan. 2003. *Accountability in Restorative Justice*. Oxford University Press.

Zehr, Howard. 2002. *The Little Book of Restorative Justice*. Good Books.

Reverberating Effects

Mark Zeitoun, Caroline Pellaton, and David Kaelin

Reverberating effects are those effects that extend beyond the physical damage, point of impact, and time of an attack.

The impact of an attack is known to "reverberate" well beyond the immediate vicinity of the attack and to endure long after hostilities cease. These "reverberating effects" compromise the people, hardware, and consumables, which enable the delivery of public services, and all of which are connected and dependent upon each other. Urban water systems are particularly prone to reverberating effects, for being so directly linked with public health, **food security**, and the natural **environment** (Zeitoun and Talhami 2017). The impact on people can last from days to years, leading to severe health consequences, including waterborne diseases, kidney complications, and other chronic health issues (Abbara et al. 2021).

Originally identified by humanitarian water engineers dissatisfied with body counts as measures of **war**'s toll (International Committee of the Red Cross [ICRC] 2015), the concept has since broadened to encompass levels of impact from explosive weapons (first, second, third-order) (Wille 2016; Baldo and Batault 2022), and related cascading effects (e.g., knock-on, secondary, domino effects) (ICRC 2015). Reverberating effects are particularly relevant in protracted urban conflicts (Wille 2016; Nohle and Robinson 2017; InterAction 2020; Schillinger et al. 2020; Wille and Baldo 2021) and inform emergency preparedness and recovery planning (ICRC et al. 2021).

The concept has direct implications for combatants. The ICRC asserts that IHL requires military targeters to consider the reasonably foreseeable reverberating effects of an attack when assessing its proportionality and precautions (Robinson and Nohle 2016; Giorgou and Zeith 2023). IHL also mandates special protection for repair crews working to mitigate these effects (Tignino and Irmakkesen 2020; ICRC and NorCross 2024). Given that knowledge of water systems and their dependencies increases over time (Zeitoun and Talhami 2021), combatants operating on their own territory, in partnered operations, or while besieging or occupying territory are generally better positioned to anticipate such effects and mitigate civilian harm. As a result, they are expected to uphold a higher standard of care and may face greater scrutiny for non-compliance with **international humanitarian law** notwithstanding recent and well-documented exceptions in Gaza (Zwijnenburg 2023; Geneva Water Hub 2024; Human Rights Watch 2024; Oxfam 2024) and Ukraine (Nikolaieva and Zwijnenburg 2022; Shumilova et al. 2023; International Federation of Red Cross and Red Crescent Societies et al. 2024).

UN Security Council Resolution 2573 (2021) implicitly acknowledges reverberating effects in its call to protect civilian **infrastructure**, because not doing so could "compound the spread of infectious disease" (Zeitoun and Talhami 2023). Efforts to deepen knowledge on reverberating effects (e.g. epidemiological studies (Tabor et al. 2023, United States Department of Defense [US DoD] 2023)) or improving combatants' awareness and recognition of patterns of water and sewage infrastructure (US DoD 2023), further align in spirit and intent with the Political Declaration of Explosive Weapons in Populated Areas (EWIPA), the **Protection of the Environment in Relation to Armed Conflict (PERAC)** principles (Conflict and Environment Observatory 2022), and the Global Alliance to Spare Water from Armed Conflict.

Given the enduring legacy of reverberating effects, the concept is also useful to efforts to promote **peace**. The psychological and economic impact of war (Action on Armed Violence 2021) all form part of the lived and remembered history of societies, through generations (Lederach 2005). Acknowledging the reverberating nature of the effects of war on water systems and people can thus be an entry point to redressing the impact–not least of all for critical **environmental peacebuilding.**

References

Abbara, Aula, Omar Zakieh, Diana Rayes, et al. 2021. "Weaponizing Water as an Instrument of War: Diarrheal Disease and the Syrian Conflict." *International Journal of Infectious Diseases* 108: 202–8. https://doi.org/10.1016/j.ijid.2021.05.030.

Action on Armed Violence (AOAV). 2021. *An Anatomy of a Grad Attack*. Action on Armed Violence.

Baldo, Alfredo Malaret, and Francesca Batault. 2022. *Second Menu of Indicators to Measure the Reverberating Effects on Civilians from the Use of Explosive Weapons in Populated Areas: Indicators for Water, Sanitation and Hygiene, Food Security, Environmental Degradation, and Economic Opportunities*. UN Institute for Disarmament Research.

Conflict and Environment Observatory. 2022. *The Protection of the Environment in Relation to Armed Conflicts (PERAC) Legal Framework: Frequently Asked Questions*. https://ceobs.org/perac-principles-frequently-asked-questions/

Giorgou, Eirini, and Abby Zeith. 2023. "When the Lights Go Out: The Protection of Energy Infrastructure in Armed Conflict." *Humanitarian Law & Policy*, April 20. https://blogs.icrc.org/law-and-policy/2023/04/20/protection-energy-infrastructure-armed-conflict/.

Geneva Water Hub. 2024. *Fully Foreseeable: The Reverberating Effects of War on Water and Health in Gaza*.

Human Rights Watch. 2024. *Extermination and Acts of Genocide: Israel Deliberately Depriving Palestinians in Gaza of Water*.

International Committee of the Red Cross and Norwegian Red Cross (ICRC and NorCross). 2024. *Keeping the Lights On and the Taps Running: Protecting and Facilitating Safer Access for Essential Service Providers in Armed Conflict*.

International Committee of the Red Cross, UNICEF, and International Bank for Reconstruction and Development (ICRC-UNICEF-World Bank). 2021. *Joining Forces to Combat Protracted Crises: Humanitarian and Development Support for Water Supply and Sanitation Providers in the Middle East and North Africa*.

International Committee of the Red Cross. 2015. *Urban Services during Protracted Armed Conflict: A Call for a Better Approach to Assisting Affected People*.

International Federation of Red Cross and Red Crescent Societies, International Committee of the Red Cross, German Red Cross, and Red Cross Red Crescent Climate Centre. 2024. *Navigating Fragility, Conflict and Violence to Strengthen Community Resilience: A Handbook for Disaster Risk Reduction Partners*.

InterAction. "Building the Evidence Base: Addressing the Reverberating Effects of Military Operations on Civilian Life." Report prepared for the Building the Evidence Base: Addressing the Reverberating Effects of Military Operations on Civilian Life Workshop, Virtual, October 2020. https://www.interaction.org/wp-content/uploads/2020/11/Building-the-Evidence-Base-Workshop-Report-October-2020-FINAL.pdf

Lederach, John Paul. 2005. *The Moral Imagination: The Art and Soul of Building Peace*. Oxford University Press.

Nikolaieva, Iryna, and Wim Zwijnenburg. 2022. "Risks and Impacts from Attacks on Energy Infrastructure in Ukraine." In *Environment and Conflict Alert Ukraine*, edited by Neil Hauer. PAX.

Nohle, Ellen, and Isabel Robinson. 2017. "War in Cities: The 'Reverberating Effects' of Explosive Weapons." *Humanitarian Law and Policy*. https://blogs.icrc.org/law-and-policy/2017/03/02/war-in-cities-the-reverberating-effects-of-explosive-weapons/.

Oxfam. 2024. *Water War Crimes: How Israel has Weaponised Water in its Military Campaign in Gaza*. Oxfam International.

Robinson, Isabel, and Ellen Nohle. 2016. "Proportionality and Precautions in Attack: The Reverberating Effects of Using Explosive Weapons in Populated Areas." *International Review*

of the Red Cross 98 (1 (Special Issue on War in Cities): 107–45. https:/doi.org/10.1017/S1816383116000552.
Schillinger, Juliane, Gül Özerol, Şermin Güven-Griemert, and Michiel Heldeweg. 2020. "Water in War: Understanding the Impacts of Armed Conflict on Water Resources and Their Management." *WIREs Water* 7 (6). https://doi.org/10.1002/wat2.1480.
Shumilova, Oleksandra, Klement Tockner, Alexander Sukhodolov, et al. 2023. "Impact of the Russia–Ukraine Armed Conflict on Water Resources and Water Infrastructure." *Nature Sustainability* 6 (5): 578–86. https://doi.org/10.1038/s41893-023-01068-x.
Tabor, Ruby, Naser Almhawish, Ibrahim Aladhan, et al. 2023. "Disruption to Water Supply and Waterborne Communicable Diseases in Northeast Syria: Apatiotemporal Analysis." *Conflict and Health* 17 (1): 4. https://doi.org/10.1186/s13031-023-00502-3.
Tignino, Mara, and Öykü Irmakkesen. 2020. "The Geneva List of Principles on the Protection of Water Infrastructure: An Assessment and the Way Forward." *International Water Law* 5 (2): 3-104.
United States Department of Defense (US DoD). 2023. "Civilian Harm Mitigation and Response." In *DoD Instruction 3000.17*. US DoD.
Wille, Christina. 2016. "The Implications of the Reverberating Effects of Explosive Weapons Use in Populated Areas for Implementing the Sustainable Development Goals." In *UNIDIR Resources*. UN Institute for Disarmament Research.
Wille, Christina, and Alfredo Malaret Baldo. 2021. *Reference Framework: Menu of Indicators to Measure the Reverberating Effects on Civilians from the Use of Explosive Weapons in Populated Areas*. UN Institute for Disarmament Research.
Zeitoun, Mark, and Michael Talhami. 2017. "The Impact of Explosive Weapons on Urban Services: Direct and Reverberating Effects Across Space and Time." *International Review of the Red Cross* 98 (1 (Special Issue on War in Cities)): 53-70. https://doi.org/10.1017/S1816383117000157.
Talhami, Michael, and Mark Zeitoun. 2021. "The Impact of Attacks on Urban Services II: Reverberating Effects of Damage to Water and Wastewater Systems on Infectious Disease." *International Review of the Red Cross* . 102 (915): 1293-1325. https://doi.org/10.1017/S1816383121000667.
Zeitoun, Mark, and Michael Talhami. 2023. "Armed Conflicts Spread Contaminated Water and Disease: Here's How to Better Protect Civilians." March 17. https://www.justsecurity.org/85514/armed-conflicts-spread-contaminated-water-and-disease-heres-how-to-better-protect-civilians/.
Zwijnenburg, Wim. 2023. *Uninhabitable? The Reverberating Public Health and Environmental Risks from the War in Gaza*. PAX.

Rights of Nature

Cormac Cullinan and Elena Kavanagh

Rights of Nature refers to the inherent rights of natural entities, such as ecosystems, rivers, forests, animals and plants, and of Earth as a whole, including the rights to exist, regenerate, and thrive. The recognition of these rights positions the holders as legal subjects rather than objects and creates corresponding human duties to respect and protect all inherent rights, including those of humans.

Rights of Nature is an application of the legal philosophy known as "Earth jurisprudence" which proposes that the primary purpose of legal systems should be to guide humans to contribute to the ecological integrity, health, and functioning of the Earth community, and not to legitimize and facilitate its exploitation by humans (Cullinan 2002). This perspective is reflected in the Universal Declaration of the Rights of Mother Earth, adopted in Cochabamba, Bolivia on April 22, 2010, which provides a context for species-specific articulations of rights such as the Universal Declaration of Human Rights.

This concept gained prominence in the 21st century, particularly after Ecuador incorporated Rights of Nature in its 2008 Constitution as a means of achieving the societal aspiration of "good living" (*buen vivir* in Spanish and *sumak kawsay* in Quechua). This means living a full life of dignity, plenitude, balance, in harmony with Nature and people, based on unity, solidarity, and empathy (Gudynas 2011). Indigenous Peoples have been key advocates for Rights of Nature, which they see as largely consistent with their understanding that humans are part of Earth and have responsibilities to the other beings and ecological communities that sustain all life.

Within **environmental peacebuilding**, the Rights of Nature movement offers an ecocentric perspective, i.e., that lasting **peace** requires promoting harmonious coexistence within Nature that focuses not only on relationships with other people, but also with the other beings that constitute the Earth community (Allen et al. 2019; Gladwin et al. 1995; Purser and Montuori 1996; Purser et al. 1995). The recognition that humans are members of an indivisible, living community of interrelated and interdependent beings gives rise to a relational approach and a focus on **restorative justice**, which seeks to heal dysfunctional relationships that give rise to exploitation and **conflict** (Llewellyn 2012). This approach challenges the anthropocentric legal objectification of Earth as a collection of **natural resources** available for exploitation, resists capitalist commodification of Nature, and prioritizes the flourishing of life over increasing economic growth.

In practice, this rights-based approach clarifies why it is inappropriate for decisions that have a significant impact on the vitality of Earth to be made on economic criteria by markets, property owners, and those who have acquired great wealth through exploiting Earth. It also reinforces the rights of Indigenous and local communities to protect their territories from exploitation, and positions activists as defenders of fundamental inherent rights instead of criminalizing them for infringing on property rights (e.g., trespass). The soft **power** that comes with adopting Nature's legal standing can also serve as an "institutionalized bridge," translating symbolism into institutionalization (Putzer et al. 2022). This can significantly promote ecological justice and pave the way for a more profound and lasting peace through the established nexus between ecosystems and **peacebuilding** (Weinthal et al. 2011; Krampe 2017). For example, the legal recognition of the Whanganui River in Aotearoa New Zealand as a living entity with rights has not only advanced environmental protection but also contributed to reconciliation processes between the Crown and the Māori, affirming Indigenous cosmologies and strengthening **Indigenous Rights** and **social cohesion** (Cribb et al. 2024). This case illustrates how Rights of Nature can operate as a transformative legal and political tool that links ecological **sustainability** with peacebuilding and justice. At the same time, important challenges remain. Translating recognition into effective enforcement is often undermined by weak institutions, political resistance, and competing development priorities. Tensions also arise when ecocentric approaches confront entrenched anthropocentric legal systems or state-driven economic agendas, creating risks of tokenism and co-optation. Yet despite these obstacles, the incorporation of Rights of Nature into peacebuilding processes broadens the horizon of possibility, reframing peace as dependent not only on human coexistence but also on the flourishing of ecosystems that sustain life.

Applications across disciplines include law and legal theory, political theory and change, **climate change**, ecology and the protection of **biodiversity**, and Indigenous rights. Recognizing the rights of Nature provides both philosophical understandings and practical tools to heal dysfunctional relationships between humans and the ecological communities that sustain us. It also contributes to the protection and flourishing of ecosystems, which is essential to long-term peace and stability.

References

Allen, Stephen, Ann L. Cuniffe, and Mark Easterby-Smith. 2019. "Understanding Sustainability Through the Lens of Ecocentric Radical-Reflexivity: Implications for Management Education." *Journal of Business Ethics* 154 (3): 781–95. https://doi.org/10.1007/s10551-016-3420-3.

Cribb, Miriama, Elizabeth Macpherson, and Axel Borchgrevink. 2024. "Beyond Legal Personhood for the Whanganui River: Collaboration and Pluralism in Implementing the Te Awa Tupua Act." *The International Journal of Human Rights* February: 1–24. https://doi.org/10.1080/13642987.2024.2314532.

Cullinan, Cormac. 2002. *Wild Law*. Siberink.

Gladwin, Thomas N., James J. Kennelly, and Tara-Shelomith Krause. 1995. "Shifting Paradigms for Sustainable Development—Implications for Management Theory and Research." *Academy of Management Review* 20 (4): 874–907. https://doi.org/10.2307/258959.

Gudynas, Eduardo. 2011. "Buen Vivir: Today's Tomorrow." *Development* 54 (4): 441–47. https://doi.org/10.1057/dev.2011.86.

Krampe, Florian. 2017. "Toward Sustainable Peace: A New Research Agenda for Post-Conflict Natural Resource Management." *Global Environmental Politics* 17 (4): 1–8. https://doi.org/10.1162/GLEP_a_00431.

Llewellyn, Jennifer. 2012. "Integrating Peace, Justice and Development in a Relational Approach to Peacebuilding." *Ethics and Social Welfare* 6 (3): 290–302. https://doi.org/10.1080/17496535.2012.704386.

Purser, Ronald E., and Alfonso Montuori. 1996. "Ecocentrism Is in the Eye of the Beholder." *Academy of Management Review* 21 (3): 611–13. https://www.jstor.org/stable/258993.

Purser, Ronald E., Changkil Park, and Alfonso Montuori. 1995. "Limits to Anthropocentrism: Toward an Ecocentric Organization." *Academy of Management Review* 20 (4): 1053–89. https://doi.org/10.2307/258965.

Putzer, Alex, Tineke Lambooy, Ignace Breemer, and Aafje Rietveld. 2022. "The Rights of Nature as a Bridge between Land-Ownership Regimes: The Potential of Institutionalized Interplay in Post-Colonial Societies." *Transnational Environmental Law* 11 (3): 501–23. https://doi.org/10.1017/S2047102522000334.

Weinthal, Erika, Jessica Troell, and Mikiyasu Nakayama. 2011. "Water and Post-Conflict Peacebuilding: Introduction." *Water International* 36 (2): 143–53. https://doi.org/10.1080/02508060.2011.561772.

S

Securitization

Meryam El Bouhati and Toni-Giovanni Pegurri

Securitization is a process where issues are framed primarily through a security lens, often emphasizing military or "hard" security approaches.

The original theory of securitization, as developed by the Copenhagen School, primarily views securitization as a discursive practice, designed to question its performative aspects (Waever 1995). In its early formulation, securitization is regarded more as framing issue than as a fully developed theory. According to Copenhagen School scholars, securitization happens when actors invoke the language of existential threats, turning a policy area not traditionally linked to **security** into a security concern. This process marks a transition from the sphere of routine politics to one characterized by emergency measures and exceptions (Buzan et al. 1998; Fischhendler 2015). Such a shift gives securitization a "post-political" dimension (Balzacq 2018), which effectively depoliticizes the issue. In turn, this suggests the possibility of a reverse process, known as desecuritization, which would lead to (re)politicization (Floyd 2010).

Significant critiques have emerged in response to this interpretation. The supposed dichotomy between securitization and politicization, which parallels the distinction between exceptionalism and routine politics, is generally not backed by empirical evidence (Bourbeau 2014). In authoritarian contexts, for example, security rhetoric is often employed to reinforce or maintain political control (Vuori 2008). As a result, contemporary scholarship frequently reverses this relationship, viewing securitization as a form of politicization and desecuritization as a means of **depoliticization** (Bourbeau 2014; Balzacq 2011). This revised understanding is prominent in studies focused on the "securitization of the **environment**," where securitization is seen not only as a rhetorical framing but also as a strategic tool. It is often used to influence the political agenda on environmental issues (Trombetta 2008; Balzacq et al. 2016) or to secure development aid for specific policy sectors (Brow and Grävingholt 2016).

Critics argue that framing **climate change** in terms of security may lead to the militarization of the issue, overshadowing its social, environmental, and developmental dimensions in favor of security-focused interventions (Daoudy et al. 2022; Oels 2012). In regions like West Africa and the Middle East, **natural resources** are

DOI: 10.4324/9781003604143-18

frequently viewed as national security concerns, leading to the securitization of resources and their management (Weinthal et al. 2015, Massé and Lunstrum 2016; Charbonneau 2022). For instance, in the Middle East and North African (MENA) region, fears about the strain on already scarce water resources caused by external migrants, including refugees, have further driven the securitization of the links between migration and water in countries like Israel, Jordan, and Syria (Weinthal et al. 2015). The same dynamics are observed in the West African Sahel, a region facing armed **conflicts**, insurgencies, jihadist terrorist groups, coups d'état, and the failure of international military intervention. It has emerged as one of the most prominent case studies in **climate security** analyzes, standing at the crossroads of debates on the roots of conflict, climate change, and security (Charbonneau 2022).

Securitization, rather than being a simple shift from routine politics to emergency measures, often serves as a strategic tool for political control and agenda-setting, especially in authoritarian regimes and regions facing resource scarcity. As securitization justifies extraordinary measures, it raises concerns about the potential for militarization and the marginalization of more comprehensive approaches to environmental challenges.

References

Balzacq, Thierry, ed. 2011. *Securitization Theory: How Security Problems Emerge and Dissolve*. Routledge.

Balzacq, Thierry, Sarah Léonard, and Jan Ruzicka. 2016. "'Securitization' Revisited: Theory and Cases." *International Relations* 30 (4): 494–531. https://doi.org/10.1177/0047117815596590.

Balzacq, Thierry. 2018. "Théories de la Sécuritisation, 1989–2018." *Etudes Internationales* 49 (1): 7–24. https://doi.org/10.7202/1050542ar.

Bourbeau, Philippe. 2014. "Moving Forward Together: Logics of the Securitization Process." *Millennium* 43 (1): 187–206. https://doi.org/10.1177/0305829814541504.

Brow, Stephen, and Jörn Grävingholt. 2016. "Security, Development and the Securitization of Foreign Aid." In *The Securitization of Foreign Aid*, edited by Stephen Brow and Jörn Grävingholt. Palgrave Macmillan. https://dx.doi.org/10.1007/978-1-137-56882-3_1.

Brown, Stephen, Jörn Grävingholt, and Rosalind Raddatz. 2016. "The Securitization of Foreign Aid: Trends, Explanations and Prospects." In *The Securitization of Foreign Aid*, edited by Stephen Brow and Jörn Grävingholt. Palgrave Macmillan. http://dx.doi.org/10.1007/978-1-137-56882-3_11.

Buzan, Barry, Ole Waever, and Jaap de Wilde. 1998. *Security: A New Framework for Analysis*. Lynne Rienner.

Charbonneau, Bruno. 2022. "The Climate of Counterinsurgency and the Future of Security in the Sahel." *Environmental Science and Policy* 138: 97–104. https://doi.org/10.1016/j.envsci.2022.09.021.

Daoudy, Marwa, Jeannie Sowers, and Erika Weinthal. 2022. "What Is Climate Security? Framing Risks Around Water, Food, and Migration in the Middle East and North Africa." *WIREs Water* 9 (3). https://doi.org/10.1002/wat2.1582.

Fischhendler, Itay. 2015. "The Securitization of Water Discourse: Theoretical Foundations, Research Gaps and Objectives of the Special Issue." *International Environmental Agreements* 15: 245–55. https://doi.org/10.1007/s10784-015-9277-6.

Floyd, Rita. 2010. *Security and the Environment: Securitisation Theory and us Environmental Security Policy*. Cambridge University Press.

Massé, Francis, and Elizabeth Lunstrum. 2016. "Accumulation by Securitization: Commercial Poaching, Neoliberal Conservation, and the Creation of New Wildlife Frontiers." *Geoforum* 69: 227–37. https://doi.org/10.1016/j.geoforum.2015.03.005.

Oels, Angela. 2012. "From 'Securitization' of Climate Change to 'Climatization' of the Security Field: Comparing Three Theoretical Perspectives." In *Climate Change, Human Security and*

Violent Conflict, edited by Jürgen Scheffran, Michael Brzoska, Hans Günter Brauch, Peter Michael Link, and Janpeter Schilling. Springer. https://doi.org/10.1007/978-3-642-28626-1_9.
Trombetta, Maria. 2008. "Environmental Security and Climate Change: Analysing the Discourse." *Cambridge Review of International Affairs* 21 (4): 585–602. https://doi.org/10.1080/09557570802452920.
Vuori, Juha A. 2008. "Illocutionary Logic and Strands of Securitization: Applying the Theory of Securitization to the Study of Non-Democratic Political Orders." European Journal of International Relations 14 (1): 65–99 https://doi.org/10.1177/1354066107087767.
Waever, Ole. 1995. "Securitization and Desecuritization." In *On Security*, edited by Ronnie Lipschutz. Columbia University Press.
Weinthal, Erika, Neda Zawahri, and Jeannie Sowers. 2015. "Securitizing Water, Climate, and Migration in Israel, Jordan, and Syria." *International Environmental Agreements: Politics, Law and Economics* 15 (3): 293–307. https://doi.org/10.1007/s10784-015-9279-4.

Security

Aliakbar Shams and Eugenia Tsatsos

Security is the condition of being protected against hazards, threats, risks, or loss.

Security has conventionally been defined exclusively in terms of the ability of the state to defend its territory and its principal values against military threats. From a military-political point of view, "security," an objective of states to be achieved primarily (but not exclusively) by military policy, is about survival. The narrow focus on external military threats and using political, legal, or coercive instruments has complemented ideas about **power** and interests. Bellany notes, "Security [...] is a relative freedom from **war**, coupled with a relatively high expectation that defeat will not be a consequence of any war that should occur" (Bellany 1981, 102).

The shift from considering security as an individual to primarily a collective good ensured by military or diplomatic means began toward the end of the 18th century, when the security of individuals became subsumed under the security of the nation. The narrow definition of security has traditionally, but not always, taken the referent object—or things that are seen to be existentially threatened and have a legitimate claim to survival—as the state, incorporating government, territory, and society (Buzan and de Wilde 1998). The first move toward expanding the notion of security dates to the 1960s, when Robert McNamara suggested that security implied the freedom of a state to develop and improve its position in the future. He remarked, "Security is the development and without development, there can be no security [...] development means economic, social, and political progress. It means a reasonable standard of living, and reasonable in this context requires continual redefinition" (McNamara 1968, 149–50). In 2005, the UN Secretary-General echoed the same sentiments: "In an increasingly interconnected world, progress in the areas of development, security, and **human rights** must go hand in hand. There will be no development without security and no security without development" (Annan 2005, 6).

Security can take many different forms depending on who or what is at risk. For example, national security focuses on protecting a country from external and internal threats; **human security** centers on safeguarding individuals' rights, dignity (United Nations 2012), and well-being; food, water, and energy security ensure access to essential resources for survival and development; and gender security addresses protection from gender-based violence and inequality. Altogether, security encompasses a broad range of areas that contribute to the safety, stability, and well-being of individuals and societies, allowing people and nations to thrive without fear of harm or deprivation.

Since the end of the Cold War, the frequency of debates on the nature and meaning of security has increased. The meaning of security has widened to include issues on economics, development, **environment**, human rights, migration, etc. (Moran 1990; Crawford 1994; Roberts 1990; Weiner 1992; Homer-Dixon 1991), as well as deepened by moving the focus down to the level of the individual or up to the level of global security, with regional or societal security as possible intermediate points. However, there is still little agreement on the definition and broadened construct of security (Krause and Williams 1996).

Security transcends traditional boundaries of defense and physical protection. Security organizations, such as ASIS International and the United Nations, prioritize the protection of people (human lives), physical assets (property), and intangible assets (rights, freedoms, reputation). ASIS International approaches security through managing and mitigating risks, preventing incidents, and ensuring continuity in operations and business environments. The United Nations approaches security through the broader perspective of safeguarding people from economic, food, health, personal, and environmental threats, emphasizing protecting people's well-being and dignity and protecting individuals from economic instability, health crises, and **conflicts**.

References

Annan, Kofi. 2005. *In Larger Freedom: Towards Development, Security and Human Rights for All: Report of the Secretary-General.* United Nations Publication. https://digitallibrary.un.org/record/543857/files/A_59_2005-EN.pdf.

Bellany, Ian. 1981. "Towards a Theory of International Security." *Political Studies* 29 (1): 102. https://doi.org/10.1111/j.1467-9248.1981.tb01276.x.

Buzan, Barry, Ole Wæver, and Jaap de Wilde. 1998. *Security: A New Framework for Analysis.* Lynne Rienner Publishers.

Crawford, Beverley. 1994. "The New Security Dilemma under International Economic Interdependence." *Millennium: Journal of International Studies* 23 (1): 25–55. https://doi.org/10.1177/03058298940230010401.

Homer-Dixon, Thomas F. 1991. "On the Threshold: Environmental Changes as Causes of Acute Conflict." *International Security* 16 (2): 76–116. https://doi.org/10.2307/2539061.

Krause, Keith, and Michael C. Williams. 1996. "Broadening the Agenda of Security Studies: Politics and Methods." Mershon International Studies Review 40 (2): 229–54. https://doi.org/10.2307/222776.

McNamara, Robert. 1968. *The Essence of Security.* Hodder and Stoughton.

Moran, Theodore. 1990/1991. "International Economics and National Security." *Foreign Affairs* 69: 74–90. https://www.foreignaffairs.com/articles/united-states/1990-12-01/international-economics-and-national-security.

Roberts, Brad. 1990. "Human Rights and International Security." *Washington Quarterly* 13 (2): 65–75. https://doi.org/10.1080/01636609009477636.

United Nations (UN). 2012. *Human Security – General Assembly*, October 25. https://www.un.org/humansecurity/wp-content/uploads/2022/06/N1147622.pdf .

Weiner, Myron. 1992/1993. "Security, Stability and International Migration." *International Security* 17 (3): 91–126. https://doi.org/10.2307/2539131.

Silvopastoralism

Alan Channer, Fakunle John Aremu, and Dalmas Tiampati

Silvopastoralism is the practice of growing trees or other woody perennials and grazing livestock on the same unit of land in ways that are mutually beneficial (Intergovernmental Science-Policy Platform on Biodiversity and Ecosystem Services 2024; International Union of Agroforestry n.d.).

The practice of growing trees and tending livestock on the same unit of land is widespread across geographies and cultures and has a long history. Almost everywhere that humans have tended livestock, from cattle across the savannas of Africa to llamas in the High Andes to pigs in Iberia, forms of silvopastoralist practice are evident (Hezron et al. 2024; Varela et al. 2022; Chillo et al. 2021).

Silvopastoralism as a term, and as a form of agroforestry, came into common usage in the 1970s and 1980s, when the concept of agroforestry itself started to gain traction and developed into a systematic approach to fostering synergies between the husbandry of trees with crops and with livestock (Bene et al. 1977; Moreno and Rolo 2019; Nair et al. 2021). In the case of silvopastoralism, these synergies are multiple: trees can provide fodder, shade, and shelter for animals, enhance soil health through the contribution of organic matter and nutrients, and provide timber, firewood, fruits, nuts, and medicines, while manure from livestock can improve soil structure and fertility. Further, silvopastoral systems can increase **resilience** to climate variability, sequester and store more carbon than pasture alone, and enhance **biodiversity** (Rigueiro-Rodríguez et al. 2010; Sales-Baptista 2021).

Silvopastoralism is germane to **environmental peacebuilding** regarding **conflict** between mobile pastoralists and sedentary farmers, known as **agricultural–pastoral conflict**, and between pastoralist communities. It addresses both the environmental and the socioeconomic drivers of such conflicts and can thereby serve as a tool both for fostering the restoration and sustainable management of rangelands and building **sustainable peace**.

Silvopastoralism works as a tool for environmental peacebuilding in pastoralist-farmer and inter-pastoralist conflicts by: (1) mitigating the root causes of scarcity, notably of food for livestock, (2) fostering collaboration over the shared management of resources, thereby building trust within and between communities (Channer 2015), (3) diversifying and strengthening livelihoods, thereby reducing economic vulnerability, and (4) building agro-ecological systems that are more resilient to climate-induced stresses.

Unfortunately, the value placed by pastoralists themselves on silvopastoralism has often been underestimated, and even willfully ignored, by colonial authorities, central governments, and development agencies (Channer 2015).

The introduction of a term less than 60 years ago to embrace and define a practice almost as old as recorded human history leaves scope for ambiguity and difference of interpretation, and it needs to be used in environmental peacebuilding with care.

The divide between herders and farmers throughout human history is evidenced in the dispute between Cain and Abel in the Old Testament and, symbolically, by the Great Wall of China, built to protect settled agrarian societies from raids by livestock-rearing nomadic communities. The phrase "beyond the pale" derives from the delineation of land by English planters in Ireland—with "the pale" meaning "an area enclosed by a fence." Those "beyond the pale" were regarded as the "wylde Irishe," herding livestock in "waste wylde places" (Spenser 1997). In terms of environmental peacebuilding, silvopastoralist programs conceived and developed in capital **cities** without input from pastoralists themselves are likely to flounder.

In Nigeria, grazing reserves have been imposed by the federal government to lessen conflict between pastoralists and farmers. Although agrosilvopastoralism is part of traditional practice, with Nigerian farmers and pastoralists in many areas highly knowledgeable about indigenous tree species (Jamala et al. 2016), the grazing reserve policy has so far largely failed to integrate **Indigenous knowledge** or foster traditional reciprocal benefits, for example, the manuring of fields by pastoralists' livestock in return for access to forage in farmers' fields.

There is a lot to build on in terms of Indigenous knowledge in many traditional silvopastoral systems. For example, the *Alalili* silvopastoral **conservation** system of the Maasai in Kenya and Tanzania conserves portions of rangeland in the wet season for improved natural regeneration of vegetation for grazing during the dry season (Hezron et al. 2024). The Maasai see trees as part of a whole, providing timber for homesteads, live fencing, traditional medicine, and less tangible benefits such as locations for traditional ceremonies and rituals. Taking Indigenous knowledge fully into account can potentiate the role of silvopastoralism in fostering inter-pastoralist and pastoralist-farmer relations.

Silvopastoralism has an important role in environmental peacebuilding. Furthermore, environmental peacebuilding, with its holistic framing and inculturated approach, can widen and deepen our understanding of what silvopastoralism is.

References

Bene, J. G., H. W. Beall, and W. Cote. 1977. *Trees, Food, and People: Land Management in the Tropics*. International Development Research Centre. https://idl-bnc-idrc.dspacedirect.org/server/api/core/bitstreams/854967b7-07ad-45d9-8171-d23eee6d3d7b/content.

Channer, Alan. 2015. "Trustbuilding and Mobile Pastoralism in Africa." In *Land Restoration: Reclaiming Landscapes for a Sustainable Future*, edited by Chbay Ilan, Martin Frick, and Jennifer Helgeson. Elsevier. https://doi.org/10.1016/B978-0-12-801231-4.00012-4.

Chillo, Verónica, Ana H. Ladio, Jaime Salinas Sanhueza, et al. 2021. "Silvopastoral Systems in Northern Argentine-Chilean Andean Patagonia: Ecosystem Services Provision in a Complex Territory." In *Ecosystem Services in Patagonia*, edited by Pablo L. Peri, Guillermo Martínez Pastur, and Laura Nahuelhual. Springer. https://doi.org/10.1007/978-3-030-69166-0_6.

Hezron, Elkana, Issakwina B. Ngondya, and Linus K. Munishi. 2024. "Sustaining Indigenous Maasai Alalili Silvo-pastoral Conservation Systems for Improved Community Livelihood and Biodiversity Conservation in East African Rangelands." *PLoS One* 19 (5): e0303649. https://doi.org/10.1371/journal.pone.0303649.

International Union of Agroforestry. n.d.. "IUAF." Accessed August 26, 2025. https://iuaf.org/.

Intergovernmental Science-Policy Platform on Biodiversity and Ecosystem Services (IPBES). 2024. "Silvopastoralism." *IPBES Glossary*. Accessed August 26, 2025. https://www.ipbes.net/glossary-tag/silvopastoralism.

Jamala, G., D. Oke, and A. Fajemisin. 2016. "Traditional Silvopastoral Systems and Farmers' Indigenous Knowledge of Fodder Trees and Shrubs in Southern Guinea Savanna, Adamawa State, Nigeria." *Forests and Forest Products Journal* 9: 93–102.
Moreno, Gerardo, and Victor Rolo. 2019. "Agroforestry Practices: Silvopastoralism." In *Agroforestry for Sustainable Agriculture*, edited by Maria Rosa Mosquera-Losada and Ravi Prabhu. Burleigh Dodds Science Publishing. https://doi.org/10.19103/AS.2018.0041.05.
Nair, P.K. Ramchandran, B. Mohan Kumar, and Vimala D. Nair. 2021. "Historical Developments: The Coming of Age of Agroforestry." In *An Introduction to Agroforestry*, edited by P. K. Ramchandran Nair, B. Mohan Kumar, and Virmala D. Nair. Springer. https://doi.org/10.1007/978-3-030-75358-0_1.
Rigueiro-Rodríguez, A., Mercedes Rois-Díaz, and Maria Rosa Mosquera-Losada. 2010. "Integrating Silvopastoralism and Biodiversity Conservation." In *Biodiversity, Biofuels, Agroforestry and Conservation Agriculture*, edited by Eric Lichtfouse. Springer. https://doi.org/10.1007/978-90-481-9513-8_12.
Sales-Baptista, Elvira, and Maria Isabelle Ferraz-de-Oliveira. 2021. "Grazing in Silvopastoral Systems: Multiple Solutions for Diversified Benefits." *Agroforestry Systems* 95: 1–6. https://doi.org/10.1007/s10457-020-00581-8.
Spenser, Edmund. 1997. *A View of the Present State of Ireland*. Oregon University Press.
Varela, Elsa, Anna M. Olaizola, Isabel Blasco, et al. 2022. "Unravelling Opportunities, Synergies, and Barriers for Enhancing Silvopastoralism in the Mediterranean." *Land Use Policy* 118: 106140. https://doi.org/10.1016/j.landusepol.2022.106140.

Social Cohesion

Katharina Löhr, Emily Sample, and Valery Ndagijimana

Social cohesion is the degree of connectedness, trust, and solidarity across individuals within a group, between groups in a society, and between society and the state.

It has increasingly been recognized as key for development and **peace** (Hope Sr. 2020; United Nations Development Programme 2020). It is a long-standing concept that is used in various disciplines within the social and political sciences but has also risen in the global development agenda. Different definitions and conceptualizations of social cohesion exist, referring to different components including social integration, solidarity, equality, place attachment, or identity, all of which are attributable to a collective—namely a community, neighborhood, region, or society (Fearon et al. 2009; Ranci 2011). Social cohesion reflects the level of connectedness, interpersonal trust, and reciprocity within a social group, as well as its norms of behavior, codes of conduct, conventions, and habits. These institutions are central to maintaining **cooperation** in social units and avoiding their disintegration. Such institutions may be formal, state-organized, or informal, with the latter often being underappreciated.

Economic equality, trust, and identity are key determinants of social cohesion (Langer et al. 2017), which can be found on different levels: relationships among individuals of the same group (depending on the extent of perceived inequalities); relationships among individuals across groups (translating into societal trust); and relationships between individuals, groups, and the state (people's adherence to a national identity). The first two are often understood as horizontal cohesion (between members of society, e.g., communities and civil society organizations), and the latter as vertical cohesion (between state and society) (Chan et al. 2006; Cox and Sisk 2017). Improving both horizontal and vertical cohesion has come to be considered a crucial factor for fostering **sustainable development** (Brown and Zahar 2015; King et al. 2010). The State Resilience Index (Fund for Peace n.d.) provides an annual quantitative measure of social cohesion at the national level, defining its key tenets as "solidarity, kinship ties, sociocentrism, and opportunities for generosity and collaboration ... which is necessary to overcome a major crisis."

Both vertical and horizontal cohesion can have distinct effects on societal relations. For example, when horizontal cohesion is strong, communities are more resilient and better able to engage in collective actions, thereby fostering a sense of belonging and, in the end, social cohesion. Vertical cohesion is equally crucial as it underpins **governance** by fostering legitimacy, which in turn promotes compliance and reduces conflicts (Aall and Crocker 2019; Cox et al. 2023; United Nations Development Programme 2020). When both forms of cohesion are strong, societal relationships are characterized by mutual trust, not only within individual communities but also between those communities and the state.

In the context of climate **resilience** and natural resource management, particularly the **conservation** of **biodiversity** and ecosystem restoration, a growing body of literature emphasizes the importance of strong social cohesion for long-term success, identifying it as a key leverage point (Cebrián-Piqueras et al. 2023). Social cohesion plays a vital role in promoting sustainable natural resource management by facilitating collective action and the joint stewardship of resources (Bodin and Crona 2009; Löhr et al. 2021). Furthermore, strong social cohesion can greatly improve the effectiveness of environmental initiatives by promoting cooperation, trust, and shared goals among the **stakeholders** (Ostrom 2009; Pretty 2003; Schiefer and Van der Noll 2017). While social cohesion is considered key for development and peace, research also highlights possible converse impacts, i.e., from natural **resource governance** and management to social cohesion. Conca and Beevers (2018) identified the **environment** as a key ingredient for building **sustainable peace** in **war**-torn societies. Evidence illustrates the profound impact of forest landscape restoration on social cohesion. Through the establishment of community-based organizations and active involvement of local volunteers, forest landscape restoration promotes collaboration in forest management, facilitates **conflict** resolution, and empowers communities (Ullah 2024). Thus, it is paramount to critically reflect on social cohesion when working in the **environmental peacebuilding** context to strengthen approaches that have positive effects on societal well-being and reduce the chances of unintended negative effects. Cultivating trust and addressing systemic inequalities are crucial for environmental peacebuilding. Failing to do so can exacerbate **violence**, especially when **natural resources** are scarce (Colletta and Cullen 2002; Mohamed 2003).

References

Aall, Pamela, and Chester A. Crocker. 2019. "Building Resilience and Social Cohesion in Conflict." *Global Policy* 10: 68–75. https://doi.org/10.1111/1758-5899.12681.

Bodin, Örjan, and Beatrice I. Crona. 2009. "The Role of Social Networks in Natural Resource Governance: What Relational Patterns Make a Difference?" *Global Environmental Change* 19 (3): 366–74. https://doi.org/https://doi.org/10.1016/j.gloenvcha.2009.05.002.

Brown, Michael J., and Joëlle Zahar. 2015. "Social Cohesion as Peacebuilding in the Central African Republic and Beyond." *Journal of Peacebuilding and Development* 10 (1): 10–24. https://doi.org/https://doi.org/10.1080/15423166.2015.1008349.

Cebrián-Piqueras, Miguel A., Ignacio Palomo, Veronica B. Lo, et al. 2023. "Leverage Points and Levers of Inclusive Conservation in Protected Areas." *Ecology & Society* 28 (4). https://doi.org/10.5751/ES-14366-280407.

Chan, Joseph, Ho-Pong To, and Elaine Chan. 2006. "Reconsidering Social Cohesion: Developing a Definition and Analytical Framework for Empirical Research." *Social Indicators Research* 75: 273–302. https://doi.org/10.1007/s11205-005-2118-1.

Colletta, Nat J., and Michelle L. Cullen. 2002. "Social Capital and Social Cohesion: Case Studies from Cambodia and Rwanda." In *The Role of Social Capital in Development: An Empirical Assessment*, edited by Christiaan Grootaert and Thierry van Bastelaer. Cambridge University Press. https://doi.org/10.1017/CBO9780511492600.011.

Conca, Ken, and Michael D. Beevers. 2018. "Environmental Pathways to Peace." In *Routledge Handbook of Environmental Conflict and Peacebuilding*, edited by Ashok Swain and Joakim Öjendal. Routledge.

Cox, Fletcher D., Charlotte Fiedler, and Karina Mross. 2023. *Strengthening Social Cohesion in Conflict-Affected Societies: Potential, Patterns and Pitfalls*. IDOS Policy Brief 3/2023. German Institute of Development and Sustainability (IDOS). https://doi.org/10.23661/ipb3.2023.

Cox, Fletcher D., and Timothy D. Sisk. 2017. "Peacebuilding: A Social Cohesion Approach." In *Peacebuilding in Deeply Divided Societies: Toward Social Cohesion?*, edited by Fletcher D. Cox and Timothy D. Sisk. Palgrave Macmillan. https://doi.org/10.1007/978-3-319-50715-6_2.

Fearon, James D., Macartan Humphreys, and Jeremy M. Weinstein. 2009. "Can Development Aid Contribute to Social Cohesion after Civil War? Evidence from a Field Experiment in Post-conflict Liberia." *American Economic Review* 99 (2): 287–91. https://doi.org/10.1257/aer.99.2.287.

Fund for Peace. n.d. "State Resilience Index." Accessed March 21, 2025. https://fundforpeace.org/SRI/about.html.

Hope Sr., Kempe Ronald. 2020. "Peace, Justice and Inclusive Institutions: Overcoming Challenges to the Implementation of Sustainable Development Goal 16." *Global Change, Peace and Security* 32 (1): 57–77. https://doi.org/10.1080/14781158.2019.1667320.

King, Elisabeth, Cyrus Samii, and Birte Snilstveit. 2010. "Interventions to Promote Social Cohesion in Sub-saharan Africa." *Journal of Development Effectiveness* 2 (3): 336–70. https://doi.org/10.1080/17449057.2010.504552.

Langer, Arnim, Frances Stewart, Kristien Smedts, and Leila Demarest. 2017. "Conceptualising and Measuring Social Cohesion in Africa: Towards a Perceptions-Based Index." *Social Indicators Research* 131 (1): 321–43. https://doi.org/10.1007/s11205-016-1250-4.

Löhr, Katharina, Bujar Aruqaj, Daniel Baumert, et al. 2021. "Social Cohesion as the Missing Link Between Natural Resource Management and Peacebuilding: Lessons from Cocoa Production in Côte D'Ivoire and Colombia." *Sustainability* 13 (23): 13002. https://doi.org/10.3390/su132313002.

Mohamed, Abdul Latif. 2003. "Genocide in Rwanda: The Interplay of Human Capital, Scarce Resources and Social Cohesion." MA thesis. Naval Postgraduate School, Monterey, CA. https://hdl.handle.net/10945/6165.

Ostrom, Elinor. 2009. "A General Framework for Analyzing Sustainability of Social-Ecological Systems." *Science* 325 (5939): 419–22. https://doi.org/10.1126/science.1172133.

Pretty, Jules. 2003. "Social Capital and the Collective Management of Resources." *Science* 302 (5652): 1912–14. https://doi.org//10.1126/science.1090847.

Ranci, Costanzo. 2011. "Competitiveness and Cocial Cohesion in Western European Cities." *Urban Studies* 48 (13): 2789–804. https://doi.org/10.1177/0042098010394688.

Schiefer, David, and Jolanda van der Noll. 2017. "The Essentials of Social Cohesion: A Literature Review." *Social Indicators Research* 132: 579–603. https://doi.org/10.1007/s11205-016-1314-5.

United Nations Development Programme (UNDP). 2020. *Strengthening Social Cohesion: Conceptual Framing and Programming Implications*. UNDP. https://www.undp.org/publications/strengthening-social-cohesion-conceptual-framing-and-programming-implications.

Ullah, Ayat. 2024. "Forest Landscape Restoration and its Impact on Social Cohesion, Ecosystems, and Rural Livelihoods: Lessons Learned from Pakistan." *Regional Environmental Change* 24: 26. https://doi.org/10.1007/s10113-024-02198-4.

Stakeholder

Rachel Weaver, Jill Baggerman, and Larry A. Swatuk

A stakeholder is an individual or group who is impacted by, has an interest in, and/or plays a role in a particular decision, intervention, or context (Bruch and Woomer 2023).

The concept of stakeholders originated in business management theory but has since been adopted across numerous fields, including development, public policy, and international relations (Wheeler and Sillanpää 1998; Buchholz and Rosenthal 2004; Sady 2023). A person or group may be a stakeholder because of their characteristics (such as because of their age, **gender**, or faith), position in society (especially how much influence they have over a situation), location, or interests. Stakeholder identification and engagement require thoughtful attention to **power** dynamics and to who has less influence or agency in decision-making processes. It is essential to be context- and conflict-sensitive when defining and engaging stakeholders, as both identification and engagement can be challenging (Eskerod et al. 2016; Bruch and Woomer 2023).

Stakeholders are typically identified through analyzes, most commonly stakeholder mapping and needs assessments, with the recognition that any organization, project, or policy operates within a web of relationships with various parties who can affect or be affected by its actions (Brugha and Varvasovszky 2000; Franklin 2006; Gaur 2013). Needs assessments can find significant value in incorporating stakeholder perspectives early at the scoping stage to identify what environmental and **conflict** issues exist and how different groups experience these challenges (Bruch and Woomer 2023). Stakeholder analyzes help distinguish between stakeholders who have direct interests or immediate impacts compared to stakeholders with more indirect relationships or broader societal concerns (Riahi 2017). Understanding these relationships, managing competing interests, and seeking solutions that consider multiple perspectives and/or parties is important for any intervention, since it has implications for decision-making and **participatory processes**. Typically speaking, the stakeholders who have a relatively high interest in an intervention or who would experience the most direct impacts of it are considered key stakeholders, both for moral reasons and because their insights increase the likelihood of an intervention's success. Including key stakeholders in decision-making processes ensures that solutions are grounded in local realities and needs, can build legitimacy and trust for interventions, and can draw on various forms of knowledge, including Indigenous Knowledge (Wheeler and Root-Bernstein 2020).

Identifying stakeholders increased in importance in the **peace** field in the early 2000s as part of the push for "local ownership" of peace processes. This understanding has adapted toward the recognition that the vast diversity of peoples involved in or impacted by conflict makes stakeholder identification and analyzes extremely important throughout **peacebuilding** interventions (Connolly and Baggerman 2014). Likewise, in **environmental peacebuilding**, stakeholders encompass a diverse array of actors whose lives, livelihoods, or interests intersect with both environmental resources and conflict dynamics. Identification of stakeholders in environmental peacebuilding includes those whose rights, well-being, and futures are directly tied to environmental or peacebuilding outcomes. For instance, communities that depend on natural resources for survival, Indigenous Peoples with ancestral connections to land, and government agencies responsible for environmental management would each be stakeholders in their respective contexts (Morales-Muñoz et al. 2021). Although engineers and technical professionals are often key stakeholders in environmental conflicts, the issues are never purely technical problems, but rather complex socio-ecological challenges rooted in competing values, needs, and worldviews (Irahola et al. 2022).

Engagement with stakeholders in environmental peacebuilding plays a crucial role in fostering **social cohesion** and trust across conflict divides. Cooperative and inclusive stakeholder engagement over environmental resources can create opportunities for cooperation around shared interests, transform relationships from competition to collaboration, and create positive momentum for broader peacebuilding (Ide 2018; Carius 2006, 2007; Conca et al. 2005). As the field of environmental peacebuilding and its terminology continues to evolve, differing and potentially conflicting understandings of terms can have practical implications on stakeholder inclusion and stakeholder-specific communication barriers (Weaver 2025).

In recent decades, the term "stakeholder" has become contested as a concept (Miles 2012; Mitchell 2012; Reed et al. 2024). Given its etymological roots in the colonizing act of driving a stake into land to declare one's right of possession (and by extension another's dispossession), there is growing debate about whether terms such as "rights-holder," "collaborator," or "partner" might be more appropriate when identifying and drawing together groups and individuals involved in resource use decisions.

Acknowledgment

This definition has been adapted from Bruch and Woomer (2023).

References

Bruch, Carl, and Amanda Woomer. 2023. *Toolkit on Monitoring and Evaluation of Environmental Peacebuilding*. Environmental Law Institute, Environmental Peacebuilding Association, and United States Institute of Peace. https://m-and-e.environmentalpeacebuilding.org/toolkit/.

Brugha, Ruairí, and Zsuzsa Varvasovszky. 2000. "Stakeholder Analysis: A Review." *Health Policy and Planning* 15 (3): 239–46. https://doi.org/10.1093/heapol/15.3.239.

Buchholz, Rogene A., and Sandra B. Rosenthal. 2004. "Stakeholder Theory and Public Policy: How Governments Matter." *Journal of Business Ethics* 51: 143–53. https://doi.org/10.1023/B:BUSI.0000033608.61005.1f.

Carius, Alexander. 2006. "Environmental Cooperation as an Instrument of Crisis Prevention and Peacebuilding: Conditions for Success and Constraints." *adelphi*. https://adelphi.de/system/files/mediathek/bilder/us_503_-_carius_environmental_peacemaking_06-07-02.pdf.

Carius, Alexander. 2007. "Environmental Peacebuilding: Conditions for Success." *Environmental Change and Security Program Report* 12: 59–75. http://nclive.org/cgi-bin/nclsm?url=http://search.proquest.com/scholarly-journals/environmental-peacebuilding-conditions-success/docview/218424864/se-2..

Conca, Ken, Alexander Carius, and Geoffrey D. Dabelko. 2005. "Building Peace Through Environmental Cooperation." In *State of the World 2005*, edited by Worldwatch Institute. Routledge. https://doi.org/10.4324/9781315065748.

Connolly, Lesley, and Jill Baggerman. 2014. "Getting Coherence and Coordination Right: Principles for the Peacebuilding Policy Community." The African Centre for the Constructive

Resolution of Disputes, August. https://www.accord.org.za/publication/getting-coherence-and-coordination-right/
Eskerod, Pernille, Martina Huemann, and Claudia Ringhofer. 2016. "Stakeholder Inclusiveness: Enriching Project Management with General Stakeholder Theory." *Project Management Journal* 46 (6): 42–53. https://doi.org/10.1002/pmj.21546.
Franklin, Aimee. 2006. "Involving Stakeholders in Organization Processes." *International Journal of Public Administration* 24 (4): 385–403. https://doi.org/10.1081/PAD-100000714.
Ide, Tobias. 2018. "Does Environmental Peacemaking Between States Work? Insights on Cooperative Environmental Agreements and Reconciliation in International Rivalries." *Journal of Peace Research* 55 (5): 351–65. https://doi.org/10.1177/0022343317750216.
Irahola, Dennis Avilés, Alejandro Mora-Motta, Aline Barbosa Pereira, et al. 2022. "Integrating Scientific and Local Knowledge to Address Environmental Conflicts: The Role of Academia." *Human Ecology* 50: 911–23. https://doi.org/10.1007/s10745-022-00344-2.
Gaur, Vidhu. 2013. "Understand Your Stakeholders." *International Journal of Advancements in Research and Technology* 2 (1). https://www.academia.edu/92176555/Understand_Your_Stakeholders#:~:text=Examples%20are%20parents%2C%20children%2C%20customers,may%20be%20and%20so%20on.
Miles, Samantha. 2012. "Stakeholders: Essentially Contested or Just Confused?" *Journal of Business Ethics* 108: 285–98. https://doi.org/10.1007/s10551-011-1090-8.
Mitchell, Ronald K. 2012. "Book Review: Stakeholder Theory: Impacts and Prospects." *Organization Studies* 33 (10): 1407–11. https://doi.org/10.1177/0170840612454490.
Morales-Muñoz, Héctor, Katharina Löhr, Michelle Bonatti, Luca Eufemia, and Stefan Sieber. 2021. "Assessing Impacts of Environmental Peacebuilding in Caquetá, Colombia: A Multistakeholder Perspective." *International Affairs* 97 (1): 179–99. https://doi.org/10.1093/ia/iiaa175.
Reed, Mark S., Bethann Garramon Merkle, Elizabeth J. Cook, et al. 2024. "Reimagining the Language of Engagement in a Post-Stakeholder World." *Sustainability Science* 19: 1481–90. https://doi.org/10.1007/s11625-024-01496-4.
Riahi, Youssra. 2017. "Project Stakeholders: Analysis and Management Process." *SSRG International Journal of Economics and Management Studies* 4 (3): 37–42. https://doi.org/10.14445/23939125/IJEMS-V4I3P108.
Sady, Monika. 2023. "The Role of Stakeholders in Sustainable Development." In *Organizing Sustainable Development*, edited by Aneta Kuźniarska, Karolina Mania, and Monika Jedynak. Routledge. https://doi.org/10.4324/9781003379409-19.
Wheeler, David, and Maria Sillanpää. 1998. "Including the Stakeholders: The Business Case." *Long Range Planning* 31 (2): 201–10. https://doi.org/10.1016/S0024-6301(98)00004-1.
Wheeler, Helen, and Meredith Root-Bernstein. 2020. "Informing Decision-Making with Indigenous and Local Knowledge and Science." *Journal of Applied Ecology* 57: 1634–43. https://doi.org/10.1111/1365-2664.13734.
Weaver, Rachel. 2025. "Tracing the Evolution of Terms and Concepts in Environmental Peacebuilding: A Critical Analysis." Master's Thesis, Nicholas School of the Environment, Duke University. https://hdl.handle.net/10161/32242.

Sustainability

Martina Jaskolski and Elena Kavanagh

Sustainability refers to a state or condition that reflects the long-term maintenance of systems and processes. The United Nations defines sustainability as meeting the needs and aspirations of the present without compromising the ability of future generations to meet their own needs (WCED 1987).

The concept can be traced back to an 18th-century forestry concept that limited tree logging to a rate that would allow **forests** to regenerate (Renn et al. 1999). In the context of human-natural resource relationships, sustainability refers to the avoidance of depletion of **natural resources** to ensure the long-term integrity of ecological systems as well as the social and economic well-being of human populations. The concept is anchored in questions on harmonious human-**environment** relations in the context of exponential population growth, economic growth, and limited global resources (Meadows et al. 1972). Sustainability has slightly differing interpretations across disciplines, such as ecology, economics, and social sciences. Achieving sustainability has been applied to numerous global challenges like **climate change**, **biodiversity** loss, and economic inequality.

In the context of global development, the World Commission on Environment and Development, in its 1987 report *Our Common Future,* coined a new approach to development that incorporates present needs and those of future generations (WCED 1987). (The monumental document is also referred to as "the Brundtland Report.") This definition's future dimension warns against the short-term exploitation of finite natural resources to ensure the long-term sustainability of human well-being and prosperity by promoting balanced development of environmental, economic, and social dimensions, often referred to as the three pillars of sustainability: environmental protection, economic growth, and social equity. The concept of **sustainable development** has since shaped the global development agenda driven by the United Nations, starting with the signing of Agenda 21 at the 1992 Earth Summit in Rio de Janeiro, and eventually leading to the Millennium Development and Sustainable Development Goals.

While the terms sustainability and sustainable development are often used interchangeably, they reflect important political differences and deep contestations over the degree and types of transformations and trade-offs required to achieve sustainability. The term sustainable development is often associated with conservative, technocratic, or human-centric approaches to change, while the term sustainability is preferred by those promoting alternative, revolutionary, or ecocentric transformations (Davidson 2000; Hopwood and Mellor 2005; Jacobs 1999). Proponents of the term sustainability point to a potential tension within the terms sustainable and development, with development often being associated with unlimited economic growth, which would be unsustainable. Moreover, postcolonial critiques refer to a lack of integration of local and **Indigenous knowledge** into notions of sustainability, i.e., long-standing practices by Indigenous populations now seen as dominated by northern science and political agendas (Nygren 1999).

Within the field of **environmental peacebuilding**, sustainability is a crucial concept from both the environmental and **conflict** perspectives. When it comes to interdependent **power** relations, a durable **peace** and environmental sustainability are mutually reinforcing (Davis et al. 2023). It is important to manage natural resources in a way that promotes long-term peace, preventing resource-based conflicts and fostering cooperative relationships between communities. As such, the concept is fundamental to creating systems that both protect the environment and contribute to social and economic stability. Examples include transboundary **water cooperation** initiatives that promote peace through shared environmental stewardship, as well as sustainable **agriculture** practices that lower the risk of conflict by reducing resource scarcity. There are, however, concerns over the "dark side" of environmental peacebuilding (Ide 2020) when some **peacebuilding** efforts result in coordinated resource exploitation (Conca 2001), and other **environmental degradation** resulting from the unsustainable political and economic order (Chandler 2020). In environmental peacebuilding, both the malleability of the term sustainability and the way it is pursued may lead to differential peace outcomes. For example, socio-ecological sustainability as a political and practical goal differs from sustaining natural resources in a way that may exclude or dominate others.

In practical terms, sustainability entails the implementation of policies and programs that reduce the vulnerability of livelihoods to environmental challenges (Barnett 2019), prioritizing **renewable energy**, **conservation** efforts, and equitable resource distribution to prevent future environmental degradation and related conflicts. Institutions and communities are encouraged to adopt sustainable practices to reduce competition over scarce resources, thus serving as a catalyst for peacebuilding. However, **political ecology**, structural inequality, and power dynamics affect and shape the ways in which different social groups can implement sustainable practices.

References

Barnett, Jon. 2019. "Global Environmental Change I: Climate Resilient Peace?" *Progress in Human Geography* 43 (5): 927–36. https://doi.org/10.1177/0309132518798077.

Chandler, David. 2020. "Security through Societal Resilience: Contemporary Challenges in the Anthropocene." *Contemporary Security Policy* 41 (2): 195–214. https://doi.org/10.1080/13523260.2019.1659574.

Conca, Ken. 2001. "Environmental Cooperation and International Peace." In *Environmental Conflict*, edited by Paul F. Diehl and Nils Petter Gleditsch. Routledge. https://doi.org/10.4324/9780429500794-11.

Davidson, Julie L. 2000. "Sustainable Development: Business as Usual or a New Way of Living?" *Environmental Ethics* 22 (1): 25–42. https://doi.org/10.5840/enviroethics200022147.

Davis, Katy, Laura E. Peters, Jamon Van Den Hoek, and Ken Conca. 2023. "Power in Environmental Peacebuilding." *World Development Sustainability* 3: 100110. https://doi.org/10.1016/j.wds.2023.100110.

Hopwood, Bill, and Mary Mellor. 2005."Sustainable Development: Mapping Different Approaches." *Sustainable Development* 13 (1): 38–52. https://doi.org/10.1002/sd.244.

Ide, Tobias. 2020. "The Dark Side of Environmental Peacebuilding." *World Development* 127: 104777. https://doi.org/10.1016/j.worlddev.2019.104777.

Jacobs, Michael. 1999. "Sustainable Development as a Contested Concept." In *Fairness and Futurity: Essays on Environmental Sustainability and Social Justice*, edited by Andrew Dobson. Oxford University.

Meadows, Donella H., Dennis L. Meadows, Jørgen Randers, and William W. Behrens III. 1972. *The Limits to Growth: A Report for the Club of Rome's Project on the Predicament of Mankind*. Universe Books.

Nygren, Anja. 1999. "Local Knowledge in the Environment-Development-Discourse from Dichotomies to Situated Knowledges." *Critique of Anthropology* 19 (3): 267–88. https://doi.org/10.1177/0308275X9901900304.

Renn, Ortwin, Anja Knaus, and Hans Kastenholz. 1999. "Wege in die Nachhaltige Zukunft." In *Agenda 21: Vision: Nachhaltige Entwicklung*, edited by Birgit Breuel. Campus Verlag.

World Commission on Environment and Development (WCED). 1987. *Our Common Future*. Oxford University Press.

Sustainable Development

Daniëlla Dam-de Jong

Sustainable development refers to development that "meets the needs of the present without compromising the ability of future generations to meet their own needs" (World Commission on Environment and Development 1987, 8). This definition aptly captures the essence of sustainable development, which is to ensure that all human beings live a life in dignity, both in the present and the future. Sustainable development's principal method for achieving this is integrated policymaking that treats economic, social, and environmental objectives, i.e., the three "pillars" or dimensions of sustainable development, as indivisible and mutually reinforcing (United Nations [UN] 2002; United Nations General Assembly 2015; Barral and Dupuy 2015).

It has long been recognized that sustainable development and **peace** are intrinsically connected. The 1992 Rio Declaration on Environment and Development already stated in Principle 25 that "peace, development and environmental protection are interdependent and indivisible," while Agenda 2030 emphasized that "[t]here can be no sustainable development without peace and no peace without sustainable development" (UN 1992). Indeed, development processes that discriminate between groups in society or that cause serious **environmental degradation** affecting communities will plant the seeds for grievances and social unrest that may become violent (Collier and Hoeffler 2004; Rustad 2016). Conversely, due to their highly destructive and disruptive nature, armed conflicts affect all three dimensions of sustainable development: the economy suffers from additional costs for military expenditure, loss of labor force due to mobilization, and wartime destruction of buildings and **infrastructure**; people may lose access to **basic services**, such as clean water and sanitation, healthcare, and education, and experience food insecurity; and the **environment** suffers from the effects of hostilities, including (toxic) remnants of **war**.

Since a durable peace is premised on socioeconomic progress that is achieved alongside environmental protection and not to its detriment (Das 2023), sustainable development is a key objective of **environmental peacebuilding**. Sustainable development can furthermore provide a framework to support environmental peacebuilding. For instance, a key challenge for environmental peacebuilding revolves around the mining of high-value **natural resources**. Their revenues are much needed to kick-start the economy of **conflict**-affected states and to provide **peace dividends** to win the hearts and minds of the people. However, when poorly managed, mining will cause new tensions and spark (localized) conflicts (Bruch et al. 2016). Sustainable development can provide guidance on how to approach this challenge through its underlying principles. These include sustainable use, inter- and intra-generational equity (solidarity between and within generations), **cooperation**, public **participation**, good **governance**, and interdependence and integration (International Law Association 2002; Das 2023). While these principles may be helpful in guiding environmental peacebuilding, sustainable development also suffers from deficiencies, most importantly a lack of guidance on how the weighing of the three dimensions of sustainable development should be reflected in the outcomes of the decision-making process (Viñuales 2013).

There is thus an urgent need to provide further substance to sustainable development principles to make them more relevant for environmental peacebuilding.

References

Barral, Virginie, and Pierre-Marie Dupuy. 2015. "Principle 4: Sustainable Development through Integration." In *The Rio Declaration on Environment and Development: A Commentary*, edited by Jorge E. Viñuales. Oxford University Press.

Bruch, Carl, Carroll Muffett, and Sandra S. Nichols. 2016. "Natural Resources and Post-Conflict Governance: Building a Sustainable Peace." In *Governance, Natural Resources, and Post-Conflict Peacebuilding*, edited by Carl Bruch, Carroll Muffett, and Sandra S. Nichols. Routledge.

Collier, Paul, and Anke Hoeffler. 2004. "Greed and Grievance in Civil War." *Oxford Economic Papers* 56 (4): 563–95. https://www.jstor.org/stable/3488799.

Das, Onita. 2023. "Sustainable Development and Environmental Peacebuilding." In *Research Handbook on International Law and Environmental Peacebuilding*, edited by Daniëlla Dam-de Jong and Britta Sjöstedt. Edward Elgar Publishing.

International Law Association. 2002. *New Delhi Declaration of Principles of International Law Relating to Sustainable Development.* UN Doc. A/57/329.

Rustad, Siri A. 2016. "Socioeconomic Inequalities and Attitudes toward Violence: A Test with New Survey Data in the Niger Delta." *International Interactions* 42 (1): 106–39. https://doi.org/10.1080/03050629.2015.1048856.

United Nations. 1992. *Rio Declaration on Environment and Development. Declaration of the United Nations Conference on Environment and Development.* Rio de Janeiro, June 13. 31 I.L.M. 874.

United Nations. 2002. *Johannesburg Declaration on Sustainable Development. Annex to Resolution 1 Political Declaration.* UN Doc. A/CONF.199/20.

United Nations General Assembly. 2015. *Transforming our World: The 2030 Agenda for Sustainable Development (Agenda 2030).* UN Doc. A/RES/70/1.

Viñuales, Jorge E. 2013. "The Rise and Fall of Sustainable Development." *Review of European, Comparative and International Environmental Law* 22 (1): 3–13. https://doi.org/10.1111/reel.12021.

World Commission on Environment and Development. 1987. *Our Common Future.* Oxford University Press.

Sustainable Peace

Nancy Boyer

Sustainable peace is an enduring state of the absence of armed conflict, a state made possible by political, economic, cultural, and societal practices adequately addressing the root causes of conflict. This condition, which can also be referred to as durable peace, is generally referred to within specified political boundaries.

The term sustainable peace is widely used in literature to denote conditions in which strong institutions and societal, cultural, and economic practices adequately promote harmony and address potential causes of armed **conflict**, so that enduring **peace** exists within a defined region (Brauch et al. 2016; United Nations Sustainable Development Group 2017; Kuwali 2022). Pursuant to the United Nations Charter, the term was notably used in a concurrent set of resolutions by the United Nations Security Council and the United Nations General Assembly, adopted on April 27, 2016. The goals and concepts are reinforced by other United Nations agreements and reports, including the 2030 Agenda for Sustainable Development. Some scholars, practitioners, and policymakers add the condition that, in addition to societal institutions and practices, adequate attention should be given to preserving the **environment**. Under this notion, sustainable peace encompasses conditions in which environmental **sustainability** is an equally important consideration to strategically developing and maintaining conditions in which material, interpersonal, and societal causes of conflict are adequately addressed. Several handbooks and numerous articles include the term sustainable peace in their titles (Brauch et al. 2016; Krampe 2017; Justino 2019; Kuwali 2022). The Earth Institute at Columbia University hosts the Sustaining Peace Project, which studies the complexity of building and sustaining peace.

A major theme in **environmental peacebuilding** literature is the management of **natural resources** in post-conflict settings with the goal of establishing conditions for maintaining a society without armed conflict. Two such management goals are the following: (1) to equitably distribute the nation's wealth, and (2) to meet development goals. In the words of Krampe (2017), the research agenda in environmental peacebuilding should include sound theory-building on how "to build peace that is socially and politically relevant and desirable as well as ecologically sensitive and viable." Krampe also finds that there are two strands of literature on the theme of building sustainable peace through attention to **natural resources**, one a **cooperation** approach and the other with a resource risk focus.

Focusing on the roles of natural resources in sustaining peace in post-conflict settings and recognizing the unique challenges resource-rich countries face to build and sustain peace after conflict, the **Environmental Law Institute**, the **United Nations Environment Programme**, McGill University, and the University of Tokyo collaborated to collect and edit six volumes on post-conflict peacebuilding. In one of the volumes, *High-Value Natural Resources and* Peacebuilding, Lujala and Rustad (2012) identify opportunities and strategies for converting resource revenues to a peaceful future in the context of "a concise theoretical and practical framework that policy makers, researchers, practitioners, and students can use to understand and address the complex interplay between the management of high-value resources and peace."

References

Brauch, Hans Günter, Úrsula Oswald Spring, John Grin, and Jürgen Scheffran, eds. 2016. *Handbook on Sustainability Transition and Sustainable Peace*. Springer International Publishing.

Justino, Patricia, Katharine Hagerman, Julius Jackson, Indira Joshi, Ilaria Sisto, and Asha Bradley. 2019. "Pathways to Achieving Food Security, Sustainable Peace and Gender Equality: Evidence from Three FAO Interventions." *Development Policy Review* 38 (1): 29–49. https://doi.org/10.1111/dpr.12413.

Krampe, Florian. 2017. "Toward Sustainable Peace: A New Research Agenda for Post-conflict Natural Resource Management." *Global Environmental Politics* 17 (4): 1–8. https://doi.org/10.1162/GLEP_a_00431.

Kuwali, Dan, ed. 2022. *The Palgrave Handbook of Sustainable Peace and Security in Africa*. Palgrave Macmillan.

Lujala, Päivi, and Siri Aas Rustad, eds. 2012. *High-Value Natural Resources and Post-Conflict Peacebuilding*. Routledge. https://doi.org/10.4324/9781849775786.

United Nations Sustainable Development Group (UNSDG). 2017. *What Does 'Sustaining Peace' Mean?* UN Peacebuilding Support Office, January 17. https://unsdg.un.org/sites/default/files/Guidance-on-Sustaining-Peace.170117.final_.pdf.

Systems Approach

Juliane Schillinger and Rejoice Papka

A systems approach reflects and embraces the complexity and interconnectedness of the world. A core tenet of systems approaches is that problems in a system cannot be treated in isolation. Instead, they need to be solved with the entire system in mind, carefully considering all relevant components and interactions. They therefore encourage holistic approaches to problem analysis and solving (Ackoff 1994).

Conceptually, a "system" describes a set of interconnected components that serve a certain purpose. While each individual component in the system holds a function, the overall system also contains its interactions, making the system "more than the sum of its parts" (Meadows 2008; Stephens and Hess 1999). Consider, for example, the human body and its many different organs that fulfill individual functions and collectively keep us alive. In the realm of **environmental peacebuilding**, social-ecological systems depict human-**environment** interactions, such as natural resource management. Such a system comprises the natural resource in question, as well as its users, the institutional framework governing natural resource management, and any physical structures required, such as resource extraction **infrastructure** (Anderies et al. 2004).

When using systems approaches, it is important to clearly define the relevant system and its connections with the "outside world." Systems thinkers distinguish between "closed systems," which are clearly delineated and unaffected by external conditions, and "open systems," which are in steady interaction with their environment and whose boundaries are more ambiguous. While closed systems are easier to depict and analyze, most natural and social systems are, in practice, open (Stephens and Hess 1999). Systems approaches, therefore, generally come with a certain level of abstraction to create a mental model of the system in question. Crucially, systems approaches should be iterative processes, in which this mental model may be updated as new empirical insights come to light.

A system analysis commonly yields system diagrams, which depict the different system components and their interactions. Such interactions may be positive/reinforcing (e.g., an increase in **environmental degradation** increases an area's exposure to climate hazards or intensifies competition for limited resources) or negative/mitigating (e.g., an increase in environmental restoration activities decreases the same area's exposure to climate hazards or decreases competition). Visualizing them in a system diagram helps to identify important processes and feedback loops, which can lead to mutually reinforcing change in a system.

Within the environmental peacebuilding field, systems approaches are used for a range of inquiries that examine the relationship between the environment, peace, and **conflict**. For example, systems analysis by Schillinger et al. (2022) of water resources management during armed conflict shows the role that institutions, financial systems, and other components beyond physical infrastructure can play in enabling or disrupting management processes, and the different ways in which conflicts can affect water resources. Systems analysis by Houiellebecq et al. (2023) of the impacts of conflict on essential services highlights relevant connections between different sectors and the potential for conflict shocks to reverberate through different systems. Both analyzes identified a range of reinforcing feedback loops that can drive vicious cycles of natural resource and service deterioration in the absence of **peace**.

In the context of environmental and **climate security**, systems approaches can be used to identify the local dynamics that link environmental or climate shocks to local tensions. While these are rarely direct linkages, they are connected through disruptions of natural resource-dependent livelihoods such as **agriculture** or fisheries and **food security**. Systems approaches can also help to analyze the drivers of climate- and insecurity-related migration, both locally and on a regional scale across borders.

Through these examples, systems approaches have proved their worth in analyzing environment-climate-conflict interactions. These approaches reject reductionism by enabling a holistic perspective that considers cross-scale dynamics and feedback loops within and across systems. The identification of root causes of environmental challenges, climate vulnerability, and insecurity can inform adaptive and transformative **governance** approaches to environmental peacebuilding.

References

Ackoff, Russell L. 1994. "Systems Thinking and Thinking Systems." *System Dynamics Review* 10 (2–3): 175–88. https://doi.org/10.1002/sdr.4260100206.

Anderies, John M., Marco A. Janssen, and Elinor Ostrom. 2004. "A Framework to Analyze the Robustness of Social-Ecological Systems from an Institutional Perspective." *Ecology and Society* 9 (1): 18. http://www.ecologyandsociety.org/vol9/iss1/art18/.

Houiellebecq, Emma, Kristen MacAskill, and Federico Sittaro. 2023. "Using Systems-Thinking Approaches to Evaluate Impacts to Essential Services in Fragile Contexts: A Case Study on Venezuela." *Civil Engineering and Environmental Systems* 40 (3): 91–128. https://doi.org/10.1080/10286608.2023.2288611.

Meadows, Donella H. 2008. *Thinking in Systems: A Primer*. Chelsea Green Publishing.

Schillinger, Juliane, Gül Özerol, and Michiel Heldeweg. 2022. "A Social-Ecological Systems Perspective on the Impacts of Armed Conflict on Water Resources Management: Case Studies from the Middle East." *Geoforum* 133 (2–3): 101–16. https://doi.org/10.1016/j.geoforum.2022.05.001.

Stephens, William, and Tim Hess. 1999. "Systems Approaches to Water Management Research." *Agricultural Water Management* 40 (1): 3–13. https://doi.org/10.1016/S0378-3774(98)00097-3.

T

Theory of Change

Leonardo Medina and Amanda Woomer

A theory of change (ToC) is a summary of how and why a specific intervention is expected to lead to desired results by mapping the causal relationships between activities and their expected outputs, outcomes, and impacts. A ToC "conveys what you are doing, what results or affects you seek, how and why you will achieve those results, who will be affected, and the assumptions or risks involved" (Bruch and Woomer 2023, 70).

A ToC is used broadly in many intervention types and is usually formed as part of an intervention's **monitoring, evaluation**, and **learning** plan. Even if a ToC is not explicitly formulated, it is the basic logic underlying why an intervention is expected to produce a desired outcome or set of outcomes. There are many guides and resources for forming a ToC (Baggerman et al. 2025; Global Environment Facility 2019). In the context of **environmental peacebuilding**, a ToC serves as a structured framework that articulates how (i.e., the causal pathways) specific interventions related to natural resource management, environmental protection, or climate action can contribute to **peace** and **conflict** resolution or vice versa. Given that environmental peacebuilding work often takes place in contexts where **environmental degradation,** competition for resources, and climate impacts are recognized to exacerbate conflicts (or vice versa), these ToCs should incorporate both the pathways of change and the risks and assumptions associated with key contextual factors.

In practical terms, a ToC in environmental peacebuilding typically begins by mapping out the key environmental, climate, or conflict drivers that the intervention seeks to address. It then outlines the desired changes in both environmental **governance** and sociopolitical relationships that could mitigate or prevent conflict while also protecting the **environment**, ensuring good natural resource management, and/or contributing to climate **resilience**. For example, interventions could focus on improving equitable access to **natural resources**, strengthening local capacities for environmental management, or fostering collaborative governance across different social groups; the ToC would be the logical description underlying how such activities work toward reducing grievances and thereby preventing **violence**.

The ToC specifies the short-, medium-, and long-term outcomes expected from these interventions. For instance, a short-term outcome might be the establishment of trust

DOI: 10.4324/9781003604143-19

between communities through joint resource management, while a medium-term outcome could be reduced tensions or conflicts over access to resources and more sustainable use of those resources. In the long term, these actions may contribute to broader peacebuilding goals, such as the stabilization of conflict-affected regions or the restoration of state legitimacy through more inclusive environmental governance. They can also contribute to sustainable ecosystems and climate resilience.

Considering the complex nature of peace, ToCs in environmental peacebuilding demand approaches that are iterative and adaptive, based on continuous learning from implementation. As such, ToCs should be considered flexible, regularly reviewed, and adjusted in accordance with lessons and developments from an intervention or the context in which it takes place. This can help practitioners use monitoring information to reflect on what works and why, adjusting strategies when certain interventions fail to deliver expected results or the context indicates a different approach. By linking environmental management or climate action to peace outcomes, the ToC becomes a powerful tool for designing, monitoring, evaluating, and learning from environmental peacebuilding programs, ensuring that they are conflict-sensitive and attuned to key social dynamics.

In conflict-affected settings, the process of developing a ToC is critical. Environmental peacebuilding theory and practice underscore the importance of creating ToCs rooted in conflict-sensitive strategies—approaches that integrate an understanding of conflict dynamics and drivers to prevent exacerbating conflict risks and to actively mitigate them. Participatory ToC development, based on conflict analysis informed by local perspectives and everyday experiences, can be particularly effective for navigating complex, conflict-affected environments and for building important, trustful relationships.

Acknowledgment

This work was carried out with support from the CGIAR Initiative on Climate Resilience, ClimBeR, and the CGIAR Initiative on Fragility, Conflict, and Migration. We would like to thank all funders who supported this research through their contributions to the CGIAR Trust Fund: https://www.cgiar.org/funders/.

References

Baggerman, Jill, Esra Cuhadar, and Kristen Wall. 2025. *Monitoring Dialogue in Conflict-Affected Contexts: A Guide for Practitioners*. United States Institute of Peace. https://www.allianceforpeacebuilding.org/afp-publications/monitoring-dialogue-conflict-affected-contexts.

Bruch, Carl, and Amanda Woomer. 2023. *Toolkit on Monitoring and Evaluation of Environmental Peacebuilding*. Environmental Law Institute, Environmental Peacebuilding Association, and United States Institute of Peace. https://m-and-e.environmentalpeacebuilding.org/toolkit/.

Global Environment Facility. 2019. "Theory of Change Primer." December 11. https://www.thegef.org/sites/default/files/council-meeting-Documents/EN_GEF_STAP_C.57_Inf.04_Theory%20of%20Change%20Primer_0.pdf.

Threat Multiplier

Pauline Baudu and Sherri Goodman

Threat multiplier refers to the ways in which climate change effects interact with and exacerbate pre-existing threats and other drivers of instability to contribute to human and national security risks (Goodman and Baudu 2023).

"Threat multiplier" has become a widely used term by scholars and practitioners to describe **climate change** implications for **security** in both the policy realm and climate-security literature. The term was coined in 2007 by the CNA Military Advisory Board under the leadership of Sherri Goodman (CNA Military Advisory Board 2007). The concept has been characterized as "definitional" in setting "a baseline for how to talk about the issue" (Burke 2016) and shaping "the way in which people studying climate policy think about risks" (Sutter 2021). Its use has also been described as "one of the most prominent ways in which the security implications of climate change have been understood" (Greaves 2022).

The term refers to the way the physical impacts of climate change and environmental stress—such as drought, flooding, or extreme weather events—can intensify existing social, political, or economic tensions and exacerbate vulnerability, especially in fragile or post-**conflict** regions. For example, competition over scarce water or arable land can heighten tensions between already fragile communities or nations.

As defined by the authors of the CNA report in 2007:

> Projected climate change will seriously exacerbate already marginal living standards, ... causing widespread political instability and the likelihood of failed states. ... Economic and environmental conditions in already fragile areas will further erode as food production declines, diseases increase, clean water becomes increasingly scarce, and large populations move in search of resources. Weakened and failing governments, with an already thin margin for survival, foster the conditions for internal conflicts, extremism, and movement toward increased authoritarianism and radical ideologies.
>
> (CNA Military Advisory Board 2007)

The use of the term "threat multiplier" led to three ways in which **climate security** risks were better understood (Goodman 2024). First, by framing risks as emanating not from climate change per se but from how it interacts with and aggravates other geopolitical, environmental, economic, and social stressors that can threaten national security, the term helped explain the systemic nature of climate risks and move away from siloed thinking. In doing so, it allowed for the rise of a broader and more comprehensive security approach to climate risks, with responses integrating defense, development, and diplomacy. Second, by highlighting the role of and implications on the military, it emphasized the necessity to incorporate climate change in every aspect of military planning. It consequently brought together the climate and defense communities and got multiple actors engaged in efforts toward increased climate **resilience** of communities and bases. Third, by recognizing that climate change is not only an environmental issue but also a national security concern, it helped broaden the bipartisan coalition of policymakers and practitioners in the United States interested in addressing climate change around military bases and **infrastructure** (Center for Climate and Security 2022; Conger 2022; Werrell and Femia 2017), and highlighted the transnational security aspects of climate risks requiring collective action.

By illustrating how climate change intersects with other drivers of **fragility**, the concept reinforces the need to apply a climate and **environment**-informed lens to address the root causes of conflict in a more holistic way. This also highlights the value of **environmental peacebuilding** strategies—such as the sustainable management of **natural resources**, **environmental cooperation**, and **sustainable development**—as essential tools for preventing or mitigating conflict, promoting **peace**, and building long-term resilience.

Acknowledgment

This entry has been adapted from Goodman and Baudu (2023).

References

Burke, Sharon. 2016. In *The Age of Consequences*, directed by Jared P. Scott. PF Pictures.

Center for Climate and Security. 2022. "Backgrounder – Climate Change and the National Defense Authorization Act." Council on Strategic Risks. https://councilonstrategicrisks.org/wp-content/uploads/2022/06/NDAA-CC-Backgrounder-2022.pdf.

CNA Military Advisory Board. 2007. *National Security and the Threat of Climate Change*. June 27. https://www.cna.org/archive/CNA_Files/pdf/national%20security%20and%20the%20threat%20of%20climate%20change.pdf.

Conger, John. 2022. "Mainstreaming Climate Security: The FY22 National Defense Authorization Act." Council on Strategic Risks, January 3. https://councilonstrategicrisks.org/2022/01/03/mainstreaming-climate-security-the-fy22-national-defense-authorization-act-ndaa/.

Goodman, Sherri, and Pauline Baudu. 2023. *Climate Change as a "Threat Multiplier": History, Uses and Future of the Concept*. The Center for Climate and Security. https://councilonstrategicrisks.org/wp-content/uploads/2023/01/38-CCThreatMultiplier.pdf.

Goodman, Sherri. 2024. *Threat Multiplier: Climate, Military Leadership, and the Fight for Global Security*. Island Press.

Greaves, Will. 2022. "Every Region of Canada is Vulnerable to Climate Insecurity – An Interview with Will Greaves." Interview by CDA Institute. CDA Institute Expert Series, August 8. https://www.youtube.com/watch?v=TfuheUj_hjc.

Sutter, John. 2021. "The Godmother of Climate Security." Heat of the Moment (podcast), *Foreign Policy,* December 2. https://foreignpolicy.com/podcasts/heat-of-the-moment-climate-change/the-godmother-of-climate-security/.

Werrell, Caitlin, and Francesco Femia. 2017. "Defense Bill Passes with Climate Change and National Security Provision." *The Center for Climate and Security,* Effective November 22. https://climateandsecurity.org/2017/11/defense-bill-passes-with-climate-change-and-national-security-provision/.

Toxic Remnants of War

Doug Weir and Iryna Nikolaieva

Toxic remnants of war (TRW) are defined as any toxic or radiological substances resulting from military activities that form a hazard to humans or ecosystems. The term was developed and popularized by the civil society Toxic Remnants of War Project (TRW Project) between 2011 and 2016 (Ghalainey 2013). The term was subsequently recognized in the International Law Commission's principles on the protection of the environment in relation to armed conflicts (Weir 2017a). The term TRW is used interchangeably with the term "conflict pollution."

A key motivation for developing the TRW framing was that, as with the more widely used Explosive Remnants of War (ERW), the means of characterizing and communicating risks, and the potential responses to different forms of harm, are broadly similar between devices or substances within these categories. Prior to its development, no overarching framing was available that covered different types of **conflict** pollution, such as oil spills, depleted uranium weapons, or the dioxin contamination from Agent Orange herbicides.

The period during which the framing was developed was marked by major conflict pollution incidents; this included urban warfare in Syria (Zwijnenburg 2015) and oil well fires in northern Iraq (Zwijnenburg 2017). However, it had also seen increasing concern over the toxic legacy of several historical and contemporary armed conflicts (Kellay 2014). A lack of obligations on either aggressor or affected states to address TRW contamination, or its health or ecological risks, had meant that post-conflict interventions were rare. TRW can undermine livelihoods, such as agricultural resources (Zwijnenburg and Nikolaieva 2023), and concern over exposure risks can be acute among affected communities, feeding distrust of local and national authorities and undermining the enjoyment of environmental **human rights**. Decisions over post-conflict TRW remediation have at times taken on a political component, as in the United States' delayed remediation of dioxin hotspots in Vietnam (Kellay 2014) and are thus relevant to wider **environmental peacebuilding** considerations.

The TRW Project characterized TRW as either direct or indirect. Direct TRW includes munitions constituents and military equipment, chemical releases, and hazardous constituents in conflict debris such as asbestos. Indirect TRW refers to factors associated with conflicts that exacerbated TRW risks, such as limited assessment capacity and weak government regulation. TRW can occur and be a risk at all phases of the **conflict life cycle**; indeed, peacetime domestic regulation around military sites, such as firing ranges, can be a greater source of environmental risk data for munitions constituents than was available in conflict contexts.

The TRW framing sought to connect humanitarian and environmental harm and contribute to greater attention to the problem (Weir 2017b). In 2017, the UN Environment Assembly passed a resolution entitled "Pollution Mitigation and Control in Areas Affected by Armed Conflict or Terrorism," which captured several of the issue's dimensions (United Nations Environment Assembly 2017).

The TRW framing was influenced by the ERW framing of humanitarian disarmament but ended up in turn influencing humanitarian disarmament. The TRW concept was used in the development of the Treaty on the Prohibition of Nuclear Weapons' positive obligations on victim assistance and environmental remediation (International Human Rights Clinic and The Conflict and Environment Observatory 2020). Arguably, it has also played a role in encouraging environmental mainstreaming across the mine action sector, which by the early 2020s had seen mine action operators facilitating the collection of TRW samples in conflict-affected areas of Ukraine (Cottrell 2024). In advancing humanitarian-oriented framings to promote a treaty ban on anti-personnel landmines in the 1990s, civil society activists had inadvertently decoupled landmines from the historical and more holistic "remnants of war," which had emerged following the Vietnam War and encompassed environmental harm and non-explosive remnants (Bolton and Weir 2015). The TRW framing has helped remedy this historic decoupling to some extent.

References

Cottrell, Lindsey. 2024. "Landmines, Climate Change and New Environmental Guidance." The Conflict and Environment Observatory and Norwegian People's Aid, July. https://ceobs.org/landmines-climate-change-and-new-environmental-guidance/.

Ghalainey, Mohamed. 2013. "Toxic Harm: Humanitarian and Environmental Concerns from Military-origin Contamination." *Toxic Remnants of War Project*. Accessed June 6 2025. https://ceobs.org/wp-content/uploads/2024/10/TRWP-2013-Toxic_Harm_humanitarian_and_environmental_concerns_from_military_origin_contamination.pdf.

International Human Rights Clinic (IHRC) and Conflict and Environment Observatory (CEOBS). 2020. "Confronting Conflict Pollution: Principles for Assisting Victims of Toxic Remnants of War." https://humanrightsclinic.law.harvard.edu/wp-content/uploads/2022/10/Confronting-Conflict-Pollution.pdf.

Kellay, Aneaka. 2014. "Pollution Politics: Power, Accountability and the Toxic Remnants of War." *The Toxic Remnants of War Project*. https://ceobs.org/wp-content/uploads/2018/01/WEB_READY%E2%80%93TRW_Pollution_Politics_Report.pdf.

United Nations Environment Assembly (UNEA). 2017. *Pollution Mitigation and Control in Areas Affected by Armed Conflict or Terrorism*. UNEP/EA.3/RES.1. UNEA. https://digitallibrary.un.org/record/3976949.

Weir, Doug. 2017a. "Reframing the Remnants of War: The Role of the International Law Commission, Governments, and Civil Society." In *Environmental Protection and Transitions from Conflict to Peace: Clarifying Norms, Principles, and Practices*, edited by Carsten Stahn, Jens Iverson, and Jennifer S. Easterday. Oxford Academic. https://doi.org/10.1093/oso/9780198784630.003.0019.

Weir, Doug. 2017b. "Conflict Pollution and the Toxic Remnants of War: A Global Problem that Receives Too Little Attention." UN Environment Perspectives No. 24. https://www.unep.org/resources/perspective-series/issue-no-24-conflict-pollution-and-toxic-remnants-war-global-problem.

Zwijnenburg, Wim. 2015. *Amidst the Debris: A Desktop Study on the Environmental and Public Health Impact of Syria's Conflict*. PAX. https://paxforpeace.nl/wp-content/uploads/sites/2/import/import/pax-report-amidst-the-debris-syria-web.pdf.

Zwijnenburg, Wim. 2017. *Living Under a Black Sky: Conflict Pollution and Environmental Health Concerns in Iraq*. PAX. https://paxvoorvrede.nl/wp-content/uploads/import/import/pax-report-living-under-a-black-sky.pdf.

Zwijnenburg, Wim, and Iryna Nikolaieva. 2023. *Attacks on Agro-Industrial Sites in Ukraine. Environment and Conflict Alert Ukraine*. PAX. https://paxforpeace.nl/news/research-bombing-of-agro-industry-in-ukraine-poses-serious-environmental-health-risks.

Transboundary Cooperation

Aleksandra Balinskaia and Tobias Ide

Transboundary cooperation is the coordination of actions, whether formal or informal, related to cross-border environmental resources or areas for mutual benefit across political borders.

Transboundary **cooperation** can take various forms, ranging from information sharing (e.g., flood early warning) to policy coordination (e.g., on **conservation**), the construction of joint **infrastructure** (e.g., dams), and the creation of joint institutions (e.g., the Permanent Indus Commission). Likewise, the cooperating parties can be state representatives and institutions as well as non-state actors. Substantial **environmental cooperation** between Israel, Jordan, and Palestine, for instance, is driven by local communities, NGOs, and scientists (Kedem et al. 2024). Transboundary cooperation and **conflict** are not mutually exclusive but can occur in parallel between the same actors. Commonly, transboundary cooperation refers to the coordination of actions across international borders, even though political borders can also exist within states.

Environmental peacebuilding research has identified environmental issues (such as water or conservation) as promising entry points for cooperation even between countries with tense relations. Environmental issues are usually less contested than economic integration, offer opportunities for mutual benefits and positive-sum interactions, are inherently cross-border, and often require long-term engagement. Once such transboundary cooperation is established, it can improve the strategic climate between states (for instance, by building trust) or even strengthen "post-Westphalian **governance**" (for instance, by building trans-societal linkages and establishing shared identities) (Conca and Dabelko 2002).

Four pathways can connect transboundary environmental cooperation to **peacebuilding** between countries. First, it can improve the environmental situation and hence address environmental drivers of conflicts, such as resource scarcity and inequality. Second, by cooperating over environmental issues, state and non-state actors build trust and understanding for each other. Third, transboundary environmental cooperation can cultivate interdependence between countries because they depend on each other to retain the benefits of cooperation. In such a situation, escalating a conflict is more costly. Fourth, transboundary cooperation often results in the establishment of joint institutions, which manage conflicts and pave the way for further cooperation (Ide 2019).

Emerging empirical evidence shows that cross-border cooperation in the joint management of **natural resources** fosters conditions conducive to mitigating conflict and establishing or reinforcing **peace**. Central to this process is the development of enduring, transparent, and inclusive institutions that connect governments, societies, and environmental practices. By acknowledging the crucial role of these institutions, it becomes possible to refine existing peacebuilding strategies by emphasizing the most effective approaches during the early phases of transboundary cooperation. Shared management of territories or resources is essential in creating stable political environments and nurturing trust and confidence among nations (Balinskaia 2024). At the same time, existing research indicates that **sustainability** initiatives, such as ecological restoration and landscape rewilding, can indirectly contribute to the **securitization** of environmental issues. The creation of protected areas often intensifies states' focus on natural resources in border regions, potentially exacerbating competition over their control and management (Constantinou et al. 2020).

While cooperative environmental management may produce mixed results, it has the potential to foster transboundary partnerships, facilitate the negotiation of new agreements, and strengthen regional frameworks. Consequently, promoting transboundary environmental cooperation increasingly emerges as a practical strategy for preventing conflicts and enhancing diplomatic relations. However, such transboundary cooperation must be pursued with a keen awareness of the intricate political dynamics involved (Balinskaia 2024).

References

Balinskaia, Aleksandra. 2024. "Interstate Environmental Peacebuilding: A Systematic Review." *Environment and Security* 3 (3): 265–98. https://doi.org/10.1177/27538796241287366.

Conca, Ken, and Geoffrey Dabelko, eds. 2002. *Environmental Peacemaking*. Johns Hopkins University Press and Woodrow Wilson Center Press.

Constantinou, Costas M., Maria Hadjimichael, and Evi Eftychiou. 2020. "Ambivalent Greenings, Collateral Conservation: Negotiating Ecology in a United Nations Buffer Zone." *Political Geography* 77 (1): 102096. https://doi.org/10.1016/j.polgeo.2019.102096.

Ide, Tobias. 2019. "The Impact of Environmental Cooperation on Peacemaking: Definitions, Mechanisms and Empirical Evidence." *International Studies Review* 21 (3): 327–46. https://www.jstor.org/stable/48557133.

Kedem, Rina, Suleiman Halasah, Tobias Ide, and Eran Feitelson. 2024. "The Role of Environmental Professionals in Post-Conflict Transboundary Cooperation: The Cases of Israel and Jordan." *Environmental Science and Policy* 151 (1): 103623. https://doi.org/10.1016/j.envsci.2023.103623.

Transdisciplinarity

Gül Özerol, Katharina Löhr, and Erica Key

Transdisciplinarity is a descriptor for efforts that blend different knowledge types from across two or more disciplines with lived experience and expertise outside of academia (Lang et al. 2012).

Transdisciplinary activities typically involve multiple participants who provide both their knowledge and involvement, as well as the context or an articulated need, requiring a holistic or **systems approach**. The term "**stakeholders**" has been commonly used to refer to the participants. However, due to the connotation of this term with colonial narratives and systemic inequities (Reed et al. 2024), there is a call to adopt more inclusive and context-specific terms, such as rightsholders, partner clients, or implementing collaborators.

A transdisciplinary process typically follows three stages (Lang et al. 2012; Mauser et al. 2013): (1) co-design and co-development stage that focuses on identifying the problem and the desired outcome, as well as possible pathways to achieve the outcome, (2) knowledge co-production stage in which expertise and experience are shared and blended to determine the path forward, with an emphasis on shared roles for all participants and a balance of **power** relations, and (3) co-implementation stage that involves collaborative action or informed decision-making with internal and external stakeholders. All stages are supported by transparent communication in language and formats accessible to the participants. The process may also include an **evaluation** framework that considers the impacts of the action or decision.

Transdisciplinarity is like interdisciplinarity and multidisciplinarity, as they all bring together knowledge from multiple domains. However, the integration of societal, and often nonacademic, knowledge and lived experience distinguishes transdisciplinarity from interdisciplinarity and multidisciplinarity. **Sustainability** science, for example, could be characterized as an interdisciplinary field of research, whereas **sustainable development** is transdisciplinary because of the inclusion of nonacademic expertise. The integration of knowledge, perspectives, and problem-solving from beyond academia introduces a practical, responsible frame that moves sometimes fundamental interdisciplinary research into an applied space. Other fields use the word transdisciplinarity to encompass different aspects of research, including use in the medical (Bewer 2017) and educational sciences (Daneshpour and Kwegyir-Afful 2022).

Transdisciplinarity links to **environmental peacebuilding** on at least three scales. First, it can contribute toward an enhanced understanding of complex **conflict** and **peacebuilding** scenarios due to cross-scale and cross-sectoral integration of diverse rightsholders/stakeholders and knowledge systems. Second, transdisciplinary processes, for instance, through co-creation of tools or co-design of innovations, open space for enhanced stakeholder dialogue that facilitates trust building, communication, and collaboration among diverse actors. Power relations need to be considered carefully, particularly with regard to the involvement of rightsholders/stakeholders with different interests or goals, and integration of different knowledge systems. Third, co-designed solutions have more long-term success potential due to participatory development of place-based innovations and the attention paid to the **evaluation** of the impacts of the transdisciplinary process.

Transdisciplinary approaches have been a cornerstone of many fields contributing to Environmental Peacebuilding, including but not limited to **conservation**, natural resource management (Gorelick 2015), and aid and development activities (Özerol et al. 2018).

References

Bewer, Vanessa V. 2017. "Transdisciplinarity in Health Care: A Concept Analysis." *Nursing Forum* 52 (4): 339–47. https://doi.org/10.1111/nuf.12200.

Daneshpour, Hosein, and Ebo Kwegyir-Afful. 2022. "Analysing Transdisciplinary Education: A Scoping Review." *Science and Education* 31 (4): 1047–74. https://doi.org/10.3389/fpubh.2023.1090725.

Gorelick, Steven. 2015. *Water Security in Jordan: A Key to the Future of the Middle East*. The Brookings Institution.

Lang, Daniel J., Arnim Wiek, Matthias Bergmann, et al. 2012. "Transdisciplinary Research in Sustainability Science: Practice, Principles, and Challenges." *Sustainability Science* 7: 25–43. https://doi.org/10.1007/s11625-011-0149-x.

Mauser, Wolfram, Gernot Klepper, Martin Rice, et al. 2013. "Transdisciplinary Global Change Research: The Co-Creation of Knowledge for Sustainability." *Current Opinion in Environmental Sustainability* 5 (3–4): 420–31. https://doi.org/10.1016/j.cosust.2013.07.001.

Özerol, Gül, Juliane Schillinger, and Maher Abu-Madi. 2018. "Transdisciplinary Research and Development Cooperation: Insights from the First Phase of the Palestinian-Dutch Academic Cooperation Programme on Water." *Water* 10 (10): 1449. https://doi.org/10.3390/w10101449.

Reed, Mark S., Bethann Garramon Merkle, Elizabeth J. Cook, et al. 2024. "Reimagining the Language of Engagement in a Post-Stakeholder World." *Sustainability Science* 19: 1481–90. https://doi.org/10.1007/s11625-024-01496-4.

Triple Bottom Line

Jocelyn Wright and John Hardman

The "triple bottom line" is a comprehensive accounting framework to guide decision-making and measure performance and impact centered on sustainable development.

In business, the "bottom line" refers to the reported net earnings or losses at the end of a corporate financial statement, while emphasizing the importance of maximizing profits. The "triple bottom line" (TBL or 3BL), coined by John Elkington in 1994, was put forth as an alternative to this narrow agenda and initially aimed to encourage companies to rethink, reprioritize, commit, invest, and take action and hold them not simply financially but morally accountable for their social and environmental as well as economic performance and impact (Elkington 2004, 2018). The **sustainability** framework Elkington had in mind, based on the Brundtland Report (World Commission on Environment and Development 1987), focused on systemic change (transforming capitalism) to promote economic prosperity, social equity, and environmental preservation (Elkington 2018) for a better future. Elkington (1994) envisaged a business agenda of "win-win-win" strategies. In 2004, he proposed that this agenda would be driven by seven sustainability revolutions related to markets, values, transparency, life-cycle technology, time, partnerships, and corporate **governance**.

In recognizing the imperative role for business to play in balancing concerns for people, planet, and profits (Elkington's 3Ps), the TBL framework has made an important contribution to **environmental peacebuilding**. Notably, by pressuring companies to assume corporate responsibility for and take a lead in reducing or eliminating social and **environmental degradation**/degeneration, such as consequential poverty, hunger, illness, illiteracy, inequality, deteriorated working conditions, crime, mass migration, **conflict**, waste, pollution, and depletion of **natural resources**, the TBL advances a paradigm shift not only in business but also in non-profit and governmental sectors (Lodder et al. 2014).

Companies have been identified as both the main contributors to **environmental violence** and potentially important actors in environmental peacebuilding (Marcantonio 2024). However, recent research in environmental peacebuilding shows that they mostly contribute to **negative peace**, the absence of physically violent conflicts, through short-term compromises aimed at **mitigation** (Dresse et al. 2021; Marcantonio 2024). A focus on TBL prompts them toward a more comprehensive **peace** approach (Galtung 1969), which additionally requires **positive peace**, an understanding of root causes, inclusive dialogue, and collaboration that leads to structural transformation that prevents and transforms inevitable conflict when it occurs (Dresse et al. 2021), and may better help to durably achieve the environmental peacebuilding aims of delivering **basic services**, rebuilding trust and relationships, generating employment and livelihoods, redistributing **power**, and establishing safety and **security** across spaces and the **conflict life cycle** while governing and managing natural resources and dealing with environmental issues (Bruch et al. 2022).

References

Bruch, Carl, David Jensen, and Monica Emma. 2022. "Defining Environmental Peacebuilding." In *Routledge Handbook of Environmental Security*, edited by Crystal Murphy, Evgenia Nizkorodov, and Richard Matthew. Routledge. https://doi.org/10.4324/9781315107592-2.

Dresse, Anaïs, Jonas Østergaard Nielsen, and Itay Fischhendler. 2021. "From Corporate Social Responsibility to Environmental Peacebuilding: The Case of Bauxite Mining in Guinea." *Resources Policy* 74: 102290. https://doi.org/10.1016/j.resourpol.2021.102290.

Elkington, John. 1994. "Towards the Sustainable Corporation: Win-Win-Win Business Strategies for Sustainable Development." *California Management Review* 36 (2): 90–100. https://doi.org/10.2307/41165746.

Elkington, John. 2004. "Enter the Triple Bottom Line." In *The Triple Bottom Line: Does It All Add Up?*, edited by Adrian Henriques and Julie Richardson. Earthscan.

Elkington, John. 2018. "25 Years Ago I Coined the Phrase 'Triple Bottom Line.' Here's Why It's Time to Rethink It." *Harvard Business Review*, June 25. https://hbr.org/2018/06/25-years-ago-i-coined-the-phrase-triple-bottom-line-heres-why-im-giving-up-on-it.

Galtung, Johan. 1969. "Violence, Peace, and Peace Research." *Journal of Peace Research* 6 (3): 167–91. https://doi.org/10.1177/002234336900600301.

Lodder, Marleen, Roebin Lijnis Huffenreuter, Michael Braungart, and Diana den Held. 2014. "Regenerative Sustainable Development: Towards a Triple Top Line Approach and Increasing Positive Externalities." Paper presented at The 5th International Sustainability Transitions Conference, August 27–29. https://www.dianadenheld.nl/wp-content/uploads/2014/10/Lodder_Huffenreuter_Braungart_den-Held_Regenerative-Sustainable-Development-final.pdf.

Marcantonio, Richard. 2024. "Environmental Violence and Enterprise: The Outsized Role of Business for Environmental Peacebuilding." *Business Horizons* 67 (6): 685–98. https://doi.org/10.1016/j.bushor.2024.07.003.

World Commission on Environment and Development (WCED). 1987. *Our Common Future*. Brundtland Report. United Nations. https://digitallibrary.un.org/record/139811?ln=en&v=pdf.

Triple Nexus

Jeannie Shroads, Héctor Morales-Muñoz, and Adam Savelli

Triple nexus, also referred to as the "Humanitarian-Development-Peace Nexus" and "HDP Nexus," is a policy concept that seeks to acknowledge and operationalize the interlinkages between the humanitarian, development, and peace sectors to create synergies and deliver on collective outcomes that address root causes of conflict and encourage resilient communities.

In his remarks upon taking the office of UN Secretary in 2016, António Guterres described humanitarian response, **sustainable development**, and sustaining peace as "three sides of the same triangle" (Caparini and Reagan 2019). This approach offers an opportunity for the humanitarian, development, and **peace** sectors to collectively address the deeper causes of **conflict**, humanitarian need, and lagging development through strengthening collaboration and coherence across HDP programming policy (Organisation for Economic Co-operation and Development [OECD] 2019). The OECD provided a roadmap for the HDP sectors to coordinate their efforts for collective outcomes, prioritizing **conflict sensitivity** and preventive programming using evidence-based and local expertise, and developing coordinated and flexible financing schemes in support of the nexus (OECD 2019).

The Triple Nexus approach brought some cultural shifts to the humanitarian and development sectors, including increased emphasis on collective outcomes and local **participation**, but progress has been slow and uneven (Inter-Agency Standing Committee 2021; Fanning and Fullwood-Thomas 2019). Despite rhetorical support, significant challenges to implementing the approach remain. Funding for HDP initiatives remains largely siloed and focused on short-term outcomes, hampering coordination and effectiveness (OECD 2024). HDP actors express concern about reconciling their individual missions while attempting to operate within the nexus. For instance, humanitarian actors worry their work may become politicized, negatively impacting their core mission of assisting those in need (Norman and Mikhael 2023). At times, working to achieve one nexus goal can undermine other nexus goals. For example, providing humanitarian support by coordinating with conflict actors (Active Learning Network for Accountability and Performance in Humanitarian Action 2023). However, the neutrality required for humanitarian efforts may at times reinforce **power** dynamics and inequalities, potentially hampering peacebuilding or development efforts. This argument leads some to argue for all actors engaging in conflict-affected regions to determine an appropriate level of coherence rather than assuming coherence and coordination, like for the Triple Nexus, should be absolute (Connolly and Baggerman 2014). Another challenge is that the siloed nature of organizations continues to hamper Triple Nexus progress. "[A] segmented mindset prevents the inter-linkages between these different actors when HDP nexus approaches are about coherence and complementarity" (OECD 2024, 28).

As **climate change** and **environmental degradation** present multilayered humanitarian, development, and **peacebuilding** challenges, the Triple Nexus approach ostensibly provides a useful framework for **environmental peacebuilding**. As such, environmental peacebuilding interventions offer valuable opportunities to reconcile and coordinate the implementation of the HDP nexus approach, ensuring a more integrated response to the intersecting challenges posed by climate change and environmental degradation, especially in conflict-affected areas. Responses to the climate crisis should intersect across HDP sectors. The development and **security** risks caused by climate change (environmental degradation, food insecurity, natural resource competition, migration, and livelihood loss, for example) create reinforcing loops of violent conflict, **fragility**, and social vulnerability (Morales-Muñoz et al. 2022). Integrating environmental peacebuilding strategies into HDP programming can generate significant **co-benefits**. For instance, complementing aid relief with climate-smart livelihood strategies within robust natural **resource governance** frameworks enhances community **resilience** and reduces dependence on short-term external aid. This can, over time, be a pathway for a more rapid transition toward sustainable, climate-neutral development. For example, partnerships between UN agencies, such as the Food and Agriculture Organization of the United Nations (FAO) or the World Food Programme (WFP), promote "durable solutions" for refugees and vulnerable communities by fostering resilience against food insecurity from conflict, climate, and economic shocks (FAO 2024). Prior to the Taliban's takeover in Afghanistan, WFP's **infrastructure** interventions, such as food aid recovery and construction of canals, from repeated shocks enhanced the resilience of communities (WFP and Konterra Group 2016). In refugee camps in Jordan and Kenya, agencies support community development and entrepreneurship through sustainable energy access for productive uses, such as solar water pumps and vocational training, viewing camps as potential permanent settlements where communities can overcome total aid dependence (Nascimento and Pureza 2024; Morales-Muñoz 2017). Uganda's government's "Comprehensive Refugee Response Framework" prevents dependency by allocating land to refugees, enabling participation in rural markets and integrating refugees into national and local development plans, focusing on self-reliance through skills training and market inclusion (Nascimento and Pureza 2024). Ultimately, such an approach can strengthen broader peacebuilding and long-term climate **adaptation** efforts and foster **social cohesion** (Morales-Muñoz et al. 2022; Mena et al. 2022).

References

Active Learning Network for Accountability and Performance in Humanitarian Action (ALNAP). 2023. "The Humanitarian-Development-Peace Nexus: Current Status and Discourse." Accessed August 11, 2025. https://alnap.cdn.ngo/media/documents/ALNAP-EXplain-ALNAP-EssentialBriefings-LongRead-HDP-Nexus-Updated.pdf.

Caparini, Marina, and Anders Reagan. 2019. *Connecting the Dots on the Triple Nexus*. Stockholm International Peace Research Institute, November 29. https://www.sipri.org/commentary/topical-backgrounder/2019/connecting-dots-triple-nexus.

Connolly, Lesley, and Jill Baggerman. 2014. *Getting Coherence and Coordination Right: Principles for the Peacebuilding Policy Community*. African Centre for the Constructive Resolution of Disputes, August. https://www.accord.org.za/publication/getting-coherence-and-coordination-right/

Fanning, Emma, and Jessica Fullwood-Thomas. 2019. *The Humanitarian-Development-Peace Nexus: What Does it Mean for Multi-Mandated Organizations?* Oxfam. https://policy-practice.oxfam.org/resources/the-humanitarian-development-peace-nexus-what-does-it-mean-for-multi-mandated-o-620820/.

Food and Agriculture Organization of the United Nations (FAO). 2024. "Agricultural Livelihoods as a Pathway for Durable Solutions." https://globalcompactrefugees.org/good-practices/agricultural-livelihoods-pathway-durable-solutions.

Inter-Agency Standing Committee. 2021. "Mapping Good Practices in the Implementation of Humanitarian-Development-Peace Nexus Approaches, Country Briefs and Synthesis Report." November 10. https://interagencystandingcommittee.org/humanitarian-development-collaboration/mapping-good-practices-implementation-humanitarian-development-peace-nexus-approaches-country-briefs.

Mena, Rodrigo, Summer Brown, Laura E. R. Peters, Ilan Kelman, and Hyeonggeun Ji. 2022. "Connecting Disasters and Climate Change to the Humanitarian-Development-Peace Nexus." *Journal of Peacebuilding and Development* 17 (3): 324–40. https://doi.org/10.1177/15423166221129633.

Morales-Muñoz, Héctor, Arwen Bailey, Katharina Löhr, et al. 2022. "Co-benefits through Coordination of Climate Action and Peacebuilding: A System Dynamics Model." *Journal of Peacebuilding and Development* 17 (3): 304–23. https://doi.org/10.1177/15423166221132149.

Morales-Muñoz, Héctor, Samantha Bidwood-ölz, Silvia Cabriolupuddu, et al. 2017. *The Role of Sustainable Energy Access in the Migration Debate*. European Union Energy Initiative, Partnership Dialogue Facility. https://doi.org/10.13140/RG.2.2.22066.76485.

Nascimento, Daniela, and José Manuel Pureza. 2024. "Refugees as Entrepreneurs? A Challenge to HDP Programmes." *The International Spectator* 59 (3): 42–57. https://doi.org/10.1080/03932729.2024.2326215.

Norman, Julie M., and Drew Mikhael. 2023. "Rethinking the Triple-Nexus: Integrating Peacebuilding and Resilience Initiatives in Conflict Contexts." *Journal of Peacebuilding and Development* 18 (3): 248–63. https://doi.org/10.1177/15423166231200210.

Organisation for Economic Co-operation and Development. 2019. "DAC Recommendation on the Humanitarian-Development-Peace Nexus." February 22. https://legalinstruments.oecd.org/en/instruments/OECD-LEGAL-5019.

The Organisation for Economic Co-operation and Development. 2024. *Report on the Implementation, Dissemination and Continued Relevance of the DAC Recommendation on the Humanitarian-Development-Peace Nexus*, March 11. https://one.oecd.org/document/DCD/DAC/INCAF(2023)1/FINAL/en/pdf.

World Food Programme (WFP) and Konterra Group. 2016. "Afghanistan, Operation Evaluation: PRRO 200447 Assistance to Address Food Insecurity and Under-Nutrition." https://one.wfp.org/operations/current_operations/project_docs/200447.pdf

V

Violence

Bryan Fok

According to the World Health Organization, violence is "the intentional use of physical force or power, threatened or actual, against oneself, another person, or against a group or community, that either results in, or has a high likelihood of, resulting in injury, death, psychological harm, maldevelopment or deprivation" (World Health Organization, Regional Office for the Eastern Mediterranean. n.d.)

Violence can take on many forms, including those that do not inflict direct, physical harm on an individual. Instances of direct violence, such as murder or rape, can cause bodily harm and deprive individuals of their capability to achieve complete well-being (Galtung 1969). However, while direct violence tends to have an immediate effect, the harm that violence inflicts can also take years or decades to materialize. The concept of slow violence (Nixon 2011), which occurs gradually and out of sight, is related to the concepts of structural violence and **environmental violence**, the latter of which is "direct and indirect harm experienced by humans due to toxic and non-toxic pollutants put into a local—and concurrently the global—ecosystem through human activities and processes" (Marcantonio and Fuentes 2023, 861). Toxic pollution is now the leading cause of premature mortality in the world (Landrigan et al. 2018), but the effects of toxic pollution are often not seen until years after the action when cancer rates and other health **indicators** spike.

The perpetrators of violence can also be the social, economic, and political institutions that ground societies. Structural violence occurs when social arrangements put individuals and groups in harm's way (Farmer et al. 2006). For example, the social practice of redlining in the United States led to residential segregation and relegated communities of color to reside in less desirable locations within urban centers. Because people of color were systematically excluded from purchasing property in specific areas by realtors and lenders, they had to resort to living in areas that today have a 3% higher exposure to flood risk and a 0.07-degree Fahrenheit higher air temperature, leading to disparate health impacts, including increased deaths during heat waves (Conzelmann et al. 2022). Other enduring effects of redlining include higher poverty rates and less funding for public schools, which prevent those living in previously redlined areas from achieving their full potential.

DOI: 10.4324/9781003604143-20

In many instances of violence, the perpetrators deem the action or behavior acceptable, or at least "not wrong." Cultural violence is the aspect of culture that "can be used to justify or legitimize direct or structural violence" (Galtung 1990, 291). Beliefs and schemas that cause one group to view another group as inferior are an example of cultural violence, as those beliefs can lead to actions that cause harm to the other group. The deep antisemitism that entrenched Nazi Germany led to the Holocaust and the extermination of approximately 6 million Jews, demonstrating that a belief can lead to direct violence. The caste system is a deeply entrenched aspect of Hindu culture, which legitimizes structural violence as members of lower castes are prevented from advancing economically and politically (Goghari and Kusi 2023).

The opposite of violence is **peace**. However, there is a distinction between a **negative peace** and a **positive peace**. Galtung (1969, 183) defines a negative peace as the "absence of personal [direct] violence" and a positive peace as the "absence of structural violence." **Environmental peacebuilding** focuses on removing ways in which states and individual actors can use the **environment** to harm others. Some environmental peacebuilding activities aim to prevent a population from being harmed by an artificially induced environment. **Peace agreements** that stipulate the removal of land mines, for example, prevent accidental deaths from unexploded ordnance. The harm that environmental peacebuilding prevents can also be indirect. For example, transboundary water management between Turkey and Armenia ensures that both countries have access to water from the Arpacay River, preventing any one country from denying access to the other to an essential resource (Altingoz and Ali 2019). For a comprehensive and **sustainable peace** to occur, all forms of violence must be eradicated so that every person can achieve complete well-being.

References

Altingoz, Mehmet, and Saleem H. Ali. 2019. "Environmental Cooperation in Conflict Zones: Riparian Infrastructure at the Armenian–Turkish Border." *The Journal of Environment and Development* 28 (3). https://doi.org/10.1177/1070496519859680.

Conzelmann, Claire, Arianna Salazar-Miranda, Toan Phan, and Jeremy Hoffman. 2022. "Long-Term Effects of Redlining on Environmental Risk Exposure." *Federal Reserve Bank of Richmond Working Papers* 22 (09): 1–18. https://doi.org/10.21144/wp22-09.

Farmer, Paul E, Bruce Nizeye, Sara Stulac, and Salmaan Keshavjee. 2006. "Structural Violence and Clinical Medicine." *PLoS Medicine* 3 (10): 1686–91. https://doi.org/10.1371/journal.pmed.0030449.

Galtung, Johan. 1969. "Violence, Peace, and Peace Research." *Journal of Peace Research* 6 (3): 167–91. https://doi.org/10.1177/002234336900600301.

Galtung, Johan. 1990. "Cultural Violence." *Journal of Peace Research* 27 (3): 291–305. https://doi.org/10.1177/0022343390027003005.

Goghari, Vina M., and Mavis Kusi. 2023. "An Introduction to the Basic Elements of the Caste System of India." *Frontiers in Psychology* 14. https://doi.org/10.3389/fpsyg.2023.1210577.

Landrigan, Philip J., Richard Fuller, and Nereus J. R. Acosta, et al. 2018. "The Lancet Commission on Pollution and Health." *The Lancet* 391 (10119): 462–512. https://doi.org/10.1016/S0140-6736(17)32345-0.

Marcantonio, Richard, and Agustín Fuentes. 2023. "Environmental Violence: A Tool for Planetary Health Research." *The Lancet Planetary Health* 7 (10): e859–67. https://doi.org/10.1016/S2542-5196(23)00190-0.

Nixon, Rob. 2011. *Slow Violence and the Environmentalism of the Poor*. Harvard University Press.
World Health Organization, Regional Office for the Eastern Mediterranean. n.d. "Violence." https://www.emro.who.int/violence-injuries-disabilities/violence/. https://www.emro.who.int/violence-injuries-disabilities/violence/.

W

War

Alexander Belyakov and Aliakbar Shams

War is a large-scale, organized armed conflict conducted by states, nations, or other structured groups to secure political, economic, or territorial goals. At its core, war often originates from a state of misery (Freedman 2012). War generally emerges when diplomatic negotiations and other non-violent methods of resolving disputes fail, leading to the strategic deployment of military forces and producing sustained, systematic violence over a defined period.

Modern international law does not define the term "war." Instead, the legal understanding of the term has evolved to indicate an escalation in the intensity of hostilities within the paradigm of "armed **conflict**" (International Committee of the Red Cross 2024). von Clausewitz recognized that war is "a true political instrument, a continuation of political intercourse, carried on with other means" (von Clausewitz 1989). War can occur in land, sea, air, and outer space and may involve conventional armies, guerrilla forces, or hybrid combatants, including private **security** contractors. Hybrid warfare supplements traditional military action with other tactics, such as disinformation and attacking computer systems.

Under the Rome Statute of the International Criminal Court, deliberate environmental destruction can be classified as a war crime (Office of the Prosecutor, International Criminal Court 2024). However, demonstrating that harm is excessive relative to military advantage remains challenging, as the war in Ukraine illustrates (Belyakov 2024).

In some circumstances, countries seek to avoid labeling military action as a "war," and instead characterize it as a "police action," "special military operation," or other term. Actors increasingly employ alternative terms to minimize the appearance of aggression, mold public opinion, or bypass **accountability**—including domestic legislation that may require legislative approval before declaring war. This rhetorical maneuver directly affects **environmental peacebuilding** efforts by masking the severity of the conflict.

Exemplary evidence is Russia's annexation of Crimea and the assertions of Russian officials. Russian propagandists labeled the war in Ukraine a "special military operation," implying a limited intervention rather than a full-scale war. Terms like

DOI: 10.4324/9781003604143-21

"restoration of constitutional order," "liberation mission," and "stabilization measures" recast violent incursions as "**governance**," hindering the recognition of environmental harm and complicating post-conflict recovery (United Nations 2022). Also, framing China's militarized expansions in the South China Sea as "maritime law enforcement" or "island reclamation" downplays their environmental impacts and impedes **peacebuilding** efforts (Yamaguchi 2021). Similarly, NATO's 2011 intervention in Libya, labeled a "no-fly zone enforcement" and "humanitarian intervention" (Isaac 2012), masked ecological damage caused by airstrikes.

Environmental degradation resulting from war encompasses deforestation, soil erosion, and contamination of water sources. Such consequences undermine local livelihoods and exacerbate resource scarcity, thereby escalating conflict (Winkel and Hermans 2024). The environmental consequences of war can lead to long-term economic and social challenges (Austin and Bruch 2000). Destroying **infrastructure** and ecosystems can hinder recovery efforts and perpetuate cycles of poverty and conflict (Conca and Wallace 2009). In this context, clarity and accountability in defining "war" in international law are essential for protecting environments and communities during and after conflict and establishing distinct boundaries between **peace**, conflict, and the transitional phases.

Preparations for war, armed conflict, wartime coping strategies, and post-conflict reconstruction cause significant environmental damage. Environmental harm has long been an underrecognized consequence of war. The environmental peacebuilding field must grapple with the full scope of conflict-related damage. In fact, the origins of the environmental peacebuilding field trace back to environmental damage assessments conducted by UNEP in the 1990s.

Understanding war's environmental legacy is critical, including recognizing how environmental issues can contribute to, exacerbate, or even trigger violent conflict. EPB needs to explore the mutual relationship between war and the **environment**, examining how conflict impacts ecosystems and how environmental stressors can influence the outbreak and intensity of war. This analysis must span different types of conflict—civil wars, interstate wars (both of which are declining), and the increasingly prevalent internationalized wars. Each raises complex questions around responsibility, liability, and long-term implications.

Environmental peacebuilding research frequently relies on conflict datasets from institutions such as the Peace Research Institute Oslo and Stockholm International Peace Research Institute, which form the backbone of its research infrastructure. These datasets enable analysts to identify patterns, trends, and outliers that inform fieldwork and policymaking. However, they are not without limitations. Since much violent conflict goes unreported, particularly in under-covered regions, media-reliant data can present a skewed picture (Shaver et al. 2022). Further research should investigate the intertwined dynamics of war and environmental harm, with a focus on shaping policies that deliver real impact.

References

Austin, Jay, and Carl Bruch, eds. 2000. *The Environmental Consequences of War: Legal, Economic, and Scientific Perspectives*. Cambridge University Press.

Belyakov, Alexander. 2024. "War's Hidden Victim: The Environmental Toll of Russia's Invasion of Ukraine." *Peace Magazine*. October–December, pp. 17–19. https://peacemagazine.org/archive/warshiddenvictim.htm.
Conca, Ken, and Jennifer Wallace. 2009. "Environment and Peacebuilding in War-torn Societies: Lessons from the UN Environment Programme's Experience with Postconflict Assessment." *Global Governance* 15 (4): 485–504. https://www.jstor.org/stable/27800777.
Freedman, Lawrence. 2012. "Defining War." In *The Oxford Handbook of War*, edited by Yves Boyer and Julian Lindley-French. https://doi.org/10.1093/oxfordhb/9780199562930.013.0002
International Committee of the Red Cross. 2024. "How Is the Term "Armed Conflict" Defined in International Humanitarian Law?" https://www.icrc.org/sites/default/files/document_new/file_list/armed_conflict_defined_in_ihl.pdf.
Isaac, Sally Khalifa. 2012. "NATO's Intervention in Libya: Assessment and Implications." Political Science Department, Cairo University; KFG Associate, Free University of Berlin. https://www.iemed.org/publication/natos-intervention-in-libya-assessment-and-implications.
Office of the Prosecutor, International Criminal Court. 2024. "Draft Policy on Environmental Crimes under the Rome Statute." https://www.icc-cpi.int/sites/default/files/2024-12/2024-12-18-OTP-Policy-Environmental-Crime.pdf.
Shaver, Andrew, Kately Arriaga, Anne Blackburn, etc. 2022. *News Media Reporting Patterns and Our Biased Understanding of Global Unrest*. Empirical Studies of Conflict Project. Princeton University. Working Paper Series No. 32. https://esoc.princeton.edu/WP32.
United Nations (UN). 2022. "Security Council Discusses Climate Change as Threat to International Peace, Security." UN Press, November 22. https://press.un.org/en/2022/sc14803.doc.htm.
von Clausewitz, Carl. 1989. *On War*. Translated by Michael Howard and Peter Paret. Indexed eBook ed. Princeton University Press.
Winkel, Georg, and Job Hermans. 2024. *Nature in the Crossfire: Examining the Environmental Effects of War*. Wageningen University, Forest and Nature Conservation Policy Group. https://edepot.wur.nl/677013.
Yamaguchi, Shinji. 2021. "Strategies of China's Maritime Actors in the South China Sea: A Coordinated Plan under the Leadership of Xi Jinping?" *China Perspectives* 2021 (4): 23–31. https://doi.org/10.4000/chinaperspectives.7022.

Water Cooperation

Susanne Schmeier, Melissa McCracken, Zoe Rosenblum, and Erika Weinthal

Water cooperation refers to formal and informal interactions, processes, mechanisms, and practices between actors that aim to achieve common goals and objectives in the use, management, and/or protection of shared water resources. Water cooperation is important for all dimensions of peaceful, equitable, and sustainable development, as the shared nature of water resources holds an inherent potential either for conflict or for jointly fostering human well-being, environmental protection, and socioeconomic development.

While water **cooperation** can occur at all **governance** levels, this entry separates the term into transboundary water cooperation between countries and national/subnational water cooperation within a country.

With 313 transboundary surface water basins (McCracken and Wolf 2019; Transboundary Freshwater Diplomacy Database 2024), almost 426 aquifers (International Groundwater Resources Assessment Centre and UNESCO-Intergovernmental Hydrological Programme 2025), and more than 300 wetlands (Rosenblum and Schmeier 2022), cooperation between states sharing these resources is essential. While the discourse at the transboundary level has often been dominated by **conflict** (Starr 1991; Bulloch and Darwish 1993; Gleick 1993; Zeitoun and

Warner 2006), more recent research shows that cooperation is more common (Wolf 1998; De Stefano et al. 2010; Bernauer and Bohmelt 2020; Turgul et al. 2024), even if recent studies highlight a potential future increase in conflict risks (Kåresdotter et al. 2023; de Bruin et al. 2024).

At the interstate level, transboundary water cooperation refers to formal and informal interactions, processes, mechanisms, and practices between actors in transboundary rivers, lake basins, or aquifers that aim to achieve common goals and objectives and attain mutual benefits relating to or derived from the transboundary nature of water resources.

The interactions, processes, and practices encompass formal, legal, and institutional aspects, all of which could have intangible outcomes such as relational dynamics and practical outcomes. Transboundary water cooperation can take many forms, including, *inter alia,* the establishment of treaties or joint basin institutions, the development of cooperative action to address environmental challenges or achieve socioeconomic objectives, the development and operation of joint water **infrastructure**, and informal cooperation practices.

Formal examples are the more than 600 treaties over shared waters, including the 1960 Indus Waters Treaty and the 1995 Agreement on the Cooperation for the Sustainable Development of the Mekong River Basin. There are over 120 basin organizations, for example, the International Commission for the Protection of the Danube River (ICPDR) and the Permanent Okavango River Basin Water Commission (OKACOM). Often, basin organizations support joint activities such as the development of a transboundary flood forecasting system (e.g., the International Sava River Basin Commission's flood forecasting and warning program) or the management of wetlands (e.g., **Nile Basin Initiative**'s Wetlands Management Strategy). Joint infrastructure is another form of cooperation, e.g., the Dostluk Reservoir shared by Iran and Turkmenistan and the Manantali and Diama dams in the Senegal River Basin, which are managed jointly by the riparians, Mali, Mauritania, and Senegal through the OMVS to share the benefits of improved navigability, irrigation capacity, and hydropower production.

Transboundary water cooperation is also understood through guiding legal principles, such as those codified in the two water conventions: 1997 **UN Convention on the Law of the Non-Navigational Uses of International Watercourses** (referred to as the UN Watercourses Convention) and 1992 UNECE Convention on the Protection and Use of Transboundary Watercourses and International Lakes (referred to as the Water Convention). Transboundary aquifers not connected to surface waters currently lack a global legal framework, as the 2008 Draft Articles on the Law of Transboundary Aquifers have not been adopted by the UN General Assembly and remain in draft status (McCaffrey 2011).

On the international stage, water cooperation is high on the international agenda, particularly within the Sustainable Development Goals (SDGs). Under goal 6, transboundary water cooperation is measured by SDG indicator 6.5.2 as the "proportion of transboundary basin area with an operational arrangement," emphasizing the procedural and institutional dimensions (McCracken and Meyer 2018) through treaties, joint bodies, regular and formal communications or meetings, joint management plans, and the regular exchange of data and information (United Nations Water 2020).

Despite the recognition the term has received in academic and policy debates, there remains a lack of clarity regarding its definition, largely because it is often used interchangeably with other terms, such as transboundary water management and water diplomacy. While related, it is important to differentiate between them to achieve peaceful, equitable, and sustainable outcomes over shared resources. Water diplomacy refers to "deliberative political processes and practices of preventing, mitigating, and resolving disputes over transboundary water resources … by applying foreign policy means" (Sehring et al. 2022). Transboundary water management, instead, emphasizes the use of technical means to achieve water-specific outcomes or respond to water-related challenges that arise from their transboundary nature (Kittikhoun and Schmeier 2021). Transboundary water management and diplomacy are two means—at the technical and political levels, respectively—to achieve transboundary water cooperation, but their presence does not automatically indicate that cooperation is also occurring.

Domestically, multiple actors share water at the local, provincial, and national levels; here, too, conflict can occur as a result, yet cooperation seems to prevail. Cooperation among different users is necessary to ensure that competing interests have access to water for drinking, agricultural, industrial, and many other purposes.

National and subnational water cooperation occurs within countries and refers to the interactions, processes, and practices between different actors—including individuals, communities, economic sectors, and civil society representatives—that aim for common goals and objectives on attaining mutual benefits with the shared water resources.

Much like at the international level, these benefits can range from social, environmental, and economic, including formal, informal, and relational aspects. For example, water cooperation can consist of the allocation of water resources from a small stream between farmers based on local water users or irrigation associations, traditional mechanisms such as "mirabs," informal arrangements over a local watershed between farmers and herders, or the formal water sharing agreements between provinces or states within a federal state. At national and subnational levels, research has highlighted the importance of **monitoring** and enforcement mechanisms to foster joint management of shared water resources (Ostrom 1990). Widespread **stakeholder** engagement, furthermore, helps to foster sustainable cooperation and build governance institutions for managing shared water. Incorporating **Indigenous knowledge** and water governance, for example, is increasingly vital for reshaping water cooperation and historical relationships of colonialism within many states (Neville and Coulthard 2019).

Regardless of the level, water cooperation should be effective and produce benefits for people, ecosystems, and countries. The challenge is determining what effective cooperation is and how it can be achieved (McCracken 2022). Understanding whether institutions governing shared water resources are effective in attaining benefits and what makes them effective are other challenges (Schmeier 2013). Drawing from the preliminary environmental regime literature, effectiveness can be determined based on different characteristics such as compliance, goal achievement, and problem-solving (Young 2001; Miles et al. 2002; Underdal 2002). These conceptualizations tend to focus on formal mechanisms, such as treaties and institutions (Schmeier 2013). Emerging research finds that informal aspects, such as the relationships developed and the outcomes achieved, are just as relevant to effective water cooperation (McCracken 2022).

Therefore, effective water cooperation can be defined as the interactions, processes, mechanisms, and practices that achieve equitable and sustainable outcomes through positive, trusting relationships between actors. This requires that the benefits of cooperation are identifiable, such as water quality improvement, flood management, **food security**, and regional stability (Sadoff and Grey 2002).

Water cooperation contributes to preventing conflicts between different actors that might have diverging interests over how to use, develop, or protect their water resources and to generating benefits for individuals, communities, economic sectors, and entire countries. It can also be a means for building **peace** beyond the water sector through **environmental peacebuilding**. Identifying shared interests and opportunities for iterated interactions provides incentives for **peacebuilding** (Conca and Dabelko 2002; Weinthal 2002; Swain 2016; Turgul et al. 2023). Water cooperation at transboundary, national, and subnational levels is critical for environmental peacebuilding because cooperative institutions and governance mechanisms build trust and confidence among adversaries, resolve conflicts, and address new challenges, such as **climate change** impacts on water availability and quality. Subnationally, for instance, restoring access to basic water services for livelihoods across water users and building institutions to manage those services can support peacebuilding efforts in conflict-affected countries (Troell and Weinthal 2014; Burt and Keiru 2011). At the transboundary level, the need to cooperate over water can bring together countries otherwise in conflict or just emerging from conflict, as the example of the Sava River Basin shows (Stec et al. 2011).

References

Bernauer, Thomas, and Tobias Böhmelt. 2020. "International Conflict and Cooperation over Freshwater Resources." *Nature Sustainability* 3.5: 350–56. https://doi.org/10.1038/s41893-020-0479-8.

Bulloch, John, and Adel Darwish. 1993. *Water Wars: Coming Conflicts in the Middle East*. Victor Gollancz.

Burt, Murray, and Bilha Joy Keiru. 2011. "Strengthening Post-Conflict Peacebuilding Through Community Water-Resource Management: Case Studies from Democratic Republic of Congo, Afghanistan and Liberia." *Water International* 36 (2): 232–41. https://doi.org/10.1080/02508060.2011.558885.

Bruin, Sophie Pieternel de, Susanne Schmeier, Rens van Beek, and Marijn Gulpen. 2024. "Projecting Conflict Risk in Transboundary River Basins by 2050 Following Different Ambition Scenarios." *International Journal of Water Resources Development* 40 (1): 7–32. https://doi.org/10.1080/07900627.2023.2184650.

Conca, Ken, & Geoffrey Dabelko (eds.). 2002. Environmental Peacemaking. Johns Hopkins University Press and Woodrow Wilson Center Press.

De Stefano, Lucia, Peter Edwards, Lynette de Silva, and Aaron T. Wolf. 2010. "Tracking Cooperation and Conflict in International Basins: Historic and Recent Trends." *Water Policy* 12 (6): 871–84. https://doi.org/10.2166/wp.2010.137.

International Groundwater Resources Assessment Centre (IGRAC) and the United Nations Educational, Scientific, and Cultural Organization (UNESCO)-Intergovernmental Hydrological Programme (IHP). 2025. *Transboundary Aquifers of the World. Edition 2025*. United Nations. https://un-igrac.org/latest/news/new-transboundary-aquifers-of-the-world-map-2025-is-available-for-download/.

Kåresdotter, Elisie, Gustav Skoog, Haozhi Pan, and Zahra Kalantari. 2023. "Water-related Conflict and Cooperation Events Worldwide: A New Dataset on Historical and Change Trends with Potential Drivers." *Science of The Total Environment* 868 (161555). https://doi.org/10.1016/j.scitotenv.2023.161555.

Kittikhoun, Anoulak, and Susanne Schmeier. 2021. *River Basin Organizations in Water Diplomacy*. Routledge.

McCaffrey, Stephen. 2011. "The International Law Commission's flawed Draft Articles on the Law of Transboundary Aquifers: The Way Forward." *Water International* 36 (5): 566–72. https://doi.org/10.1080/02508060.2011.597094.

McCracken, Melissa. 2022. *Defining Effective Transboundary Water Cooperation*. Routledge. https://doi.org/10.4324/9781003126249.

McCracken, Melissa, and Chloé Meyer. 2018. "Monitoring of Transboundary Water Cooperation: Review of Sustainable Development Goal Indicator 6.5.2 Methodology." *Journal of Hydrology* 563: 1–12. https://doi.org/10.1016/j.jhydrol.2018.05.013.

McCracken, Melissa, and Aaron T. Wolf. 2019. "Updating the Register of International River Basins of the World." *International Journal of Water Resources Development* 35 (5): 732–82. https://doi.org/10.1080/07900627.2019.1572497.

Miles, Edward L., Steinar Andresen, Elaine M. Carlin, Jon Birger Skjærseth, Arild Underdal, and Jørgen Wettestad, eds. 2002. *Environmental Regime Effectiveness: Confronting Theory with Evidence*. MIT Press.

Neville, Kate J., and Glen Coulthard. 2019. "Transformative Water Relations: Indigenous Interventions in Global Political Economies." *Global Environmental Politics* 19 (3): 1–15. https://doi.org/10.1162/glep_a_00514.

Ostrom, Elinor. 1990. *Governing the Commons: The Evolution of Institutions for Collective Action*. Cambridge University Press.

Rosenblum, Zoe H., and Susanne Schmeier. 2022. "Global Wetland Governance: Introducing the Transboundary Wetlands Database." *Water* 14 (19): 3077. https://doi.org/10.3390/w14193077.

Sadoff, Claudia W., and David Grey. 2002. "Beyond the River: The Benefits of Cooperation on International Rivers." *Water Policy* 4 (5): 389–403. https://doi.org/10.1016/S1366-7017(02)00035-1.

Schmeier, Susanne. 2013. *Governing International Watercourses. River Basin Organizations and the Sustainable Governance of Internationally Shared Rivers and Lakes*. Routledge.

Sehring, Jenniver, Susanne Schmeier, Rozemarijn ter Horst, Alyssa Offutt, and Bota Sharipova. 2022. "Diving into Water Diplomacy–Exploring the Emergence of a Concept." *Diplomatica* 4 (2): 200–221. https://doi.org/10.1163/25891774-bja10082.

Starr, Joyce R. 1991. "Water Wars." *Foreign Policy* 82: 17–36. https://doi.org/10.2307/1148639.

Stec, Stephen, Jana Kovandžić, Marija Filipović, and Amar Čolakhodžić. 2011. "A River Ran Through it: Post-Conflict Peacebuilding on the Sava River in Former Yugoslavia." *Water International* 36 (2): 186–96. https://doi.org/10.1080/02508060.2011.558886.

Swain, Ashok. 2016. "Water and Post-Conflict Peacebuilding." *Hydrological Sciences Journal* 61 (7): 1313–22. https://doi.org/10.1080/02626667.2015.1081390.

Transboundary Freshwater Diplomacy Database (TFDD). 2024. "Transboundary Freshwater Diplomacy Database." College of Earth, Ocean and Atmospheric Sciences, Oregon State University. Accessed September 12, 2025. https://transboundarywaters.ceoas.oregonstate.edu/.

Troell, Jessica, and Erika Weinthal. 2014. "Harnessing Water Management for More Effective Peacebuilding: Lessons Learned." In *Water and Post-Conflict Peacebuilding*, edited by Erika Weinthal, Jessica Troell, and Mikiyasu Nakayama. Routledge. https://doi.org/10.4324/9781849775809.

Turgul, Alexandra, Melissa McCracken, Susanne Schmeier, Zoe H. Rosenblum, Lynette de Silva, and Aaron T. Wolf. 2024. "Reflections on Transboundary Water Conflict and Cooperation Trends." *Water International* 49 (3–4): 274–88. https://doi.org/10.1080/02508060.2024.2321727.

Turgul, Alexandra, Zoe H. Rosenblum, Melissa McCracken, Susanne Schmeier, and Aaron T. Wolf. 2023. "Can Water Help Quench the Flames of Hostility? How Shared Waters Can Promote Dialogue During Conflict." *SAIS Review of International Affairs* 43 (2): 69–94. https://dx.doi.org/10.1353/sais.2023.a918646.
Underdal, Arild. 2002. "One Question, Two Answers." In *Environmental Regime Effectiveness: Confronting Theory with Evidence*, edited by Edward L. Miles, Steinar Andresen, Elaine M. Carlin, Jon Birger Skjærseth, Arild Underdal, and Jørgen Wettestad. MIT Press.
United Nations Water. 2020. *Step-by-Step Monitoring Methodology for SDG Indicator 6.5.2: Proportion of Transboundary Basin Area with an Operational Arrangement for Water Cooperation*. United Nations Water, January 25. https://www.unwater.org/sites/default/files/app/uploads/2020/02/SDG_652_Step-by-step_methodology_2020_ENG.pdf.
Weinthal, Erika. 2002. *Statemaking and Environmental Cooperation: Linking Domestic and International Politics in Central Asia*. MIT Press.
Wolf, Aaron T. 1998. "Conflict and Cooperation Along International Waterways." *Water Policy* 1 (2): 251–65. https://doi.org/10.1016/S1366-7017(98)00019-1.
Young, Oran R. 2001. "Inferences and Indices: Evaluating the Effectiveness of International Environmental Regimes." *Global Environmental Politics* 1 (1): 99–121. https://doi.org/10.1162/152638001570651.
Zeitoun, Mark, and Jeroen Warner, 2006. "Hydro-Hegemony: A Framework for Analysis of Transboundary Water Conflicts." *Water Policy* 8 (5): 435-60. https://doi.org/10.2166/wp.2006.054.

Water Diplomacy

Jill Baggerman

Water diplomacy is the political process of using diplomatic mechanisms without the use of force to advance interests, resolve disputes, or otherwise conduct international relations related to transboundary water resources. Diplomatic mechanisms include formal and informal processes such as persuasion, dialogue, negotiation, and confidence-building measures relating to transboundary water access and use.

Water diplomacy is a subset of broader diplomacy between riparians (states sharing a water resource either as upstream and downstream from one another or sharing a water body along their borders). As a political process, water diplomacy uses diplomatic instruments for issues related to transboundary water resources, including consultations with state and non-state actors, persuasion, negotiations around water allocation, dialogue to foster water cooperation, or official decision-making processes for jointly managed transboundary water institutions.

Like diplomacy more broadly, water diplomacy often occurs on multiple levels, referred to as tracks. The different levels of tracks typically occur in parallel but are distinguished primarily by the type of actors engaged and on what level (Fisher et al. 2023). Track 1 diplomacy is official, government-to-government relations operating through formal channels, such as between ambassadors or through treaty implementation. For instance, officials from the United States and Mexico regularly negotiate how to address water-related challenges related to the implementation of treaty obligations over their transboundary rivers through the International Boundary and Water Commission (Buono et al. 2021). Track 2 diplomacy is unofficial, typically occurring between leaders (such as civil society and nongovernmental organization leaders or people with religious influence) who are not a part of a government. For instance, through the Arava Institute, scientists and civil society leaders from Israel and Palestine have collaborated across conflict lines to improve water supply and wastewater treatment (Arava Institute n.d.). Track 1.5 diplomacy occurs in between these, as unofficial platforms for government officials to communicate with their counterparts in their personal capacities and engage with non-government leaders (Fisher et al. 2023; Staats et al. 2019). Track 1.5 sometimes occurs as "shuttle diplomacy," especially when actors on a track 1 level are unwilling to meet face-to-face with one another, and so a mediator or other intermediary who is engaged with **stakeholders** on opposing sides shares perspectives and ideas from one forum to another. (Track 3 diplomacy is a more nebulous term, typically referring to national dialogues or community-level efforts.) Since water resources are cross-sectoral—including energy, **agriculture**, development, private industry, public health, ecosystems, disaster prevention—and have significant implications from the household to the international level, water diplomacy is often a multitrack diplomacy process.

Water diplomacy is distinct but interconnected with water management and **water cooperation** (Sehring et al. 2022; Schmeier 2018; Troell and Weinthal 2014). The emphasis of water diplomacy is on the policy mechanisms and political relations between riparians, although water diplomacy relies in many ways on technical expertise, and many water diplomats have at least some technical understanding of water-related sciences. The same skill sets and techniques used in water diplomacy can also be employed by actors sharing water subnationally, but these contexts are not typically referred to as water diplomacy. Water management relates to the technical and administrative functions around water, such as water regulation, distribution, or monitoring hydrometeorological conditions. Water management occurs across all levels of water use, from municipal to regional. Water cooperation refers to the combination of technical means, diplomatic relations, and broader governance for collaborating with riparians or other subnational groups sharing a water resource. Water diplomacy often entails the use of transboundary water resources as a means to foster water cooperation, make decisions around water management, or for broader goals such as conflict prevention or peacebuilding. For instance, downstream countries must often negotiate with their upstream neighbors a schedule for releasing water from hydroelectric dams, important for both their domestic management around flood prevention and drought mitigation, and contributing to long-term cooperation on water and energy. However, a state may not necessarily use water diplomacy as part of **conflict prevention** or cooperation over water resources. For instance, India's suspension of the Indus Water Treaty, an agreement which served more to divide water access between the states rather than a platform for water cooperation (Michel 2025), qualifies as water diplomacy, though the suspension is escalatory rather than peace-oriented.

Water diplomacy can be a strategic element of an **environmental peacebuilding** process. **Water security** is typically heavily connected to national security (Vishwanath 2018) and water allocation between riparians and domestically often reflects underlying **power** dynamics. However, even when one or more riparians do not engage in water diplomacy with the goal of cooperation or peacebuilding, actors taking a **conflict transformation** lens can still promote environmental peacebuilding via water diplomacy. Water diplomacy has the potential to build trust and develop multiple, incremental **confidence-building measures**. When conducted with conflict analyzes, **conflict sensitivity**, a **systems approach**, and **participatory processes**, the same linkages that make water part of national security and power dynamics—combined with the direct implications the water sector has on **human security** from international to household levels—peacebuilding efforts can leverage water diplomacy as an entry point toward addressing conflict drivers in a practical and strategic way.

References

Arava Institute for Environmental Studies. n.d. "About the Arava Institute." Accessed October 3, 2025. https://arava.org/about-our-community/about-arava/.

Buono, Regina, Jill Baggerman, and Lindsay C. Sansom. 2021. "Got a Minute for the Future of the Rio Grande? Considering the Prospects for a Sustainability Minute in the Wake of the Colorado's Minute 323." *Water International* 46 (4): 543–66. https://doi.org/10.1080/02508060.2021.1915562.

Fisher, Ronald, Margarita Tadevosyan, and Esra Cuhadar. 2023. "Track 2 Dialogues." United States Institute of Peace, April. https://www.usip.org/sites/default/files/Track-2-Dialogues-Evidence-Review-Paper.pdf.

Michel, David. 2025. "Can India Cut Off Pakistan's Indus River Lifeline?" Center for Strategic and International Studies, May 1. https://www.csis.org/analysis/can-india-cut-pakistans-indus-river-lifeline.

Schmeier, Susanne. 2018. "What is Water Diplomacy and Why Should You Care?" Global Water Forum, August 31. https://www.globalwaterforum.org/2018/08/31/what-is-water-diplomacy-and-why-should-you-care/.

Sehring, Jenniver, Susanne Schmeier, Rozemarijn ter Horst, Alyssa Offutt, and Bota Sharipova. 2022. "Diving into Water Diplomacy–Exploring the Emergence of a Concept." *Diplomatica* 4 (2): 200–21. https://doi.org/10.1163/25891774-bja10082.

Staats, Jennifer, Johnny Walsh, and Rosarie Tucci. 2019. "A Primer on Multi-track Diplomacy: How Does it Work?" United States Institute of Peace, archived January 26, 2024. https://web.archive.org/web/20240126103157/https://www.usip.org/publications/2019/07/primer-multi-track-diplomacy-how-does-it-work.

Troell, Jessica, and Erika Weinthal. 2014. "Harnessing Water Management for More Effective Peacebuilding: Lessons Learned." In *Water and Post-Conflict Peacebuilding*, edited by Erika Weinthal, Jessica Troell, and Mikiyasu Nakayama. Routledge.

Vishwanath, Ambika. 2018. "Paddling Upstream: Transboundary Water Politics in South Asia." Carnegie India, October. https://carnegie-production-assets.s3.amazonaws.com/static/files/Viashwanath_GBM_Basin_October_2018.pdf.

Water-Energy-Food Nexus

Larry A. Swatuk

The Water-Energy-Food (WEF) Nexus is a conceptual framework devised for understanding the interrelationships among water, energy, and food in human systems of production and consumption so as to minimize negative cross-sector and ecosystem effects and maximize opportunities for resource use efficiency, policy coherence, sustainable development, and overall security (Allouche et al. 2019; Al-Saidi and Elagib 2017; Swatuk and Cash 2018). The WEF Nexus framework can be applied at different scales, from local to national and from regional to global. Recognizing its interlinkages with biodiversity and ecosystems, programs seeking to understand trade-offs or identify synergies across sectors, the nexus is often broadened to the Water-Energy-Food-Ecosystems (WEFE) Nexus approach (United Nations Economic Commission for Europe 2025).

The idea of using an integrated approach to governing and managing natural resource use is not new. Allouche and colleagues (2019) note that its first use seems to be in 1983 with the United Nations University's Food-Energy Nexus Program. Over the ensuing four decades, a variety of different nexuses have emerged, most often in global **governance** contexts, to help inform national and interstate policymaking and resource use practice. The WEF iteration of this basic nexus "emerged at a critical moment against the backdrop of cascading global crises in energy, food and global finance in 2008, and a sense of scarcity and insecurity that accompanied it, especially among transnational corporations and the Organisation for Economic Co-operation and Development (OECD) country governments" (Allouche et al. 2019, 4–5; World Economic Forum [WEF] 2011a; WEF 2011b).

As a conceptual framework, the WEF Nexus has been widely adopted across national and intergovernmental agencies, from German and United Kingdom development agencies to the United Nations (through the Sustainable Development Goals (SDGs)), the OECD, FAO, and European Union, and to regional entities such as the Southern African Development Community (Swatuk et al. 2025). Unlike earlier nexus formulations, the private sector, through the World Economic Forum and the SDGs, has taken on a significant role in nexus modeling and implementation. Utilizing complementary concepts such as the green economy and ecological modernization, private sector actors emphasize the role the WEF Nexus can play in maximizing economic and social benefits while minimizing **environmental degradation** (Hoff 2011). Investment and lending organizations such as the World Bank, the Asian Development Bank, and the African Development Bank are working alongside major donors to support WEF Nexus research, planning, and action. Depending on one's entry point or expertise, the framework is often presented as the energy-food-water (EFW), food-energy-water (FEW), or, most recently, the water-energy-food-ecosystems (WEFE) nexus.

As with any concept whose effects are material, the WEF nexus is highly contested. On one hand, coalitions of influential actors across states, civil society, and the private sector have supported or directly undertaken numerous projects devoted to modeling and operationalizing the WEF nexus at a variety of scales: urban, river basin, national, and regional (Al-Zu'bi and Keough 2018; Carvalho et al. 2022; Chen and Chen 2021; Mabhaudhi et al. 2021; Nhamo et al. 2018; Nhamo et al. 2022; Simpson et al. 2022). On the other hand, an extensive set of critiques has been leveled at the WEF Nexus, in particular its dominant, macro-focused, and market-oriented framing where trade-offs are calculated in terms of resource efficiencies, and resource efficiencies are understood as best use only when they have a high return-on-investment (e.g., Leese and Meisch 2015; Swatuk and Cash 2018; Allouche et al. 2019). For Allouche et al. (2019, 130), the WEF nexus needs to be:

> reconceptualized ... away from a technical solution to natural resource scarcity, which is apolitical in origin and intent, towards a clear and articulated political choice about allocation and trade-offs between resources and the imagining of the future of water-food-energy systems and their interlinkages.

By promoting large-scale, market-oriented approaches to resource optimization and economic efficiency, the dominant discourse encourages "the commodification of resources, neglecting other issues such as environmental externalities, poverty alleviation, and everyday realities at the local level" (Allouche et al. 2019, 33).

In relation to **environmental peacebuilding,** the WEF Nexus is a discursive space, and it draws together a broad coalition of **stakeholders** interested in a particular range of socio-ecological, political, or economic outcomes. At minimum, the WEF Nexus encourages integrated approaches to resource use, governance, planning, and management. This suggests the possibility for collaboration and constructive engagement. It is necessary for creative coalitions of actors interested in **peace, justice,** equity, and **sustainability** to push back against persistent, special interest-based attempts to narrow the understanding of the WEF Nexus.

References

Al-Saidi, Mohammad, and Nadir A. Elagib. 2017. "Towards Understanding the Integrative Approach of the Water, Energy and Food Nexus." *Science of the Total Environment* 574: 1131–39. https://doi.org/10.1016/j.scitotenv.2016.09.046.

Al-Zu'bi, Maha, and Noel Keough. 2018. "Water, Energy and Food: The Problematic Aspects of the Transition from 'Silo Approach' to 'Nexus Approach' in the Arab Region." In *Water, Energy, Food and People Across the Global South: 'The Nexus' in an Era of Climate Change*, edited by Larry A. Swatuk and Corrine Cash. Palgrave Macmillan. https://doi.org/10.1007/978-3-319-64024-2.

Allouche, Jeremy, Carl Middleton, and Dipak Gyawali. 2019. *The Water–Food–Energy Nexus: Power, Politics, and Justice*. Routledge. https://doi.org/10.4324/9781315209067.

Carvalho, Pedro N., David C. Finger, Fabio Masi, et al. 2022. "Nature-Based Solutions Addressing the Water-Energy-Food Nexus: Review of Theoretical Concepts and Urban Case Studies." *Journal of Cleaner Production* 338 (1): 130652. https://doi.org/10.1016/j.jclepro.2022.130652.

Chen, Weizhong, and Yan Chen. 2021. "Pre-Warning Measurement of Water Resources Security in the Yangtze River Basin from the Perspective of Water-Energy-Food Symbiosis." *Water* 13 (4): 475. https://doi.org/10.3390/w13040475.

Hoff, Holger. 2011. "Understanding the Nexus." Background Paper for the Bonn 2011 Conference: The Water Energy and Food Security Nexus, November. Stockholm Environment Institute. https://www.sei.org/mediamanager/documents/Publications/SEI-Paper-Hoff-UnderstandingTheNexus-2011.pdf.

Mabhaudhi, Tafadzwanashe, Luxon Nhamo, Tendai P. Chibarabada, et al. 2021. "Assessing Progress towards Sustainable Development Goals through Nexus Planning." *Water* 13 (9): 1321. https://doi.org/10.3390/w13091321.

Meisch, Simon, and Matthias Leese. 2015. "Securitising Sustainability? Questioning the 'Water, Energy and Food-Security Nexus.'" *Water Alternatives* 8 (1): 695-709. https://www.water-alternatives.org/index.php/alldoc/articles/vol8/v8issue1/272-a8-1-5/file.

Nhamo, Luxon, Bekithemba Ndlela, Charles Nhemachena, Tafadadzwanashe Mabhaudhi, Sylvester Mpandeli, and Greenwell Matchaya. 2018. "The Water-Energy-Food Nexus: Climate Risks and Opportunities in Southern Africa." *Water* 10 (567): 1–18. https://doi.org/10.3390/w10050567.

Nhamo, Luxon, Sylvester Mpandeli, Stanley Kiphadzi, and Tafadzwanashe Mabhaudhi. 2022. "Securing Land and Water for Food Production through Sustainable Land Reform: A Nexus Planning Perspective." *Land* 11: 974. https://doi.org/10.3390/land11070974.

Simpson, Gareth B., Graham P. W. Jewitt, William Becker, et al. 2022. "The Water-Energy-Food Nexus Index: A Tool to Support Integrated Resource Planning, Management and Security." *Frontiers in Water* 4: 825854. https://doi.org/10.3389/frwa.2022.825854.

Swatuk, Larry A., and Corrine Cash, eds. 2018. *Water, Energy, Food and People across the Global South: 'The Nexus' in an Era of Climate Change*. Springer. https://doi.org/10.1007/978-3-319-64024-2.

Swatuk, Larry A., Caroline Cash, and Tang Kai Nicole. 2025. "Leverage Points for Achieving Integrated Socio-Ecological Security across the Water, Energy, Food (WEF) Nexus." In *The Elgar Companion to Food System Transformation for Sustainable Development,* edited by Prajal Pradhan and Laxmi Prasad Pant. Edward Elgar Publishing.
United Nations Economic Commission for Europe. 2025. "Water-Food-Energy-Ecosystem Nexus." Last modified May 22, 2025. https://unece.org/environment-policy/water/areas-work-convention/water-food-energy-ecosystem-nexus.
World Economic Forum (WEF). 2011a. *Water Security: The Water–Food–Energy–Climate Nexus,* edited by Dominic Waughray. Island Press.
World Economic Forum (WEF). 2011b. *Global Risks 2011 6th Edition.* WEF. https://www3.weforum.org/docs/WEF_Global_Risks_Report_2011.pdf.

Water Security

Alessandra Vaselli and Emmanuel Valax

Water security refers to people's capacity and freedom to pursue human security through sustainable access to adequate quantities and quality of water, along with protection against water-related risks and ecosystem preservation.

Water security emerged in the 1990s and has undergone several conceptual iterations since (Cook and Bakker 2013). Initially, the concept developed within the broader **human security** discourse that had swept through the field of international development **cooperation** toward the end of the 20th century. According to Cook and Bakker, in the 2000 World Water Forum, the Global Water Partnership "introduced an integrative definition of water security that considered access and affordability of water, as well as human needs and ecological health" (Cook and Bakker 2013). That marked a turning point, as the concept quickly entered scholarly and policy debates, specifically around water availability, exposure to water-related hazards, livelihoods linked with **food security**, and the more general **sustainability** discourse (Cook and Bakker 2013). A little over a decade later, in one of the most comprehensive efforts to conceptualize this term, the United Nations University defined "water security" as:

> the capacity of a population to safeguard sustainable access to adequate quantities of acceptable quality water for sustaining livelihoods, human well-being, and socio-economic development, for ensuring protection against water-borne pollution and water-related disasters, and for preserving ecosystems in a climate of **peace** and political stability.
>
> (United Nations University Institute for Water, Environment and Health 2013)

Though different definitions of "water security" exist, its core idea is fundamentally linked to multiple human (in)security dimensions, including economic, food, health, environmental, personal, community, and national security. These are interconnected with broader social-ecological systems and depend on preserving ecosystem functionality. In line with human security's twofold protection and **empowerment** nature (United Nations General Assembly 2012), water security encompasses both freedom from insecurity and freedom to development. While the former informs risk-based approaches addressing social-hydrological thresholds for water-related risks, such as scarcity and hazards, freedom-to-development approaches view water security as the capacity to achieve well-being in social-hydrological systems (Grey and Sadoff 2007; Garrick and Hall 2014; Jepson et al. 2017; Mishra et al. 2021; Zawahri and Weinthal 2019).

More broadly, the concept of water security is linked to national and international **security** concerns, as the imperative to "secure" water intersects with strategic domestic and geopolitical priorities. Within the water, peace, and **conflict** nexus, national efforts for water security are seen both as possible triggers of international conflict and, more commonly, as catalysts for strategic cooperation (Gleick 1993; Link et al. 2016; MacQuarrie and Wolf 2013). At the subnational level, competition over water resources intersecting with **governance** and socioeconomic tensions may also fuel localized conflicts (Gizelis and Wooden 2010; Ravnborg et al. 2012). Conversely, efforts to achieve water security in fragile, conflict-affected, or post-conflict settings are also often seen as opportunities to build political stability and peace through improved water management and governance (Grey and Sadoff 2007; Link et al. 2016; Swain 2016; Septon et al. 2019; UNESCO World Water Assessment Programme 2024). In this sense, water security interventions can help prevent conflicts and build **positive peace** by enhancing **social cohesion** and inclusive institutions. There is an ongoing debate about how to sustain peace through developing water security and whose water security this entails. Water **justice** and socio-hydrological approaches highlight the subjective and socially constructed nature of water security, which unfolds in relation to shared or contested needs, preferences, and values, shaping diverse definitions of the "security" dimension of water (Zeitoun 2011; Jepson et al. 2019; Döring et al. 2024). In line with these critical perspectives, pursuing water security alongside **environmental peacebuilding** efforts also requires challenging the uneven distribution of water-related vulnerabilities and benefits within social-hydrological systems.

Acknowledgment

This work was carried out with support from the CGIAR Science Programs on Climate Action and Food Frontiers and Security. We would like to thank all funders who supported this research through their contributions to the CGIAR Trust Fund: https://www.cgiar.org/funders/.

References

Cook, Christina, and Karen Bakker. 2013. "Debating the Concept of Water Security." In *Water Security*. Routledge.

Döring, Stefan, Kyungmee Kim, and Ashok Swain. 2024. "Integrating Socio-Hydrology and Peace and Conflict Research." *Journal of Hydrology* 633: 131000. https://doi.org/10.1016/j.jhydrol.2024.131000.

Garrick, Dustin, and Jim W. Hall. 2014. "Water Security and Society: Risks, Metrics, and Pathways." *Annual Review of Environment and Resources* 39 (1): 611–39. https://doi.org/10.1146/annurev-environ-013012-093817.

Gizelis, Theodora-Ismene, and Amanda E. Wooden. 2010. "Water Resources, Institutions, and Intrastate Conflict." *Political Geography* 29 (8): 444–53. https://doi.org/10.1016/j.polgeo.2010.10.005.

Gleick, Peter H. 1993. "Water and Conflict: Fresh Water Resources and International Security." *International Security* 18 (1): 79–112. https://doi.org/10.2307/2539033.
Grey, David, and Claudia W. Sadoff. 2007. "Sink or Swim? Water Security for Growth and Development." *Water Policy* 9 (6): 545–71. https://doi.org/10.2166/wp.2007.021.
Jepson, Wendy, Jessica Budds, Laura Eichelberger, et al. 2017. "Advancing Human Capabilities for Water Security: A Relational Approach." *Water Security* 1: 46–52. https://doi.org/10.1016/j.wasec.2017.07.001.
Jepson, Wendy, Amber Wutich, and Leila M. Harris. 2019. "Water-Security Capabilities and the Human Right to Water." In *Water Politics*, edited by Farhana Sultana and Alex Loftus. Routledge.
Link, P. Michael, Jürgen Scheffran, and Tobias Ide. 2016. "Conflict and Cooperation in the Water-Security Nexus: A Global Comparative Analysis of River Basins Under Climate Change." *WIREs Water* 3 (4): 495–515. https://doi.org/10.1002/wat2.1151.
MacQuarrie, Patrick, and Aaron T. Wolf. 2013. "Understanding Water Security." In *Environmental Security*, edited by Rita Floyd and Richard Matthew. Routledge.
Mishra, Binaya Kumar, Pankaj Kumar, Chitresh Saraswat, Shamik Chakraborty, and Arjun Gautam. 2021. "Water Security in A Changing Environment: Concept, Challenges and Solutions." *Water* 13 (4): 490. https://doi.org/10.3390/w13040490.
Ravnborg, Helle Munk, Rocio Bustamante, Abdoulaye Cissé, et al. 2012. "Challenges of Local Water Governance: The Extent, Nature and Intensity of Local Water-Related Conflict and Cooperation." *Water Policy* 14 (2): 336–57. https://doi.org/10.2166/wp.2011.097.
Septon, Tal, Nidhi Nagabhatla, and Caner Sayan. 2019. "Peace as a Strategy for Planning Water Secure Futures." *OIDA International Journal of Sustainable Development* 12 (12): 51–66. https://oidaijsd.com/wp-content/uploads/2020/04/12-12-4-160.pdf.
Swain, Ashok. 2016. "Water and Post-Conflict Peacebuilding." *Hydrological Sciences Journal* 61 (7): 1313–22. https://doi.org/10.1080/02626667.2015.1081390.
UNESCO World Water Assessment Programme. 2024. *Water for Prosperity and Peace: The United Nations World Water Development Report 2024*. UNESCO. https://www.unwater.org/publications/un-world-water-development-report-2024.
United Nations General Assembly. 2012. Resolution 66/290, Follow-up to Paragraph 143 on Human Security of the 2005 World Summit Outcome. A/RES/66/290, 25 October. https://www.un.org/humansecurity/wp-content/uploads/2022/06/N1147622.pdf.
United Nations University Institute for Water, Environment and Health (UNU-INWEH). 2013. *Water Security and the Global Water Agenda: A UN Water Analytical Brief*. United Nations University Institute for Water, Environment, and Health.
Zawahri, Neda A., and Erika Weinthal. 2019. "The Intersecting Dimensions of Water Security." In *Understanding New Security Threats*, edited by Michel Gueldry, Gigi Gokcek, and Lui Hebron. Routledge.
Zeitoun, Mark. 2011. "The Global Web of National Water Security." *Global Policy* 2 (3): 286–96. https://doi.org/10.1111/j.1758-5899.2011.00097.x.

Weaponization of Natural Resources

Tobias von Lossow and Nancy Boyer

The weaponization of natural resources is the practice of using natural resources as instruments and tools to deliberately create a threat or cause harm in violent conflicts to achieve strategic political and/or tactical military goals, and thereby also unfolding strong psychological impact (von Lossow 2016).

The concept of weaponizing natural resources requires definitions of both "the environment and natural resources" as well as "weapon" and "weaponization." The **environment** and **natural resources** include water, land/soil, timber, **biodiversity**, and minerals, which are found in the environment and used beneficially by and for human beings. In the context of weaponization, the focus is mostly on resources that are essential for human well-being and survival, such as water or food, although manipulating other resources, such as minerals and **fossil fuels**, can also be included under certain circumstances (International Institute for Strategic Studies 2022).

While a clear, uniform, and internationally agreed and accepted definition of a weapon does not exist, essential characteristics of weapons include being instruments or tools used to threaten, fight, or cause harm to beings, thus affecting **human security** or **security** more broadly. This entails, for example, defeating, injuring, or killing others as well as destroying resources, facilities, and **infrastructures** fundamental for human well-being and survival.

The term weaponization describes that a tool, practice, resources, or act of a generally innocuous, civil, and often common context is twisted and used to create a threat, harm, or damage. Weaponization can be used for defensive purposes, for example, to prevent opponent troops from entering certain territories, as well as for offensive purposes, for example, to attack territories, deprive people of necessities for survival, or force civilians to flee (von Lossow 2016). In the context of the environment and natural resources, weaponization goes beyond the simple destruction of resources, driven by more complex, strategic thinking and logics related to an ongoing violent **conflict**. Such acts are usually thoroughly planned and executed rather than an ad hoc, spontaneous action or reaction.

While being also driven by strategic political goals, the weaponization of the environment and natural resources does not cover (1) purely economic considerations, for example, increasing revenues by implementing tolls or changing main trade partners and volumes, which is rather a weaponization of trade. The weaponization of the environment and natural resources neither covers (2) the inability to prevent disasters caused by natural resources, such as floods, nor (3) unintentional action, accidents, and collateral damage when parties demonstrably tried to avoid the harm and did not declare the actions or inactions an act of **war**.

Still, (1) economic considerations can turn into a weaponization of the environment and natural resources if they deliberately threaten the well-being, livelihoods, or even survival of civilians, for example, by disrupting oil and gas supplies during wintertime. Similarly, it can be considered a weaponization of the environment if (2) natural disasters remain deliberately unaddressed, for example, floods that might expel people from certain territories, or if (3) accidents and collateral damage are frequently tolerated or willingly accepted, for example, burning forests or polluted waters as a consequence of missile attacks.

The numerous examples of weaponizing the environment and natural resources include the withholding and diverting water by the so-called Islamic State (IS) at major dams in Syria and Iraq 2014–2017 (von Lossow 2020), the blowing up of Kakhovka dam in Ukraine 2023 (Schmeier and von Lossow 2023), the Assad government's starving towns of Foah, Kefraya, Madaya, and Zabadani in Syria 2015–2016 (Shaheen and Graham-Harrison 2016), Revolutionary Armed Forces of Colombia's (FARC's) blowing up oil pipelines in Colombia 2015 (Brodzinsky 2015), the United States's use of Agent Orange in the years 1962–1971 during the Vietnam War, the confiscation of agricultural land in the West Bank by Israeli settlers since 1967, or forced displacements and resettlements of Indigenous peoples in certain territories by the French colonial rulers in Algeria.

The weaponization of the environment and natural resources describes regularly occurring political and military practices in violent conflicts that date back millennia, such as the legendary salting of the earth around Carthage by victorious Romans in 146 BCE or the rerouting of the river in Babylon by Cyrus the Great in 539 BCE (von Lossow 2020). The weaponization of the environment and natural resources has never fully disappeared, and some practices and forms, such as targeting dams, starving people, or cutting off water supplies, have gained new momentum in recent years (von Lossow 2016).

Generally, the practice of weaponizing the environment and natural resources—and also the damage to and destruction of the environment and natural resources—is a serious violation of international law, **international humanitarian law**, and international norms and standards (Bothe et al. 2010). The Hague Convention of 1907 and the additional protocols to the Geneva Conventions of 1977, for example, address the issue through the protection of civilians (Office of the United Nations High Commissioner for Human Rights 1977a). The Convention on the Prohibition of Military and Other Hostile Use of Environmental Modification Techniques (ENMOD) of 1976 prohibits modifying weather patterns and causing tsunamis and earthquakes (United Nations Office for Disarmament Affairs 1976). The Rome Statute of 1998, the founding document of the International Criminal Court (ICC), is another example that lists excessive damage to the natural environment as a serious violation of international law (ICC 2021).

In 2018, the **United Nations Security Council** held an Arria-Formula meeting on "The Protection of the Environment During Armed Conflict," which also reflects the relevance of this topic (United Nations Security Council 2018). In addition to current international law and protocols to limit damage to the environment and prohibit the weaponization of the environment and natural resources, the **International Law Commission** (ILC) is working to develop additional legal protections (ILC 2022). Beyond the focus on harm and damage, it is also important to identify entry points and develop strategies for environmental peacebuilding practices that make the environment and natural resources contribute to pathways of peace, stability, and cooperation instead of being weaponized (Ahmadnia et al. 2022).

References

Ahmadnia, Shaadee, Agathe Marie Christien, Phoebe Spencer, Tracy Hart, and Caio Cesar de Araujo Barbosa. 2022. *Defueling Conflict: Environment and Natural Resource Management as a Pathway to Peace*. The World Bank.

Bothe, Michael, Carl Bruch, Jordan Diamond, and David Jensen. 2010. "International Law Protecting the Environment During Armed Conflict: Gaps and Opportunities." *International Review of the Red Cross* 92 (879). https://doi.org/10.1017/S1816383110000597.

Brodzinsky, Sibylla. 2015. "Farc Rebels Bomb New Section of Colombian Oil Pipeline." *The Guardian*, June 29. https://www.theguardian.com/world/2015/jun/29/colombia-farc-rebels-bomb-oil-pipeline.

International Criminal Court (ICC). 2021. Rome Statute. https://www.icc-cpi.int/sites/default/files/2024-05/Rome-Statute-eng.pdf.

International Institute for Strategic Studies (IISS). 2022. "Climate Change and the Weaponization of Natural Resources." *Online Event at IISS*, February 8. https://www.iiss.org/events/2022/02/climate-change-and-weaponization-of-natural-resources/.

International Law Commission (ILC). 2022. "Draft Principles on Protection of the Environment in Relation to Armed Conflicts. Texts, Instruments or Final Reports Adopted by the Commission." https://legal.un.org/ilc/texts/8_7.shtml.

Office of the United Nations High Commissioner for Human Rights (OHCHR). 1977a. "Protocol Additional to the Geneva Conventions of 12 August 1949, and Relating to the Protection of Victims of International Armed Conflicts (Protocol I)." https://www.ohchr.org/en/instruments-mechanisms/instruments/protocol-additional-geneva-conventions-12-august-1949-and.

Schmeier, Susanne, and Tobias von Lossow. 2023. "Key Insights into the Nova Kakhovka Dam Destruction: An Initial Analysis." *WPS Blog*, June 8. https://waterpeacesecurity.org/info/blog-08-06-2023-key-insights-into-the-nova-kakhovka-dam-destruction-an-initial-analysis.

Shaheen, Kareem, and Emma Graham-Harrison. 2016. "Madaya: Residents of Besieged Syrian Town say they are Being Starved to Death." *The Guardian*, January 6. https://www.theguardian.com/world/2016/jan/06/residents-besieged-syrian-town-madaya-starved-to-death.

United Nations Office for Disarmament Affairs. 1976. *Convention on the Prohibition of Military or Any Other Hostile Use of Environmental Modification Techniques (ENMOD)*. https://disarmament.unoda.org/en/our-work/weapons-mass-destruction/convention-prohibition-military-or-any-other-hostile-use-environmental-modification-techniques.

United Nations Security Council. 2018. "The Protection of the Environment During Armed Conflict Arria-Formula Meeting." https://www.securitycouncilreport.org/whatsinblue/2018/11/protection-of-the-environment-during-armed-conflict-arria-formula-meeting.php.

von Lossow, Tobias. 2016. "The Rebirth of Water as a Weapon: IS in Syria and Iraq." *The International Spectator* 51 (3): 82–99. https://www.tandfonline.com/doi/full/10.1080/03932729.2016.1213063.

von Lossow, Tobias. 2020. "Weaponizing Water in the Middle East: 'Lessons Learned' from IS." In *Water and Conflict in the Middle East*, edited by M. King. Hurst Publishers, 151–170. https://doi.org/10.1093/oso/9780197552636.003.0007.

Y

Youth

Natalija Vojno, Ambreen Tour Ben-Shmuel, and Elsa Barron

The definition of a "youth" varies across projects, policies, and research (Bennett et al. 2003). We define "youth" or "young people" to be people 18 to 35 years of age who come from a diversity of backgrounds—many of whom face intersecting social, economic, and political marginalization.

Generalizations that homogenize this diverse generation can portray them as perpetrators or heroic victims (Drummond-Mundal and Cave 2007). Young people have been depicted as protagonists of protests or threats to the status quo, especially in economic downturns (Huntington 1996). They are also often ascribed the potential to change the future. Both views are evident in the critique or praise for the work of young activists like Sweden's Greta Thunberg, who speaks out for climate **justice**, and Anishinaabe water rights advocate Autumn Peltier. Yet "youth" cannot be associated with one political position or policy preference (Thew et al. 2020). This category includes significant diversity, with different experiences, opportunities, capacities, responsibilities, and worldviews.

United Nations (UN) resolutions have called upon states to recognize youth as positive actors who can participate meaningfully in peace processes (UN Security Council 2015, 2018). These resolutions have implications for youth involvement in **environmental peacebuilding** by emphasizing the importance of youth **participation** in **peace**, **security**, justice, and **sustainable development** more broadly. Yet, contextual and individual differences motivate young people to apply different strategies to achieve goals that may or may not align with the objectives of institutional actors engaging with youth (Wehn et al. 2018). In transboundary water **governance**, youth exert agency by adopting roles that fit highly politicized and securitized water governance arenas; this results in an ecosystem of young leaders engaged in **water cooperation** and **conflict** as champions, advisers, strategists, researchers, cross-boundary connectors, capacity-builders, entrepreneurs, or caretakers of well-being (Vojno et al. 2022). How young people engage is dependent on the financial and institutional support available.

Youth are increasingly mobilizing around environmental issues, utilizing social media and grassroots organizing in awareness-raising efforts and demands for

DOI: 10.4324/9781003604143-22

climate action. Youth-led movements, specifically around climate, are shaping public discourse and policy agendas (Fisher 2019). We are beginning to see this intersect with **peacebuilding** efforts as well. For example, the 2023 COP28 Global Youth Statement, which combined the inputs of 700,000 young people from across 150 countries, included a section on climate, peace, and security with recommendations on topics including equity, resource management, **conflict prevention**, migration and **displacement**, youth engagement in peacebuilding, and diplomacy (Official Youth and Children's Constituency of the UNFCCC 2023). In their framing of peace and security, this coalition notably includes climate justice topics and bridges between both negative and **positive peace**, while embracing intersectionality and other advocacy movements. Within these movements, youth operate as both educators and learners (Trott 2024). Their involvement as educators attempts to build peace and justice in society. At the same time, youth organizing is a form of **empowerment** through **learning** and skill-building and an **indicator** of positive peace.

Intergenerational collaboration in environmental peacebuilding involves youth working alongside older generations to address environmental challenges and promote peace. While generations are conceptualized in various ways, empirical studies focusing on intergenerational engagement in environmental peacebuilding have noted that youth bring unique perspectives and skills to problem-solving and decision-making processes, which are shaping the field (Ben-Shmuel 2024; Kronich et al. 2023). This approach recognizes the value of diverse age-related experiences and competencies in addressing complex environmental and political issues. Intergenerational initiatives can foster knowledge exchange, build mutual understanding, and create more inclusive and sustainable solutions to environmental conflicts (UN 2021). By bridging generational gaps, these collaborations can lead to more comprehensive and effective strategies for environmental peacebuilding.

Acknowledgment

This work was supported by the Water Youth Network (financed through the IHE Delft Water and Development Partnership Programme), Directorate-General for International Cooperation, and the Truman Institute and Azrieli Foundation.

References

Bennett, Andy, Mark Cieslik, and Steven Miles. 2003. *Researching Youth*. Palgrave Macmillan.

Ben-Shmuel, Ambreen. 2024. "Intergenerational Confrontation and Organizational Change in the Israeli-Palestinian Peace Context." *Academy of Management Proceedings* (1). https://doi.org/10.5465/AMPROC.2024.19057abstract.

Drummond-Mundal, Lori, and Guy Cave. 2007. "Young Peacebuilders: Exploring Youth Engagement with Conflict and Social Change." *Journal of Peacebuilding and Development* 3 (3): 63–76. https://doi.org/10.1080/15423166.2007.620722997322.

Fisher, Dana R. 2019. "The Broader Importance of #FridaysForFuture." *Nature Climate Change* 9 (6): 430–31. https://doi.org/10.1038/s41558-019-0484-y.

Huntington, Samuel P. 1996. *The Clash of Civilizations and the Remaking of World Order*. Simon and Schuster.

Kronich, Shira, Liel Maghen, and Mohammad Al-Azraq. 2023. *Addressing the Intergenerational Tension in a Rapidly Changing Climate: A Case Study of the Mena Region Young Professionals Group (YPG) and Policy Recommendations*. IEMed, European Institute of the Mediterranean: https://www.euromesco.net/wp-content/uploads/2023/06/EuroMeSCo-Paper-62.pdf.

Official Youth and Children's Constituency of the UNFCCC. 2023. "Global Youth Statement: Declaration for Climate Justice." https://coy17eg.com/globalyouthstatement/.
Thew, Harriet, Lucie Middlemiss, and Jouni Paavola. 2020. "Youth Is Not a Political Position": Exploring Justice Claims-Making in the UN Climate Change Negotiations." *Global Environmental Change* 61. https://doi.org/10.1016/j.gloenvcha.2020.102036.
Trott, Carlie D. 2024. "Activism as Education in and through the Youth Climate Justice Movement." *British Educational Research Journal*. https://doi.org/10.1002/berj.4082.
United Nations. 2021. *Our Common Agenda – Report of the Secretary-General*. https://www.un.org/en/content/common-agenda-report/assets/pdf/Common_Agenda_Report_English.pdf
United Nations Security Council. 2015. Resolution 2250. S/RES/2250. December 9.
United Nations Security Council. 2018. Resolution 2419. S/RES/2419. June 6.
Vojno, Natalija, Rozemarijn ter Horst, Hussam Hussein, et al. 2022. "Beyond Barriers: The Fluid Roles Young People Adopt in Water Conflict and Cooperation." *Water International* 47 (3): 480–505. https://doi.org/10.1080/02508060.2021.2021481.
Wehn, Uta, Kevin Collins, Kim Anema, Laura Basco-Carrera, and Alix Lerebours. 2018. "Stakeholder Engagement in Water Governance as Social Learning: Lessons from Practice." *Water International* 43 (1): 34–59. https://doi.org/10.1080/02508060.2018.1403083.

Annex I
Institutions in Environmental Peacebuilding

adelphi

Janani Vivekananda and Benjamin Pohl

adelphi is a Berlin-based think tank and policy consultancy specializing in **sustainable development** and environmental policy. adelphi has worked on **environmental peacebuilding** since 2001. Through research, supporting policy and **capacity development**, accompanying implementation processes, and raising public awareness of the risks that **environmental change** entails for **peace** and stability, adelphi works to advance context-informed action addressing these risks.

adelphi conducts research to understand the complex interaction effects between environmental change on the one hand, and peace and **conflict** on the other hand. adelphi's research has analyzed this nexus at global—e.g., *A New Climate for Peace* (Climate Diplomacy 2015), *10 Insights into Climate and Peace* (Berlin Climate and Security Conference 2020), *The Rise of Hydro-Diplomacy* (Pohl et al. 2014), *The Nature of Conflict and Peace* (Rüttinger et al. 2022)—as well as regional, national, and local levels, e.g., by serving as Secretariat for the Climate Security Expert Network (Climate Diplomacy 2025a), as well as numerous research projects in the context of Weathering Risk (2025a), Climate Diplomacy (2025b), the Organization for Security and Co-operation in Europe (OSCE) Climate and Security project (2025) and many others. These analyzes have covered **environment**-related **fragility** across the world.

adelphi also supports policy development, working with international organizations, national governments, and local **stakeholders** to shape policies that integrate environmental considerations into **peacebuilding** strategies (examples include the **African Union** (Weathering Risk 2024) and the European Union, the OSCE (OSCE 2025) and the Pacific Island Forum (United Nations Development Programme and Pacific Islands Forum Secretariat 2023), South Asia (Vishwanath et al. 2024), Sudan (Foong et al. 2020), Haiti (Mosello et al. 2023), Yemen (Barry et al. 2024) and many more). adelphi's policy development work aims to support making environmental **sustainability** both **conflict-sensitive** and a core component of peacebuilding efforts. It complements policy development with training and capacity development for policymakers, practitioners, and community leaders who seek to support individuals and organizations with the skills and knowledge needed to implement environmental peacebuilding initiatives effectively.

One critical aspect of environmental peacebuilding is a granular and localized contextual understanding. adelphi brings diverse stakeholders together to address environmental

DOI: 10.4324/9781003604143-23

challenges in conflict-affected areas, partnering with experts from local communities and academic institutions to develop trust and capacities and benefiting from local knowledge and insights. One example of this approach is the first-ever integrated climate fragility assessment, *Shoring Up Stability* (Vivekananda 2019). The initiative analyzed dynamics in the Lake Chad basin, integrating satellite observation-based hydrological analysis with qualitative conflict analysis based on over 200 interviews across communities in the four basin countries conducted by local researcher teams.

adelphi also supports the integration of climate and environmental considerations into peacebuilding and **mediation** work. Leading the Weathering Risk Peace Pillar (Weathering Risk 2025b), it is supporting peace programming in regions severely affected by conflict and climate risks. As the Secretariat for Climate for Peace (Climate Diplomacy 2025c), a multilateral initiative, adelphi promotes, coordinates, and advances concrete projects on the ground to ensure **climate security** is a local and national priority. Since 2019, adelphi has organized the Berlin Climate and Security Conference (2025) together with the German Foreign Office, the global forum connecting governments, international organizations, experts, and practitioners working to promote interlinkages between different approaches within the climate and **security** agenda.

Environmental peacebuilding continues to face numerous obstacles: notably, there is too little understanding of the way environmental and security risks intersect; such understanding does not reach relevant policymakers; and stove-piped sectoral approaches make it hard to deliver solutions that simultaneously address environmental, livelihood, political, and security risks. By combining research, policy, and capacity development, collaborative projects, and advocacy, adelphi seeks to combine different entry points to help overcome these obstacles.

References

adelphi and Climate Diplomacy Initiative. 2025a. *The Climate Security Expert Network (CSEN)*. Climate Diplomacy. https://climate-diplomacy.org/climate-security-expert-network.

Barry, Sinéad, Beatrice Mosello, Spencer McMurray, and Lucas Destrijcker. 2024. "Undercurrents: How Conflict, Climate Change and the Environment Intersect in Yemen." *adelphi*, December 6. https://weatheringrisk.org/en/publiation/undercurrents-how-conflict-climate-change-and-environment-intersect-yemen.

https://climate-diplomacy.org/magazine/conflict/undercurrents-how-conflict-climate-change-and-environment-intersect-yemen

Berlin Climate and Security Conference. 2020. "10 Insights on Climate Impacts and Peace: A Summary of What We Need to Know." adelphi and Potsdam Institute for Climate Impact Research. https://berlin-climate-security-conference.de/system/files/document/10_insights_on_climate_impacts_and_peace_key_facts_0.pdf

Berlin Climate and Security Conference. 2025. "Berlin Climate and Security Conference 2025." https://berlin-climate-security-conference.de/en.

Climate Diplomacy. 2015. "A New Climate for Peace: Taking Action on Climate and Fragility Risks." Accessed August 25, 2025. https://climate-diplomacy.org/magazine/conflict/new-climate-peace.

Climate Diplomacy. 2025a. "Climate Diplomacy." https://climate-diplomacy.org/.

Climate Diplomacy. 2025b. "Climate for Peace Initiative." https://climate-diplomacy.org/climate-for-peace-initiative.

Foong, Adrian, Benjamin Pohl, and Lukas Rüttinger. 2020. *Climate-Fragility Risk Brief: Sudan*. https://climate-diplomacy.org/magazine/conflict/climate-fragility-risk-brief-sudan.

Mosello, Beatrice, Lucas Destrijcker, and Spencer Adrian McMurray. 2023. *Roots for Peace: Uncovering Climate Security Challenges in Haiti and What to Do about Them*. https://

weatheringrisk.org/sites/default/files/document/Roots_for_peace_uncovering_climate_security_challenges_in_Haiti_3.pdf.
Organization for Security and Co-operation in Europe (OSCE). 2025. *Strengthening Responses to Security Risks from Climate Change in South-Eastern Europe, Eastern Europe, the South Caucasus and Central Asia.* https://www.osce.org/node/521965.
Pohl, Benjamin, Alexander Carius, Ken Conca, et al. 2014. *The Rise of Hydro-Diplomacy. Strengthening Foreign Policy for Transboundary Waters.* https://adelphi.de/en/publications/the-rise-of-hydro-diplomacy.
Rüttinger, Lukas, Raquel Munayer, Pia van Ackern, and Florian Titze. 2022. *The Nature of Conflict and Peace: The Links Between Environment, Security and Peace and Their Importance for the United Nations.* https://adelphi.de/en/publications/the-nature-of-conflict-and-peace.
United Nations Development Programme (UNDP) and Pacific Islands Forum Secretariat (PIFS). 2023. *Pacific Climate Security Assessment Guide.* https://forumsec.org/sites/default/files/2023-12/Pacific-Climate-Security-Assessment-Guide.pdf.
Vishwanath, Ambika, Priyanka Bhide, Namrata Kabra, Harsh Vasani, Benjamin Pohl and Mary Potts. 2024. *Building a Climate for Peace: Climate Security in South Asia.* https://climate-diplomacy.org/magazine/cooperation/building-climate-peace-climate-security-south-asia.
Vivekananda, Janani, Martin Wall, Florence Sylvestre, Chitra Nagarajan, and Oli Brown. 2019. *Shoring Up Stability: Addressing Climate and Fragility Risks in the Lake Chad Region.* https://shoring-up-stability.org/.
Weathering Risk. 2024. *Africa Climate Security Risk Assessment.* https://weatheringrisk.org/en/publication/africa-climate-security-risk-assessment.
Weathering Risk. 2025a. *Weathering Risk.* https://weatheringrisk.org/en.
Weathering Risk. 2025b. *The Weathering Risk Peace Pillar.* https://weatheringrisk.org/en/peacepillar.

African Union (AU)

Ousseyni Kalilou

The African Union (AU) launched in 2002 and succeeded the Organization of African Unity (OAU), which originated from the Charter signed in Addis Ababa in May 1963. The AU, as a continuation of the pan-African vision, seeks to build on the principles of African unity and **cooperation** at the state level to mitigate polity fragmentation and promote a people-centered approach to **peace** and development (Maphosa et al. 2014). Before the 1992 Earth Summit, African states took significant steps to address **biodiversity** disturbance and ecological degradation. Between 1920 and 1990, many African states established national parks and protected areas, such as Kruger and Serengeti, to combat pollution, deforestation, and poaching. In 1968, the independent African states signed the African Convention on the Conservation of Nature and Natural Resources, which was later revised in July 2003 to become a governing treaty. The Sirte Declaration of 1999, which led to the establishment of the African Union, reflected the continent's growing recognition of collective action in addressing emerging transboundary challenges, including environmental degradation and conflict. Building on this foundation, the AU and its internal and external partners collaborate with the UN on the principle of subsidiarity to advance the UN's interface with regional organizations, a significant achievement in the field of **environmental peacebuilding**.

Many other collective efforts underscore the shared responsibility and unity in achieving the AU's **sustainability** goals. In 2002, in Durban, South Africa, the OAU Council of Ministers designated March 3rd as Africa Environment Day. In January 2012, the AU adopted a decision calling for the joint commemoration of Africa Environment Day and Wangari Maathai Day on March 3 (AU 2023). In 2004, the AU Peace and Security Council (PSC) emerged to fall into the scope of environmental **peacebuilding** by promoting **conflict prevention**, ongoing **conflict** situation peacemaking and peacekeeping, and post-conflict reconstruction and peacebuilding (Baregu 2011). In 2007, the AU launched the Great Green Wall (GGW) Initiative for the Sahara and the Sahel to enhance ecosystem and climate **resilience** (Bernard 2023). On January 31, 2015, in Addis Ababa, the AU launched Agenda 2063, with an important aspiration, the Comprehensive African Agriculture Development Programme (CAADP), to help eliminate hunger and reduce poverty through agricultural development (AU 2023). The African Union Development Agency-New Partnership for African Development (AUDA-NEPAD), the AU Inter-African Bureau for Animal Resources (AU-IBAR), the African Union Semi-Arid Food Grain Research and Development (SAFGRAD), the Department of Agriculture, Rural Development, Blue Economy, and Sustainable Environment (ARBE), the African Risk Capacity (ARC) Group, and AU Wildlife Strategy (AU 2015, 2023) are among the AU institutions working collaboratively to implement the AU sustainable **environment** management priorities, policies, and frameworks set in the CAADP and other programs.

More recently, on July 15, 2021, the AU launched the comprehensive continental Green Recovery Action Plan. This plan, along with the AU Climate Change and Resilient Development Strategy and Action Plan (2022–2032), represents the AU's commitment to concretizing the pan-African vision of peacebuilding, climate cooperation, and climate-resilient development for the future.

African environmental peacebuilding is uniquely rooted in African colonial history, the neo-colonial state-centric approaches to peace, the Eurocentric view of peacebuilding, and the lessons learned from the OAU (Zondi 2017). The unique innovations and efforts of AU environmental peacebuilding should shift from the political, economic, social, and environmental dimensions of modernity inherited from colonization to a reforming decolonial peace. In African conflict situations, decolonial peace is a pursuit of peace that abandons the continued colonial order in post-colonial Africa, including neocolonialism, to achieve full liberation, lasting peace, and prosperity (Grosfoguel 2009; Ndlovu-Gatsheni 2015; Zondi 2017). This shift is crucial and should involve a more substantial consideration of the roles of Indigenous peoples, communities, and civil society actors in peacebuilding processes.

However, the lack of technical and climate finance self-reliance undermines the very purpose of the AU (Zondi 2017). The AU decolonial efforts offer ways to push multinational companies for carbon credit integrity, the sustainability of the environment, and the local communities beyond the transfer of technology. Competition over depleting **natural resources** due to **climate change** and **environmental degradation** (Kalilou 2021) is compounding other conflict factors such as ethnicity, proliferation of arms, politico-religious ideologies, and auto-determination (Bujra 2002).

It is crucial to note that the AU's commitment to environmental peacebuilding is not just a historical legacy, but also a contemporary reality. The AU's current approach to environmental peacebuilding is deeply rooted in the dominant trend of environmental protection ideology, encompassing agro-ecology, ecotourism, circular economy, green economy, green energy, energy efficiency, sustainable **cities**, and the like. This modern approach, coupled with the AU's support for member states in climate **adaptation**, loss and damage financing, biological **security**, and **conservation**, reinforces its position in mainstream environmental peacebuilding.

References

African Union (AU). 2013. *Revised African Convention on the Conservation of Nature and Natural Resources.* 2nd Ordinary Session of the Assembly Maputo, Mozambique. July 11.

African Union (AU). 2015. *African Strategy on Combating Illegal Exploitation and Illegal Trade in Wild Fauna and Flora in Africa.* https://au.int/sites/default/files/documents/33796-doc-african_strategy_strategy_africaine_au.pdf.

African Union (AU). 2023. *Africa Environment and Wangari Maathai Day.* March 3. https://au.int/en/wangari-maathai-day.

Baregu, Mwesiga, ed. 2011. *Understanding Obstacles to Peace: Actors, Interests, and Strategies in Africa's Great Lakes Region.* Fountain Publishers.

Bernard, Florence. 2022. *Regional Policy Coherence Analysis to Scale Up Action and Achievements of the Great Green Wall Initiative.* Centre for International Forestry Research (CIFOR) and World Agroforestry. https://core.ac.uk/download/552858395.pdf.

Bujra, Abdullah. 2002. *African Conflicts: Their Causes and Their Political and Social Environment.* DPMF Occasional Paper 4. Development Policy Management Forum.

Grosfoguel, Ramon. 2009. "A Decolonial Approach to Political-Economy: Transmodernity, Border Thinking and Global Coloniality." Kult 6 – Special Issue: Epistemological Transformation. http://www.postkolonial.dk/artikler/kult_6/GROSFOGUEL.pdf.

Kalilou, Ousseyni. 2021 "Climate Change and Conflict in the Sahel: the Acacia Gum Tree as a Tool for Environmental Peacebuilding." *International Affairs* 97 (1): 201–18. https://doi.org/10.1093/ia/iiaa178.

Maphosa, Sylvester B., Laura DeLuca, and Alphonse Keasley, eds. 2014. *Building Peace From Within: An Examination of Community-Based Peacebuilding and Transitions in Africa.* Africa Institute of South Africa. https://doi.org/10.2307/j.ctvh8r4g3.

Ndlovu-Gatsheni, Sabelo J. 2015. "Decoloniality in Africa: A Continuing Search for a New World Order." *The Australasian Review of African Studies* 36 (2): 22–50. https://search.informit.org/doi/epdf/10.3316/informit.640531150387614.

Zondi, Siphamandla. 2017. "African Union Approaches to Peacebuilding: Efforts at Shifting the Continent Towards Decolonial Peace." *ACCORD, AJCR 2017/1.* https://www.accord.org.za/ajcr-issues/african-union-approaches-peacebuilding/.

Alliance for Peacebuilding (AfP)

Nick Zuroski

The Alliance for Peacebuilding (AfP) is an award-winning non-profit and non-partisan network of 235 organizations working in 181 countries to prevent and reduce violent **conflict** and build **sustainable peace**. AfP's members span non-profit and for-profit organizations working in **peacebuilding**, humanitarian assistance, development, civilian protection, and social impact, as well as university institutions, including Search for Common Ground, Nonviolent Peaceforce, and the University of Notre Dame's Keough School of Global Affairs. AfP cultivates a network to strengthen and advance the peacebuilding field, enabling peacebuilding organizations to achieve greater impact—tackling issues too large for any one organization to address alone.

AfP hosts numerous working groups and communities of practice, including the Digital Peacebuilding Community of Practice, the Global Fragility Act (GFA) Coalition, and the Prevention and Protection Working Group. Specifically, regarding **environmental peacebuilding**, AfP convenes the Environment, Climate, and Conflict Working Group, which works to develop the evidence base on the integration of **conflict prevention** and environmental action in policy, programming, and funding; generate guidance and collate tools on that integration; and elevate discourse on the connections between **environmental degradation** and violent conflict. The working group has hosted in-depth discussions of key environmental peacebuilding research, such as the Institute for Economics and Peace's *Ecological Threat Report*, facilitated dialogues before and after international conferences, and contributed to key policy and advocacy documents, such as AfP's policy brief on climate and conflict integration (AfP 2022) and AfP's transition memo for the Trump Administration (AfP 2024).

References

Alliance for Peacebuilding (AfP). 2022. *The Climate Change Crisis: Ensuring Integration of Conflict Prevention and Climate Change in U.S. Foreign Policy and Development Assistance.* Policy brief by Elizabeth Hume, Megan Corrado, and Nick Zurosk. Alliance for Peacebuilding. https://www.allianceforpeacebuilding.org/afp-publications/ensuring-integration-conflict-climate-april22.

Alliance for Peacebuilding (AfP). 2024. *Status Quo No More: How the New Administration Can Prevent and Resolve Conflict.* Policy Brief. https://www.allianceforpeacebuilding.org/afp-publications/2024-transition-memo.

Alliance of Bioversity and the International Center for Tropical Agriculture (CIAT)

Grazia Pacillo

The Alliance of Bioversity International and the International Center for Tropical Agriculture (CIAT) delivers research-based solutions that address the global crises of malnutrition, **climate change, biodiversity** loss, and **environmental degradation.** The Alliance focuses on the nexus of **agriculture,** nutrition, and the **environment.** It works with local, national, and multinational partners across Africa, Asia, Latin America, and the Caribbean, and with the public and private sectors and civil society. With novel partnerships, the Alliance generates evidence and mainstreams innovations to transform food systems and landscapes so that they sustain the planet, drive prosperity, and nourish people in a climate crisis. The Alliance is part of the Consultative Group on International Agricultural Research (CGIAR), a global research partnership for a food-secure future dedicated to transforming food, land, and water systems in a climate crisis (The Alliance of Bioversity International and CIAT n.d.).

The CGIAR FOCUS Climate Security team, within the Alliance of Bioversity International and CIAT, leads the climate, **peace,** and **security** research agenda. The CGIAR FOCUS Climate Security Team is a team of over 70 researchers based in 20 countries and operating from 7 global hubs (Cali, Cairo, Dakar, Hanoi, Nairobi, Pretoria, and Rome). Its expertise includes migration and **displacement** analysis; peace and **conflict** analysis; policy and **governance** analysis; geospatial, **remote sensing** analysis; early warning, early action, anticipatory action; systems and complexity analysis; climate risk and vulnerability assessments; spatial-econometric analysis; digital platforms and news media analysis; and machine **learning,** natural language processing, and other big data approaches (CGIAR FOCUS Climate Security n.d.a).

The team works on assessing the role of food, land, and water systems as an intermediary space where climate impacts may contribute to human insecurity and increase the risk of conflict, as well as how building resilient and inclusive food, land, and water systems may contribute to **conflict prevention** and sustaining peace. This includes examining the intersection of climate with livelihood security, human mobility, peace, and **social cohesion**, assessing how, where, and for whom climate exacerbates root causes of human insecurity and peace, and how climate **adaptation** efforts can become sensitive to these interactions and proactively contribute to conflict prevention, peace, and stability. This also includes supporting humanitarian actors in developing climate-sensitive strategies, programs, and plans. Through secondments in key UN organizations, the team also aims to develop a "Surge Capacity" to deploy staff members and strategic science to support humanitarian crisis response.

Specifically, the team offers includes the following:

I. Mixed-method, cutting-edge research on the climate, food, land, and water systems, conflict, and fragility nexus (Pacillo et al. 2024; Craparo et al. 2024; CSO n.d.a)
II. Deployment of research and researchers to support timely emergency and crisis response (the CGIAR SURGE Capacity)
III. Tools and methodologies for implementing peace-positive food, land, and water systems transformations across the climate, conflict, **fragility**, and displacement nexus (Climate Security Observatory (CSO) n.d.b)
IV. Policy and governance analyzes for mainstreaming peace and security into key policies and programs (Schapendonk et al. 2023)
V. Supporting United Nations conventions, frameworks, conferences, and approaches to building climate-resilient livelihoods in conflict-affected areas (International Center for Tropical Agriculture 2023; Switzerland UN 2023; Carneiro et al. 2023)
VI. De-risking climate adaptation investments in high-risk areas (CGIAR FOCUS Climate Security n.d.b)
VII. Integrating climate and peace into social protection approaches and across the humanitarian-development-peace (HDP) nexus
VIII. Localizing and decolonizing research through the climate, peace, and displacement partnership (Desai et al. 2024)

This research agenda covers the full fragility and conflict continuum, including relatively stable and less fragile countries such as Kenya, Senegal, Mauritania, Benin, Zambia, Madagascar, Vietnam, the Philippines, India, and Guatemala, to more fragile, conflict-affected and displacement hotspots such as Sudan, South Sudan, Jordan, Mali, Burkina Faso, Nigeria, Niger, Libya, Benin, Côte d'Ivoire, Ghana, Guinea, Togo, Somalia, Ethiopia, Democratic Republic of the Congo, Zimbabwe, Mozambique, Pakistan, Bangladesh, Honduras, Haiti, and Papua New Guinea.

Acknowledgment

This work was carried out with support from the CGIAR Initiative on Climate Resilience, ClimBeR, and the CGIAR Initiative on Fragility, Conflict, and Migration. We would like to thank all funders who supported this research through their contributions to the CGIAR Trust Fund: https://www.cgiar.org/funders/.

References

Carneiro, Bia, Niklas Sax, and Grazia Pacillo. 2023. *Exploring the Research Gaps in Climate Security for Somalia.* Working Paper, Alliance of Bioversity International and CIAT, CGIAR. https://hdl.handle.net/10568/137400.

CGIAR FOCUS Climate Security. n.d.a. "Home Page." Accessed August 26, 2025. https://climatesecurity.cgiar.org/.

CGIAR FOCUS Climate Security. n.d.b. *Climate Security Programming Dashboard (CSPDxCF) Landing Page.* Accessed August 26, 2025. https://cspd.cso.cgiar.org/.

International Center for Tropical Agriculture (CGIAR). 2023. "Climate Security Challenges in the Arab Region: Salma Kadry's Historic Address at the UN Security Council." *CGIAR News & Events*, June 14. https://www.cgiar.org/news-events/news/unveiling-the-arab-regions-climate-security-challenges-salma-kadrys-historic-address-at-the-un-security-council/.

Climate Security Observatory (CSO). n.d.a. "Home Page." Accessed August 26, 2025. https://cso.cgiar.org/#/LandingPage.

Climate Security Observatory (CSO). n.d.b. "Global Tools: Climate Security Tools." Accessed August 26, 2025. https://cso.cgiar.org/#/GlobalTools.

Craparo, Alessandro, Ashleigh Basel, Onivola Minoarivelo, Grazia Pacillo, and Peter Laderach. 2024. *Compound Risk Framework: Approach Overview.* Alliance of Bioversity International and CIAT, CGIAR. https://hdl.handle.net/10568/159330.

Desai, Bhavya, Peter Läderach, and Grazia Pacillo. 2024. *Climate Peace and Displacement Partnership: IPSR Innovation Profile.* 1st ed. CGIAR System Organization. https://hdl.handle.net/10568/151834.

Pacillo, Grazia, Leonardo Medina, Theresa Liebig, et al. 2024. "Measuring the Climate Security Nexus: The Integrated Climate Security Framework." *PLoS Climate* 3 (10): e0000280. https://doi.org/10.1371/journal.pclm.0000280.

Schapendonk, Frans, Carolina Sarzana, Adam Savelli, Ignacio Madurga-Lopez, Grazia Pacillo, and Peter Läderach. 2023. "Are Climate and Environment- and Peace- and Security-Related Policy Outputs Coherent? A Policy Coherence and Awareness Analysis for Climate Security." *International Development Planning Review* 45 (4): 437–67. https://doi.org/10.3828/idpr.2023.

Switzerland UN (@swiss_un). 2023. "Switzerland, Mozambique & United Arab Emirates co-hosted a discussion on Bridging Science-Policy Gap: Fostering Solutions to Address Climate, Peace & Security. Thanks to the UN Climate Security Mechanism for their perspective & @CGIAR for presenting the Climate Security Observatory." *Twitter* (now X), October 12. https://x.com/swiss_un/status/1712582518610833717.

The Alliance of Bioversity International and the International Center for Tropical Agriculture (CIAT). n.d. "Home Page." Accessed August 26, 2025. https://alliancebioversityciat.org/.

Antarctic Treaty

Christianne Zakour

The Antarctic Treaty of 1959 is the main regulatory agreement for the protection of Antarctica, signed in 1959 and effective since 1961. It designates Antarctica and the area south of 60° South Latitude as a scientific preserve, bans military activity, and ensures freedom of scientific investigation (Lord 2020). Importantly, it suspends any territorial sovereignty claims, making it a unique example of collaborative management and environmental protection. The history of the Treaty lies in the tense climate of the Cold War between the Soviet Union and the United States. It was negotiated by states that were already present in Antarctica and were concerned about the threat of armed **conflict** encroaching on the region (Haward and Jackson 2023; Lord 2020).

The Antarctic Treaty anchors the Antarctic Treaty System (ATS), the whole complex of arrangements made for the purpose of regulating relations among states in the Antarctic (Yermakova 2021). Some key treaties within the ATS include the Protocol on Environmental Protection to the Antarctic Treaty (Madrid Protocol) of 1991, designating Antarctica as a natural reserve; the Convention for the Conservation of Antarctic Seals (1972), which aims to regulate and manage the sealing industry; and the Convention on the Conservation of Antarctic Marine Living Resources (1980), which focuses on preserving marine life and ecosystems in the Southern Ocean (Yermakova 2021). The ATS has demonstrated remarkable **resilience** thus far against internal and external stressors due to shared member state commitment to the cardinal values of **peace**, science, and **cooperation**, along with a stable legal framework built on consensus decision-making (Haward and Jackson 2023). Its ability to effectively respond to the challenges posed by illegal fishing and whaling, **climate change**, commercial tourism, energy, and **human security** will be tested in the coming years. If the ATS continues its trajectory of dynamic growth, it can be extrapolated that member states will tackle these challenges with flexibility and the political will to maintain their success (Triggs 2011).

The Antarctic Treaty has served as a model for other treaties; for example, it directly inspired and influenced many of the provisions in the Outer Space Treaty of 1967 (Race 2011). The similarities can be seen as Antarctica and outer space are both harsh, remote locations with vast untapped resources, with many countries interested in making territorial claims (Kerrest 2011). The Treaty and the ATS have faced criticisms and skepticism about their effectiveness and limitations. Membership in the Treaty has expanded since the 1960s, to mitigate any claims of exclusion and imperialism (Lord 2020). Two problems result from this: geopolitical tensions between members about consensus decision-making practices or interests over territorial claims (Haward and McGee 2021), and the main influences on Treaty **governance**. Since the key currency of Antarctic governance remains scientific output, the ability to participate in Antarctic scientific research remains in the hands of a few wealthy states, and thus political agendas are shaped by the member states that lead research on the continent (Lord 2020). The ATS has also been slow to address the impacts of climate change on the Antarctic ecosystem (Haward and McGee 2021). For example, the Antarctic Treaty Consultative Parties have engaged with climate change on a vague basis by committing to reducing greenhouse gas emissions from Antarctic research operations, rather than taking a holistic stance against global climate change.

The Treaty was enacted well before the official dawn of the field of **peacebuilding**, yet it remains firmly in place and sets a significant precedent for international cooperation on environmental protection. At the forefront is the Treaty's considerations for environmental protection. Notably, the basis of the Antarctic Treaty is the use of scientific collaboration as a tool of peacebuilding by encouraging freedom of scientific investigation and data sharing among partner nations. The Treaty suspends territorial sovereignty claims and bans military activities, thereby setting a precedent for demilitarized zones and peaceful coexistence. Multiple nations are involved in the decision-making processes, demonstrating vast potential for collaborative management of global commons. The Antarctic Treaty and the ATS are important contributors to **environmental peacebuilding** and global governance because they encourage nations to work together and promote a sense of shared environmental responsibility and stewardship.

References

Haward, Marcus, and Andrew Jackson. 2023. "Antarctica: Geopolitical Challenges and Institutional Resilience." *The Polar Journal* 13 (1): 31–48. https://doi.org/10.1080/2154896X.2023.2205237.

Haward, Marcus, and Jeffrey McGee. 2021. "Antarctic Geopolitics: Emerging Cracks in the Ice?" *Georgetown Journal of International Affairs*. https://gjia.georgetown.edu/2021/02/03/antarctic-geopolitics-emerging-cracks-in-the-ice/.

Kerrest, Armel. 2011. "Outer Space as International Space: Lessons from Antarctica." *Science Diplomacy: Antarctica, Science, and the Governance of International Spaces*. http://dx.doi.org/10.5479/si.9781935623069.133.

Lord, Thomas. 2020. "The Antarctic Treaty System and the Peaceful Governance of Antarctica: the Role of the ATS in Promoting Peace at the Margins of the World." *The Polar Journal* 10 (1): 3–21. https://doi.org/10.1080/2154896X.2020.1757821.

Race, Margaret S. 2011. "Policies for Scientific Exploration and Environmental Protection: Comparison of the Antarctic and Outer Space Treaties." In *Science Diplomacy: Antarctica, Science, and the Governance of International Spaces*, edited by Paul Arthur Berkman, Michael A. Lang, David W. H. Walton, and Oran R. Young. Smithsonian Contributions to Knowledge. http://dx.doi.org/10.5479/si.9781935623069.143.

Triggs, Gillian. 2011. "The Antarctic Treaty System: A Model of Legal Creativity and Cooperation." In *Science Diplomacy: Antarctica, Science, and the Governance of International Spaces*, edited by Paul Arthur Berkman, Michael A. Lang, David W. H. Walton, and Oran R. Young. Smithsonian Contributions to Knowledge. http://dx.doi.org/10.5479/si.9781935623069.39.

Yermakova, Yelena. 2021. "Legitimacy of the Antarctic Treaty System: Is It Time for a Reform?" *The Polar Journal* 11 (2): 342–59. https://doi.org/10.1080/2154896X.2021.1977048.

The Center for Climate and Security (CCS)

Noah Fritzhand and Tom Ellison

The Center for Climate and Security (CCS) is a solutions-oriented, bridge-building research institute focused on tackling systemic climate and ecological **security** risks. CCS works to break down silos across disciplines to ensure **climate change** considerations are integrated into security policies and security concerns are integrated into climate policy, both in the United States and countries around the world. CCS facilitates interdisciplinary policy development, conducts groundbreaking research and analysis, and hosts resources and communities of practice on the intersection of climate change and security challenges.

Founded in 2010 with a highly **stakeholder**-centric approach, CCS engages in a broad range of United States and international work focused on interdisciplinary environmental challenges through three main lines of effort:

I. *Research and Analysis*: CCS conducts forward-leaning research and publishes analysis to examine the impact of climate change on a range of national and international security issues. This policy-relevant research seeks to translate between disciplinary divides, including how climate change is exacerbating **conflict** risk; compounding food, water, and human insecurity; straining institutions, and reshaping geopolitics. For example, CCS and Woodwell Climate Research Center have conducted joint climate modeling and security analysis on countries and regions key to international security, including Turkey, Iran, North Korea, India, China, and the Arctic (Center for Climate and Security n.d.a)

II. Policy Engagement: CCS develops policies for governments and other stakeholders (i.e., multilateral institutions) to better address the climate-security nexus, educate policymakers, and track implementation. CCS convenes and engages policymakers and practitioners publicly and privately for discussions, events, and tabletop exercises. CCS is pragmatic without losing sight of the transformative changes required by climate change. For example, in the ongoing Feeding Resilience project, CCS provided a set of actionable policy recommendations to the US government to mitigate the national security risks posed by food insecurity and climate change (Center for Climate and Security n.d.b). In Phase II of Feeding Resilience, policy recommendations on this topic are being developed for European and multilateral institutions. CCS has also conducted several tabletop and scenario exercises with high-level stakeholders on topics such as ecological risks in a future Southeast Asia, renewable energy in Japan and Korea, and climate security in the Balkans

III. Professional Community Development: CCS offers training, hosts resources, and cultivates professional communities and networks to develop the security workforce required by a changing climate. These include, for example, the Climate Security Working Group (CSWG), the International Military Council on Climate and Security (an international consortium of security research institutions and leaders), and fellowship programs for young people in climate and ecological security. CCS also acts as a resource hub for key climate security documents from governments and international institutions

CCS has a robust history of research, writing, and convening on the topics of **environmental peacebuilding**, climate and conflict-affected settings, and the humanitarian and development sectors. Some examples of recent key projects include *Security and Climate Change Implications for Humanitarian, Development and Peace-building Programmes in Somalia* (ODI et al. 2024), *Climate Change and Fragility: Improving Early Warning and Climate-proofing Development and Conflict Interventions* (Sikorsky et al. 2023), *Breaking Silos: Climate Change, Security, and Humanitarian Action—Roundtable Summary* (Barron et al. 2023), and *Climate Change and Human Security Implications for Humanitarian, Development, Disaster and Climate Risk Management in Mozambique* (USAID Bureau of Humanitarian Assistance et al. 2024).

References

Barron, Elsa, Brigitte Hugh, and Alejandra Portillo-Taylor. 2023. *Breaking Silos: Climate Change, Security, and Humanitarian Action—Roundtable Summary*. CCS and ICRC. https://councilonstrategicrisks.org/2023/01/31/briefer-breaking-silos-climate-change-securi ty-and-humanitarian-action-roundtable-summary/.

Center for Climate and Security (CCS). n.d.a. "Climate Risks: Fusing Security, Science, and Storytelling." Council on Strategic Risks. https://councilonstrategicrisks.org/ccs/climaterisks/#:~:text=Since%202021%2C%20CCS%20and%20Woodwell,Climate%20Security%20Crises%20in%20Iran

Center for Climate and Security (CCS). n.d.b. "Feeding Resilience: Food, Climate Change, Global Security." Council on Strategic Risks. https://councilonstrategicrisks.org/ccs/feedingresilience/.

ODI, USAID's Bureau of Humanitarian Assistance, and the Center for Climate and Security (CCS). 2024. "Security and Climate Change Implications for Humanitarian, Development and Peacebuilding Programmes in Somalia." Report prepared for the Catalysing Cohesive Action on Climate and Security Workshop, Nairobi, Kenya, January 2024. https://odi.org/en/publications/meeting-report-security-and-climate-change-implications-for-hdp-programmes-in-somalia/

USAID Bureau of Humanitarian Assistance, ODI, the Center for Climate and Security (CCS), and the CGIAR FOCUS Climate Security Team at the Alliance of Bioversity International and International Center for Tropical Agriculture (CIAT). 2024. "Climate Change and Human Security Implications for Humanitarian, Development, Disaster and Climate Risk Management in Mozambique." Report prepared for the Catalysing Cohesive Action on Climate and Security Workshop, Mozambique, March 2024. https://odi.org/en/publications/climate-change-human-security-implications-mozambique/.

Sikorsky, Erin, Francesco Femia, and Brigitte Hugh. 2023. "Climate Change and Fragility: Improving Early Warning and Climate-Proofing Development and Conflict Interventions." In *Handbook of Fragile States*. Edward Elgar Publishing. https://www.e-elgar.com/shop/usd/handbook-of-fragile-states-9781800883468.html.

The Climate Security Mechanism (CSM)

Meryam El Bouhati and Matti Lehtonen

The Climate Security Mechanism (CSM) is a joint initiative by the United Nations Department of Political and Peacebuilding Affairs (DPPA), the United Nations Development Programme (UNDP), the United Nations Environment Programme (UNEP), and the United Nations Department of Peace Operations (DPO). Established in 2018, the CSM is a flagship United Nations (UN) initiative aimed at integrating climate, **peace**, and **security** (CPS) considerations across the UN system. Acknowledging the intricate interconnections between these domains, the CSM promotes cross-sectoral collaboration to enhance the UN's capacity to assess and respond to **climate-related security risks**. Drawing on the expertise of its four constituent entities, it develops comprehensive approaches that address the impacts of **climate change** on stability while fostering **conflict sensitivity** in climate action (CSM n.d.).

The CSM operates through a Joint Programme financed by a flexible pooled funding mechanism administered by the UN Multi-Partner Trust Fund Office, with support from the governments of Austria, Canada, Denmark, France, Germany, Ireland, Norway, Slovenia, Sweden, Switzerland, the United Arab Emirates, and the United Kingdom (UN Multi-Partner Trust Fund n.d.).

The Joint Programme builds on a long-standing legacy of UN collaboration in the field of **environmental peacebuilding**, where **cooperation** among agencies at the country level has helped shape emerging guidance. Moreover, this operational groundwork has complemented the growing discursive consolidation—shifting from references to "climate and security" and "climate-related security risks" to the more integrated framing of climate, peace, and security—a transition reflected in Security Council deliberations and field-level practice. At the same time, the institutionalization of the climate-security agenda within the UN evolved alongside broader efforts to promote the humanitarian-development-peace nexus and to strengthen integrated responses to environmental challenges. In this process, Security Council debates played a key role, beginning with the United Kingdom's 2007 initiative to place **climate security** on the Council's agenda (Hardt et al. 2023). Later, Germany's 2011 thematic debate culminated in a presidential statement that called on the Secretary-General to assess the effects of climate change in contexts where it could affect peace and security (Lehtonen 2022).

During its 2017–2018 Security Council tenure, Sweden played a pivotal role in shifting the Council's approach to the climate-security nexus by promoting country-specific resolutions that embedded climate, peace, and security considerations within concrete geopolitical contexts. This move expanded the scope of the debate and laid the groundwork for more context-sensitive interventions (Lehtonen 2022).

Over time, discourse around climate and security evolved—moving from fragmented references to more structured, systemic recognition of climate, peace, and security as interlinked priorities. This framing, now central to the CSM's approach, reflects a growing recognition of climate change as a risk multiplier and a challenge that transcends sectoral boundaries (Maertens 2021; United Nations University Centre for Policy Research 2023).

The CSM has evolved to include targeted support to UN missions, country teams, and regional actors, strengthening their capacity to manage climate-related security risks. In addition to deploying Climate, Peace, and Security Advisers, the CSM advances systemic change by promoting knowledge sharing, capacity building, and multistakeholder collaboration. Leveraging the convening **power** of its member entities, it facilitates coordinated action across **governance** levels (CSM n.d.).

References

Climate Security Mechanism (CSM). n.d. "Climate, Peace and Security | Climate Security Mechanism." Accessed February 17, 2025. https://www.un.org/climatesecuritymechanism/en.

Hardt, Judith Nora, Cameron Harrington, Franziskus Von Lucke, Adrien Estève, and Nicholas P. Simpson, eds. 2023. *Climate Security in the Anthropocene: Exploring the Approaches of United Nations Security Council Member-States*. The Anthropocene: Politik—Economics—Society—Science 33. Springer International Publishing.

Lehtonen, Matti. 2022. "Les Risques Sécuritaires Climatiques à l'ONU (1990–2020)." *Les Champs de Mars* 35 (2): 211–25. https://doi.org/10.3917/lcdm.035.0211.

Maertens, Lucile. 2021. "Climatizing the UN Security Council." *International Politics* 58 (4): 640–60. https://doi.org/10.1057/s41311-021-00281-9.

United Nations Multi-Partner Trust Fund. n.d. "JP Climate Security Mechanism." Accessed February 17, 2025. https://mptf.undp.org/fund/jxe00.

United Nations University Centre for Policy Research (UNU CPR). 2023. *Thematic Review on Climate Security and Peacebuilding 2023*. https://www.un.org/peacebuilding/content/thematic-review-climate-security-and-peacebuilding-2023.

The Conflict and Environment Observatory (CEOBS)

Doug Weir and Oli Brown

The Conflict and Environment Observatory (CEOBS) is a specialized charity that launched in 2018 and undertakes research and advocacy on the environmental dimensions of armed **conflict** and military activities. CEOBS's primary focus areas include remote environmental assessments of conflict-affected areas, research and advocacy on the contributions that conflicts and **militarism** make to the climate crisis, and supporting environmental mainstreaming in the humanitarian mine action sector. CEOBS also has long-standing interests in the progressive development and implementation of the legal framework protecting the **environment** in relation to armed conflicts (CEOBS 2022a).

CEOBS is based in West Yorkshire, UK, and was inspired by a six-year research project initiated by the International Coalition to Ban Uranium Weapons and the Dutch NGO IKV Pax Christi (later **PAX**). The Toxic Remnants of War Project identified, among other things, that increasing the availability of environmental data gathered during armed conflicts was vital to help inform international policy processes aimed at reducing or mitigating environmental and humanitarian harm (CEOBS 2023).

Since launch, CEOBS has explored and developed a wide range of methodologies for remote environmental data collection in conflict contexts, for example, to assess groundwater in Yemen (CEOBS 2021) and conflict and vegetation loss in Tigray (CEOBS 2022b). CEOBS monitors environmental trends and incidents, and in early 2022, launched an incident database in response to the full-scale invasion of Ukraine. The assessment methodology that it developed was subsequently applied to Sudan, in support of the United Nations Environment Programme (UNEP). In 2023, CEOBS and PAX were jointly awarded the **Environmental Peacebuilding Association**'s Al-Moumin Award for Thought Leadership in Environmental Peacebuilding for "their ongoing leadership in **monitoring,** quantifying, and redressing the environmental impacts of **war**" (Environmental Peacebuilding Association 2023). This recognized the criticality of robust environmental data to not only drive **environmental peacebuilding,** but all actions necessary to reduce or mitigate harm.

CEOBS and its predecessor organization have long engaged in advocacy initiatives across both **peace** and **security** and environmental fora. They were instrumental in the development and successful passage of UN Environment Assembly Resolutions 2/15, 3/1, and 6/12 (CEOBS 2024), which have helped to secure and enhance UNEP's mandate to work on conflicts. They were also heavily involved in the **International Law Commission**'s decade-long process to develop its Principles on the **Protection of the Environment in Relation to Armed Conflicts (PERAC)**, providing data and argumentation to legal and state **stakeholders** (CEOBS 2022a).

At the UNFCCC COP26 in 2021, CEOBS launched its *Military Emissions Gap* project, which aims to enhance the reporting transparency of military greenhouse gas emissions—a sector that had hitherto been neglected and excluded from national reporting obligations (The Military Emissions Gap n.d.). CEOBS and Scientists for Global Responsibility would later calculate that militaries were likely responsible for around 5.5% of global emissions.

Since launch, CEOBS has worked closely with the humanitarian organization Norwegian People's Aid to promote environmental mainstreaming in the humanitarian mine action sector. This not only seeks to reduce the impact of clearance activities but also to explore the environmental role that mine action operators can play in the areas where they work, for example, in data collection (CEOBS n.d.). Such efforts align closely with CEOBS' core interests in improving data collection and assessment methodologies in fragile and conflict contexts, which stretch from satellite **remote sensing** to participatory "civilian science" (Weir et al. 2019).

References

The Conflict and Environment Observatory (CEOBS). 2021. "Groundwater Depletion Clouds Yemen's Solar Energy Revolution." https://ceobs.org/groundwater-depletion-clouds-yemens-solar-energy-revolution.

The Conflict and Environment Observatory (CEOBS). 2022a. "Explore our PERAC Publications." https://ceobs.org/explore-our-perac-publications.

The Conflict and Environment Observatory (CEOBS). 2022b. "The War in Tigray is Undermining its Environmental Recovery." https://ceobs.org/the-war-in-tigray-is-undermining-its-environmental-recovery.

The Conflict and Environment Observatory (CEOBS). 2023. "Commentary: Who Cares about the Environment in War?" https://ceobs.org/commentary-who-cares-about-the-environment-in-war.

The Conflict and Environment Observatory (CEOBS). 2024. "UNEA-6 Passes Resolution on Environmental Assistance and Recovery in Areas Affected by Armed Conflict." https://ceobs.org/unea-6-passes-resolution-on-environmental-assistance-and-recovery-in-areas-affected-by-armed-conflict.

The Conflict and Environment Observatory (CEOBS). n.d. "Environment in Humanitarian Disarmament." Accessed June 9, 2025. https://ceobs.org/projects/project-one.

Environmental Peacebuilding Association (EnPAx). 2023. "Al-Moumin Lectures." https://www.environmentalpeacebuilding.org/events/al-moumin-lectures.

The Military Emissions Gap (MEG). n.d. "The Military Emissions Gap Project." Accessed June 9, 2025. https://militaryemissions.org.

Weir, Doug, Dan McQuillan, and Robert A. Francis. 2019. "Civilian Science: The Potential of Participatory Environmental Monitoring in Areas Affected by Armed Conflicts." *Environmental Monitoring and Assessment* 191 (618). https://doi.org/10.1007/s10661-019-7773-9.

Conservation International (CI)

Janet Edmond and Shaadee Ahmadnia

Founded in 1987, Conservation International (CI) is a United States-based non-governmental organization working to spotlight and secure the critical benefits that nature provides to humanity. Building upon a strong foundation of science, partnership, and field demonstration, CI empowers societies to responsibly and sustainably care for nature, our global **biodiversity**, and the well-being of humanity. CI imagines a healthy, prosperous world in which societies are forever committed to caring for and valuing nature, for the long-term benefit of people and all life on Earth (CI n.d.a). CI's institutional priorities are to stabilize our climate by protecting and restoring nature, doubling ocean protection, and expanding nature-positive economies.

Conservation International recognizes that conserving nature requires engagement with the local communities who depend on it, but **conflict** can arise between **stakeholders** over values, priorities, and other factors. If carried out with care, however, **conservation** efforts can encourage collaboration, transform relationships, and support **peace**. Since 2012, CI has advanced **conflict sensitivity** and **environmental peacebuilding** in its work and has committed to understanding and fostering nature's role in resolving conflict—for nature and human well-being (CI n.d.b).

Since 2012, CI has championed conflict sensitivity and environmental peacebuilding by integrating these approaches across organizational policies, practices, and programming. The organization has identified and nurtured staff champions while sharing practical resources and insights with the broader development and conservation community working at the **environment**-conflict nexus. This includes the following:

I. Publishing *Conflict Sensitivity & Environmental Peacebuilding in Conservation: Lessons Learned from Conservation International*
II. Developing and implementing the *Environmental Peacebuilding Training Manual* in three languages across hundreds of CI staff in global and country programs (Ajroud et al. 2017)
III. Improving conflict analysis and resolution skills in program design and delivery across diverse contexts (Peru's Alto Mayo Protected Area forest and buffer zone, Brazil's forest rangers, global coastal fisheries programs, etc.)
IV. **Monitoring** and **evaluation** of environmental peacebuilding through conflict-sensitive pilot study in rural Kenya (Donatti 2024)

References

Ajroud, Brittany, Nathalie Al-Zyoud, Lydia Cardona, Janet Edmond, Danny Pavitt, and Amanda Woomer. 2017. *Environmental Peacebuilding Training Manual.* Conservation International. https://d2iwpl8k086uu2.cloudfront.net/docs/default-source/s3-library/publication-pdfs/environmental-peacebuilding-training-manual-eng.pdf?sfvrsn=eec8748a_2.

Conservation International. n.d.a. "Home Page." Accessed September 19, 2025. https://www.conservation.org/about.

Conservation International. n.d.b. "Conservation and Peacebuilding." Accessed September 19, 2025. https://www.conservation.org/projects/conservation-and-peacebuilding.

Donatti, Camila I., Josephat Nyongesa, Maureen Mwanzia, Janet Edmond, and Giacomo Fedele. 2024. "Grassland Restoration Impacts Human-wildlife and Social Conflicts in the Chyulu Hills, Kenya." *Frontiers in Environmental Science* 12: 1431316. https://doi.org/10.3389/fenvs.2024.1431316.

Convention on Biological Diversity (CBD)

Celestine Procter and Alexander Belyakov

The Convention on Biological Diversity (CBD) is an international treaty adopted at the Earth Summit in Rio de Janeiro in 1992, aimed at conserving biological diversity, ensuring the sustainable use of its components, and promoting the fair and equitable sharing of benefits arising from the utilization of genetic resources. It serves as a global framework for **biodiversity** conservation and **sustainable development.**

The CBD holds significant relevance in **environmental peacebuilding** due to its foundational role in promoting **cooperation** and sustainable management of **natural resources** across nations. Biodiversity, a cornerstone of ecological stability, is often a shared resource that transcends national boundaries. The treaty emphasizes collaboration, equity, and **sustainability**, fostering frameworks that prevent resource-based conflicts and promote **peace.**

Within environmental peacebuilding, the CBD's three objectives—the **conservation** of biodiversity, sustainable use, and benefit-sharing—resonate deeply. Its protocols, such as the Nagoya Protocol on Access and Benefit-Sharing (2010), address contentious issues like biopiracy and unequal resource distribution, creating pathways for equitable resource management. These measures can reduce grievances and foster trust among communities and states, mitigating potential conflicts.

The 16th Conference of the Parties to the CBD, held under the theme Peace with Nature, highlights the Convention's commitment to integrating **peacebuilding** and biodiversity conservation. Central to this effort is the Peace and Biodiversity Dialogue Initiative (PBDI), launched by the Republic of Korea during its Presidency of the 12th meeting of the Conference of the Parties to the CBD (Convention on Biological Diversity 2024). The PBDI promotes **transboundary cooperation** by facilitating the exchange of knowledge and best practices among Parties to the CBD, strengthening their efforts in managing shared conservation areas. As part of its capacity building initiatives, the PBDI also offers the Massive Open Online Course on Peace Park Development and Management, available in five languages, equipping **stakeholders** with the tools needed to advance peace and biodiversity goals. In recent times, the PBDI has focused on technical cooperation and supported transboundary conservation projects in partnership with other organizations.

The CBD's participatory mechanisms, which emphasize inclusivity by engaging Indigenous peoples and local communities, enhance its role in conflict resolution (Morgera and Tsioumani 2010; Swiderska and Argumedo 2006; Bavinck 2014). Recognizing traditional ecological knowledge as vital to biodiversity conservation, the treaty aligns with environmental peacebuilding's focus on integrating diverse stakeholders. The CBD also underscores the importance of restoring degraded ecosystems, which can serve as **confidence-building measures** between conflicting parties by demonstrating shared commitment to sustainable practices.

Despite its achievements, the CBD faces challenges, including uneven implementation among countries and limited enforcement mechanisms. However, its role as a platform for international dialogue and cooperation makes it indispensable for addressing global environmental challenges, which are increasingly interlinked with issues of peace and security.

References

Bavinck, Maarten, Lorenzo Pellegrini, and Erik Mostert. 2014. *Conflicts over Natural Resources in the Global South: Conceptual Approaches*. CRC Press.

Convention on Biological Diversity. 2024. "Peace and Biodiversity Dialogue Initiative." https://www.cbd.int/peace.

Morgera, Elisa, and Elsa Tsioumani. 2010. "The Evolution of Benefit Sharing: Linking Biodiversity and Community Livelihoods." *Review of European, Comparative and International Environmental Law* 19 (2): 150–70. https://doi.org/10.1111/j.1467-9388.2010.00674.x.

Swiderska, Krystyna, and Alejandro Argumedo. 2006. *Biocultural Heritage in the CBD: Valuing Traditional Knowledge and Genetic Resources*. International Institute for Environment and Development (IIED).

EcoPeace Middle East

Gidon Bromberg and Danni Reches

EcoPeace Middle East is a unique Jordanian-Palestinian-Israeli organization seeking to advance efforts to protect our shared **environment**, promote sustainable regional development and **climate security**, and create conditions for lasting **peace** in our region.

For three decades, EcoPeace has adopted top-down and bottom-up approaches to **conflict transformation**. The methods developed have been designed, tested, and redesigned to meet enormous challenges faced in a protracted and bloody **conflict**. Sustaining internal organizational trust amid intensive **violence**, responding to populist accusations of being traitors, and the threat to the actual safety of staff and participants are just a few examples of the challenges faced. Despite the challenges, EcoPeace has demonstrated that regional **cooperation** on shared environmental issues can advance climate **resilience**, promote regional peace, and provide concrete solutions that benefit all our communities.

EcoPeace's top-down advocacy strategies focus on engaging local, regional, and international decision-makers and **stakeholders**, the private sector, and civil society. Through these engagements, EcoPeace secures buy-in, fosters cooperation, and promotes regional **environmental cooperation** and climate security. Our advocacy efforts have resulted in:

I. *Water-Energy Nexus*: The Middle East faces severe freshwater scarcity but has desert areas ideal for **renewable energy** production. A 2017 EcoPeace study identified Israel and Palestine's comparative advantage of proximity to the sea, enabling them to produce desalinated water and supply it to Jordan, whereas Jordan benefits from vast deserts that could supply Palestine and Israel with renewable energy (Katz and Sharfran 2017). Based on the study, in 2021, Israel and Jordan signed a Letter of Intent to create a water-energy exchange, building climate resilience. EcoPeace identified John Kerry as the project's champion and the UAE as a trusted investor, leading to a $1 billion investment MoU signed at the Climate COP in 2022. EcoPeace continues to promote Palestine's inclusion in the exchange (Udasin 2021; EcoPeace Middle East 2025)
II. Mitigating Gaza's Sanitation Crisis: In 2015, a five-year-old Palestinian boy died after swimming in raw sewage off Gaza's coast. International agencies approached EcoPeace on how to overcome the challenge of Israel's security concerns around importing construction materials into Gaza for wastewater treatment plants (WWTPs) (De Bruyne 2018). By 2022, EcoPeace advocacy framing Gaza's sewage outflow as a national security threat to Israel had led to the construction of four WWTPs, representing an investment of $250 million, and taking sewage treatment from 0 to 65% in four years (Hermesh et al. 2019; Serfaty 2022)
III. Jordan River Rehabilitation: Due to conflict, the Jordan River sees only 5% of its historical flow, is heavily polluted, and has lost 50% of its biodiversity. Following a decade of EcoPeace-led community action (Mayoral MoUs, a regional master plan for the Jordan Basin (Kool 2015), etc.), at the 2022 COP, Israel and Jordan signed a Letter of Intent to rehabilitate the river (Surkes 2022). EcoPeace identified mutual interests to create political will and ministers willing to champion the effort. Still needed is the inclusion of Palestinians in these regional efforts (Bromberg et al. 2020)

EcoPeace's bottom-up strategy involves educating **youth** and young professionals on environmentalism and diplomacy.

I. Climate Diplomats Young Professionals (CDYP): CDYP brings together young Palestinian, Jordanian, and Israeli professionals (ages 21–35) to learn about climate and water diplomacy and become agents of change (EcoPeace Middle East n.d.a). The combination of national and regional events enables them to develop personal relationships and constructively engage in in-depth conversations. Each cohort attracts over 150 professionals

II. Teacher Training: EcoPeace facilitates national, bi-national, and regional teacher training seminars in Jordan, Israel, and Palestine on EcoPeace-developed water diplomacy and biology curricula (EcoPeace Middle East n.d.b). EcoPeace offers 30-hour training courses on integrating the lessons into existing curricula. Teachers participate in regional meetings to share best practices and experience transboundary cooperation. Each regional engagement includes over 40 educators

III. Good Water Neighbors (GNW): EcoPeace's award-winning GWN project engages youth to support concrete environmental solutions. The school program has educated thousands of Palestinians, Israelis, and Jordanians about the interdependent nature of water resources, the cross-border environmental impact of policies, and the need for, and self-interest in, cooperation. Tours expose participants to shared water realities, inspiring them to plan and implement community projects

EcoPeace's 30 years of experience help provide valuable lessons for the **environmental peacebuilding** community. The EcoPeace model shows that it is meaningful and impactful to create a single organization consisting of members of conflicted communities, working through consensus decision-making, to deliver real benefits to communities and champion policy change. Particularly in protracted conflict zones, the combined efforts of bottom-up and top-down activities help maintain the credibility and relevance of action. The environmental entry point to **peacebuilding** is often better understood by the public as an issue that requires cooperation, especially when linked to critical daily needs such as water and energy supply. Advocacy is most effective when based on scientific research and economic and geopolitical analysis that, by design, tries to highlight the win-win potential for all parties involved.

References

Bromberg, Gidon, Nada Majdalani, and Yana Abu Taleb. 2020. *A Green Blue Deal for the Middle East*. EcoPeace Middle East. https://ecopeaceme.org/wp-content/uploads/2022/03/a-green-blue-deal-for-the-middle-east.pdf.

De Bruyne, Charlotte. 2018. *Report on the Status of the Hebron-Besor-Wadi Gaza Basin*. EcoPeace Middle East. https://ecopeaceme.org/wp-content/uploads/2024/04/nahal-hevron_web_ENG.pdf.

EcoPeace Middle East. n.d.a. "EcoPeace Education: Climate Diplomacy for Young Professionals (CDYP)." Accessed September 8, 2025. https://ecopeaceme.org/wp-content/uploads/2025/03/cdyp-one-pager.pdf.

EcoPeace Middle East. n.d.b. "EcoPeace Education: Specialized Teacher Training Programs." Accessed September 8, 2025. https://ecopeaceme.org/wp-content/uploads/2025/03/TEACHER-TRAININGS-ONE-PAGER.pdf.

EcoPeace Middle East. 2025. *Our New Path to Sustainability: The IMEC Peace Triangle*. https://ecopeaceme.org/2025/01/28/the-imec-peace-triangle/.

Hermesh, Barak, Ma'ayan Maya, and Nadav Davidovitch. 2019. *Health Risk Assessment for the Israeli Population Following the Sanitary Crisis in Gaza*. EcoPeace Middle East. https://ecopeaceme.org/wp-content/uploads/2024/04/Gaza-Health-Report-Eng.-Mar.-2019.pdf.

Katz, David, and Arkadiy Shafran. 2017. *Water Energy Nexus. A Prefeasibility Study for Mid-East Water-Renewable Energy Exchanges*. EcoPeace Middle East and Konrad-Adenauer-Stiftung. https://ecopeaceme.org/wp-content/uploads/2022/03/water-energy-nexus-a-pre-feasibility-study-for-mid-east-water-renewable-energy-exchanges.pdf.

Kool, Jeroen. 2015. *Regional NGO Master Plan for Sustainable Development in the Jordan Valley*. EcoPeace Middle East. https://ecopeaceme.org/wp-content/uploads/2022/03/Regional_NGO_Master_Plan_Final.pdf.
Serfaty, Sara. 2022. "Gaza Beaches Made Safe for Swimming with Help of Environmental Peace Group." *The Times of Israel*, June 10. https://www.timesofisrael.com/gaza-beaches-made-safe-for-swimming-with-help-of-environmental-peace-group/.
Surkes, Sue. 2022. "Israel and Jordan Pen Agreement to Clean up Jordan River." *The Times of Israel*, November 17. https://www.timesofisrael.com/israel-and-jordan-pen-agreement-to-clean-up-jordan-river/.
Udasin, Sharon. 2021. "Israel, Jordan, UAE Sign Pivotal Deal to Swap Solar Energy, Desalinated Water." *The Hill*, November 23. https://thehill.com/policy/equilibrium-sustainability/582750-israel-jordan-uae-sign-pivotal-deal-to-swap-solar-energy/.

Energy Peace Partners (EPP)

David Mozersky

Energy Peace Partners (EPP) is built on the idea that **renewable energy** can serve as a building block for peace. Since 2017, EPP has developed innovative solutions to catalyze new renewable energy development in fragile, energy-poor regions. EPP seeks system-level change that will promote greater investment in climate solutions to support peace and development in fragile settings (EPP n.d.a). EPP works across three main areas:

I. *Peace Renewable Energy Credit (P-REC)*: EPP developed the P-REC as a tool to increase funding for renewable energy projects in **conflict**-affected countries (EPP n.d.b). By mid-2025, P-REC deals had unlocked over $1.5 million for high-impact projects, benefiting over 230,000 people, including through first-time electrification, public streetlights, and hospital electrification in Chad, the Democratic Republic of the Congo, Nigeria, Somalia, South Sudan, and Ethiopia. P-REC buyers to date include Microsoft, Google, Block, and Oriflame. EPP works with a range of partners in support of the P-REC: on the demand side (local renewable energy companies), the supply side (multinational buyers and climate advisory firms), and the issuance side (international standards bodies). EPP is also developing the P-REC Aggregation Facility (PAF) to scale up the P-REC market (EPP n.d.c). The PAF will unlock over $130 million in new renewable energy investment across a dozen fragile countries in Africa. EPP currently serves as the I-REC authorized country issuer in seven countries and will continue to expand I-REC and P-REC architecture to additional fragile states

II. *Powering Peace*: Powering Peace is a research and advocacy initiative, run jointly with the Stimson Center, that examines the role of energy in different conflict settings and the opportunities for renewable energy to support peace in each context (EPP n.d.a, n.d.d). It further aims to catalyze the system-wide transition of United Nations (UN) peacekeeping operations from diesel generators to renewables. These missions are often the largest energy consumers in their host countries and represent the single best entry point for catalyzing new renewable energy projects at scale where they are deployed, with benefits for both the UN and local communities. Since 2019, Powering Peace has published a range of reports, including five in-depth country case studies, analyzing the opportunities and pathways for renewable energy development to advance **peace** efforts in different countries. It has succeeded in raising awareness of this issue, reorienting UN policy and priorities, and creating a coalition of partners, including Norway, the United Arab Emirates, the International Renewable Energy Agency (IRENA), and the UN, who have been working to move this agenda forward

III. Peace Impacts Research Initiative: EPP developed an impact measurement framework to quantify the social and peace impacts of P-REC supported renewable energy projects based on the idea that increasing renewable electricity access results in improved short-term social and economic outcomes and long-term peace impacts. The framework is based on the Positive Peace framework, developed by the Institute for Economics and Peace (EPP n.d.e)

References

Energy Peace Partners (EPP). n.d.a. “Innovation for Peace.” Accessed September 12, 2025. https://www.energypeacepartners.com/mission.

Energy Peace Partners (EPP). n.d.b. “Peace Renewable Energy Credit.” Accessed September 12, 2025. https://www.energypeacepartners.com/prec.

Energy Peace Partners (EPP). n.d.c. “P-REC Aggregation Facility.” Accessed September 12, 2025. https://www.energypeacepartners.com/p-rec-fund.

Energy Peace Partners (EPP). n.d.d. “Powering Peace.” Accessed September 12, 2025. https://www.energypeacepartners.com/powering-peace.

Energy Peace Partners (EPP). n.d.e. “Peace Impacts of Renewable Energy.” Accessed September 12, 2025. https://www.energypeacepartners.com/peace-impacts.

Environmental Change and Security Program, Wilson Center

Geoffrey Dabelko, Lauren Herzer Risi, and Meaghan Parker

The Environmental Change and Security Program (ECSP) at the Woodrow Wilson International Center for Scholars has actively pursued the connections between the environment, health, population, development, **conflict**, and **security** since 1994. ECSP brought together scholars, policymakers, media, and practitioners through events, research, publications, multimedia content, and its award-winning blog, *New Security Beat.*

The ECSP at the Wilson Center had three primary focus areas:

I. Climate and Environment: Environmental dynamics, including climate change, natural resource management, and water quality and availability, can pose risks to security, but can also be used as a tool for building cooperation. ECSP worked with policymakers, practitioners, and scholars to explore new research and develop policy responses that linked the environmental, diplomatic, development, and security realms, with a focus on water security, critical minerals, and the energy transition, and improving predictive capabilities for climate security risks

II. Population Dynamics: Global population dynamics such as migration, youth bulges, and urbanization can affect political stability, conflict trends, and environmental vulnerability. ECSP served as a forum for discussing new research, practical rights-based policies, and programs on population-health-environment connections and demographic security

III. Global Health and Gender: Global health and gender are often addressed in silos, but where these issues connect is where targeted investments can produce real change. ECSP worked to help policymakers understand how issues related to gender equality, maternal and newborn health, and global health are critical to unlocking a sustainable, prosperous, and stable future for populations

ECSP was founded in 1994 to provide a non-partisan, non-advocacy forum where scholars and practitioners could explore the conceptual and practical dimensions of the environment's links to conflict, **peace**, and security after the Cold War. ECSP's meetings (ranging from 20 to more than 40 a year) convened public and private sector actors and institutions who didn't often work together to engage in dialogue that led to new understandings, connections, and partnerships. ECSP was also home to focused efforts such as the prominent forum for debate and discovery on China's environmental conditions and efforts, the China Environment Forum (founded in 1998), the Maternal Health Initiative (originally the Global Health Initiative), and the Wilson Center's urban-focused programming.

The *ECSP Report*, first published in 1995, was the only regularly published journal dedicated to **environmental security** until 2023 (when the **Environmental Peacebuilding Association** began publishing *Environment and Security* with Sage). In 2007, ECSP launched its award-winning blog *New Security Beat,* which is the only online publication covering the intersection of environment, health, population, and security (Environmental Change and Security Program, Wilson Center. n.d.). Multimedia productions include documentary films; a short, animated film on water, conflict, and cooperation; and various podcast series. In 2020, ECSP launched a new series of short-form policy briefs called the *New Security Brief.*

In the late 1990s, the Environmental Change and Security Program facilitated dialogue among scholars and practitioners around the peacemaking and **peacebuilding** potential of the environment and **natural resources.** These efforts produced the 2002 volume *Environmental Peacemaking*, edited by Ken Conca and Geoffrey Dabelko (Conca and Dabelko 2002). As a direct response to the overwhelming focus on environment and conflict that dominated the 1990s, this mix of case study evidence and deductive arguments for the confidence-building properties of environmental management laid the groundwork for subsequent advances in **environmental peacebuilding** theory and practice.

Beginning in 2007, ECSP facilitated dialogue among scholars and practitioners on a neglected dimension of climate-related conflict and peacebuilding that ECSP staff labeled "**backdraft.**" The 2013 issue of the ECSP Report, *Backdraft: The Conflict Potential of Climate Mitigation and Adaptation*, focused on identifying how well-intentioned climate **mitigation** and **adaptation** programs might spur conflict and instability (Dabelko et al. 2013).

In 2015, ECSP, as part of an international think tank collaboration, published *A New Climate for Peace: Taking Action on Climate and Fragility Risks*, an independent report commissioned by the G7 Foreign Ministers. The report identified seven climate-**fragility** risks that pose serious threats to stability and resulted in the elevation of climate-fragility risks as a foreign policy priority and the establishment of a G7 task force on climate and security (Rüttinger et al. 2015). The report laid the groundwork for countries working to integrate climate, security, and peacebuilding into their policy frameworks, including President Obama's 2016 Memorandum on Climate Change and National Security. Further work by ECSP focused on the connections between climate change, migration, conflict, and peace, and the development of a decision-support framework on climate security risks (Herzer Risi ad Carius 2021; Null and Herzer Risi 2023).

Following the de facto closure of the Wilson Center in 2025, several components of the Environmental Change and Security Program, including *New Security Beat*, transitioned to the Stimson Center.

References

Conca, Ken, and Geoffrey Dabelko, eds. 2002. *Environmental Peacemaking*. Johns Hopkins University Press and Woodrow Wilson Center Press.

Dabelko, Geoffrey D., Lauren Herzer, Schuyler Null, Meaghan Parker, and Russell Sticklor, eds. 2013. "Backdraft: The Conflict Potential of Climate Change Adaptation and Mitigation." *Environmental Change and Security Program Report* 14: 2.

Environmental Change and Security Program, Wilson Center. n.d. *New Security Beat*. Accessed February 23, 2025. https://www.newsecuritybeat.org.

Herzer Risi, Lauren, and Alexander Carius. 2021. *About 21st Century Diplomacy*. Woodrow Wilson Center Press.

Null, Schuyler, and Lauren Herzer Risi. 2023. *Navigating Complexity: Climate, Migration, and Conflict in a Changing World*. Woodrow Wilson Center Press.

Rüttinger, Lukas, Dan Smith, Gerald Stang, et al. 2015. *A New Climate for Peace: Taking Action on Climate and Fragility Risks*. adelphi, International Alert, Wilson Center, and European Institute for Security Studies. [Report prepared at the request of the Group of 7 Foreign Ministers].

Environmental Law Institute (ELI)

Susannah Dibble and Carl Bruch

The Environmental Law Institute (ELI) is a non-profit, non-partisan environmental institution based in Washington, DC. Established in 1969, ELI has played a pivotal role in shaping the development and implementation of environmental **governance**, domestically in the United States and globally. ELI works with partners to deliver insightful and impartial analyzes, provide evidence-based technical assistance, build capacity, and convene diverse **stakeholders** to develop solutions to pressing problems.

ELI is a leading institution shaping the theory and practice of **environmental peacebuilding**. For more than 25 years, this work has been grounded in experience helping **conflict**-affected countries and partners in Africa, Latin America, the Middle East, and Asia to more effectively manage **natural resources** to support the transition to **peace**. In the early 1990s, ELI facilitated **cooperation** in the Gulf of Aqaba on shared interests around water and the **environment** (Sandler et al. 1993). Since 2004, ELI has provided legal analysis, drafting assistance, and capacity building to support forestry sector reform as Liberia emerged from conflict. In the 2010s, ELI supported the development of a feedback and grievance redress mechanism to peacefully resolve forest-related disputes (especially for **REDD+**) and helped to develop national standards to ensure that communities' land and resource rights are protected effectively. In South Sudan, ELI supported the development of forestry law and policy, ensuring harmonization with other natural resource management policies.

With the **United Nations Environment Programme** (**UNEP**), the University of Tokyo, and McGill University, ELI led a global initiative that analyzed experiences in post-conflict **peacebuilding** and natural resource management; identified lessons; and raised the awareness of practitioners and researchers. This initiative resulted in six edited books between 2012 and 2016 arranged around six topics: high-value **natural resources**, assessment and restoration, land, water, livelihoods, and governance (Lujala and Rustad 2012; Jensen and Lonergan 2012; Unruh and Williams 2013; Weinthal et al. 2014; Young and Goldman 2015; Bruch et al. 2016).

ELI has led and coordinated several important research and policy projects on the environment, conflict, and peace. In the late 1990s, ELI convened a global conference and published a foundational book on the *Environmental Consequences of War: Legal, Economic, and Scientific Perspectives* (Austin and Bruch 2000) with leading academics and practitioners. The book informed how the UN Compensation Commission addressed the environmental claims against Iraq for its 1990 invasion of Kuwait. ELI staff also coauthored a report for the **UN Environment Programme** on *Protecting the Environment during Armed Conflict: An Inventory and Analysis of International Law* (Maruma Mrema et al. 2009). This report led the UN **International Law Commission** to take up the topic of **Protection of the Environment in Relation to Armed Conflict**, articulating principles of international law protecting the environment before, during, and after conflict (United Nations International Law Commission 2022).

Much of ELI's work focuses on building knowledge and capacity for environmental peacebuilding. With the **Environmental Peacebuilding Association** and the US Institute of Peace, ELI developed a toolkit that equips practitioners with tools and approaches tailored to navigating key issues of **monitoring** and **evaluation** of environmental peacebuilding (Bruch and Woomer et al. 2023). In partnership with UNEP, Columbia University, Duke University, UC-Irvine, and the UN Development Program, ELI coordinated the development of a massive open online course (MOOC) on Environmental Security and Sustaining Peace in which more than 10,000 people from 170 countries enrolled in the first running (2018) and more than 7,000 people from 175 countries enrolled in the second running (in 2019). As of 2025, it is in its fifth running.

ELI is dedicated to cultivating the environmental peacebuilding community. Initially drawing on the Post-conflict Peacebuilding and Natural Resource Management book series, ELI and UNEP launched the Environmental Peacebuilding Knowledge Platform in 2013 as a comprehensive international clearinghouse on environmental peacebuilding and established an Environmental Peacebuilding Community of Practice (which includes more than 4,300 people from 150 countries, as of December 2024). ELI was a founding institutional member of the Environmental Peacebuilding Association and currently serves as the Secretariat for the Association.

References

Austin, Jay, and Carl Bruch, eds. 2000. *The Environmental Consequences of War: Legal, Economic, and Scientific Perspectives*. Cambridge University Press. https://doi.org/10.1017/CBO9780511522321.

Bruch, Carl, and Amanda Woomer. 2023. *Toolkit on Monitoring and Evaluation of Environmental Peacebuilding*. Environmental Law Institute, Environmental Peacebuilding Association, and United States Institute of Peace. https://m-and-e.environmentalpeacebuilding.org/toolkit/.

Bruch, Carl, Carroll Muffett, and Sandra S. Nichols, eds. 2016. *Governance, Natural Resources, and Post-Conflict Peacebuilding*. Routledge. https://doi.org/10.4324/9780203109793.
Jensen, David, and Steve Lonergan, eds. 2012. *Assessing and Restoring Natural Resources in Post-Conflict Peacebuilding*. Routledge. https://doi.org/10.4324/9780203550199.
Lujala, Päivi, and Siri Aas Rustad, eds. 2012. *High-Value Natural Resources and Post-Conflict Peacebuilding*. Routledge. https://doi.org/10.4324/9781849775786.
Maruma Mrema, Elizabeth, Carl Bruch, and Jordan Diamond. 2009. *Protecting the Environment during Armed Conflict: An Inventory and Analysis of International Law*. United Nations Environment Programme. https://www.unep.org/resources/report/protecting-environment-during-armed-conflict-inventory-and-analysis-international.
Sandler, Deborah, Emad Adly, Mahmoud A. Al-Khoshman, Philip Warbug, and Tobie Berstein, eds. 1993. *Protecting the Gulf of Aqaba: A Regional Environmental Challenge*. Environmental Law Institute.
United Nations International Law Commission (ILC). 2022. *Draft Principles on Protection of the Environment in Relation to Armed Conflicts*. UN ILC. https://legal.un.org/ilc/texts/instruments/english/draft_articles/8_7_2022.pdf.
Unruh, Jon, and Rhodri C. Williams, eds. 2013. *Land and Post-Conflict Peacebuilding*. Routledge. https://doi.org/10.4324/9781849775793.
Weinthal, Erika, Jessica Troell, and Mikiyasu Nakayama, eds. 2014. *Water and Post-Conflict Peacebuilding*. Routledge. https://doi.org/10.4324/9781849775809.
Young, Helen, and Lisa Goldman, eds. 2015. *Livelihoods, Natural Resources, and Post-Conflict Peacebuilding*. Routledge. https://doi.org/10.4324/9781849775816.

Environmental Peacebuilding Association (EnPAx)

Carl Bruch and Erika Weinthal

The Environmental Peacebuilding Association® (EnPAx®) is the premier global association dedicated to bringing together researchers, practitioners, and decision-makers working on issues of environment, **conflict**, and peace. Founded in 2018, the Association is a multidisciplinary forum with three main aims: knowledge development and consolidation, capacity building, and fostering dialogue.

EnPAx seeks to identify promising research avenues and best policy practices, as well as foster the exchange of knowledge and data on **environmental peacebuilding**. Toward this end, EnPAx has partnered with organizations to develop a series of tools, flagship reports, and other resources. These include, among others, a *Supplementary Note to the African Union Transitional Justice Policy on Environment and Natural Resources in Transitional Justice Processes and Mechanisms* (Mutuku 2022), a *Toolkit on Monitoring and Evaluation of Environmental Peacebuilding* (Bruch and Woomer 2023), and an associated *Primer* (Bruch et al. 2023) (both with the **Environmental Law Institute** and the **United States Institute of Peace**, available in English, French, and Spanish); and a flagship report on *Digital Technologies for Environmental Peacebuilding: Horizon Scanning of Opportunities and Risks* (Martinez Sequero et al. 2024) (with the **United Nations Environment Programme**). The glossary in which this entry occurs was a showcase effort by the Association. The Association also supported the development of an **International Union for Conservation of Nature** report on *Planet on the Move: Reimagining Conservation at the Intersection of Migration, Environmental Change, and Conflict* (Hsiao et al. 2024) and a book with ELI and the **Global Environment Facility** on *Conflict-Sensitive Conservation: Lessons from the Global Environment Facility* (Bruch et al. 2024). In partnership with Sage Publishing, EnPAx launched *Environment and Security* in 2023, a quarterly journal that is the leading global outlet for peer-reviewed research and debates on issues at the intersection of **environment** and **security**.

EnPAx builds capacity and awareness among practitioners to advance the field and increase impact. It builds capacity through a massive open online course (MOOC) on Environmental Security and Sustaining Peace (offered through the SDG Academy), webinars and skills-building courses, mentorship arrangements for young professionals, and training courses at its biennial conferences. It has also supported the development of an online training on conflict-sensitive **conservation** (led by the PeaceNexus Foundation). EnPAx builds awareness through its biweekly *Environmental Peacebuilding Update* (since 2013, circulated to the Environmental Peacebuilding Community of Practice) and monthly *Newsletter on Gender, Natural Resources, Climate, and Peace* (since 2019). To honor outstanding contributions to environmental **peacebuilding**, EnPAx sponsors awards for leaders, practitioners, and researchers, including, among others, the Al-Moumin Award for Thought Leadership in Environmental Peacebuilding.

EnPAx fosters interactions among scholars, practitioners, decision-makers, and others across disciplines, genders, geographical locations, and stages of professional development. EnPAx has more than a dozen Interest Groups that provide spaces for collaborative work, networking, mentoring, and sharing knowledge and expertise. Although specific activities will vary depending on the Interest Group, activities may include, among others: organizing panels, webinars, and other events; preparing technical papers and other publications; selecting and maintaining a group of experts that can be contacted for information by individuals and groups outside the Association; and organizing sessions at the global EnPAx conferences.

EnPAx brings together scholars, practitioners, decision-makers, and students at its biennial conferences and thematic symposia, as well as the consultative processes leading up to the conferences. The EnPAx conferences emphasize interchange between researchers, practitioners, and decision-makers; an emphasis on different modalities for engagement (including through the arts); sessions that seek to collaboratively workshop ideas; and showcasing practical initiatives that advance understanding and practice of environmental peacebuilding.

References

Bruch, Carl, Geeta Batra, Anupam Anand, Shehla Chowdhury, and Sierra Killian. 2024. *Conflict-Sensitive Conservation: Lessons from the Global Environment Facility*. Routledge.

Bruch, Carl, and Amanda Woomer. 2023. *Toolkit on Monitoring and Evaluation of Environmental Peacebuilding*. Environmental Law Institute, Environmental Peacebuilding Association, and United States Institute of Peace. https://m-and-e.environmentalpeacebuilding.org/toolkit/.

Bruch, Carl, Amanda Woomer, Buket Altınçelep, and Erica Key. 2023. *Primer on Monitoring and Evaluation of Environmental Peacebuilding*. Environmental Law Institute, Environmental Peacebuilding Association, and United States Institute of Peace.

Hsiao, Elaine, Richard Matthew, Philippe Le Billon, and Galeo Saintz, eds. 2024. *Planet on the Move: Reimagining Conservation at the Intersection of Migration, Environmental Change, and Conflict*. International Union for Conservation of Nature.

Martinez Sequero, Albert, Asim Zia, Carl Bruch, David Jensen, and Silja Halle. 2024. *Digital Technologies for Environmental Peacebuilding: Horizon Scanning of Opportunities and Risks*. United Nations Environment Programme and Environmental Peacebuilding Association.

Mutuku, Munini. 2022. *Supplementary Note to the African Union Transitional Justice Policy on Environment and Natural Resources in Transitional Justice Processes and Mechanisms*. Bosch Alumni Network and Environmental Peacebuilding Association.

The Fund for Peace (FFP)

Emily Sample and Juliet Chiamaka Osueke

The Fund for Peace (FFP) is a globally recognized non-profit organization committed to reducing **conflict** and fostering sustainable **security** through innovative approaches and partnerships. Since its inception in 1957, FFP has evolved to address the multifaceted challenges of fragile states and violent conflict by equipping policymakers, practitioners, and local communities with the tools and data needed to identify risks and implement solutions. FFP's mission is to work toward a world where conflicts are mitigated before they escalate into crises, prioritizing partnerships, local capacity building, and data-driven approaches. FFP's programs encompass early warning systems, **conflict prevention**, and promoting sustainable business practices in fragile states. FFP integrates **environmental peacebuilding** into programming by addressing the intersection of environmental stressors, **governance** challenges, and community **resilience**. FFP is headquartered in Washington, DC, and collaborates with diverse **stakeholders** across the globe to create resilient societies through collective dialogue and actionable insights.

FFP is renowned for its flagship tool, the Fragile States Index (FSI), which is published annually. Since its inception in 2005, the FSI has evaluated the pressures affecting the national stability of 179 countries and territories. Through a mixed methods analysis, the FSI ranks the countries based on 12 **indicators**, including group grievance, economic decline, **human rights**, political factionalization, and demographic pressures, providing insight into the structural vulnerabilities contributing to **fragility**. This data-driven tool provides a comprehensive, evidence-based overview of patterns of state fragility.

The State Resilience Index (SRI) stands alongside the FSI as a new tool to identify capacities and capabilities in countries under stress. This tool measures the extent to which a country can prepare, manage, and recover from a crisis, relative to the severity of that crisis. Through a better understanding of resilience factors, stakeholders can assess how and where to leverage those indicators to build future resilience.

FFP's research and programs target several critical areas (The Fund for Peace n.d.):

I. *Early Warning Systems*: FFP uses data-driven methodologies to identify potential conflict triggers, enabling timely interventions that prevent **violence** and instability
II. *Fragility and Resilience Assessments*: These analyzes explore a country's vulnerabilities and strengths, offering strategies to enhance resilience against crises
III. *Climate Security*: Recognizing the growing impact of environmental factors, FFP examines how **climate change** exacerbates insecurity and works on strategies for **adaptation** and **mitigation**. FFP has mainstreamed environmental assessment across programs to provide actionable and holistic recommendations
IV. *Risk Assessment and Mitigation*: Through tailored tools and technologies, FFP helps stakeholders navigate complex risks, providing actionable guidance for reducing threats

V. *Human Rights and Business*: The organization fosters corporate responsibility by training businesses to engage ethically in fragile settings, reducing their impact on conflict dynamics. FFP is a signatory to the Voluntary Principles on Security and Human Rights, which guide land-intensive industry to mitigate risks and support community stability

The Fund for Peace's indices, research, and programs provide critical, data-driven insights that illuminate the intersection of **environmental degradation**, fragility, and conflict, enabling more targeted and effective **peacebuilding** interventions. By combining early warning systems, community resilience initiatives, and innovative conflict assessments, FFP advances environmental peacebuilding approaches that are inclusive, locally led, and designed to mitigate both human and ecological insecurity.

Reference

The Fund for Peace. n.d. "Programs." Accessed September 12, 2025. https://fundforpeace.org/programs/.

Global Environment Facility (GEF)

Susanne Schmeier

The Global Environment Facility (GEF) is a multilateral funding mechanism dedicated to addressing key environmental challenges and generating environmental benefits in developing countries. It places particular emphasis on meeting the objectives of international environmental agreements. The GEF was established in 1991 ahead of the 1992 Rio Earth Summit. Its establishment was based on the recognition that significantly more financial means were required for developing countries to achieve international environmental objectives.

The GEF provides funding—through grants and blended financing mechanisms—through 18 GEF agencies—to national government agencies, civil society organizations, private sector companies, research institutions, and other partners. Activities and engagement topics vary widely, including **biodiversity**, chemicals and waste, **land degradation**, large marine ecosystems, ozone layer depletion, illegal wildlife trade, freshwater resources, and **mitigation** and **adaptation** to **climate change**, among others. The GEF is also the financial mechanism for currently five international environmental agreements: The Minamata Convention on Mercury, the Stockholm Convention on Persistent Organic Pollutants, the **United Nations Convention on Biological Diversity (UNCBD)**, the **United Nations Convention to Combat Desertification (UNCCD)**, and the **United Nations Framework Convention on Climate Change (UNFCCC)**. This has been expanded to also include the 2023 Agreement on the Conservation and Sustainable Use of Marine Biological Diversity of Areas beyond National Jurisdiction (Biodiversity Beyond National Jurisdiction (BBNJ) Agreement), an implementing agreement to the 1982 United Nations Convention on the Law of the Sea (UNCLOS).

Funding to the GEF is provided by donor countries, who make financial contributions to the different GEF Trust Funds. Those Trust Funds are administered by the **World Bank** as the GEF Trustee and serviced by the GEF Secretariat, housed at the World Bank but independent. GEF Trust Funds include the GEF Trust Fund, the main funding mechanism, the Global Biodiversity Framework Fund (GBFF), the Least Developed Countries Fund (LDCF), the Special Climate Change Fund (SCCF), the Nagoya Protocol Implementation Fund (NPIF), and the Capacity-building Initiative for Transparency Trust Fund (CBIT). The funding mechanism for the 2023 BBNJ Agreement will soon be included as well.

After some political challenges right after the GEF's establishment (Sjöberg 1999; Boisson de Chazournes 2005), the GEF has effectively mobilized and disseminated funding to developing countries and successfully addressed important environmental problems (Clemencon 2006; Mee et al. 2008; Nyekwere 2017). Its engagement did not only focus on generating environmental benefits within its partner countries but also increasingly focused on socioeconomic **co-benefits** (Scientific and Technical Advisory Panel [STAP] 2022; GEF 2024), acknowledging the inherent interlinkages between human well-being and development on the one hand and environmental protection and **sustainability** on the other. Moreover, the GEF has effectively supported **cooperation** between countries, especially those that share ecosystems or **natural resources**.

In recent years, the GEF has had to face newly emerging challenges, relating both to new or intensifying environmental challenges that require its attention, as well as highly effective and efficient management and increasing competition over financial resources and donor commitment (Nyekwere 2017).

With increasing GEF financing going to fragile and **conflict**-affected countries (GEF 2023; Independent Evaluation Office 2024), the GEF has increasingly acknowledged the linkages between natural resources, the **environment**, conflict, and cooperation and engages in environmental **conflict prevention** and management, the promotion of cooperation over the environment, and **environmental peacebuilding** (Bruch et al. 2024; STAP 2018, 2024a). This concerns both the negative effects that **fragility** and conflict can have on the environment (and thus also the outcomes of GEF projects) and, more recently, the acknowledgment of the **peacebuilding** potential in environmental management and cooperation. In international waters, for example, the increasing importance of linking the management of shared freshwater and marine resources to conflict and cooperation considerations has been highlighted (STAP 2024b). In land and forest management, participatory ecosystem management by local communities has been used to strengthen relations between communities in post-conflict areas, contributing to peacebuilding (GEF-United Nations Development Programme 2021).

The GEF thus now plays a role not only in addressing key environmental challenges of global importance but also in preventing and mitigating conflict risks pertaining to the shared nature of natural resources and the environment at different **governance** levels and leveraging the potential for environmental peacebuilding inherent to natural resources and the environment.

References

Boisson de Chazournes, Laurence. 2005. "The Global Environment Facility (GEF): A Unique and Crucial Institution." *Review of European Community and International Environmental Law (RECIEL)* 14 (3): 193–201. https://doi.org/10.1111/j.1467-9388.2005.00441.x.

Bruch, Carl, Geeta Batra, Anupam Anand, Shehla Chowdhury, and Sierra Killian. 2024. *Conflict-sensitive Conservation. Lessons from the Global Environment Facility*. Routledge.

Clemencon, Raymond. 2006. "What Future for the Global Environment Facility." *The Journal of Environment and Development* 15 (1): 50–74. https://doi.org/10.1177/1070496506286438.
Global Environment Facility (GEF) Secretariat. 2023. *Gap Analysis of GEF-Funded Activity and Engagement in Fragility, Conflict, and Violence-Affected States*. GEF/C.64/Inf.13, 64th GEF Council Meeting. GEF.
Global Environment Facility (GEF) Secretariat. 2024. *Tracking and Measuring the Socio-Economic Co-Benefits of GEF Investments*. GEF/C.66/12, 66th GEF Council Meeting, January 8. GEF.
Global Environment Facility-United Nations Development Programme. 2021. *Participatory Conservation and Peacebuilding in Dry Forest as Production Landscape. Nature-based Solutions for Local and Global Benefits in Columbia*. https://www.thegef.org/sites/default/files/publications/GEF_GoodPracticesBriefs_Colobia_CRA.pdf.
Independent Evaluation Office (IEO). 2024. *GEF Support in Fragile and Conflict-Affected Situations*. Evaluation Report No. 151. January. Global Environment Facility (GEF).
Mee, Laurence D., Holly T. Dublin, and Anton A. Eberhard. 2008. "Evaluating the Global Environment Facility: A Goodwill Gesture or a Serious Attempt to Deliver Global Benefits?" *Global Environmental Change* 18 (4): 800–10. https://doi.org/10.1016/j.glonvcha.2008.07.005.
Nyekwere, Empire Hechime. 2017. "International Environment Financing: A Review of the Global Environment Facility." *Groningen Journal of International Law* 5 (2): 2780297.https://doi.org/10.21827/5a6afa321fa06.
Scientific and Technical Advisory Panel (STAP). 2018. *Environmental Security: Dimensions and Priorities*. Global Environment Facility.
Scientific and Technical Advisory Panel (STAP). 2022. *Refining the Tracking of Co-benefits in Future GEF Investments*. A STAP Information Brief. Global Environment Facility (GEF).
Scientific and Technical Advisory Panel (STAP). 2024a. "Environmental Security: Achieving Durable Outcomes in Fragile and Conflict-Affected Situations. GEF/STAP/C.66/Inf.03." 66th GEF Council Meeting. Global Environment Facility.
Scientific and Technical Advisory Panel (STAP). 2024b. "Fostering Cooperation and Managing Conflict: A STAP Information Note on GEF Transboundary Water Projects. GEF/STAP/C.67/Info.04." 67th GEF Council Meeting. Global Environment Facility.
Sjöberg, Helen. 1999. *Restructuring the Global Environment Facility*. Working Paper 13. Global Environmental Facility (GEF).

Green Climate Fund (GCF)

Milen Yishak

The Green Climate Fund (GCF) is the world's largest climate financial mechanism to mobilize climate finance to support scaled-up **mitigation** and **adaptation** action in developing countries (Basak and Karlsson-Vinkhuyzen 2022; Climate Focus 2016). Established in 2010 at the **United Nations Framework Convention on Climate Change (UNFCCC)** 16th session of the Conference of the Parties (COP 16), it is headquartered in the Republic of Korea. The GCF expedites climate action through a country-owned partnership approach and employs financing solutions and climate investment expertise by allocating its resources to low-emission and climate-resilient projects and programs in developing countries (GCF n.d.; United Nations Environment Programme n.d.).

The purpose of the Fund is to make a significant and ambitious contribution to the global efforts toward realizing goals established by the international community to respond to the challenges posed by **climate change** (GCF 2023). The two long-term strategic visions of the Fund, outlined in the Strategic Plan for 2024–2027, are to "(1) promote a paradigm shift toward low-emission and climate resilient development pathways in the context of **sustainable development** and (2) support developing countries in the implementation of the UNFCCC and the Paris Agreement within the evolving climate finance landscape" (GCF 2023, 2). As of mid-2025, the Fund operates 314 projects in 133 countries with a budget of US$18 billion. It aims for a 50-50 ratio of investment between mitigation and adaptation, and the ability to designate 50% of the adaptation allocation for particularly vulnerable countries, comprising the least developed countries (LDCs), Small Island Developing States (SIDS), and African states. After its establishment to operate as the financial mechanism of the UNFCCC in 2010, the Fund was operationalized in 2015 (UN Climate Change 2025; Treichel et al. 2024; Hammersley et al. 2023).

Historically, the formal discourse on the Fund's inception and purpose in climate financing can be traced to the 2009 Copenhagen Accord, at COP15, during which developed countries committed to provide "new and additional" financial resources of US$30 billion to fast-start funding between 2010 to 2012 to support developing country plans on reducing deforestation, deploying clean technologies, and adaptation and set the goal of US$100 billion per year by 2020 to address the needs of developing countries (Basak and Karlsson-Vinkhuyzen 2022; Cui et al. 2014; Union of Concerned Scientists 2009; Organization for Economic Co-operation and Development n.d.).

Although the Fund's portfolio includes vulnerable populations in LDCs, SIDS, African states, and fragile and conflict states, an analysis of integrated **climate security** programming in the Fund concluded that it does not fully incorporate **conflict sensitivity** or **peace** responsiveness in its project cycle (Scartozzi 2023). Reviewing publicly available literature on risks and opportunities for projects in fragile and conflict-affected settings in **infrastructure** and built environments, Scartozzi et al. (2023) identified a range of **security** risks for GCF-supported projects. The risks include destruction of infrastructure and the built **environment**, occupation by **conflict** parties, political instability, uncertain legal and regulatory frameworks, and cross-border and regional implications. The projects can also produce security risks—exacerbating or reinforcing conflict dynamics—by exacerbating competition over **natural resources**, causing **displacement** and **land tenure** issues, and interfering with local **livelihoods**, and through lack of community engagement and consultation, leading to mistrust and resistance. Concluding their analysis, Scartozzi et al. recommended that the GCF integrate conflict sensitivity throughout projects that it supports and include **peacebuilding** actions.

Concerns regarding **conflict sensitivity** are not unique to GCF-supported projects, as climate financing has yet to mainstream the benefits of conflict sensitivity programming as a **co-benefit** of mitigation and climate resiliency goals and fully understand the complexities that limit climate finance access to conflict-affected and fragile populations (Chambers and Kyed 2024). In general, climate finance tends to be risk-averse and state-based, prioritizing large projects in safe operating environments, which have the potential to contribute to inequality and vulnerability (Environment, Climate, Conflict and Peace Community of Practice 2024; Reda and Wong 2021). Studies and analyzes, including by the United Nations Development Programme (UNDP) and **World Bank**, have indicated that fragile and conflict-affected states are receiving noticeably less from the climate financing portfolio despite the vital support climate financing can provide to mitigation and adaptation efforts (Reda and Wong 2021; Jones et al. 2024; Gulati et al. 2024).

The lack of conflict sensitivity in climate finance can result in operational and reputational risks and negative impacts on the mobilization of funding (especially in conflict-affected regions) and undermine the potential peacebuilding co-benefits (Scartozzi 2023). Integrating conflict sensitivity in climate finance projects can help to manage risk, realize peace **co-benefits**, prevent harmful side effects of climate finance, and incentivize investment (Environment, Climate, Conflict and Peace Community of Practice 2024; Reda and Wong 2021).

One of the GCF programming priorities is to "address urgent and immediate adaptation and **resilience** needs for particularly vulnerable countries" (GCF 2023, 5). This will expand coverage of climate information and early warning systems in accordance with the early warning for all initiatives and supporting integrated risk management approaches.

Environmental peacebuilding integrates conflict sensitivity into climate projects as well as integrating considerations of climate change into peacebuilding (Nielsen et al. 2022; Peters et al. 2025; Albakjaji et al. 2025; Gianesello and Desmidt 2025; Tarif 2024; Gaston et al. 2023).

References

Albakjaji, Mohamad, Soumaya Khammassi, and Yusra Alshanqityi. 2025. "The Role of the International Environmental Cooperation in Achieving Sustainable Peacebuilding in International Relations: A Comprehensive and Critical Analysis." *Journal of Posthumanism* 5 (4): 1257–68. https://doi.org/10.63332/joph.v5i4.1222.

Basak, Rishi, and Sylvia Karlsson-Vinkhuyzen. 2022. "The Green Climate Fund: History, Status and Legitimacy." In *Handbook of International Climate Finance*. Edward Elgar Publishing. https://doi.org/10.4337/9781784715656.00013.

Chambers, Justine, and Helene M. Kyed. 2024. *Bridging the Gap in Climate Change Financing to Violent Conflict Affected Areas*. Danish Institute for International Studies (DIIS). DIIS Policy Brief. https://www.diis.dk/en/research/bridging-the-gap-in-climate-change-financing-to-violent-conflict-affected-areas.

Climate Focus. 2016. "Briefing Note: Green Climate Fund and the Paris Agreement." https://climatefocus.com/wp-content/uploads/2022/06/GCF-and-Paris-Brief-2016.new_.pdf

Cui, Lian-biao, Lei Zhu, Marco Springmann, and Ying Fan. 2014. "Design and Analysis of the Green Climate Fund." *Journal of Systems Science and Systems Engineering*. 23: 266–99. https://doi.org/10.1007/s11518-014-5250-0.

Environment, Climate, Conflict and Peace Community of Practice. 2024. "Conflict Sensitive Climate-Action in Practice." https://medium.com/@ecosystemforpeace/conflict-sensitive-climate-action-in-practice-fe3252316bc6.

Gaston, Erica, Oliver Brown, Nadwa al-Dawsari, Cristal Downing, Adam Day, and Raphael Bodewig. 2023. *Climate-Security and Peacebuilding: A Thematic Review*. United Nations University Centre for Policy Research. https://www.un.org/peacebuilding/content/thematic-review-climate-security-and-peacebuilding-2023.

Gianesello, Sara, and Sophie Desmidt. 2025. *Environmental Peacebuilding as Climate-Security Priority for the EU- Examples from the Sahel*. The European Centre for Development Policy Management. https://ecdpm.org/application/files/5817/5205/9636/Environmental-Peacebuilding-Climate-Security-Priority-EU-Examples-Sahel-ECDPM-Briefing-Note-197-2025.pdf.

Green Climate Fund. 2023. "Strategic Plan for the Green Climate Fund 2024–2027." https://www.greenclimate.fund/document/strategic-plan-green-climate-fund-2024-2027.

Green Climate Fund. n.d. "Green Climate Fund." Accessed September 12, 2025. https://www.greenclimate.fund.

Gulati, Manisha, Yue Cao, Amir Khouzam, and Mauricio Vazquez. 2024. *Climate Change Conflict and Fragility: A Recipe for Disaster*. ODI Global. https://odi.org/en/insights/climate-change-conflict-fragility-recipe-for-disasters/.

Hammersley, Georgia, Melanie Pill, and Roland Rajah. 2023. *Revitalising the Green Climate Fund*. Lowy Institute. https://www.lowyinstitute.org/publications/revitalising-green-climate-fund.

Jones, Lindsey Paul, Josue Banga, Benjamin Savely Notkin, and Arthur Guillaume Julien Brochen. 2024. *Closing the Gap: Trends in Adaptation Finance for Fragile and Conflict-affected Settings - Executive Brief.* World Bank Group. http://documents.worldbank.org/curated/en/099071924093054876.

Nielsen, Flemming, Atila Uras, Eissa Yagoub Musa, and Awadalla Hamid Mohamed Osman. 2022. "Natural Resources Management, Environmental Governance, and Peacebuilding in Darfur: Creating a Win-Win Situation when Farmers and Pastoralists were Incentivized to Find Shared Solutions." In *The Future of Environmental Peacebuilding: Nurturing an Ecosystem for Peace: A White Paper and Compendium.* https://www.ecosystemforpeace.org/compendium/natural-resources-management-environmental-governance-and-peacebuilding-in-darfur.

Organization for Economic Co-operation and Development (OECD). n.d. "Climate Finance and the USD 100 Billion Goal." https://www.oecd.org/en/topics/sub-issues/climate-finance-and-the-usd-100-billion-goal.html.

Peters, Laura E. R., Ken Conca, Aaron T. Wolf, Ari Lippi, and Jamon Van Den Hoek. 2025. "The Many Voices of Environmental Cooperation: A Relational Analysis of 30 Years of Environmental Peacebuilding Over Shared Waters in Israel, Jordan, and Palestine." *Environment and Security.* https://doi.org/10.1177/27538796251347820.

Reda, Dima, and Catherine Wong 2021. *Climate Finance for Sustaining Peace: Making Climate Finance Work for Conflict-Affected and Fragile Contexts.* United Nations Development Programme. https://www.undp.org/publications/climate-finance-sustaining-peace-making-climate-finance-work-conflict-affected-and-fragile-contexts.

Scartozzi, Cesare M. 2023. "Conflict Sensitive Climate Finance: Lessons from the Green Climate Fund." *Climate Policy* 24 (3): 297–313. https://doi.org/10.1080/14693062.2023.2212640.

Scartozzi, Cesare M., Giula Caroli, Carolina Sarzana, et al. 2023. *Guidance Note for Peace-Informed Programming at the Green Climate Fund: Infrastructure and Built Environment.* CGIAR and InterPeace. https://cgspace.cgiar.org/items/a5a706ec-2e8c-4114-aba9-dbf923a4cde6.

Tarif, Kheira. 2024. *From Conflict to Collaboration: Co-Funding Environmental Peacebuilding in South-Central Somalia.* SIPRI. https://doi.org/10.55163/KSJV5069.

Treichel, Pia, Michai Robertson, Emily Wilkinson, and Jack Corbett. 2024. "Scale and Access to the Green Climate Fund: Big Challenges for Small Island Developing States." *Global Environmental Change* 89: 102943. https://doi.org/10.1016/j.gloenvcha.2024.102943.

UN Climate Change. 2025. "Green Climate Fund." https://unfccc.int/process/bodies/funds-and-financial-entities/green-climate-fund.

United Nations Environment Programme (UNEP). n.d. "Green Climate Fund." Accessed September 12, 2025. https://www.unep.org/about-un-environment/funding-and-partnershis/green-climate-fund.

Union of Concerned Scientists. 2009. "The Copenhagen Accord: Not Everything We Wanted, but Something to Build on." https://www.ucs.org/resources/copenhagen-accord#:~:text=Climate%20financethe%20Accord%20commits,the%20impacts%20of%20climate%20change.

Intergovernmental Panel on Climate Change (IPCC)

Stephen Lonergan

The Intergovernmental Panel on Climate Change (IPCC) is an intergovernmental body established in 1988 by the **United Nations Environment Programme (UNEP)** and the World Meteorological Organization (WMO) to inform governments on the state of knowledge of human-induced **climate change**. Its mandate is to prepare a comprehensive review and set of recommendations concerning the science of climate change, the likely social and economic impacts of climate change, and the potential strategies available to respond to climate change.

The UN General Assembly officially endorsed the creation of the IPCC in December 1988 to "provide internationally coordinated scientific assessments of the magnitude, timing, and potential environmental and socio-economic impacts of climate change and realistic response strategies" (United Nations General Assembly Resolution 43/53 1988). *The First Assessment Report*, published in 1990, was important not only in recognizing climate change as a major international challenge but also in setting the foundation for the creation of the **UN Framework Convention on Climate Change (UNFCCC)**, the key international treaty on reducing greenhouse gas emissions and dealing with the consequences of climate change. Since its inception, the IPCC has produced six assessment reports, along with numerous special reports and technical papers. These assessments—which provide a synthesis of the latest climate change research—are authored and reviewed on a voluntary basis by thousands of experts from around the world. At present (2024), the IPCC has 195 member countries and is based in Geneva, Switzerland.

The IPCC is divided into three working groups, covering (1) the physical science of climate change, (2) impacts, **adaptation**, and vulnerability, and (3) **mitigation**. There is also a task force on national greenhouse gas inventories. In addition to the assessment reports, the IPCC prepares greenhouse gas emission scenarios for use in the general circulation models (GCMs) of the Earth's atmosphere. These, in turn, are used to project changes in future climate conditions at local, regional, and global levels.

Working group 2, Impacts, Adaptation, and Vulnerability, assesses "the vulnerability of socio-economic and natural systems to climate change, negative and positive consequences of climate change, and options for adapting to it" (IPCC 2024). Although there is no direct reference to **environmental peacebuilding**, various IPCC reports do address the potential threats that climate change poses for **human security**, including "undermining livelihoods, compromising culture and identity, increasing migration, and challenging the ability of states to meet the needs of people" (Adger et al. 2014; IPCC 2022). While there is no evidence of a direct link between climate change and violent **conflict**, there is acceptance that negative social and economic impacts from climate change could, in turn, increase the risk of violent conflict. There is also agreement that people living in places affected by violent conflict are particularly vulnerable to climate change.

References

Adger, W. Neil, Juan M. Pulhin, Jon Barnett, et al. 2014. "Human Security." In *Climate Change 2014: Impacts, Adaptation, and Vulnerability. Part A: Global and Sectoral Aspects. Contribution of Working Group II to the Fifth Assessment Report of the Intergovernmental Panel on Climate Change*, edited by Core Writing Team, Rajendra Pachauri, and Leo Meyer. IPCC.

Intergovernmental Panel on Climate Change (IPCC). 2022. *Climate Change 2022: Impacts, Adaptation, and Vulnerability*. Contribution of Working Group II to the Sixth Assessment Report of the Intergovernmental Panel on Climate Change (Summary for Policymakers), edited by Hans-Otto Pörtner, Debra C. Roberts, Melinda Tignor, et al. Cambridge University Press.

Intergovernmental Panel on Climate Change (IPCC). 2024. *Intergovernmental Panel on Climate Change (home page)*. Accessed October 16, 2024. https://www.ipcc.ch.

United Nations General Assembly (UNGA). 1988. *Resolution 43/53: Protection of Global Climate for Present and Future Generations of Mankind*, December 6.

International Court of Justice (ICJ)

Zerrin Savaşan

The International Court of Justice (ICJ) is the oldest permanent international court in operation (United Nations 1945) and a principal organ of the United Nations (UN). The

Court possesses general jurisdiction, meaning its jurisdiction is limited neither to a specific field of international law nor to a certain type of dispute (ICJ Statute, Article 36.1). This broad scope makes it feasible to bring environmental disputes before the Court. It can either settle environmental legal disputes submitted by parties in accordance with international law (contentious jurisdiction) or provide advisory opinions on environmental legal questions when requested by the UN General Assembly or other authorized UN organs (advisory jurisdiction) (UN Charter, Articles 65, 96). Crucially, it has the authority to indicate provisional measures (ICJ Statute, Article 41.1-2), which are especially vital in environmental cases due to the "irreparability of environmental damage" (Sands 1996, 77). In 1993, the ICJ even established an ad hoc Chamber for Environmental Cases (ICJ Statute, Article 26.1). However, because no case was ever brought to this Chamber, the Court decided not to hold elections for a Bench for it in 2006.

Since its inception, the ICJ has handled several cases involving environmental issues. Some were settled outside the Court, e.g., the *Nuclear Tests* case (ICJ 1974a, 1974b). Others, such as the *Gabcikovo-Nagymaros* case (ICJ 1997), saw environmental issues considered indirectly. More directly, in cases like the *Whaling* case (ICJ 2014) and the *Pulp Mills* case (ICJ 2010), the Court based its decisions on relevant international treaties' obligations. Similarly, in the *Costa Rica v Nicaragua* case (ICJ 2018) and the *Nicaragua v Costa Rica* case (ICJ 2015), the Court referenced states' obligations under customary international law. An important step in environmental jurisprudence was taken with the Whaling case, where the ICJ implicitly acknowledged an *erga omnes* obligation, indicating that Japan had breached an obligation owed to all states party to the *International Convention for the Regulation of Whaling* (ICRW) (ICJ 2014). Building on this, on March 23, 2023, the UNGA adopted Resolution (77/276), requesting the ICJ to render an advisory opinion on states' obligations with respect to **climate change** (UN General Assembly 2023).

Despite its potential, the Court is often perceived as having limited or no effect in specialized areas like environmental protection. This is largely due to its slow, expensive, and confrontational proceedings (Aust 2010; Downes and Penhoet 1999; Kolari 2002; Wang and Wiser 2002). Moreover, while its judgments are binding on the parties and final without appeal apart from exceptional situations (ICJ Statute, Articles 59–60) (Article 61.3 ICJ Statute), the Court lacks compulsory jurisdiction independent of state consent (Article 34.1 ICJ Statute). Its success depends, therefore, heavily on the willingness of individual states (Petersmann 1999). Lastly, non-state actors cannot initiate proceedings before the Court, even if they have a legal interest. This limitation often hinders their potential to contribute to the "**depoliticization** of international jurisdiction" and public awareness (Petersmann 1999, 783).

Although often criticized for its perceived failure to resolve environmental disputes (Hempel 1996; Schwabach 2005), a combined assessment of the ICJ's decisions and advisory opinions demonstrates its significant contribution to guiding the peaceful resolution of environmental disputes. The Court establishes the applicable rules of international environmental law (IEL) (Coulon 2022) and clarifies environmental obligations, thereby promoting **environmental peacebuilding**. Looking ahead, this proactive role in the development of IEL is vital to prevent environmental tensions from escalating into **conflict**.

References

Aust, Anthony. 2010. *Handbook of International Law*. Cambridge University Press.
Coulon, Anne. 2022. "The International Court of Justice and the Protection of the Environment." In *The Environment Through the Lens of International Courts and Tribunals*, edited by Edgardo Sobenes, Sarah Mead, and Benjamin Samson. T. M.C. Asser Press. https://doi.org/10.1007/978-94-6265-507-2_2.
Downes, David, and Bethanie Penhoet. 1999. *Effective Dispute Resolution, A Review of Options, For Dispute Resolution Mechanisms and Procedures*. Center for International Environmental Law.
Hempel, Lamont C. 1996. *Environmental Governance: The Global Challenge*. Island Press.
International Court of Justice (ICJ). 1945. *Statute of the International Court of Justice*.
International Court of Justice (ICJ). 1974a. *Nuclear Tests (Australia v France)*, Judgment of 20 December 1974.
International Court of Justice (ICJ). 1974b. *Nuclear Tests (New Zealand v France)*, Judgment of 20 December 1974.
International Court of Justice (ICJ). 1997. *Gabčíkovo-Nagymaros Project (Hungary v. Slovakia)*, Judgment of 25 September 1997.
International Court of Justice (ICJ). 2010. *Pulp Mills on the River Uruguay (Argentina v. Uruguay)*, Judgment of 20 April 2010.
International Court of Justice (ICJ). 2014. *Whaling in the Antarctic (Australia v. Japan: New Zealand intervening)*, Judgment of 31 March 2014.
International Court of Justice (ICJ). 2015. *Construction of a Road in Costa Rica along the San Juan River (Nicaragua v Costa Rica)*, Judgment of 16 December 2015.
International Court of Justice (ICJ). 2018. *Certain Activities Carried Out by Nicaragua in the Border Area (Costa Rica v Nicaragua)*, Judgment of 2 February 2018.
Kolari, Tuomas. 2002. "Promoting Compliance with International Environmental Agreements-An Interdisciplinary Approach with Special Focus on Sanctions." Pro Gradu Thesis, Faculty of Social Sciences, Department of Law, University of Joensuu.
Petersmann, Ernst-Ulrich. 1999. "Constitutionalism and International Adjudication: How to Constitutionalize the U.N. Dispute Settlement System?" *New York University Journal of International Law and Politics* 31 (4): 753–90. https://heinonline.org/HOL/LandingPage?handle=hein.journals/nyuilp31&div=33&id=&page=.
Sands, Philippe. 1996. "Compliance with International Environmental Obligations: Existing International Legal Arrangements." In *Improving Compliance with International Environmental Law*, edited by James Cameron, Jacob Werksman, and Peter Roderick. Earthscan.
Schwabach, Aaron. 2005. *International Environmental Disputes: A Reference Handbook*. Santa Barbara.
United Nations (UN). 1945. *Charter of the United Nations*. San Francisco: United Nations.
United Nations General Assembly (UNGA) 2023. Resolution 77/276. Request for an Advisory Opinion of the International Court of Justice on the Obligations of States in Respect of Climate Change. Seventy-seventh session, Agenda item 70: Report of the International Court of Justice.
Wang, Xueman, and Glenn Wiser. 2002. "The Implementation and Compliance Regimes under the Climate Change Convention and Its Kyoto Protocol." *Review of European Community and International Environmental Law* 11 (2): 181–98. https://doi.org/10.3233/CL-2010-006.

International Institute for Sustainable Development (IISD)

Alec Crawford

The International Institute for Sustainable Development (IISD) is an independent policy research think tank. IISD has a long history of working on the **environment, conflict**, and **peacebuilding** (IISD n.d.a). This work—which spans more than twenty years—has traditionally focused on understanding and promoting the role of ecosystem management and natural **resource governance** in moving fragile and conflict-affected societies onto pathways of **peace** and **resilience** (Matthew et al. 2002). Working with policymakers, practitioners, and academics, among other partners, IISD's work in this space has traditionally focused on three things: research, policy, and practice.

Within the first pillar, IISD conducts research and analysis to better understand and communicate to decision-makers the links among nature, climate resilience, and peace and conflict. In the policy space, IISD identifies entry points and strategies for better integrating peace and conflict considerations into national and international planning and policy processes. Finally, working with practitioners, IISD builds capacities to design, implement, monitor, evaluate, and learn from conflict-sensitive approaches to **conservation** and climate change **adaptation** programming.

Thematically, IISD's work has focused largely on **climate change** and conservation. With regard to climate change, IISD made important early contributions to the research based on the extent to which climate change was or would be a contributing factor in conflicts and **violence**. This included research on these links in West Africa (Brown and Crawford 2008) and the Middle East (Brown and Crawford 2009).

Since that early research, IISD's climate change work has shifted toward conflict-sensitive climate change adaptation planning and programming and climate-resilient peacebuilding in conflict-affected contexts (Crawford et al. 2015). This is largely focused on the National Adaptation Planning (NAP) process. Through the NAP Global Network, for which IISD runs the Secretariat, the organization has developed guidance and worked directly with the governments of conflict-affected states on integrating conflict and peacebuilding considerations into their NAP processes (Crawford et al. 2023). This work is helping to ensure that national and subnational adaptation actions are well aligned with peacebuilding agendas, and that they are conflict-sensitive—meaning that they do not create new or exacerbate existing conflict risks, but rather that they support peace, trust, and **cooperation**.

Within the climate change space, IISD undertook early research to understand how the transition to a low-carbon economy—and the minerals and metals required to make that shift—could affect **fragility**, conflict, and violence dynamics in mineral-rich states (Church and Crawford 2018). This award-winning research provided an early illustration of how and where the extraction and trade of **critical minerals** could influence peace and conflict.

IISD also has a long track record on conflict-sensitive conservation (IISD n.d.b). Working with partners in Central, East, and West Africa, IISD developed a groundbreaking approach to conflict-sensitive conservation that is still in use today (Hammill et al. 2009). Guidance and toolkits on conflict-sensitive conservation have been complemented by extensive research, training, and application of the approach by IISD and partners across critical landscapes and ecosystems in many of the world's most biodiverse regions.

References

Brown, Oli, and Alec Crawford. 2008. *Assessing the Security Implications of Climate Change for West Africa: Country Case Studies of Ghana and Burkina Faso*. International Institute on Sustainable Development. https://www.iisd.org/publications/assessing-security-implications-climate-change-west-africa-country-case-studies-ghana.

Brown, Oli, and Alec Crawford. 2009. *Rising Temperatures, Rising Tensions: Climate Change and The Risk of Violent Conflict in the Middle East*. International Institute on Sustainable Development. https://www.iisd.org/publications/report/rising-temperatures-rising-tensions-climate-change-and-risk-violent-conflict.
Church, Clare, and Alec Crawford. 2018. *Green Conflict Minerals: The Fuels of Conflict in the Transition to a Low-carbon Economy*. International Institute on Sustainable Development. https://www.iisd.org/publications/report/green-conflict-minerals-fuels-conflict-transition-low-carbon-economy.
Crawford, Alec, Angie Daze, Anne Hammill, Jo-Ellen Parry, and Alicia Natalia Zamudio. 2015. *Promoting Climate-Resilient Peacebuilding in Fragile States*. International Institute on Sustainable Development. https://www.iisd.org/publications/report/promoting-climate-resilient-peacebuilding-fragile-states.
Crawford, Alec, Anne Hammill, and Anne Tadgell. 2023. *Peace, Conflict, and National Adaptation Plan (NAP) Processes*. International Institute on Sustainable Development. https://napglobalnetwork.org/resource/peace-conflict-and-national-adaptation-plan-nap-processes/.
Hammill, Anne, Alec Crawford, Robert Craig, Robert Malpas, and Richard Matthew. 2009. *Conflict-Sensitive Conservation: Practitioners' Manual*. International Institute on Sustainable Development. https://www.iisd.org/publications/guide/conflict-sensitive-conservation-practitioners-manual.
International Institute on Sustainable Development. n.d.a. "Environment, Conflict and Peacebuilding." Accessed September 24, 2025. https://www.iisd.org/topics/environment-conflict-and-peacebuilding.
International Institute on Sustainable Development. n.d.b. "Conflict-Sensitive Conservation." Accessed September 24, 2025. https://www.iisd.org/projects/conflict-sensitive-conservation.
Matthew, Richard, Mark Halle, and Jason Switzer, eds. 2002. *Conserving the Peace: Resources, Livelihoods and Security*. International Institute for Sustainable Development. https://www.iisd.org/system/files/publications/envsec_conserving_peace.pdf.

International Law Commission (ILC)

Stavros Evdokimos Pantazopoulos, Marie G. Jacobsson, and Marja Lehto

The International Law Commission (ILC) of the United Nations was established as a subsidiary organ of the UN General Assembly (UNGA) in 1947 to promote the progressive development and codification of international law. It is composed of 34 individuals recognized for their expertise and qualifications in international law, who are elected by the UNGA every five years.

The ILC enjoys significant authority within international law-making because of its thorough procedures, which entail an iterative process of interaction with States and, increasingly, with other actors. The informal relations between the ILC and the **International Court of Justice** (ICJ) are revealed through the work of the ILC and the reasoning of the Court (Sender 2024). In this respect, mention can be made of the International Court of Justice (ICJ) recent Advisory Opinion on *Legal Consequences arising from the Policies and Practices of Israel in the Occupied Palestinian Territory, including East Jerusalem*, which refers to the ILC Principles on the **Protection of the Environment in Relation to Armed Conflicts (PERAC)** (ICJ 2024, 124).

The ILC included the topic of the *Protection of the Environment in Relation to Armed Conflicts* in its long-term program of work in 2011, acting upon a recommendation included in a 2009 report of the **UN Environment Programme**. The topic was eventually included in the ILC's program of work in 2013, when Ambassador Marie G. Jacobsson was appointed as the Special Rapporteur. Special Rapporteur Jacobsson submitted three reports on PERAC throughout her term with the ILC, which ended in 2016. Ambassador Marja Lehto succeeded her in 2017 as the second Special Rapporteur on the topic. Special Rapporteur Lehto submitted three additional reports until the successful conclusion of the ILC's work on the topic in 2022, when the Commission adopted 27 principles on PERAC (ILC 2022), following nearly a decade of work (ILC 2024).

The work that led to the adoption of the PERAC principles involved extensive consultation with States, the ICRC, international organizations, academicians, and other experts. Following the first reading of the text in 2019, the ILC invited input from States, international organizations, and other **stakeholders**. In the same vein, more than 70 States expressed their views on the finalized text of the PERAC principles and their commentaries at the Sixth Committee of the United Nations General Assembly during its 77th session in 2022. Subsequently, the UNGA adopted a resolution without a vote, through which it took note of the PERAC principles, annexed them and their commentaries to the said resolution, brought them to the attention of States, international organizations, and all who may be called upon to deal with the subject, and encouraged their widest possible dissemination (UNGA 2022).

The PERAC principles aim to enhance environmental protection in relation to armed **conflict** by preventing, mitigating, and remediating environmental harm (Principle 2). The principles adopt a comprehensive, temporal approach, covering the entire **conflict life cycle**—before, during, and after armed conflict, including situations of occupation (Principle 1) (Jacobsson and Lehto 2020, 32). Their application spans multiple phases: principles of general application, those specific to armed conflict, those addressing occupation, and those addressing post-conflict recovery.

In terms of personal scope, the principles are primarily directed at States but also address international organizations, non-state actors, and corporate entities involved in conflict zones. Notably, the principles do not differentiate between international armed conflicts (IACs) and non-international armed conflicts (NIACs), bringing the environmental conduct of non-state armed groups and other actors within their scope.

While the principles themselves do not generally distinguish between IACs and NIACs, their accompanying commentaries offer clarity, particularly regarding how specific laws apply in each context (Pantazopoulos 2022). The principles also draw from various branches of international law (Dam-de Jong and Sjöstedt 2021), such as the law of armed conflict (e.g., Part 4), international environmental law (e.g., Principle 21), and international **human rights** law (e.g., Principle 19).

In sum, the ILC's PERAC principles offer a holistic approach to protecting the **environment** during all stages of conflict, integrating insights from multiple legal frameworks and addressing a wide array of actors involved in both state and non-state armed conflicts. The ILC Principles could prove to be among the building blocks of PERAC for the years to come, meaning that attention should shift to their implementation (Weir and Pantazopoulos 2021). The IUCN Specialist Group on Environmental Security and Conflict Law has already published implementation guidance on ILC Principles 10 and 11 (IUCN-WCEL Specialist Group on Environmental Security and Conflict Law 2024).

References

Dam-de Jong, Daniëlla, and Britta Sjöstedt. 2021. "Enhancing Environmental Protection in Relation at Armed Conflict: An Assessment of the ILC Draft Principles." *Loyola of Los Angeles International and Comparative Law Review* 44 (2): 129–56. https://digitalcommons.lmu.edu/ilr/vol44/iss2/1.

International Court of Justice (ICJ). 2024. *Legal Consequences Arising from the Policies and Practices of Israel in the Occupied Palestinian Territory, Including East Jerusalem*. Advisory Opinion, July 19. https://www.un.org/unispal/document/report-of-the-secretary-general-icj-19dec24/.

International Law Commission (ILC). 2022. *Report of the International Law Commission on the Work of Its Seventy-Third Session (18 April–3 June and 4 July–5 August 2022)*. UN Doc. A/77/10, Chapter V "Protection of the Environment in Relation to Armed Conflicts." A/CN.4/L.968.

International Law Commission (ILC). 2024. *Analytical Guide to the Work of the International Law Commission: Protection of the Environment in Relation to Armed Conflicts*. https://legal.un.org/ilc/guide/8_7.shtml#dcommrep.

International Union for Conservation of Nature (IUCN) – World Commission on Environmental Law (WCEL) Specialist Group on Environmental Security and Conflict Law. 2024. *Corporate Due Diligence and Liability for Natural Resources Supply Chains in/from Conflict-Affected Regions: Implementation Guidance for Principles 10 and 11 of the ILC Principles on Protection of the Environment in Relation to Armed Conflict*. https://iucn.org/sites/default/files/2024-10/guidance-ilc-principles-10-and-11.pdf.

Jacobsson, Marie, and Marja Lehto. 2020. "Protection of the Environment in Relation to Armed Conflicts – An Overview of the International Law Commission's Ongoing Work." *Goettingen Journal of International Law* 10 (1): 27–46. https://doi.org/10.3249/1868-1581-jacobsson-lehto.

Pantazopoulos, Stavros. 2022. *The ILC Draft Principles on Protection of the Environment in Armed Conflict*. https://lieber.westpoint.edu/ilc-protection-environment-armed-conflict.

Sender, Omri. 2024. *International Law-Making by the International Court of Justice and the International Law Commission*. Cambridge Studies in International Law. Cambridge University Press.

United Nations General Assembly (UNGA). 2022. *Protection of the Environment in Relation to Armed Conflicts*. UN Doc. A/77/104, December 7.

Weir, Doug, and Stavros Pantazopoulos. 2021. "If Not Us, Who? How States – with the Help of Civil Society – Can Implement the Legal Framework Protecting the Environment from Armed Conflict." *Humanitarian Law and Policy*, June 10. https://blogs.icrc.org/law-and-policy/2021/06/10/states-civil-society-protecting-environment-armed-conflict.

International Military Council on Climate and Security (IMCCS)

Timothy Balag'kutu

The International Military Council on Climate and Security (IMCCS) is an international network of senior military and **security** experts drawn from governments and the non-governmental sector. The Council on Strategic Risks' (CSR) Center for Climate and Security (CCS) founded and launched the IMCCS on February 19, 2019, in the Hague, Netherlands. The IMCCS came into being in response to rising **climate security** concerns and demands for expert information and best practices to address security and military implications of **climate change**.

Membership in the IMCCS is by invitation and comprises: (1) participants, those committed to the objective of the organization; and (2) observers, individuals with an interest in the subject but who are unable to provide the same level of commitment as participants. The IMCCS is committed to providing an open marketplace of ideas on the military and security dimensions of climate change and has a mission to anticipate, analyze, and address security implications of a changing climate (IMCCS 2025). The organization and its members are driven and bound by a common concern about the challenges of climate change with respect, especially to global peace, stability, and security, and hence the role of military missions in global peacekeeping and **peacebuilding**. The IMCCS emerged from concerns among several militaries and national security communities about the climate-security nexus.

Three institutions lead the IMCCS. First, the Secretariat, which is housed in the CCS, provides day-to-day **governance** in consultation with and execution of the organization's agenda set by the Executive Council. Second, an Expert Group consists of the following specialized organizations: CCS; French Institute for International and Strategic Affairs (IRIS); The Hague Centre for Strategic Studies (HCSS); and the Planetary Security Initiative of the Netherlands Institute of International Relations (also known as Clingendael Institute). Third, an Executive Committee consists of senior members representing all the organizations of the Expert Group consortium and is co-led by the Chair and Secretary General.

The IMCCS facilitates strategic and policy conferences, workshops, and other platforms that explore and share ideas on climate security and **environmental peacebuilding**. The Council maintains an accessible online repository of regularly updated scientific publications and policy outputs, as well as information on various activities and programs by its members, partners, and donor organizations, think tanks, non-governmental organizations, and regional and global institutions. Through its Expert Group, the IMCCS publishes timely research information that provides relevant policy guidance on environmental peacebuilding. The network's annual World Climate and Security Report, which began in 2020, a year after the launch of the IMCCS, assesses and recommends measures for addressing security risks of climate change. Individual experts of the IMCCS regularly appear on media and on other platforms to engage in discussions and provide expert policy advice to governments at all levels. The IMCCS is a resource hub on climate security and environmental peacebuilding. For example, among others, the IMCCS provided expert input to a 2023 report on *Climate Security Scenarios in the Balkans* (Sikorsky and Hugh 2023); helped to place on the agenda and guide discussions on climate security at the 75th NATO Summit in Washington, DC; undertakes various **youth** climate-security leadership programs; facilitates incorporation of climate change into professional military education; and provides policy guidance to the United Nations Security Council on climate security in global peacebuilding. Among many notable individual scholarly contributions, *The Climate General: Stepping Up the Fight* (2023) and *Threat Multiplier: Climate, Military Leadership and the Fight for Global Security* (2024), respectively by the Council's Chair, Retired General and Netherlands Chief of Defence, Tom Middendorp, and Secretary General, the Honorable Sherri Goodman, are major and essential recent additions to the growing body of knowledge on climate security. In its brief history, the IMCCS has had considerable influence in global thinking, policy, and practice/action on climate security, and its impact continues to extend in membership, partnership, and global policy input.

References

Goodman, Sherri. 2024. *Threat Multiplier: Climate, Military Leadership, and the Fight for Global Security*. Island Press.

International Military Council on Climate and Security (IMCCS). 2025. "International Military Council on Climate and Security." https://imccs.org/.

Middendorp, Tom, and Antonie Van Campen. 2023. *The Climate General: Stepping Up the Fight*. La Butineuse.

Sikorsky, Erin, and Brigitte Hugh. 2023. *Weathering Risk Climate Security Scenarios in the Balkans*. The International Military Council on Climate and Security Expert Group, the Center for Climate and Security Institute of the Council on Strategic Risks, and adelphi.

International Union for Conservation of Nature (IUCN)

Karen Hulme and Galeo Saintz

The International Union for Conservation of Nature and Natural Resources (IUCN) focuses on the status of the natural world and the measures needed to safeguard it. The IUCN is a membership union, which is uniquely composed of both government and civil society organizations. The IUCN was created in 1948 and with state members, more than 1,400 member organizations, and 16,000 experts, the union is today the world's largest and most diverse environmental network. The IUCN works with a multitude of partners and supporters to implement a diverse portfolio of **conservation** projects worldwide. Its mission is to "influence, encourage and assist societies to conserve the integrity and diversity of nature and ensure that any use of **natural resources** is equitable and ecologically sustainable" (IUCN n.d.).

IUCN supports conservation, environmental protection, and **governance** through the provision of knowledge and tools to public, private, and non-governmental organizations that enable human progress, economic development, and nature conservation to take place together. It is also a leading provider of conservation data, assessments, and analysis, including through the IUCN Red List of Threatened Species—the world's most comprehensive information source on the global conservation status of animal, fungi, and plant species. Its broad membership enables IUCN to create, promote, and disseminate new tools, best practices, and international standards. IUCN is also able to provide a neutral space in which diverse **stakeholders**, including governments, NGOs, scientists, businesses, local communities, Indigenous Peoples' organizations, and others, can work together to forge and implement solutions to environmental challenges and achieve **sustainable development**.

IUCN has seven Commissions, including the World Commission on Environmental Law (WCEL), Commission on Environmental, Economic and Social Policy (CEESP), and the World Commission on Protected Areas (WCPA). Within the Commissions, there are several smaller research groupings with topics that relate to the work on **environmental peacebuilding**, including the Specialist Group of the WCEL on Environmental Security and Conflict Law. CEESP has also written several publications in the area, including *Conflict and Conservation* (IUCN 2021).

Acknowledgment

The definition of IUCN is drawn from the official IUCN publications and website.

References

International Union for Conservation of Nature. n.d. "About IUCN." Accessed September 8, 2025. https://iucn.org/about-iucn.

International Union for Conservation Nature (IUCN). 2021. *Conflict and Conservation*. International Union for Conservation of Nature. https://iucn.org/resources/publication/conflict-and-conservation.

Nile Basin Initiative (NBI)

Sarah Bebb, Ana Elisa Cascão, and Jeannie Sowers

The Nile Basin Initiative (NBI) was established in 1999 by ministers in charge of water issues from all Nile Basin countries, with the support of the World Bank and development partners. The goal is "to achieve sustainable socio-economic development through the equitable utilization of, and benefit from, the common Nile Basin water resources" (Nile Basin Initiative 2019) by facilitating information collection, transboundary dialogue, joint action, and investments. NBI member states include Burundi, the Democratic Republic of the Congo, Egypt, Ethiopia, Kenya, Rwanda, South Sudan (since gaining independence in 2011), Sudan, Tanzania, Uganda, and Eritrea (as observer). The NBI Secretariat is in Uganda, the Eastern Nile Technical Regional Office in Ethiopia, and the Nile Equatorial Lakes Subsidiary Action Program Coordination Unit in Rwanda. The NBI is financed by annual contributions from member countries and external partners.

Nile basin politics have long been dominated by the vested interests of the most powerful downstream state, Egypt, along with midstream Sudan (Waterbury 1979). Regional water development plans made under British colonial rule in the 1800s focused on providing large water storage to expand irrigation in Egypt and Sudan (Tvedt 2004). The 1929 Nile Water Agreement between Britain and Egypt allocated the majority of water to Egypt for irrigation purposes. The 1959 Agreement between Egypt and Sudan, as newly independent states, further allocated Nile water between them, enabling them to pursue major hydraulic projects, including the Aswan High Dam that epitomized Egypt's lofty development aspirations (Collins 1990). Upstream states were not party to these agreements and do not recognize their validity; they claim the right to pursue development projects without the prior consent of downstream states (Blackmore and Whittington 2008).

To work on bridging these differences, all the Nile Basin countries adopted a two-track approach from the outset, seeking to foster technical **cooperation** among its members under the auspices of the NBI alongside political negotiations toward a new Cooperative Framework Agreement (CFA). NBI members found many areas of agreement during the ten-year negotiating process, but Egypt and Sudan opposed the final CFA document as it did not recognize their historical water rights and current uses. Regardless, in May 2010, seven upstream states agreed to open the CFA for signature. Egypt and Sudan withdrew from NBI activities, although Sudan later resumed **participation**. Ethiopia, Rwanda, Tanzania, Uganda, and Burundi ratified the CFA between 2013 and 2023, and South Sudan acceded in July 2024. The CFA formally entered into force on October 13, 2024, and the NBI is expected to gradually transition to the Nile River Basin Commission. It remains to be seen how this will affect collaboration with countries that are currently active participants in the NBI process but have not signed or ratified the CFA, namely the Democratic Republic of the Congo, Kenya, and Sudan.

The NBI supports projects designed to foster win-win approaches. These are envisioned as trust-building efforts and aim to provide tangible benefits in the areas of integrated water management, energy trade, agricultural investment, drought, and flood management (Cascão et al. 2020). While the NBI has generated considerable amounts of jointly generated knowledge and built an active community of water practitioners, the pipeline of multi-purpose **infrastructure** projects developed during its initial years has been unable to bridge political differences in the region and remains largely unrealized, leading Ethiopia to pursue infrastructure projects unilaterally (Lavers 2024). In the Equatorial Lakes Region, the Rusumo Falls Hydroelectric Project presents the first major, joint dam by NBI member countries and is designed to generate shared benefits for Burundi, Rwanda, and Tanzania. The NBI exemplifies both the potentials and the challenges of **environmental peacebuilding**, serving as a platform for dialogue, trust-building, and technical cooperation amidst enduring political tensions in the Nile basin.

References

Blackmore, Don, and Dale Whittington. 2008. *Opportunities for Cooperative Water Resources Development on the Eastern Nile: Risks and Rewards*. Final Report of the Scoping Study Team to the Eastern Nile Council of Ministers. World Bank.

Cascão, Ana Elisa, Wubalem Fekade, Malte Grossmann, and Abdulkarim Seid. 2020. "Water Diplomacy and Conflict Transformation in the Nile River Basin: The Key Role of the Nile Basin Initiative over the Past 20 Years." In *River Basin Organizations in Water Diplomacy*, edited by Anoulak Kittikhoun and Susanne Schmeier. Routledge.

Collins, Robert O. 1990. *The Waters of the Nile: Hydropolitics and the Jonglei Canal, 1900–1988*. Oxford Studies in African Affairs. Clarendon Press.

Lavers, Tom. 2024. "Upending the Hydropolitics of the Nile: From Cooperation to Unilateralism." In *Dams, Power, and the Politics of Ethiopia's Renaissance*, edited by Tom Lavers. Oxford University Press. https://doi.org/10.1093/oso/9780192871213.003.0006.

Nile Basin Initiative. 2019. *A Long River Journey*. https://nilebasin.org/node/11924.

Tvedt, Terje. 2004. *The River Nile in the Age of the British: Political Ecology and the Quest for Economic Power*. I.B. Tauris.

Waterbury, John. 1979. *Hydropolitics of the Nile Valley*. Syracuse University Press.

PAX

Christina Parandii and Wim Zwijnenburg

PAX is the largest peace organization in the Netherlands, focusing on protecting civilians from the **violence** of **war**, trying to end armed **conflict**, and building inclusive peace (PAX 2025). PAX works in conflict areas worldwide together with local partners and people who believe that everyone has a right to a dignified life in a peaceful society. PAX is headquartered in Utrecht, the Netherlands, and has offices in Brussels (EU) and New York (US). PAX's projects are centered around the themes of Inclusive Peacebuilding, Protection of Civilians, Humanitarian Disarmament, and Business, Conflict, and Human Rights, and cover programs in Africa, the Balkans, the Middle East, Colombia, and Ukraine. Having its origins in the **peace** movement of the late 1940s, PAX resulted from the merging of Pax Christi, Netherlands, and the IKV (Interchurch Peace Council) in 2006 (PAX n.d.a).

Since 2001, PAX has been working on **natural resources** and armed conflict through its work on oil and **human rights** in (South) Sudan, through its Unpaid Debt campaign and coordinating the European Coalition on Oil in Sudan (PAX n.d.b); Unpaid Debt 2025). Following concerns raised by partners about conflict pollution in Iraq in 2009, the organization has been documenting and **monitoring** the environmental impact of new and ongoing conflicts, including in Ukraine, Iraq, Syria, Gaza, Libya, Yemen, South Sudan, and Sudan. Lastly, PAX is also supporting partners in Colombia on corporate **accountability** in conflict-affected areas related to natural resource extraction (PAX n.d.c).

PAX has developed a methodology of documentation of environmental damage using open-source investigation and satellite analysis, with the purpose of building better responses to environmental disasters in conflict (Zwijnenburg and Ballinger 2023). Through documentation and monitoring, and collaboration with local partners, PAX's research has examined how armed conflicts destroy ecosystems (PAX 2023a) and exacerbate **biodiversity** loss (PAX 2020), leave a trail of conflict-pollution (PAX 2023b) and **toxic remnants of war** (PAX 2024a), and lead to degradation and overexploitation of natural resources due to weakened environmental **governance** and unsustainable coping strategies by affected populations (PAX 2021). PAX's work has also been dedicated to exploring the potential of nature in **peacebuilding** through coordinating efforts around the establishment of a digital catalog of **nature-based solutions** (NbS) for peace (Ecosystem for Peace 2024).

The findings and recommendations are incorporated in PAX's research-based advocacy on local, regional, and multilateral levels, including in UN and EU platforms, where it aims to set norms around the link between **environmental degradation**, **climate change**, and armed conflict to help prevent, minimize, and mitigate the environmental damage, as well as to ensure accountability for it. PAX representatives have spoken several times at the UN, including to the Security Council, as well as initiated and provided input to numerous international discussions on the topic of environmental degradation, climate crisis, and conflict (PAX 2019, 2024b). PAX has also supported with expertise and advocacy several resolutions of the UN Environmental Assembly that aimed at addressing the environmental impacts of conflicts in the affected countries and supporting their environmental recovery.

Building capacity for local partners is part of PAX's work to support environmental investigations in conflict-affected areas and to strengthen accountability. Therefore, PAX is providing training on the use of **remote sensing** and open-source investigations on environmental and climate issues for local groups and individuals interested in these topics.

PAX received the UN Environment/UNOCHA Green Star Award in 2017 for its work in environmental emergencies (PAX 2017). In 2023, PAX was jointly awarded the Al-Moumin Award for Thought Leadership in Environmental Peacebuilding by the **Environmental Peacebuilding Association** (PAX 2023c). PAX is also an active member of the International Environmental Peacebuilding Association and the Community of Practice on Environment, Climate, Conflict, and Peace (ECCP).

References

Ecosystem for Peace. 2024. "Nature-based Solutions for Peace and Security." https://solutions.ecosystemforpeace.org/.

PAX. 2017. "PAX Wins International Award." Accessed March 1, 2025. https://paxforpeace.nl/news/pax-wins-international-award/.
PAX. 2019. "PAX Briefs the UN Security Council on Conflict and Environment." https://paxforpeace.nl/news/pax-briefs-the-un-security-council-on-conflict-and-environment/.
PAX. 2020. "Environment and Conflict Alert: Burned Area Analysis in Iraqi Kurdistan." https://paxforpeace.nl/wp-content/uploads/sites/2/import/import/pax-environment-conflict-alert-iraq.pdf.
PAX. 2021. "War, Waste, and Polluted Pastures." https://paxforpeace.nl/publications/war-waste-and-polluted-pastures/.
PAX. 2023a. "Axed and Burned: How Conflict-caused Deforestation Impacts Environmental, Socio-economic and Climate Resilience in Syria." https://paxforpeace.nl/wp-content/uploads/sites/2/import/2023-03/PAX_axed_and_burned.pdf.
PAX. 2023b. "Toxic Floods: Climate, Natural Hazards and Risks to South Sudan's Oil Infrastructure." https://paxforpeace.nl/wp-content/uploads/sites/2/import/2023-05/PAX_Toxic%20Floods.pdf.
PAX. 2023c. "PAX Receives Al-Moumin Award for Environmental Peacebuilding." https://paxforpeace.nl/news/pax-receives-al-moumin-award-for-environmental-peacebuilding/.
PAX. 2024a. "After the Blast: Mapping Environmental Risks from Explosive Weapons in Ukrainian Towns and Cities." https://paxforpeace.nl/wp-content/uploads/sites/2/2024/04/PAX_After-the-Blast.pdf.
PAX. 2024b. "Protecting the Environment Is Protecting Civilians." https://paxforpeace.nl/news/protecting-the-environment-is-protecting-civilians/.
PAX. 2025. "PAX for Peace." https://www.paxforpeace.nl.
PAX. n.d.a. "The Story of PAX." Accessed March 1, 2025. https://paxforpeace.nl/who-we-are/the-story-of-pax.
PAX. n.d.b. "Oil Extraction in South Sudan." Accessed March 4, 2025. https://paxforpeace.nl/what-we-do/programmes/oil-extraction-in-south-sudan/.
PAX. n.d.c. "Corporate Accountability in Conflict Areas: Coalmining in Northern Colombia." Accessed March 4, 2025. https://paxforpeace.nl/what-we-do/programmes/corporate-accountability-in-conflict-areas-coalmining-in-northern-colombia/.
Unpaid Debt. 2025. "Unpaid Debt Campaign on Oil and Human Rights in Sudan." https://unpaiddebt.org.
Zwijnenburg, Wim, and Ollie Ballinger. 2023. "Leveraging Emerging Technologies to Enable Environmental Monitoring and Accountability in Conflict Zones." *International Review of the Red Cross* 105 (2023): 1497–521. https://doi.org/10.1017/S1816383123000383.

South Asian Association for Regional Cooperation (SAARC)

Machane Daniel Isaac

The South Asian Association for Regional Cooperation (SAARC) was established on December 8, 1985, as a regional intergovernmental organization and geopolitical union of eight member states in South Asia with a goal of promoting regional **cooperation** and development. SAARC's member states include Afghanistan, Bangladesh, Bhutan, India, the Maldives, Nepal, Pakistan, and Sri Lanka.

South Asia is home to nearly a quarter of the world's population and is characterized by complex socioeconomic relationships, rich cultural diversity, and shared environmental challenges that transcend national boundaries. The Secretariat of SAARC is headquartered in Kathmandu, Nepal. The highest decision-making body of SAARC is the Summit, which consists of the Heads of State or Government of the member countries. In accordance with Article III of the Charter, SAARC makes decisions at all levels based on unanimity and excludes bilateral and contentious issues from being placed on the agenda.

SAARC fosters collaboration across a range of sectors, including culture, economics, **environment**, health, science, social, and technology. SAARC, further, encourages member states to share best practices, engage in joint research, and develop regional frameworks for addressing shared challenges. This collaborative spirit is essential in a region where environmental concerns often transcend national borders. Transboundary environmental issues, if unaddressed, have the potential to exacerbate existing tensions between neighboring countries in South Asia. Since the late 1990s, SAARC has placed a growing emphasis on **environmental cooperation** to mitigate **conflict** and promote **sustainable development**. SAARC recognizes that managing environmental challenges cooperatively is key to sustaining **peace** and development in the region (Takahashi and Kato 2001).

The field of **environmental peacebuilding** continues to evolve, offering theoretical frameworks and empirical insights. However, South Asia's complex geopolitical landscape presents unique challenges and opportunities for implementing these ideas. Recent scholarship has begun to explore specific avenues for environmental peacebuilding in the South Asian context. For instance, Swain (2002) examines the role of SAARC in promoting environmental peacebuilding, particularly in the context of shared water resources and its contribution to the confidence-building process in the region. Here, the author acknowledges that SAARC is struggling to achieve its full potential due to geopolitical imbalances, internal tensions, and a lack of political will among member states, especially India and Pakistan (Dash 2008; Jha 2004), as well as uneven development and resource distribution across the region. Despite these challenges, Swain (2002) has identified opportunities for environmental peacebuilding through subregional initiatives and the growing involvement of non-state actors, which could complement SAARC's efforts and foster trust among South Asian countries. Similarly, Huda and Ali (2018) outline the role of the SAARC Framework Agreement for Energy Cooperation that gave impetus to hydroelectric cooperation in the Ganges-Brahmaputra-Meghna basin, rooted in environmental peacebuilding mechanisms. Huda (2021) highlights the role of SAARC in linking tertiary education to regional **peacebuilding**, emphasizing the South Asian University (SAU) as a regional hub for excellence established by SAARC's eight member states. Its mandate explicitly aims to foster mutual understanding and strengthen regional consciousness among South Asian communities, positioning education as a tool for advancing ecological cooperation and **sustainable peace**. These initiatives suggest that while environmental cooperation faces significant challenges in South Asia, it also holds promise for advancing peacebuilding efforts by engaging a wide range of **stakeholders**. This underscores SAARC's potential to foster regional stability by prioritizing and contextualizing environment-driven cooperation initiatives.

SAARC has played a vital role in amplifying the voices of its member states in global environmental forums. By fostering regional solidarity and presenting a united front in key discussions, SAARC not only contributes to environmental cooperation within South Asia but also enhances the region's influence in global environmental **governance**. However, the region's colonial legacy, post-independence nation-building processes, and ongoing territorial disputes have profoundly shaped interstate relations, creating a complex backdrop against which environmental initiatives must navigate. This very complexity makes it even more critical to harness environmental cooperation as a tool for peacebuilding, particularly to address the transboundary environmental challenges such as shared water resources, **climate change**, **energy security**, and **biodiversity** loss, which demand collaborative solutions and foster trust among nations. Despite these challenges, a comprehensive review of the existing scholarship reveals that scholars consistently emphasize SAARC's untapped potential and continue to remain an important forum for dialogue and cooperation on critical issues, particularly in addressing shared environmental concerns in South Asia.

Acknowledgment

Introduction to SAARC has been obtained from the SAARC Secretariat portal published at https://saarc-sec.org/ as of October 24, 2024.

References

Dash, Kishore C., eds. 2008. *Regionalism in South Asia: Negotiating Cooperation, Institutional Structures*. Routledge. http://ci.nii.ac.jp/ncid/BA85848140.
Huda, Mirza Sadaqat. 2021. "An Ecological Response to Ethno-nationalistic Populism: Grassroots Environmental Peacebuilding in South Asia." *International Affairs* 97 (1): 119–38. https://doi.org/10.1093/ia/iiaa176.
Huda, Mirza Sadaqat, and Saleem H. Ali. 2018. "Environmental Peacebuilding in South Asia: Establishing Consensus on Hydroelectric Projects in the Ganges-Brahmaputra-Meghna (GBM) Basin." *Geoforum* 96: 160–71. https://doi.org/10.1016/j.geoforum.2018.08.015.
Jha, U C. 2004. "Environmental Issues and SAARC." *Economic and Political Weekly* 39 (17): 1666–71. https://www.jstor.org/stable/4414925.
Swain, Ashok. 2002. "Environmental Cooperation in South Asia." In *Environmental Peacemaking*, edited by Ken Conca and Geoffrey D. Dabelko. Woodrow Wilson Center Press.
Takahashi, Wakana, and K. Kato. 2001. *Environmental Cooperation in South Asia*. IGES Environmental Governance Project, Regional/Subregional Environmental Cooperation in Asia and the Pacific.

United Nations Convention on the Law of Non-Navigational Uses of International Watercourses

Mara Tignino

The UN General Assembly adopted the Convention on the Law of Non-Navigational Uses of International Watercourses (UNWC) on May 21, 1997. The Convention is the result of 20 years of work of the **International Law Commission (ILC)**, started in 1973 and finalized in 1994, with the adoption of the Draft Articles with commentaries and a resolution on transboundary confined groundwater (McCaffrey 2019). The UNWC entered into force on August 17, 2014, after the deposit of the thirty-fifth instrument of ratification by Vietnam (Article 36.1).

The UNWC is a framework convention enshrining the key principles of international law on the management, uses, and protection of transboundary water resources (Boisson de Chazournes et al. 2018). The Convention covers both surface and groundwater resources to the extent that they are linked to international watercourses (Article 2.a).

The **International Court of Justice (ICJ)** and arbitration tribunals have recognized the customary nature of several principles contained in the UNWC, including the principle of equitable and reasonable utilization, the obligation not to cause significant harm, and the duty to cooperate. In the cases of the Gabcikovo-Nagymaros project between Slovakia and Hungary (1997), the 2010 Pulp Mills on the Uruguay River case between Argentina and Uruguay, and the 2022 Dispute over the Status and Use of the Waters of the Silala between Chile and Bolivia, the ICJ clarified the extent to which the principles previously mentioned apply to transboundary water resources.

One of the main goals of UNWC is to provide a legal basis for the establishment of peaceful relations between States sharing transboundary water resources. To achieve this objective, the UNWC contains a solid clause on the settlement of disputes providing for the resort to negotiation, good offices, **mediation**, conciliation, or joint watercourse institutions, as well as the use of judicial means including arbitration and the ICJ (Article 33.2). The possibility to unilaterally submit a dispute to an impartial fact-finding commission is also provided under the Convention (Articles 33.3–33.9).

In addition to the peaceful management and uses of transboundary freshwaters, the protection of the **environment** is a key element of UNWC. UNWC States parties have a duty to protect and preserve freshwater ecosystems and to take measures to prevent, control, and reduce pollution (Articles 20–21). Parties are also required to prevent the introduction of invasive and new species and to take, individually or in **cooperation** with other States, measures that are necessary to protect and preserve the marine environment, including estuaries and deltas (Articles 22–23). In providing common principles and rules to states sharing an international watercourse, the UNWC contributes to the prevention and resolution of water disputes that have the potential to threaten regional **peace**. Joining the UNWC can help riparians to de-escalate water tensions and contribute to the broader aims of **environmental peacebuilding**.

References

Boisson de Chazournes, Laurence, Makane Moïse Mbengue, Mara Tignino, Komlan Sangbana, and Jason Rudall, eds. 2018. *The UN Convention on the Law of the Non-Navigational Uses of International Watercourses. A Commentary*. Oxford University Press.

Convention on the Law of the Non-Navigational Uses of International Watercourses, May 21, 1997, U.N. Doc. A/51/869.

McCaffrey, Stephen C. 2019. *The Law of International Watercourses*. Oxford University Press.

United Nations Convention to Combat Desertification (UNCCD)

Utchang Kang

The United Nations Convention to Combat Desertification (UNCCD), adopted in 1994, is the sole legally binding international agreement linking **environment** and development to sustainable land management and drought **resilience**. The mandate of the Convention and its 197 Parties is to support "a future that avoids, minimizes, and reverses desertification/**land degradation** and mitigates the effects of drought in affected areas at all levels – and to achieve a land degradation-neutral world consistent with the 2030 Agenda for Sustainable Development" (UNCCD 2018).

The Land Degradation Neutrality (LDN Sustainable Development Goals 15.3) is the guiding principle for the implementation of the UNCDD (UNCCD n.d.). It provides a flexible framework to identify and design an optimal mix of policies and interventions for sustainable land management (SLM), aligned with existing international agreements and initiatives. LDN provides an additional opportunity to forge a broader partnership to support SLM with maximized synergy and complementarity, including governments, non-state actors, NGOs, organizations, expert institutions, and private sector entities, among others. It can serve as a linchpin in addressing land-related environmental issues such as watershed management, drought, **biodiversity** loss, disaster risk management, and sand and dust storms at a landscape scale that often transcends national boundaries.

LDN is vital for **environmental peacebuilding**, as it can play a role in sustainable ecosystem management in fragile and **conflict**-affected areas. Degraded land accelerates resource competition, causing insecurity in livelihoods and social stability and, therefore, affects **peace** and **security**. Achieving LDN by jointly caring for land and **forests** can foster confidence-building and underpin collective security and peace. Building on existing mechanisms and/or initiatives, LDN can forge collaboration to mobilize further exchanges and practical actions between concerned **stakeholders**. Where **governance** systems for SLM are not in place, LDN can provide an entry point for **environmental cooperation** for building confidence and peace.

A general strength of LDN is that it emphasizes the importance of the human dimension—land-users, stakeholder groups, and institutions—in **land use** planning and development. As a participatory and inclusive process, it can be used to systematically address conflicts as they arise, while contributing to the achievement of the global restoration goals and targets, particularly during the UN Decade on Ecosystem Restoration (2021–30) (Mosello 2024).

UNCCD's initiatives, including the Peace Forest Initiative (PFI), highlight this transformative potential of land restoration and environmental peacebuilding in fostering political **cooperation** and **social cohesion**, particularly through cross-border partnerships that promote the rehabilitation of ecosystems and strengthen the community's resilience in conflict-affected and fragile regions, hence making a direct linkage between land restoration and conflict **mitigation** (Kang et al. 2024; Upla et al. 2024).

The Peace Forest Initiative (PFI), launched during the 14th Session of the Conference of the Parties of the UNCCD in 2019, is a flagship program of the UNCCD to address the nexus between land, peace, and security (Kang et al. 2024). The initiative seeks to promote exchanges, dialogues, capacity building, and cooperation on ecosystem and restoration between concerned parties in fragile and conflict-affected locations (Upla et al. 2024). It aims to provide a flexible, inclusive, and participatory platform for coordination and co-design, development, and implementation of joint activities, plans, strategies, projects, and programs on the sustainable management of **natural resources** (land, soil, water, and forest). Additionally, the PFI promotes a comprehensive approach by combining **conflict sensitivity** with a rights-based framework to address inequalities and root causes of conflicts, involving a multistakeholder platform, locally led conflict analysis, and bottom-up **peacebuilding** (Mosello et al. 2024). Integrating conflict sensitivity into land restoration is essential for lasting peace outcomes (Upla et al. 2024).

Ultimately, the PFI aims to contribute to building confidence and peace through joint activities and collaboration on restoring and developing land-based resources as well as enhancing the livelihood and resilience of the community (Kang et al. 2024). As land restoration is emerging as a crucial strategy for **conflict prevention** and long-term stability, the PFI seeks to address complex intersections between land restoration and security, demonstrating the linkages between LDN (SDG 15.3) and SDG 16 (peace and justice) (Upla et al. 2024).

References

Kang, Utchang, Pauliina Upla, and Corinna Voigt. 2024. *Peace Forest Initiative: Restoring Peace and Trust through Nature*. United Nations Convention to Combat Desertification. https://www.unccd.int/sites/default/files/202406/PFI%20brochure_final_%20june%20%28003%29.pdf.

Mosello, Beatrice, Mary Elizabeth Potts, Héctor Morales-Muñoz, Sandor Madar, and Oli Brown. 2024. *Ground for Peace: Land Restoration for International Peace and Security*. United Nations Convention to Combat Desertification. https://www.unccd.int/sites/default/files/202412/Peace%20Forest%20Initiative_Ground%20for%20Peace_full%20report_UNCCD.pdf.

United Nations Convention to Combat Desertification (UNCCD). n.d. "Land Degradation Neutrality Principles." https://www.unccd.int/land-and-life/land-degradation-neutrality/ldn-principles.

United Nations Convention to Combat Desertification (UNCCD). 2018. "The UNCCD 2018-2030 Strategic Framework." ICCD/COP(13)/21/Add.1, Decision 7/COP.13. https://www.unccd.int/resources/other/unccd-2018-2030-strategic-framework.

Upla, Pauliina, Utchang Kang, and Ani Tovmasyan. 2024. *Peace Forest Initiative: Operational Guidelines*. United Nations Convention to Combat Desertification. https://www.unccd.int/sites/default/files/2024-12/Peace%20Forest%20Initiative%20Operational%20Guidelines_UNCCD.pdf.

United Nations Department of Peace Operations (UNDPO)

Billy Agwanda and Silvia Danielak

The United Nations Department of Peace Operations (UNDPO), formerly the Department of Peacekeeping Operations (DPKO), is one of the main institutions within the United Nations (UN) framework mandated with the implementation of **UN Security Council** resolutions on international **peace** and **security** (UN n.d.; Koops et al. 2015). The Department of Peace Operations' (DPO's) predecessor was created in 1992 in response to the evolving and multidimensional nature of peacekeeping operations (UN Archives and Records Management 2007) as part of a restructuring of the UN Secretariat. Then UN Secretary-General Boutros Boutros-Ghali elevated the Department's primary role from military observation during the implementation of **peace agreements** to that of robust and multidimensional military, police, and civilian support during ongoing **conflict** and its aftermath. Since 2019, the renamed DPO provides political guidance, diplomatic liaison, and administrative support to build strong foundations for **sustainable peace** in UN member states. Institutionally, DPO is headed by an Under-Secretary-General and consists of three main entities (Office of Rule of Law and Security Institutions, Office of Military Affairs, and the Policy, Evaluation and Training Division), as well as regional divisions shared with the Department of Political and Peacebuilding Affairs (DPPA).

Already in 2012, the **UN Environment Programme (UNEP)** drew attention to the important environmental role of peacekeepers, especially vis-à-vis **natural resources** (UNEP 2012). Given the scope of DPO's mandate, and its large number of personnel (civilian staff, formed police units, troops, and military observers) deployed in the field through its peace operations, DPO has a large ecological footprint and possibly significant impacts on environmental and cultural heritage preservation, urbanization patterns, and **infrastructure** development (Danielak 2023; Maertens 2019; Ravier et al. 2016). DPO has thus stood out as one of the best-placed institutions within the UN framework to mitigate ecological harm and improve the organization's environmental footprint. Furthermore, DPO can also foster positive environmental norms and practices through climate-sensitive programming, and environmental **mediation** and **peacebuilding** within peacekeeping operations (Bakaki and Böhmelt 2021; Scartozzi 2022; Sarfati 2022).

Owing to its environmental footprint, DPO has stepped up its policy engagement and environmental risk and performance management over the last few years. In its Environmental Policy for Peacekeeping Operations published in 2022, DPO—together with DPPA and the Department of Operational Support, which has led much of the environmental engagement in peace operations—committed to limiting its overall environmental footprint of missions (UN Department of Peace Operations et al. 2022). The policy outlines measures to ensure that missions institute sound environmental practices and fully consider the environmental impact of their operations, underpinned by five core principles. These include stewardship of the environment, efficiency in resource use and operations, continuous improvement of environmental performance, **stakeholder** engagement at all levels, and **adaptation** and **resilience**. UNDPO also joined the UN **Climate Security Mechanism** in 2021 to address the links between **climate change** and security risks and has deployed **climate security** focal points and advisers to its field missions. Finally, diplomatically, DPO has many opportunities through training and **evaluation**, transition planning, financing, and even diplomatic engagement to advance the environmental agenda in peace operations.

References

Bakaki, Zorzeta, and Tobias Böhmelt. 2021. "Can UN Peacekeeping Promote Environmental Quality?" *International Studies Quarterly* 65 (4): 881–90. https://doi.org/10.1093/isq/sqab051.

Danielak, Silvia. 2023. "Peacekeeping's Conflict Urbanism: A Decade of Urban Change in Mali." *Peacebuilding* 11 (4): 446–62. https://doi.org/10.1080/21647259.2023.2219131.

Koops, Joachim A., Norrie MacQueen, Thierry Tardy, and Paul D. Williams. 2015. "Introduction: The United Nations and Peacekeeping." In *The Oxford Handbook of U.N. Peacekeeping Operations*, edited by Joachim A. Koops, Thierry Tardy, Norrie MacQueen, and Paul D. Williams. Oxford University Press. https://doi.org/10.1093/oxfordhb/9780199686049.013.78.

Maertens, Lucile. 2019. "From Blue to Green? Environmentalization and Securitization in UN Peacekeeping Practices." *International Peacekeeping* 26 (3): 302–26. https://doi.org/10.1080/13533312.2019.1579648.

Ravier, Sophie, Anne-Cecile Vialle, Russ Doran, and John Stokes. 2016. "Environmental Experiences and Developments in United Nations Peacekeeping Operations." In *Governance, Natural Resources and Post-Conflict Peacebuilding*, edited by Carl Bruch, Carroll Muffett, and Sandra S. Nichols. Routledge.

Sarfati, Agathe. 2022. *Toward an Environmental and Climate-Sensitive Approach to Protection in UN Peacekeeping Operations*. International Peace Institute. https://www.ipinst.org/2022/10/toward-an-environmental-climate-sensitive-approach-to-protection-in-un-peace-ops.

Scartozzi, Cesare M. 2022. "Climate-Sensitive Programming in International Security: An Analysis of UN Peacekeeping Operations and Special Political Missions." *International Peacekeeping* 29 (3): 488–521. https://doi.org/10.1080/13533312.2022.2084387.

United Nations. n.d. "Department of Peace Operations." https://peacekeeping.un.org/en/department-of-peace-operations.

United Nations Archives and Records Management. 2007. "Department of Peacekeeping Operations–Office of the Under-Secretary-General–UNARMS." https://search.archives.un.org/department-of-peacekeeping-operations-office-of-the-under-secretary-general.

United Nations Department of Peace Operations (UNDPO), United Nations Department of Peacebuilding Support (UNDPOS), and United Nations Department of Political and

Peacebuilding Affairs (UNDPPA). 2022. *Environmental Policy for Peacekeeping Operations and Field-Based Special Political Missions*. United Nations. https://operationalsupport.un.org/sites/default/files/dos_2022.01_environmental_policy_for_peackeeping_operations_and_special_political_missions.pdf.
United Nations Environment Programme (UNEP). 2012. *Greening the Blue Helmets: Environment, Natural Resources and UN Peacekeeping Operations*. http://postconflict.unep.ch/publications/UNEP_greening_blue_helmets.pdf.

United Nations Environment Programme (UNEP)

Oli Brown, Maria Ivanova, and Stephen Lonergan

The United Nations Environment Programme (UNEP) is the leading UN authority on the environment, the "anchor institution" designed to catalyze environmental action and foster collaboration across the United Nations system (Ivanova 2021). UNEP plays a central role in catalyzing global action on the **environment**, with a particular focus on **climate change, biodiversity,** nature loss, and pollution and waste. It is also instrumental in advancing **environmental peacebuilding** by addressing the environmental causes and impacts of **conflict.**

Established in 1972 following the Stockholm Conference on the Human Environment, UNEP has been a critical actor in shaping global environmental **governance** (United Nations General Assembly 1972). As the first international organization with a mandate to address environmental issues, UNEP was designed to catalyze action, promote global environmental norms, and coordinate international responses to pressing environmental challenges. Headquartered in Nairobi, Kenya, UNEP is a global organization with more than 2,000 personnel working in seven regional offices and 18 country offices around the world. Nairobi is the only UN headquarters in the Global South (UNEP 2024).

UNEP's contribution to environmental peacebuilding began in 1999, when it was tasked with producing a post-**conflict environmental assessment** of the NATO military campaign in Kosovo. This marked UNEP's first foray into addressing the environmental impacts of conflict, setting the stage for future work on environmental causes and consequences of warfare. Since then, UNEP has conducted more than 30 post-conflict and post-disaster environmental assessments and managed country programs in Afghanistan, Sudan, South Sudan, Haiti, Somalia, and Sierra Leone, assisting these nations to restore environmental governance and rebuild livelihoods.

UNEP's role in environmental peacebuilding extends beyond post-conflict recovery. In 2005, UNEP brokered talks between Afghanistan and Iran over shared water resources, demonstrating its capacity to mediate environmental disputes that could lead to conflict. UNEP has also integrated environmental concerns into broader **peacebuilding** strategies, highlighting the importance of natural resource management for **sustainable peace** (Matthew et al. 2009).

In addition to post-conflict efforts, UNEP has helped create key international bodies and agreements that support peacebuilding through environmental governance. UNEP was instrumental in the creation of the **Intergovernmental Panel on Climate Change (IPCC)**, the Intergovernmental Science-Policy Platform on Biodiversity and Ecosystem Services (IPBES), and the International Resource Panel (IRP). These bodies provide the scientific assessments necessary for informed decision-making on global environmental issues, laying the groundwork for long-term stability and **cooperation**.

One of UNEP's major successes in environmental peacebuilding has been its leadership in promoting the sustainable management of **natural resources**. Through efforts such as the Environmental Cooperation for Peacebuilding initiative, UNEP has demonstrated that effective resource management is crucial in preventing conflicts and fostering **resilience** in fragile states (United Nations Environment Programme 2024). By addressing resource scarcity, UNEP helps reduce the likelihood of **environmental degradation** leading to conflict, while also ensuring that post-conflict recovery efforts are sustainable and equitable.

Despite its successes, UNEP faces several challenges, including limited funding and political constraints. Nevertheless, UNEP continues to be a key player in the intersection of environmental governance and peacebuilding, advocating for the integration of environmental management into **conflict prevention** and resolution efforts.

References

Ivanova, Maria. 2021. *The Untold Story of the World's Leading Environmental Institution: UNEP at Fifty*. MIT Press. https://direct.mit.edu/glep.

Matthew, Richard, Oli Brown, and David Jensen. 2009. *From Conflict to Peacebuilding: The Role of Natural Resources and the Environment*. No. 1. United Nations Environment Programme and Earthprint. https://www.iisd.org/system/files/publications/conflict_peacebuilding.pdf

Programme and Earthprint. https://www.iisd.org/system/files/publications/conflict_peacebuilding.pdf

United Nations Environment Programme (UNEP). 2021. *Environmental Cooperation for Peacebuilding*.

United Nations Environment Programme (UNEP). 2024. "About the United Nations Environment Programme." https://www.unep.org/who-we-are/about-us.

United Nations General Assembly (UNGA). 1972. *Resolution 2997 (XXVII): Institutional and Financial Arrangements for International Environmental Cooperation*, December 15, 1972.

United Nations Framework Convention on Climate Change (UNFCCC)

Gabriela Nagle Alverio and B. Eni Owoeye

The United Nations Framework Convention on Climate Change (UNFCCC) is an international treaty designed to address the collective action challenges of **climate change** by fostering **cooperation** among countries to mitigate greenhouse gas emissions and engage in climate **adaptation**. Established in 1992, the UNFCCC provides a framework for negotiating consensus-based agreements to set greenhouse gas emissions-reduction targets and commitments for developed and developing countries to achieve these goals (UNFCCC n.d.d). In 1995, countries launched negotiations to strengthen the global response to climate change and adopted the Kyoto Protocol. The Protocol established legally binding emissions-reduction targets for developed countries, referred to as Annex I countries, based on the principle of "common but differentiated responsibilities" (UN 1997). This principle recognizes that industrialized nations, which have contributed more to climate change, should take the lead in reducing emissions. The Kyoto Protocol, however, has not proven successful because the largest emitter at the time, the United States, did not ratify it and because of its stringent targets that did not account for varying economic and political landscapes (Vihma 2011).

Subsequent agreements have sought to overcome the Kyoto Protocol's failures. The 2015 Paris Agreement introduced a more flexible approach to global climate action by allowing countries to set their own voluntary emission reduction targets, called Nationally Determined Contributions (NDCs), which must be made more ambitious every five years (UN 2015). The Agreement's central objective is to limit global warming to 1.5 degrees, and well below 2 degrees (UNFCCC 2015; UN 2015).

Key events under the UNFCCC, such as the annual Conference of the Parties (COP), which resulted in the Paris Agreement, facilitate dialogue between the 198 countries that are parties to the Convention, review progress toward the previously agreed targets, and push for stronger climate action globally (UNFCCC 2022). The COP is the largest United Nations conference and is hosted by different countries annually (UNFCCC n.d.a). The UNFCCC organizes several negotiating sessions as well as subsidiary body meetings and workshops throughout the year leading up to COP (UNFCCC n.d.a).

While only state parties can vote on agreements, **environmental peacebuilding** advocates may participate closely in the negotiations process through the nine recognized constituencies: Business and Industry NGOs (BINGO), Environmental NGOs (ENGO), Farmers, Indigenous Peoples Organizations (IPO), Local Government and Municipal Authorities (LGMA), Research and Independent NGOs (RINGO), Trade Union NGOs (TUNGO), Women and Gender (WGC), and Youth NGOs (YOUNGO) (UNFCCC n.d.b). The voices of advocates in these constituencies have been critical in advancing climate **justice** considerations within UNFCCC discussions. For instance, in 2013, the Warsaw International Mechanism for Loss and Damage was introduced to promote capacity building and knowledge sharing around the irreversible losses caused by climate change (UNFCCC n.d.c). Largely driven by continued advocacy efforts, in 2023, parties moved beyond capacity building to operationalize a fund to respond to loss and damage (UNFCCC 2023).

Through sustained collaboration among governments, civil society, and other **stakeholders**, the UNFCCC continues to evolve as a platform not only for climate policy but also for the collective advancement of global environmental **resilience** and **justice**.

References

United Nations (UN). 1997. *Kyoto Protocol to the United Nations Framework Convention on Climate Change*. https://unfccc.int/resource/docs/convkp/kpeng.pdf.

United Nations (UN). 2015. *Key Aspects of the Paris Agreement*. United Nations Climate Change. https://unfccc.int/most-requested/key-aspects-of-the-paris-agreement.

United Nations Framework Convention on Climate Change (UNFCCC). 2023. *COP 27 Report*. https://unfccc.int/documents/632319.

United Nations Framework Convention on Climate Change (UNFCCC). 2022. *Status of Ratification of the Convention*. https://unfccc.int/process-and-meetings/the-convention/status-of-ratification-of-the-convention.

United Nations Framework Convention on Climate Change (UNFCCC). n.d.a. "About the Secretariat." Accessed August 26, 2025. https://unfccc.int/about-us/about-the-secretariat#:~:text=The%20ultimate%20objective%20of%20all,naturally%20and%20enables%20sustainable%20development.

United Nations Framework Convention on Climate Change (UNFCCC). n.d.b. "Constituencies in the UNFCCC." Accessed August 26, 2025. https://unfccc.int/process-and-meetings/parties-non-party-stakeholders/non-party-stakeholders/overview/admitted-ngos#Constituencies-in-the-UNFCCC.

United Nations Framework Convention on Climate Change (UNFCCC). n.d.c. "Warsaw International Mechanism for Loss and Damage." Accessed August 26, 2025. https://unfccc.int/topics/adaptation-and-resilience/workstreams/loss-and-damage/warsaw-internationalmechanism#:~:text=Facilitating%20the%20mobilization%20and%20securing,and%20damage%20associated%20with%20climate.

United Nations Framework Convention on Climate Change (UNFCCC). n.d.d. "What Is the United Nations Framework Convention on Climate Change?" Accessed August 26, 2025. https://unfccc.int/process/the-convention/what-is-the-united-nations-framework-convention-on-climate-change.

United Nations Framework Convention on Climate Change (UNFCCC). 2015. *Paris Agreement.* https://unfccc.int/sites/default/files/english_paris_agreement.pdf.

Vihma, Antto. 2011. "The Failure of the Kyoto Protocol: A Policy Perspective." *Global Environmental Politics* 11 (2): 125–41. https://www.jstor.org/stable/43267167.

United Nations Security Council (UNSC)

Zerrin Savaşan and Daniëlla Dam-de Jong

The United Nations Security Council (UNSC) is the primary organ within the United Nations (UN) for the maintenance of international **peace** and **security**. It is also the only organ of the UN that has the right to take decisions that are coercive in nature or authorize the use of force (United Nations 1945, Articles 41, 42) and that are directly binding on all UN members (UN Charter, Articles 24, 25).

To act, the UNSC must first determine whether a situation constitutes a threat to the peace, a breach of the peace, or an act of aggression (UN Charter, Article 39). In practice, it often relies on the concept of a threat to peace. This notion has evolved considerably over time (Scott and Andrade 2011; Scott and Ku 2018). Until the 1990s, the UNSC defined threats to the peace primarily within the context of armed conflicts between states. From then on, it started to expand the notion to include situations of internal armed **conflict** with (potential) regional repercussions (UNSC 1992b). It also expanded the range of issues that could fall under the notion, broadening the scope to nonmilitary threats (UNSC 1992a). By the 2000s, the UNSC further extended its traditional jurisdiction. Relevant examples include its practice on **environment**-related issues like the consequences of the earthquake in Haiti (UNSC 2010), the impact of HIV/AIDS on international peace and security (UNSC 2011), and the deadly Ebola outbreak in Africa (UNSC 2014).

Through this practice, the UNSC made several important contributions to **environmental peacebuilding.** The first was the establishment of the UN Compensation Commission, tasked with assessing claims by states and individuals for damage suffered by Iraq's illegal invasion of Kuwait in 1990, including in relation to environmental damage (Payne and Sand 2011; UNSC 1990, 1991a, 1991b, 1991c). A second concerns the UNSC's practice in relation to **conflict resources** from the 1990s onwards (Dam-de Jong 2017). A third contribution relates to the recent question of **climate security** and **environmental degradation** more generally (Maertens 2021; Cusato 2022; Dam-de Jong 2023; Dienelt 2023). The Council developed relevant practice in this field, especially with respect to the effects of **climate change** on conflicts and instability in Lake Chad and in the Sahel region (UNSC 2017) and on the effects of ecological change on Somalia's stability (UNSC 2018). Other important contributions by the UNSC relate to addressing the relations between conflict and chemical and biological weapons (Kapoor and Lewis 2023).

Despite these developments in the UNSC's practice, its contribution to addressing environmental threats is still quite controversial, especially if there is no armed conflict (Savaşan 2020). Proposals have been made to activate Article 39 of the UN Charter and make it possible for the UNSC to consider the security effects of environmental problems as "a direct threat" to international peace (Voigt 2009); or to assume a larger role for the Council in protecting the environment through preemptive military strikes in the name of environmental defense (Eckersley 2007); or to refer to the terrorism-related Resolution 1373 and Resolution 1540 as a precedent for environmental threats (UNSC 2001, 2004; Martínez 2008; Talmon 2005). These proposals, however, do not seem to align with current political realities, as exemplified by the UN Secretary-General's failed efforts to strengthen the contribution of UN peacekeeping operations to climate security as part of general UN policy (UN 2023, 2024).

References

Cusato, Eliana. 2022. "Of Violence and (In)Visibility: The Securitisation of Climate Change in International Law." *London Review of International Law* 10 (2): 203242. https://doi.org/10.1093/lril/lrac015.

Dam-de Jong, Daniëlla A. 2017. "UN Natural Resources Sanctions Regimes: Incorporating Market-Based Responses to Address Market-Driven Problems." In *Research Handbook on UN Sanctions and International Law*, edited by Larissa van den Herik. Edward Elgar Publishing.

Dam-de Jong, Daniëlla A. 2023. "The Contribution of the UN Security Council to Environmental Peacebuilding." In *Research Handbook on International Law and Environmental Peacebuilding*, edited by Daniëlla A. Dam-de Jong and Britta Sjöstedt. Edward Elgar Publishing.

Dienelt, Anne. 2023. "The Climatization of International Peace and Security – A Missed Opportunity?" In *Climate Change and the Testing of International Law*, edited by Sandrine Maljean-Dubois and Jacqueline Peel. Centre for Studies and Research in International Law and International Relations Series. Brill Publishers.

Eckersley, Robyn. 2007. "Ecological Intervention: Prospects and Limits." *Ethics, and International Affairs* 21 (3): 292–316. https://doi.org/10.1111/j.1747-7093.2007.00101.x.

Kapoor, Radhika, and Dustin A. Lewis. 2023. "The Practice of The UN Security Council Pertaining to the Environment and Armed Conflict, 1945–2021." *International Review of the Red Cross* 105 (924): 1646–67. https://international-review.icrc.org/articles/the-practice-of-the-un-security-council-pertaining-to-the-environment-924.

Maertens, Lucile. 2021. "Climatizing the UN Security Council," *International Politics,* UN Charter (58): 640–60. https://doi.org/10.1057/s41311-021-00281-9.

Martínez, Luis Miguel Hinojosa. 2008. "The Legislative Role of the Security Council in Its Fight against Terrorism: Legal, Political and Practical Limits." *The International and Comparative Law Quarterly* 57 (2): 333–59. https://doi.org/10.1017/S0020589308000201.

Payne, Cymie, and Peter Sand, eds, 2011. *Gulf War Reparations and the UN Compensation Commission: Environmental Liability*. Oxford University Press.

Savaşan, Zerrin. 2020. "İklim Güvenliği Konusunda BM Güvenlik Konseyi'nin Olası Rolü [The Role of the UN Security Council on Climate Security]." In *Küresel Ekolojik Güvenlik,* edited by Nesrin Algan, Duygu Özel Demiralp, and Feza Sencer Çörtoğlu. Ankara Üniversitesi.

Scott, Shirley V., and Roberta C. D. Andrade. 2011. "The Global Response to Climate Change: Can the Security Council Assume a Lead Role?" *Brown Journal of World Affairs* 18 (2): 215–26. https://www.jstor.org/stable/24590874.

Scott, Shirley V., and Charlotte Ku, eds. 2018. *Climate Change and the UN Security Council.* Edward Elgar Publishing.

Talmon, Stefan. 2005. "The Security Council as World Legislature." *The American Journal of International Law* 99 (1): 175–93. https://doi.org/10.2307/3246097.

United Nations (UN). 1945. *Charter of the United Nations.*

United Nations (UN). 2023. *A New Agenda for Peace: Our Common Agenda Policy Brief 9.* https://www.un.org/sites/un2.un.org/files/our-common-agenda-policy-brief-new-agenda-for-peace-en.pdf.
United Nations (UN). 2024. *Pact for the Future, Global Digital Compact and Declaration on Future Generations.* https://www.un.org/sites/un2.un.org/files/sotf-pact_for_the_future_adopted.pdf.
UN Security Council (UNSC). 1990. Resolution 674. UN Doc.S /RES/674, October 29.
UN Security Council (UNSC). 1991a. Resolution 686. UN Doc. S/RES/686, March 2.
UN Security Council (UNSC). 1991b. Resolution 687. U.N. Doc. S/Res/687, April 3.
UN Security Council (UNSC). 1991c. Resolution 692. UN Doc. S/RES/692, May 20.
UN Security Council (UNSC). 1992a. Presidential Statement. UN Doc. S/23500, January 31.
UN Security Council (UNSC). 1992b. Resolution 794. U.N. Doc. S/RES/794, December 3.
UN Security Council (UNSC). 2001. Resolution 1373. U.N. Doc. S/RES/1373, September 28.
UN Security Council (UNSC). 2004. Resolution 1540. U.N. Doc. S/RES/1540, April 28.
UN Security Council (UNSC). 2010. Resolution 1908. U.N. Doc. S/RES/1908, January 19.
UN Security Council (UNSC). 2011. Resolution 1983. U.N. Doc. S/RES/1983, June 7.
UN Security Council (UNSC). 2014. Resolution 2177. U.N. Doc. S/RES/2177, September 18.
UN Security Council (UNSC). 2017. Resolution 2349. UN Doc. S/RES/2349, March 31.
UN Security Council (UNSC). 2018. Resolution 2408. UN Doc. S/RES/2408, March 27.
Voigt, Christina. 2009. "Security in a 'Warming World': Competences of the U.N. Security Council for Preventing Dangerous Climate Change." In *Security: A Multidisciplinary Normative Approach*, edited by Cecilia M. Bailliet. Brill Publishers.

The Women in Water Diplomacy Network (WWDN)

Elizabeth Koch

The Women in Water Diplomacy Network (WWDN) is a community of practice comprised of both formal and informal women water diplomats, decision-makers, and experts, representing a wealth of collective experience and expertise with a focus on shared waters in some of the most water-insecure and **conflict**-affected regions of the world. Originating in the Nile basin in 2017, the Network is now a global community of women water leaders and male champions supporting **gender** equality, working collectively to strengthen women's leadership in transboundary water decision-making with support from a global coalition of partners and members in the Nile, Central Asia-Afghanistan, Southern Africa, North America, the South Caucasus, and elsewhere. Members of the Women in Water Diplomacy Network are directly engaged in a range of ongoing formal and informal transboundary water and climate-related **peace** processes/agreements. Members serve in diverse official roles/functions from technical advisory roles to administrator, mediator, or negotiator roles at various scales and stages of transboundary management and negotiations in fragile and conflict-affected regions of the world. The Women in Water Diplomacy Network draws inspiration from the Dublin Statement (1992), the Women, Peace and Security agenda adopted by the United Nations Security Council through Resolution 1325 (2000), and UN Resolution 2250 on Youth, Peace and Security (2015), working to catalyze women's leadership in high-level decision-making and negotiations in transboundary basins.

Women's **participation** at all scales of development and **peacebuilding** processes, including **conflict prevention** and post-conflict reconstruction and recovery, is essential for a resilient and peaceful future—of which sustainable water for all is foundational. The Women in Water Diplomacy Network engages in coalition building and dialogue, serving as a community of practice for key actors representing decision-making roles, largely

Ministries of Water and Foreign Affairs and other related line ministries from basin states, to support each other in increasing women's participation in decision-making processes around cross-border water resource management. The Network enables members to get to know one another, learn from each other, share experiences, and build cross-border relationships and trust while collectively elevating the importance of inclusive decision-making in transboundary water **governance**.

The Women in Water Diplomacy Network is guided by its *Global Strategy, A Path Forward for Women, Water, Peace and Security* (2022), as well as the Network's Global Leadership Council. At the close of 2024, the Network has engaged more than 1,000 individuals from more than 80 countries and more than 10 Indigenous communities as part of Network in-person training and knowledge exchange workshops, online training and knowledge exchange workshops/seminars, contributed to communications or resource products (including novel research and the Network newsletter's 2,000+ subscribers), and collectively engaged at major water and climate policy and decision-making events including the UN Water Conference 2023, World Water Weeks, and World Water Forums. Hundreds of the Network's community members engage frequently across Network programming. In addition, the Network's broader community online numbers 5,000+ followers on LinkedIn (launched in November 2023). The Network also maintains several WhatsApp groups with more than 200 Network members and champions actively engaged in the sustained dialogue.

By intentionally applying gender equality principles to water diplomacy, the Women in Water Diplomacy Network works to elevate women's participation in water-related negotiations and decision-making processes, build **resilience** in water-insecure regions, and foster cooperative and inclusive governance. Through its strategies, the Network encourages decision-makers to embrace inclusive leadership as a guiding framework for equitable and sustainable water management across scales, ultimately contributing to global **peace, security**, and **sustainable development** through improved **water cooperation**. In practice, the application of the principles of gender equality to water diplomacy processes has not only contributed to a broader recognition of the benefits of inclusion within water and climate-related decision-making but has also paved the way for a transformative water diplomacy derived from women-led practices that reshape the core principles of what effective water diplomacy can achieve (Koch and Mengistu 2025).

References

A Path Forward for Women, Water, Peace and Security: Women in Water Diplomacy Network Nile and Beyond Global Strategy. 2022. Stockholm International Water Institute.

Koch, Elizabeth A., and Bethelehem Mengistu. 2025. "Evolving Principles of Women-led Water Diplomacy." *Nature Water* 3: 246–47. https://doi.org/10.1038/s44221-025-00414-x.

World Bank Group (WBG)

Elise Doumergue, Phoebe Girouard Spencer, and Tracy Hart

The World Bank Group (WBG) is made up of five institutions: the International Bank for Reconstruction and Development, the International Development Association, the

International Finance Corporation, the Multilateral Investment Guarantee Agency, and the International Centre for Settlement of Investment Disputes, working together to reduce poverty and promote **sustainable development** with knowledge, financing, and expertise. Within the context of **environmental peacebuilding**, WBG support plays a crucial role in addressing the environmental dimensions of **conflict** and **fragility**, building **resilience** to shocks, and advancing development priorities in fragile and conflict-affected settings.

The World Bank Group's involvement in environmental peacebuilding is rooted in its institutions' broader mission to reduce poverty and boost shared prosperity on a livable planet. Recognizing the intricate linkages between **environmental degradation**, natural resource management, fragility, and conflict, the WBG has developed strategies and programs aimed at mitigating these issues to foster **peace** and stability.

One of the key frameworks guiding the institutions' efforts in this area is the World Bank Group Strategy for Fragility, Conflict, and Violence (FCV) 2020–2025, which marks a shift toward a more holistic approach to **conflict prevention** (WBG 2020). Notably, the strategy underlines the importance of addressing the environmental drivers and impacts of FCV and enhancing the resilience of vulnerable populations. It also explicitly acknowledges the role of **climate change** as a **threat multiplier**, emphasizing the need for integrated environmental and conflict-sensitive interventions.

WBG environmental programs span various priority areas, including **biodiversity** protection, landscape and seascape management, climate action, pollution abatement, water and green energy provision, and agricultural support. Projects are often designed to address the root causes of environmental degradation and promote sustainable natural resource management. A portfolio analysis conducted in 2024 revealed that in the period between 2014 and 2023, among the 755 WBG development projects tagged for natural resource management activities in fragile and conflict-affected situations, 44 have been explicitly designed with the objective of reducing conflict drivers through strategic environmental interventions (Doumergue et al. 2025). These projects tend to include peace-positive components responding to local FCV drivers by promoting sustainable livelihoods, basic service provision, inclusive **governance** of resources, or seeking to strengthen horizontal and vertical **social cohesion**, for example. Participatory approaches, stakeholder engagement, and social and conflict analyzes are different tools that may feed into project design and implementation to reduce the potential for resource-based grievances and to foster the **conflict sensitivity** and **peacebuilding** outcomes of these initiatives.

While the World Bank has been operating at the intersection of environment, fragility, and conflict for decades, its connection to the environmental peacebuilding space has been strengthened with the creation of its Environmental Peacebuilding team and the publication of *Defueling Conflict: Environment and Natural Resource Management as a Pathway to Peace* (Ahmadnia et al. 2022a). This report provides a comprehensive analysis of the nexus between natural resource management, fragility, conflict, and peacebuilding, and it offers practical guidance for designing and implementing conflict-sensitive environmental initiatives. Its follow-up report, *Defueling Conflict: Notes from the Field* (Doumergue et al. 2025), dives into the operationalization of environmental peacebuilding and synthesizes tools and lessons learned for **environment**-climate-FCV analytics, project design (a **theory of change**, safeguards), **monitoring**, **evaluation**, and **learning**, as well as institutional capacity building. The World Bank Environmental Peacebuilding team has sought to mainstream an understanding of **natural resources**-FCV linkages and opportunities for peacebuilding across the institution, including in upstream analytics (Risk and Resilience Assessments;Ahmadnia et al. 2022b, Country Climate and Development Reports;Doumergue et al. 2024) and in downstream operations. Other WBG initiatives connected to the environmental peacebuilding field include work on the peace and social dividends of climate action, on resilience in FCV-impacted settings, as well as on locally led climate action.

Acknowledgment

The work of the WB Environmental Peacebuilding Team is funded by the State and Peacebuilding Umbrella Trust Fund (SPF). SPF is a global multi-donor fund administered by the World Bank that works with partners to address the drivers and impacts of fragility, conflict, and violence, and strengthen the resilience of countries and affected populations, communities, and institutions. SPF is kindly supported by Denmark, France, Germany, the Netherlands, Norway, Sweden, Switzerland, and IBRD.

References

Ahmadnia, Shaadee, Agathe Christien, Phoebe Spencer Girouard, Tracy Hart, and Caio De Araujo Barbosa. 2022a. *Defueling Conflict: Environment and Natural Resource Management as a Pathway to Peace*. World Bank Group. http://documents.worldbank.org/curated/en/099520010272224660/P1771511b38fda01e1afec11edd810d8cde.

Ahmadnia, Shaadee, Phoebe Spencer Girouard, Tracy Hart, Caio De Araujo Barbosa, and Agathe Christien. 2022b. *Environment and Natural Resources Supplementary Guidance Note for Risk and Resilience Assessments*. World Bank Group. https://documents.worldbank.org/en/publication/documents-reports/documentdetail/099040423160036586/p1771510f078dc071084fc00455b7cab874.

Doumergue, Elise, Shaadee Ahmadnia, Max Hammond, and Caio De Araujo Barbosa. 2024. *Natural Resource Management, Fragility, and Conflict Issues: Guidance Note for Country Climate and Development Reports*. World Bank Group. https://documents.worldbank.org/pt/publication/documentsreports/documentdetail/099092424121514170/p1771511ae43bc0ce19a0d1662993dda2e3.

Doumergue, Elise, Shaadee Ahmadnia, Tracy Hart, Phoebe Spencer Girouard, Sally Judson, and Amanda Woomer. 2025. *Defueling Conflict: Notes from the Field*. World Bank Group.

World Bank Group (WBG). 2020. *World Bank Group Strategy for Fragility, Conflict, and Violence 2020–2025*. World Bank. http://documents.worldbank.org/curated/en/844591582815510521/World-Bank-Group-Strategy-for-Fragility-Conflict-and-Violence-2020-2025.

Index